ere is a thorough introduction to the ndamentals of reinforced concrete nstruction. Extensively revised to conrm to the latest ACI Standard Building ode Requirements for Reinforced Conete, this essential book sets down the sic requirements of all sound construcn and outlines the practical procedures necessary for an intelligent approach to design problems. It provides understanding of the nature and operties of concrete . . . shows how to ecify proper reinforcing . . . shows how calculate stresses in members . . . and uch more. In addition, the book gives lutions to many typical problems, and raws its illustrations from actual structures.

cluded in this fourth edition is information from the latest edition of the merican Association of State Highway fficials' Standard Specifications for ighway Bridges. Chapters on compose beams and large slabs have been completely revised and those on combined ending and compression and prestressed oncrete have received major revision as ell. There has also been considerable roadening of the chapters on bending in eams, bond, shear, columns, footings, nd architectural details. In addition, orking-stress design and ultimate-trength design are treated in parallel hroughout the book.

—how to visualize just how each part of a reinforced concrete structure acts
—how to design these parts so that each will safely perform the service for which it is intended
—how to plan operations in the field so that work will progress rapidly and with a minimum of trouble

Geared to the needs of the practicing engineer, the book emphasizes the field operations and conditions under which structures are erected and shows how they strongly influence the work of the designer. It devotes special attention to the action of structures, the behavior of materials, and the proper use of all facilities for both design and construction. The theory and application of both working-stress and ultimate-strength design are described and illustrated to enable the engineer to predict the magnitude of the failure load as well as to understand the action of his structure under working loads.

ABOUT THE AUTHOR

CLARENCE W. DUNHAM is a Civil Engineer with extensive experience in both teaching and engineering. He has taught at Rensselaer Polytechnic Institute, Cooper Union, and Yale University. He worked with the Port of New York Authority and participated in the design of the George Washington Bridge, the Lincoln Tunnel, and many other major structures and highways. Mr. Dunham has been affiliated with the Bethlehem Steel Company, the Phelps Dodge Corporation, and has acted as consultant on industrial plant construction for many years. At present he is a consulting engineer. In addition, he is the author of PLANNING INDUSTRIAL STRUCTURES, FOUNDATIONS OF STRUCTURES, ADVANCED REINFORCED CONCRETE, and, with R. D. Young, CONTRACTS SPECIFICATIONS, AND LAW FOR ENGINEERS, all published by McGraw-Hill.

THE THEORY
AND PRACTICE OF
REINFORCED CONCRETE

Clarence W. Dunham

Consulting Engineer

*Formerly associated with the Civil Engineering Department, Yale University
Consulting Structural Engineer for Anaconda Co., New York
Assistant Engineer, Design Division, Port of New York Authority
Assistant Chief Draftsman, Bethlehem Steel Co.
Chief Structural Designer, Phelps Dodge Corp.
Member, American Society of Civil Engineers, American Institute of
Consulting Engineers, American Concrete Institute, Connecticut Society of
Civil Engineers, American Society for Engineering Education*

Fourth Edition

McGraw-Hill Book Company

New York   *St. Louis*   *San Francisco*
Toronto   *London*   *Sydney*

THE THEORY AND PRACTICE OF REINFORCED CONCRETE

Library of Congress Catalog Card Number 65-24523

18225

1234567890 MP 7321069876

PREFACE

This volume includes material which the undergraduate or the young engineer should master if he is to be well prepared to handle the planning and design of ordinary reinforced-concrete structures with reasonable facility. Indeterminate structures and advanced material for the graduate student and the professional engineer have been incorporated in another volume entitled "Advanced Reinforced Concrete."[1] However, even in this elementary volume it has seemed to be desirable to include an introduction to precast concrete, prestressed concrete, and composite construction as well as working-stress design and ultimate-strength design. Also included are many of the details that are essential for an understanding of practical work and for the production of safe and proper designs.

Much in the line of concrete construction is changing and developing and will undoubtedly continue to grow and to improve. This applies to theory as well as to practice. Many things are debatable even now, and not all men will have the same ideas and opinions as those expressed here. Nevertheless, the author has attempted to present the theory and the art as he sees them.

In this volume he has tried to present fundamental principles and concepts. Admittedly, it is difficult to tell where to begin and where to stop in such a vast field as reinforced concrete.

The author hopes to teach the reader to visualize how each part of a structure acts, to design these parts so that each one will perform safely the service for which it is intended, and finally, to plan the operations in the field so that the entire work will be a thing of which he is proud. Sound judgment and engineering sense are exceedingly important. These essentials are attained chiefly through hard work and long experience by the individual. However, the author hopes to expedite their attainment by presenting the subject from the viewpoint of the practicing engineer.

The reader will notice that in many cases this book does not go into extreme refinements of design and calculation. When one realizes that the assumed loads, their distribution, the allowable unit stresses or safety factors for the concrete, and even the ultimate strength of the concrete

[1] C. W. Dunham, "Advanced Reinforced Concrete," McGraw-Hill Book Company, New York, 1964.

are often rather approximate and that they are based upon experiments, experience, and judgment, it seems to be inadvisable to carry subsequent computations to a degree of refinement that is not justified by the accuracy of the fundamental data from which the calculations are started. Therefore, the use of the slide rule is sufficient for all work in this volume. In many cases, the numerical answers are rounded to the nearest important significant figure. The methods of analysis that are employed are designed to show fundamental principles and their application. They are believed to yield results that are on the side of safety and to be sufficiently accurate for all practical purposes.

The study of reinforced-concrete design should not be confined to making an acquaintance with a large number of formulas and to substituting quantities in those formulas in order to obtain a lot of numerical answers. This can lead to dangerous results because the structures built from the plans generally involve the safety of persons and property as well as the wise use of money and materials. It is essential for an engineer to develop a thorough understanding of the action of structures, the behavior of materials, and the proper use of all of the facilities at his command for both design and construction.

The theory and application of both working-stress and ultimate-strength design are described and illustrated because the engineer should understand the action of his structure under working loads, but he should also be able to predict the magnitude of its failure load. In any case, he will desire to be sure that his structure will be safe, reasonable in cost, and satisfactory in performance. He knows that failure is unforgivable.

One of the principal objectives of this edition is to make the text illustrate and conform to the ACI Standard Building Code Requirements for Reinforced Concrete (ACI 318-63), June, 1963, published by the American Concrete Institute, P.O. Box 4754, Redford Station, Detroit, Michigan, 48219. The requirements of the Standard Specifications for Highway Bridges, 7th Edition, of the American Association of State Highway Officials, 1220 National Press Building, Washington, D.C., are also included where applicable.

The tables and diagrams in the Appendix are useful for many purposes. Frequent references to them are made in order to have the reader appreciate their utility and learn when and how to use them. However, the author's chief emphasis is upon the basic principles of the analysis and design of various types of structures and upon the practical features which are so important in planning and building them.

Specific recommendations have been given for many theoretical and practical procedures. Where this has been done without reference to other authorities, the author merely attempts to provide the reader with some definite suggestions, but he does not pretend to set up unchangeable specifications.

It is desirable for a student to work on the creation of designs for complete structures, not merely the analysis or dimensioning of isolated members. For that reason a few plans of miscellaneous structures are presented in Chap. 15. Each one is based upon an actual structure, but most are simplified in detail. The instructor or the student alone can use as a backlog or project whatever problem or problems his interest and available time will permit.

The author is grateful to W. B. Sinnickson, Engineer of Tests, The Port of New York Authority, who has written most of Chap. 1. Mr. Sinnickson has endeavored to explain the characteristics and proper use of cement and aggregates and the essentials of good workmanship in making and placing concrete, and to aid the reader in understanding the particular and peculiar properties of the concrete of which the engineer will build his structures. Mr. Sinnickson wishes it stated that opinions and interpretations of fact contained in Chap. 1 are his personal opinions and interpretations and do not necessarily indicate or reflect the policies or opinions of the engineering staff of The Port of New York Authority.

The author is indebted to many associates and other friends for useful data and for helpful suggestions. He is especially grateful to The Port of New York Authority; to Samuel Potashnick, A. C. Seaman, Leon Kirsch, Walter Gadkowski, William J. Delaney, Paul F. Pape, L. A. Warner, and H. Gesund; and to Profs. William S. LaLonde, Jr., Leroy W. Clark, Bert B. Williams, Hardy Cross, Francis M. Baron, and Henry A. Pfisterer, who, along with others, assisted in various ways in connection with the previous editions.

Clarence W. Dunham

LIST OF ABBREVIATIONS

bbl = barrels
DL = dead load
fpm = feet per minute
fps = feet per second
ft = foot or feet
ft-k = foot-kip
ft-lb = foot-pound
ft^2 = square foot
ft^3 = cubic foot
gal = gallon
hr = hour
in. = inch
$in.^2$ = square inch
$in.^3$ = cubic inch or inches cubed
$in.^4$ = inches to fourth power
in.-k = inch-kip
in.-lb = inch-pound
kip = 1,000 pounds
ksf = kips per square foot
ksi = kips per square inch
lb = pound
lin in. = linear inch
LL = live load
min = minute
pcf = pounds per cubic foot
plf = pounds per linear foot
pli = pounds per linear inch
psf = pounds per square foot
psi = pounds per square inch
sec = second
SF = safety factor
USD = ultimate-strength design
WSD = working-stress design
yd = yard
yd^2 = square yard
yd^3 = cubic yard

CONTENTS

PROPERTIES AND MANUFACTURE OF CONCRETE[1]

1-1. Introduction. A concrete structure, either plain or reinforced, is unique among the many systems of modern construction. In many cases it is a type of structure that is manufactured from its component materials on the site of the work. This may be called *poured-in-place* construction. The quality of its raw materials may be decidedly variable. The compounding of its ingredients, the control of its chemical processes, and the arrangement of its parts are often performed by unskilled mechanical artisans. The inspection of its fabrication is sometimes delegated to an inexperienced member of a supervisory force and may occasionally be neglected entirely.

On the other hand, the development of *precast-concrete* and *prestressed-concrete* units for incorporation in structures has made vast strides in recent years. These members are made in special plants designed and equipped for the manufacture of these materials by skilled workmen and under excellent control of all procedures involved, including curing.

For poured-in-place concrete construction, the personal element—the care with which work is executed in the field—is of major importance. The concrete may be manufactured at the work site by whoever is available. This is in marked contrast to structures that are built partially or entirely of steel or masonry units prefabricated in factories by skilled workmen and assembled in the field by expert mechanics. The designer of reinforced-concrete structures that are built entirely on the site should know the useful properties and practical limitations of the materials with which his plan will be constructed. With this knowledge

---

[1] Contributed by W. B. Sinnickson, engineer of tests, The Port of New York Authority.

he should plan the work in such a manner that desirable results are easily and correctly attained in the field.   The study of concrete as a material is a complete subject in itself.[1]   Only some of the most important aspects of the subject—the characteristics of concrete and the factors influencing its quality—can be given in this chapter.

1-2.   Definition and Description of Concrete.   Concrete is an artificial stone that is cast in place in a plastic condition.   Its essential ingredients are cement and water, which react with each other chemically to form another material having useful strength.   A mixture of cement and water is termed *cement paste*.   Such a mixture is expensive.   To increase the volume of artificial stone produced from a prescribed amount of cement it is customary to add inert filler materials known as *aggregates*.   When a large amount of cement paste is combined with a small amount of fine aggregate, and the combination is of fluid consistency, the mixture is termed *grout*.   With the addition of somewhat more fine aggregate, such that the paste loses its fluidity and behaves as a cohesive plastic, the mixture is termed *mortar*.   With the further addition of coarse aggregate, the mixture is called *concrete*.

It has long been customary to designate these mixtures in terms of the relative volumes of cement, fine aggregate, and coarse aggregate used in their preparation.   For example, concrete proportions given as 1:2:4 mean a mixture of 1 ft³ of cement, 2 ft³ of fine aggregate, and 4 ft³ of coarse aggregate.   Another given as 1:3 is intended to mean a mixture of cement and fine aggregate, without any coarse aggregate.   The latter would be classified as mortar.   It should be observed that in each of these examples the amount of water to be used is undisclosed.

The foregoing system of indicating proportions of materials by volume is obsolescent but is still used on small projects or in connection with minor work.   Now proportions are more often given by weight, and sometimes the total water to be used is also indicated.   For example, proportions of materials for constructing the anchorages of the Bronx-Whitestone Bridge, in New York City, were determined by experiment and were given as 94:184:380 lb plus 5.6 gal of water.   It should be remembered that, invariably when proportions are given, whether by weight or by volume, the ingredients are in the same order: cement first, fine aggregate next, and coarse aggregate last.   Water is indicated separately, most often as gallons per bag of cement, sometimes as a ratio associated with the cement, and, occasionally, as the total amount in a unit volume of concrete.

Water, cement, and both fine and coarse aggregate, when mixed

[1] Edward E. Bauer, "Plain Concrete," 3d ed., McGraw-Hill Book Company, New York, 1949.

together in suitable proportions, produce concrete that is a plastic mass capable of being poured into molds. Concrete castings are made in this manner into objects of predetermined shape and size. The molds, which are actually called *forms*, must be built to restrain the plastic mass until it solidifies. Usually, the forms must be constructed in such a manner that the concrete, when it is poured, will be in its final position in the structure. This is not always necessary, however, and precast-concrete members made in forms on the ground or in specialized plants remote from the work are becoming increasingly important. Precast units are most economical when many members of identical size are required, in which case a single form or a set of forms can be used repeatedly. Forms, in addition to their primary function of restraining concrete within dimensional limits until it solidifies, serve a less obvious purpose that should not be overlooked. They support the mass until it has attained sufficient strength to support itself without undue deflection or complete collapse.

Concrete does not solidify or attain useful strength quickly. The chemical reaction of cement and water is relatively slow and requires time and favorable temperatures for its completion. The reaction requires at least several days and may require several weeks for the production of worthwhile results, and it continues thereafter for several years. It is customarily divided, for descriptive purposes, into three distinct phases. The first, designated the time of *initial set*, requires from 45 min to about 8 hr for completion. During this time the freshly mixed concrete gradually decreases in plasticity and develops pronounced resistance to flow. Disturbance of the mass or remixing during this time may cause serious damage to the concrete.

The second phase is an interval during which the concrete appears to be a relatively soft solid without surface hardness. It will support light loads without indentation; but it is easily abraded, and its surface can be scored, roughened, or otherwise marred without appreciable effort. This phase is termed the interval of *final set*. The time required for concrete to attain a condition that might be regarded as completely and finally set is very indefinite, inasmuch as the condition is in itself indefinable. However, within an interval of about 5 to 20 hr after the original mixing operation, the mass develops surface hardness to such a degree that its finish can no longer be manipulated or modified with ordinary hand tools such as trowels, floats, edgers, belts, and brooms, and when this condition prevails the concrete is said to have set.

The third phase is one of progressive hardening and increase in strength. For concrete of good quality this progressive improvement continues indefinitely. It is rapid during early ages until about 1 month after mixing, at which time the mass has attained the major portion of its potential hardness and strength. After the first month the improvement continues, but at a greatly reduced rate. This peculiar property of improvement with age will be discussed later in greater detail. It is graphically illustrated in Fig. 1-1.

1-3. Cement. For more than two thousand years man has used various cementitious materials in the building of his important structures. All can be placed in one or the other of two distinct categories: those which do not, and those which *do* set and harden in the presence of appreciable amounts of water. The latter are of major importance and are said to have *hydraulic* properties. A general classification of hydraulic cements should include *pozzolanic material* of volcanic origin, so effectively used by the Romans; *hydraulic lime* frequently used in France; *natural cement* such as that produced near Louisville, Ky., and Rosendale, N.Y., of the type used in constructing the Brooklyn Bridge; *alumina cement,* which is popular for sea-water construction in European countries; and *portland cement.*

Each of these types of hydraulic cement is in current use in some part of the world. The use of any particular type is a matter of engineering psychology and economic necessity. Engineering practice, in the United

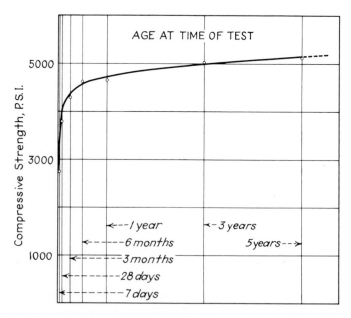

Fig. **1-1** Relation of compressive strength to age. George Washington Bridge, Riverside Drive connection to the New York approach. Each point is the average unit strength of twenty-five 6- by 12-in. cylinders mixed in the laboratory and cured in moist air at 70°F.

States, is inalterably linked to the rapid construction and early utilization of a structure, for which reason portland cement is almost universally used. Although aluminous and natural cements are sometimes used in this country for structures of a specialized nature, the further discussion of reinforced-concrete construction will be confined mainly to that in which portland cement is the binding agent.

During the year 1904 the American Society for Testing Materials adopted a standard specification for portland cement. Its requirements were definitely regulatory, but their range was such that manufacturers throughout the United States could meet them with little difficulty. During 1912, the Federal Specification Board also adopted a standard specification having requirements similar to the contemporary ASTM specification. These specifications remained similar in essential requirements even though they were revised at times. Because of their great latitude, manufacturers could solve their individual problems of composition and plant operation with little attention to specification restrictions. During the period 1914 to 1918 it became evident to many engineers and cement technologists that, although most cements conformed with then current specification requirements, they were often different in accomplishment. Differences in durability of concrete made with cements which seemed to be chemically similar and were of neighboring origin were observed particularly in structures built in sea water. Some cements attained unusual strength. Others reached adequate strength but later showed evidence of retrogression. Some produced unforeseen expansion of structures. Occasionally, some set quickly and generated much heat, while others were slow and lazy in their rate of gaining useful strength. Other less significant differences were also observed.

To understand and control this contradictive behavior, the Portland Cement Association in September, 1914, and many engineers and others about 1920 began intensive investigations of problems pertaining to concrete proportioning. Several years thereafter most of the manufacturers, by means of the Portland Cement Association Fellowship at the National Bureau of Standards, began coordinated scientific studies of the reactions occurring during manufacture, the constitution of the finished product, and the chemical and physical behavior incidental to the utilization of cement. Earlier chemical research had been carried on by governmental agencies[1] or by persons[2] not necessarily associated with cement manufacture, as well as by individual producers who were in most instances poorly equipped for pure research but were anxious to solve specific problems.

[1] U.S. Geological Survey; Carnegie Geophysical Laboratory; National Bureau of Standards.

[2] H. LeChatalier, W. Michaelis, and A. E. Tornebohm were noteworthy among scores of others as being so nearly in accord with current concepts.

Much of the early research was contradictory or inconclusive and sometimes extremely controversial.

During 1927, the International Cement Corp. placed on the market a so-called *high-early-strength* portland cement. This was the first of several important deviations from the all-purpose general-utility portland cement, the only type available until that time. There was, at the time, little apparent difference in the composition of this as compared with other portland cements, but the performance of the new material was remarkably different. Later, however, when the results of research on the constitution of cement clinker became known, its real difference became evident. Concrete made with this new cement exhibited similar plastic properties during mixing and placing, required almost identical proportions of ingredients, and set and hardened at about the same rate as did concrete made with normal cement. Nevertheless, the premium material was amazingly quick in its development of strength and attained useful values within about one-fifth the time experienced with ordinary cement. The ASTM in 1930 and the Federal government in 1936 adopted specifications for this improved type of portland cement.

Competition of other manufacturers soon provided other sources of supply of the premium material. As a consequence of this competitive activity the strength of ordinary portland cement was also improved by almost all manufacturers, and the strength differential between normal and high-early-strength cements became less pronounced. New disparities in the behavior of cement in concrete now became troublesome. Some cements were finely ground and others were coarse. Some required much more water than others to produce concrete of comparable consistency. Some produced cohesive plastic concrete of fatty texture while others encouraged segregation and bleeding of water to the surface of plastic mixtures.[1] Even though the compressive strength of ordinary concrete per pound of cement used was appreciably greater after 1926, a similar improvement of other attributes such as durability and impermeability was not evident. It became possible during the early 1930's to make concrete of good compressive strength but with insufficient cement to provide adequate resistance to the action of water, ice, and chemically destructive salts.

Great constructive activity during the 1930's on the part of state highway departments, Federal and other governmental agencies, and private industry emphasized differences in the behavior of apparently similar cements. Engineers engaged in building great irrigation and flood-control projects were anxious to minimize or control the heat evolved

---

[1] T. C. Powers, The Bleeding of Portland Cement Paste, Mortar and Concrete, *PCA Research Lab. Bull.* 2, July, 1939.

during the reaction of cement and water. Others responsible for sea-coast construction were active in searching for the inherent characteristic of cement that would assure durability of concrete exposed to tidal action and sea water. Highway engineers, and especially those of Northern states experiencing severe winter climates, were concerned about the spalling and surface disintegration of pavement slabs because repeated freezing and thawing and the chemical attack of snow-removal aids caused unpredictable deterioration of concrete pavements. During this decade many independent specifications were prepared by engineers for the purpose of controlling the quality of cement used in specialized structures.

Some of these independently prepared specifications contained logical and reasonable requirements. Others were founded largely on the faith of their proponents in some unusual testing procedure or some unstudied chemical reaction. Manufacturers of cement were faced with the problem of meeting the requirements of a score or more diverse specifications. Industrial endeavor toward standardization of behavior of cement suffered a temporary setback. Some industrial distress was caused by the increasing number and individual diversity of nonstandard specifications. Confusion was relieved in some degree, however, when in 1936 the Federal government adopted a specification for several types, and more so in 1940 when the ASTM adopted a specification for five types of portland cement.

Then during the 1950's, the Federal government had one, and the ASTM had two specifications for portland cement. These modern specifications provide not only for general-purpose cement but also for other types more suitable for specialized structures. One of the ASTM specifications covers five different types[1] of cement having specifically designated applications as described later; whereas the other[2] covers three of the five types and provides, in addition, for the entrapment of a myriad of small bubbles of air within the concrete in which the cement is used. The contemporary Federal specification[3] provides for five types of cement having applications identical with those designated by ASTM as well as for five others of similar usage but having the property of *air entrainment*. An interesting commentary on the evolution of portland-cement specifications is contained in American Concrete Institute literature.[4] In addition to the foregoing specifications for true portland cements, both ASTM[5] and Federal[6] specifications are also available for portland-

[1] Specification for Portland Cement, ASTM Designation: C 150.
[2] Specification for Air-entraining Portland Cement, ASTM Designation: C 175.
[3] Cements; Portland, Federal Specification SS-C-192.
[4] J. C. Pearson, Comments on Changes in Cement Specifications . . . , *J. ACI*, Vol. 19, No. 8, p. 705, April, 1948.
[5] Specification for Blast-furnace Slag Cement, ASTM Designation: C 205.
[6] Cement; Portland, Pozzolana, Federal Specification SS-C-208.

pozzolana types of cement. An ASTM specification[1] is also available for natural cement which, although it was a forerunner of portland, cannot be considered as a true portland cement, inasmuch as the material during manufacture never reaches sintering temperatures essential to the chemical reaction of lime with silica.

Other hydraulic cements of good quality and satisfactory performance which are not covered by nationally recognized specifications are available to the engineer. One of these that should be especially noted is Lumnite cement. It is an aluminosilicate cement that is sometimes used in concrete construction where unusually great strength or resistance to severe chemical attack may be required, or where temperatures as great as 1000°F may be encountered in service. Lumnite is not a portland cement, and concrete made with it requires somewhat different handling than does ordinary concrete. The experienced advice of its manufacturers should be solicited when its use is contemplated. *Ciment fondu* is a French equivalent of the aluminous type. Other European cements that might occasionally be used for concrete construction in this country are the iron-ore cements such as *erz* cement of Germany and *ferrari* cement made in France and Italy; the pozzolana-portlands such as *eisenportland* and *trass* cements of Germany, *gaize* and *metallurgique de fer* of France, and *silikatcement* of Sweden. True portlands from Belgium may sometimes be encountered.

Other cements such as white portland made by maintaining the iron content of raw mixtures at a very low value, waterproof portland made by incorporating organic soaps during grinding of clinker, oil-well portland made by special heat-treatment, and tinted or colored portlands made by intergrinding with limeproof pigments are all available for special uses. Such cements are generally used in monolithic concrete topping applied to ordinary concrete members. Expansive cements made by blending portland cement and metallic iron filings, sometimes accompanied by ammonium chloride, are sometimes used for grouting column bases and other parts of structures when expansion during setting is desirable. When water is present, the rusting of exposed metallic particles may produce an objectionable appearance. Other expansive cements are apparently made by combining portland cement with a sulfoaluminous cement and a stabilizing agent. One should consult the manufacturers of these special products to learn what can be accomplished.

An engineer can now select from among the several types of cement and can designate, by means of a standard specification, the most suitable cement for a special project or an unusual condition of service. The scope of his freedom of choice is best indicated by a brief description of

---

[1] Specification for Natural Cement, ASTM Designation: C 10.

the designated uses of each type of portland cement as set forth in Federal Specification SS-C-192:

Type I. For use in general concrete construction where the special properties specified for types II, III, IV, and V are not required.

Type II. For use in general concrete construction exposed to moderate sulfate action, or where moderate heat of hydration is required.

Type III. For use when high early strength is required.

Type IV. For use when low heat of hydration is required.

Type V. For use when high sulfate resistance is required.

Types I-A, II-A, III-A, IV-A, and *V-A.* For identical uses as the foregoing types of the same number where air entrainment is required.

The relative demand for each of these distinctive types of cement is indicated by our national average production[1] during the years 1945 to 1949. During this period the production of portland cement averaged 173,700,000 bbl; and the production of portland-pozzolana, masonry, and natural cements combined averaged 2,700,000 bbl per year. The relative average production of distinctive types was as follows:

	Per cent
Types I and II combined	85.1
All air-entrainment types combined	8.9
Type III high early strength	3.4
Type IV low heat of hydration	0.1
Type V sulfate-resistant	0.05
All others (white, plastic, etc.)	2.5

Although the foregoing percentages are informative they fail to disclose a very significant fact: that production of air-entrainment types of cement increased fourfold during the 5-year interval. The noteworthy performance of concrete having entrained air has almost completely reassured us about the ultimate durability of well-proportioned and properly installed concrete exposed to freezing, the leaching action of fresh water, and the chemical attack of salt water. It is quite likely that the demand of progressive engineers for air-entrainment types of cement will continue to cause increased production for an indefinite time to come.

The manufacture of cement is widespread throughout the United States. It is produced in 150 mills located in 36 states and the island of Puerto Rico. The area of most significant production is the Lehigh Valley of Pennsylvania, and this is closely followed next in importance by California. The close proximity of argillaceous as well as pure limestone and coal is responsible for the advantageous position of the Lehigh Valley. However, because of technical progress in the use of clay, shale,

---

[1] Computed from yearly production given in "Minerals Yearbook," U.S. Bureau of Mines, 1949.

chalk, and siliceous limestones accompanied by the availability of natural gas or petroleum, and because of improved plant-operating procedures, this one-time preeminent advantage is now greatly reduced. Increased use of a flotation process,[1] commonly used for ore dressing but first applied to the commercial production of cement at the Conshohocken plant of the Valley Forge Cement Co., may eliminate all factors other than cheap fuel as local advantages in the manufacture of cement. The schematic flow of material through a typical cement plant is shown in Fig. 1-2.

Cement is a conglomeration of many compounds in varying proportions. Its composition is dependent upon the impurities present in its raw materials, one of which is the fuel used in its manufacture, and upon the treatment given it during the operations of calcination and grinding. One of its most important constituents is isomeric in form, and others are suspected to be. The contribution of such dual-personality constituents to over-all quality is dependent upon their casual condition of existence. An idealized portland cement that might be assumed to exist only for the purpose of discussion is a chemical combination of lime and silica forming a mixture of tricalcium and beta dicalcium silicates. The manufacture of such a product would, because of temperature limitations, be commercially difficult and the product might be hard to manage in construction. Aluminous and iron compounds, normally present in cement raw mixtures, make it possible to approximate this idealized cement by means of practical production methods. Both the alumina and the iron compounds form effective fluxing agents, and they accelerate the reaction of lime and silica which exist largely as unmelted solids at kiln operating temperatures. These fluxing agents react at calcination temperatures with lime to form tetracalcium aluminoferrite and tricalcium aluminate. They thus isolate some of the lime and make it unavailable for reaction with silica to form other compounds of great cementing value. Tricalcium aluminate is predominantly responsible for the setting and early hardening of cement.[2] Unfortunately, this constituent is also responsible for the deterioration of cement exposed to sulfates, inasmuch as its hydration product combines with the sulfate radical to form calcium sulfoaluminate and, in doing so, exhibits an increase of approximately 227 per cent in molecular volume.[3]

---

[1] C. H. Breerwood patent 1,931,921, Oct. 24, 1933.

[2] R. H. Bogue, The Nature of the Setting and Hardening Processes in Portland Cement, *PCA Fellowship Paper* 17, October, 1928; and R. H. Bogue and W. Lerch, Hydration of Portland Cement Compounds, *PCA Fellowship Paper* 27, August, 1934.

[3] R. H. Bogue, W. Lerch, and W. C. Taylor, Influence of Composition on Volume Constancy and Salt Resistance of Portland Cement Pastes, *PCA Fellowship Paper* 28, October, 1934.

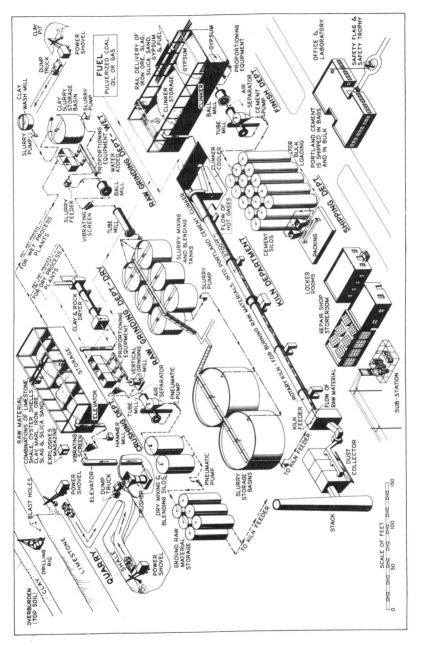

Fig. 1-2 Flow sheet of a plant for the manufacture of portland cement. (*Courtesy of the Portland Cement Association.*)

11

Other important constituents occurring in cement in lesser amounts are: uncombined calcium oxide, magnesium oxide, calcium sulfate, and poorly understood complex compounds of sodium and potassium. All these have a profound influence on the performance of cement in service. Free calcium oxide, usually trapped in the glassy structure of other major constituents, is the primary cause of unsoundness manifest by the warping or cracking of a pat of hardened cement paste exposed to low-pressure steam.[1]   Crystalline magnesia occurring as a result of low calcination temperature, or slow cooling of clinker, is most responsible for autoclave unsoundness evident as abnormal expansion of hardened paste exposed to high-pressure steam.[2]   Both the free lime and the magnesia are responsible for the long-time expansion of concrete continuously saturated with water. Calcium sulfate is deliberately added to clinker during finish grinding and is essential for retardation and establishment of a time of set that is practical for use in most concrete construction. An abnormally great amount of this constituent may indicate that the cement is of improper composition or was unsuitably calcined or cooled. Such cement, if prepared with a normal amount of retarder, might generate much heat and exhibit a *flash set* causing unwanted cracking of a structure. Sodium and potassium compounds are most certainly the cause of deterioration of concrete made with reactive aggregates, but the chemistry of the phenomenon is not yet well understood. These compounds are also believed to influence the formation of calcium silicates at kiln temperatures and to determine in some degree the amount and condition of the free lime remaining after calcination. The presence of other constituents such as compounds of manganese, phosphorus, titanium, and other elements in small amounts is of general occurrence. However, their significance in determining the quality or influencing the behavior of cement is uncertain.

The chemical processes of hydrolysis and hydration occur simultaneously during the reaction of cement and water. Hydrolysis is, briefly, the change of a compound into others as a result of the chemical action of water. Hydration is the combination of a material with water. Some of the constituents of cement disintegrate in the presence of water and, in doing so, they form other compounds which combine with water. The following outline of the reactions of cement with water should be considered as only generalities, since the reactions are not at all so simple and conclusive as they are described. Because of the ponderous names of the constituents and the unwieldiness of their common chemical symbols, they and their reaction products are usually referred to in cement

[1] W. Lerch, *Concrete*, Vol. 35, No. 1, p. 109; Vol. 35, No. 2, p. 119, 1929.
[2] W. Lerch and W. C. Taylor, Some Effects of Heat Treatment of Portland Cement Clinker, *PCA Fellowship Paper 33*, July, 1937.

technology in an abbreviated symbolic form.   For example: tetracalcium aluminoferrite is expressed in conventional symbols as $4CaO.Al_2O_3.-Fe_2O_3$; whereas, in the abbreviated form used by cement chemists, it is written as C_4AF.   In the same manner, tricalcium silicate is noted as C_3S, while hydrated tricalcium aluminate would be recorded as C_3AH_6. The abbreviated form of notation is used in the following description whenever its meaning is clear and its use is convenient.

Briefly, the C_3S disintegrates slowly by hydrolysis and the products of its dissociation then hydrate; the beta form of C_2S reacts directly with water but the reaction takes place slowly and requires many years for its completion.   Gamma C_2S, produced by the inversion of the beta form as a result of the slow cooling of clinker, hydrates very slowly and its hydration product develops insignificant strength even after several years.   Both the C_4AF and the C_3A combine actively with water and form hydrated products which contribute moderate strength to concrete within the first few hours or days of their existence.   The hydration of C_3A contributes most to the initial set and the early hardening of concrete and, unless an adequate amount of retardant in the form of a sulfate is present, the reaction may be vigorous and cause rapid set.   It may seem anomalous that sulfates are used to control the set and that they are also destructive to hardened concrete.   However, the reaction of C_3A with retarders takes place largely while the mixture is plastic or, at worst, while the concrete is weak and capable of readjustment to the expansive forces of formation of calcium sulfoaluminate.   The relatively energetic and prolonged reaction of C_3S, and the slow reaction of beta C_2S with water to form C_2SH_x is responsible for the progressive increase in strength of portland-cement concrete.[1]

The constitution, the chemistry of formation and utilization, and what might be called the *metallurgy* or phase relationships of cement are now fairly well defined.   Phenomenal gains in our knowledge of cement have been made since 1920 by research chemists and physicists. Although many persons of most diversified national background have accomplished this and each deserves part of the credit, nevertheless most of the references given here are related to research performed in the United States.   Sincere apologies are offered to the scientists of foreign lands who have contributed to our present state of knowledge but whose contributions have not been noted because of limited space.

Because of the chemical complexity of the material, it is suggested that the designing engineer should apply generally accepted standard specifications for cement and should leave improvements and departures from standard practice to the specialized fields of the cement technologist and

---

[1] R. H. Bogue and W. Lerch, Hydration of Portland Cement Compounds, *PCA Fellowship Paper* 27, August, 1934.

materials engineer.   Designers should also realize that, although special types of cement behave during mixing and placing in a manner similar to ordinary cement, they differ greatly and may be substantially slower in their rate of gaining strength.   Special types should not be used without a thorough understanding of their properties.

1-4.   Aggregate.   Aggregate, either fine or coarse, is inert filler material added to cement paste to increase its bulk.   Aggregates do have other functions and may impart beneficial properties to concrete, but a proper appreciation of their primary function as a filler makes the proportioning of concrete mixtures more easily understood.   Fillers may be of either natural or artificial origin.   Because of widespread distribution, natural sand and gravel and mechanically crushed rock are the most commonly used aggregates.   Unusual materials such as blast-furnace slag, pumice, calcined clay, diatomaceous silica, asbestos, sawdust, vegetable fiber such as seaweed, and others are sometimes used as concrete aggregates. Materials deliberately made to have a cellular structure, such as Haydite, Lelite, Perlite, Waylite, and others, as well as natural materials such as vermiculite, are used for making acoustical, thermal-insulating, and lightweight concretes.   Even the small air voids entrapped in concrete when air-entraining cement is used should properly be thought of as part of the aggregate.   It is customary to consider any sound filler material that will pass through a sieve having $\frac{1}{4}$-in.-square openings as fine aggregate.   It follows that particles larger than $\frac{1}{4}$ in. in size are classed as coarse aggregate.

It is of major importance that aggregate be nonreactive with cement[1] and water and that it be structurally sound, strong, and durable. Hardness and toughness are also desirable properties, especially in highway construction, where resistance to abrasion and impact are of functional importance.   In general, igneous and metamorphic rocks as well as most siliceous sands and gravel are usually of excellent quality for use as aggregates.   However, siliceous materials having constituents of amorphous or cryptocrystalline structure should always be suspected of being potentially reactive with high-alkali cement.   Sedimentary rocks in general should be considered with suspicion unless thorough and complete tests or extensive experience in their use have proved their worthiness.   Natural materials having pronounced planes of weakness or cleavage, such as slate, shale, and micaceous materials, are usually undesirable; whereas others having uniform shearing strength in all directions are ideal for use as aggregates.

[1] Petrographic examinations made by a competent person may reveal the probability of alkali-aggregate reactivity.   Otherwise, tests lasting 3 to 12 months may be necessary to make sure that an aggregate can be used safely.

Aggregates should be clean, since particles coated with clay, silt, organic matter, or crusher dust will not bond with the surrounding cement paste.   The interface between an aggregate particle and its cementing medium is the most critical plane through which tensile or shearing failures are likely to occur.   This interfacial area, if poorly bonded to the cement paste, may also provide channels for capillary percolation of water into the concrete mass.   Natural sands are particularly prone to surface coating, especially when obtained from near the surface of a deposit.   Material obtained from subaqueous sources is often found to be coated with algae and contaminated with marine animals and vegetation.   Substances of organic origin are usually harmful to freshly mixed concrete.   Fine, and sometimes coarse, material from sources of supply containing clay, silt, and natural overburden of disintegrated organic matter should be thoroughly washed before use.   Even after washing the material should be tested, preferably in comparison with a material of known dependability such as *Ottawa* sand.[1]   The contamination of aggregate with topsoil, humus, or earthy material containing products of organic decay even in small amounts is practically certain to cause early disintegration or complete collapse of a structure.[2]

Specifications for concrete aggregates should require that the material be "clean, hard, strong, durable, and sound material free from harmful amounts of soft, friable, thin, elongated, or laminated pieces."   It is customary to indicate the maximum amounts of harmful substances such as clay,[3] silt,[4] shale,[5] coal,[6] and organic matter[7] that will be acceptable in an aggregate.   In some instances, particularly when a material has had little background in use, it may also be desirable to require that it meet empirical performance tests giving some measure of hardness,[8] toughness,[9] or resistance to repeated freezing.[10]

Aggregate should be inert to cement and water.   It has of late been

[1] Method of Test for Measuring Mortar-making Properties of Fine Aggregate, ASTM Designation: C 87.

[2] The use of air to blow off the coating should never be trusted.

[3] Method of Test for Clay Lumps in Aggregates, ASTM Designation: C 142.

[4] Method of Test for Amount of Material Finer than No. 200 Sieve in Aggregates, ASTM Designation: C 117.

[5] Method of Test for Soft Particles in Coarse Aggregate, ASTM Designation: C 235.

[6] Method of Test for Coal and Lignite in Sand, ASTM Designation: C 123.

[7] Method of Test for Organic Impurities in Sands for Concrete, ASTM Designation: C 40.

[8] Method of Test for Abrasion of Graded Coarse Aggregate by Use of the Deval Machine, ASTM Designation: D 289.

[9] Method of Test for Abrasion of Coarse Aggregate by Use of the Los Angeles Machine, ASTM Designation: C 131.

[10] Method of Test for Soundness of Aggregate by Use of Sodium Sulfate or Magnesium Sulfate, ASTM Designation: C 88.

emphasized that some materials heretofore used with confidence as aggregates have, when used with certain cements, caused great expansion and cracking of structures.[1] Aggregate reactivity has been most often observed in Pacific Coast and Rocky Mountain states as well as in parts of the Missouri River Basin. Much more rarely have other cases been evident in New York State and in the Southern Appalachian Highlands. The mechanism of failure was first most reasonably hypothesized by Hansen,[2] and it is now factually accepted that sodium and potassium constituents of cement react with amorphous, pseudocrystalline, or microcrystalline siliceous mineral components of certain aggregates. Opal and chalcedony are serious offenders, but other siliceous minerals, and even natural and man-made glasses, may cause trouble.[3] Destruction is caused by forces resulting from the formation of hygroscopic silica gel within the concrete which, in company with hydrated cement paste acting as a semipermeable membrane, creates osmotic pressure tending to burst the concrete. Where alkali-aggregate reactivity may be encountered it is imperative that definition of types and designation of acceptable amounts of reactive minerals be stated. In some situations the danger must be circumvented by rigorously selecting low-alkali cement when none other than hazardous aggregates can be obtained. Tests for aggregate reactivity are now in the process of standardization.[4] A comprehensive discussion of the ways of evaluating aggregate reactivity is available in ASTM literature.[5]

Water is essential to alkali-aggregate reactivity, and repeated wetting and drying aggravates the rate and emphasizes the degree of failure. Water trapped in discrete crevices or small pores within aggregate particles is also troublesome in that it exerts disruptive forces upon freezing. For these reasons the use of impermeable aggregate, of poor ability to transmit or store water, is desirable. The density and absorption of aggregate[6] is a fair indication of its worthiness for use in concrete. In general, materials of greater density are most suitable for concrete

[1] T. E. Stanton, *Eng. News-Record*, Vol. 124, p. 171, 1940.

[2] W. C. Hansen, Studies Relating to the Mechanism by Which the Alkali-aggregate Reaction Produces Expansion in Concrete, *J. ACI*, Vol. 15, No. 3, p. 213, January, 1944.

[3] T. M. Kelly, L. Schuman, and F. B. Hornibrook, A Study of Alkali-aggregate Reactivity by Means of Mortar Bar Expansions, *J. ACI*, Vol. 20, No. 1, p. 57, September, 1948.

[4] Method of Test for Potential Alkali Reactivity of Cement-aggregate Combinations, ASTM Designation: C 227.

[5] Symposium on Methods and Procedures Used in Identifying Reactive Materials in Concrete, *ASTM Proc.*, Vol. 48, p. 1055, 1948.

[6] Method of Test for Specific Gravity and Absorption of Coarse Aggregate, ASTM Designation: C 127; also . . . Fine Aggregate, C 128.

exposed to weathering. This generality should, however, be applied with good judgment since some materials of dense structure contain discontinuous internal cells not likely to serve as water reservoirs, and particles of such dense materials may exhibit low *apparent* density. Porous aggregate of large pore size, even though such pores may be interconnected, is less susceptible to damage by freezing because large pores are less tenacious in their retention of water than are small pores.

Several other characteristics are also of great importance in determining the behavior of aggregates, but their consideration is often neglected in urban communities where relative utility of available materials has, in most instances, already been established by the time-consuming and sometimes expensive method of trial and error. However, in the building of structures in undeveloped areas where new and untried sources of aggregate must be used, these characteristics may be definitive in evaluating promising materials. One of these characteristics is surface texture. A material of rugged surface is—by reason of its greater likelihood of mechanically adhering to cement paste—more desirable than another of vitreous, conchoidal, or smoothly fractured surface. Materials of rougher texture are also less likely to develop continuous crevice areas beneath horizontally oriented particles from which cement paste may have been unfortunately washed by bleeding of water or sedimentation of cement.

Another significant characteristic is the coefficient of thermal expansion of the material with respect to the same property of the cement paste or, in the case of coarse aggregate, the mortar. Many rocks have lower coefficients than paste or mortar; and, if the difference is great, the aggregate will contract less on cooling and cause tensile fractures of interfacial bond, thus providing channels for the percolation of water and subsequent deterioration by repeated freezing. Other influential characteristics are the specific heat and the thermal conductivity of aggregate. These determine to a great degree the temperature gradient within a concrete mass having temperature differences on opposite faces of a section. Where the gradient is sufficiently abrupt, the concrete may be subjected to destructive internal strains.

Size and shape, as well as the relative number of particles of different size, are important in determining the suitability of a material for use as concrete aggregate. Remembering that a major function of aggregate is to act as the bulky filler in an expensive cement-water paste, and considering also that workable mixtures are necessary for ease in placement with minimum effort, it should then be obvious that particles offering least resistance to rearrangement among their kind are most desirable. Spherical particles meet this criterion of best shape. They roll against each other and their relative position in a group is easily changed. Furthermore, spheres have least surface area for a specific bulk volume and, as a consequence, less cement is required to coat their surface.

Cubes offer more resistance to rearrangement because of interferences of their edges and corners.   For equal bulk volumes of material, cubes require 25 per cent more cement to coat their surface than do spheres. Flat, elongated, and prismoidal particles interlock with each other even more than do cubes.   Their relative mobility in plastic mixtures is poor, and they behave much like a log jam.   They impart harshness and encourage oversanding of mixtures in which they are used.   Compared with spheres and cubes of equal bulk volume, they require much more cement to coat their surface.   In attaining a prescribed consistency of concrete a greater bulk volume of spheres, as compared with cubes or prisms, can be added to a paste of specific fluidity because the spheres are more easily rearranged to provide plasticity and they use less of the paste to coat their surface.   For this reason, natural gravel is usually more economical than is crushed stone in massive construction.

Size also influences the worth of a material used as an aggregate.   For example, a single cube of 1-in. size has 6 in.^{2} of surface.   If this cube is divided by planes through the center of each face, the volume of the material is still 1 in.^{3}, but the number of individual particles has become eight cubes—each of $\frac{1}{2}$-in. size.   The total surface of the eight cubes is 12 in.^{2} area.   Further subdivision of these particles produces no increase of their spatial volume, or what might be termed their bulking value, but the number of particles and the sum of their individual surface areas soon reach extremely large values.   During the mixing operation each aggregate particle must be wet by, and become intimately associated with, a companion film of paste.   The sum of these coatings is immobilized, thus leaving less of the total paste available for the separation and flotation of particles, which condition is essential for plasticity.   Because of their lesser surface area, a specific volume of large, as compared with the same volume of small, particles has less stiffening effect on a prescribed amount of paste.   Consequently, a greater volume of large particles can be used as filler for a particular consistency of mixture, and this produces a greater volumetric yield of concrete.   For this reason greater maximum sizes of aggregate produce more economical mixtures.

The relative frequency of occurrence of particles of different size is another factor influencing the utility of aggregate.   For example, consider a cubical box of 1 ft^3 capacity.   If this is filled with 1-in.-diameter spheres, each tangent to others, it will contain 12 layers, each comprising 144 spheres.   It might at first be inferred that the box is completely filled, but upon further consideration, it is evident that there are air spaces between adjacent spheres.   These 1,728 particles, each of 1-in. diameter, constitute 0.52 ft^3 of bulk volume.   The interstitial space amounts to

0.48 ft³, or 48 per cent of the apparent volume of the material. The spheres could be more closely arranged in a system of hexagonal rather than cubical packing, and in this circumstance, the space between particles would be 44 per cent of their apparent volume. This space between aggregate particles is commonly termed *voids*. These uniformly sized particles, in their most compact arrangement, would require 0.44 ft³ of paste to fill the voids and cement them into a solid mass.

If smaller particles were placed in each interstice, less paste would then be required to fill the voids. By using both fine and coarse aggregate having particles well distributed from small to large in size it is often possible to limit the interstitial space to less than 25 per cent of the apparent volume of the aggregate.[1] To produce plastic concrete of good workability, however, something more than a filling of the voids with paste is necessary. A surplus of paste must be provided to separate the particles at points of tangency or planes of contact and thus produce plasticity. Since well-graded aggregate possesses fewer voids, it requires less paste to produce a mixture of prescribed plasticity. For the same reason, a paste of specific dilution will tolerate the incorporation of more well-graded as compared with poorly graded aggregate, in preparing mixtures of suitable workability.

In almost all circumstances a statement of maximum particle size, and acceptable ranges of particle-size distribution, is necessary for procurement of suitable aggregate. When the gradation of aggregate is uncontrolled, the consistency of the concrete is almost always erratic, and the quality of the concrete is usually variable. Furthermore, if aggregate of too large a size is used, the narrow clearances between steel and the face of a form may cause segregation of mixture components, or closely spaced reinforcement may behave like a sieve and separate parts of the mixture. Aggregate having particles well graded from fine to coarse in size facilitates placement of concrete in narrow spaces and in heavily reinforced structures.

Classification of aggregate with regard to size is accomplished by separating a representative sample of the material, using a standardized series of testing sieves.[2] Until about 1936, much confusion was experienced among aggregate producers, because some consumers expressed their requirements in terms of square openings while others preferred to use round openings.[3] However, cooperative action of manufacturers' associations, individual producers, public and private consumers, and various engineering societies, under the guidance of the National Bureau

[1] Method of Test for Voids for Aggregate in Concrete, ASTM Designation: C 30.
[2] Method of Test for Sieve Analysis of Fine and Coarse Aggregates, ASTM Designation: C 136.
[3] Edmund Shaw, Sieve Testing of Aggregates, *Rock Products*, May 9, 1931.

of Standards, has done much to improve the situation particularly with regard to coarse aggregate. Now, square-mesh sieves are almost universally used, and acceptable equivalents of square as compared with round openings have been established.[1] A system of square-mesh sieves logically related one to another was first produced in 1910[2] and is described in the 1913 ASTM Proceedings. A more modern system[3] providing closer separation of sizes is one so arranged that successive sieve openings are related in size as $1 : \sqrt[4]{2}$. Only for the testing of coarse aggregate, however, are the intermediate sieves of this more closely spaced series generally used.

An exceedingly useful application of the Tyler series of sieves, which are related as $1 : \sqrt{2}$, was suggested in 1918,[4] when it was proposed that concrete aggregate be classified in terms of an abstract number called the *fineness modulus*. This number, commonly abbreviated "F.M.," is the sum of the percentages of material coarser than the following sieves: 3 in., $1\frac{1}{2}$ in., $\frac{3}{4}$ in., $\frac{3}{8}$ in., Nos. 4, 8, 16, 30, 50, and 100. Its value is an index of the average surface area of the aggregate. Aggregates of different particle-size distribution may have similar fineness moduli; and furthermore, for aggregates of a specific type it is reasonably certain that those of similar fineness moduli will, if used in similar amounts, produce concrete of similar plasticity. For this reason, the fineness modulus of an aggregate is a worthwhile adjunct to inspection and readily discloses a change in concrete-making quality of material from a particular source.

The highway departments of some states, as well as some other major consumers of aggregates, still adhere to the use of independently prepared specifications which are believed to be peculiarly suitable to their local problems. However, many progressive engineers, and most of the engineering societies by means of joint committee action in the promulgation of building-code requirements,[5] prefer to use nationally recognized standards,[6] which are in turn based to a great extent upon widely accepted simplified-practice agreements. The student is cautioned that, in some places, aggregate of best quality may be unavailable or obtainable only at great effort and expense. Local sources should be investigated if enforceable specifications are to be prepared and aggregate is to be obtained at reasonable cost. In any case, specifications should take into considera-

---

[1] U.S. Department of Commerce, Simplified Practice Recommendation R 163-48.

[2] W. S. Tyler Company, Cleveland, Ohio.

[3] Specification for Sieves for Testing Purposes, ASTM Designation: E 11.

[4] Duff A. Abrams, Design of Concrete Mixtures, Structural Materials Research Laboratory, Lewis Institute, *Bull.* 1, 1918.

[5] Building Code Requirements for Reinforced Concrete, ACI 318-63.

[6] Specifications for Concrete Aggregates, ASTM Designation: C 33; or Specifications for Lightweight Aggregates for Concrete, ASTM Designation: C 130.

tion local conditions as well as what is to be accomplished, and restrictions should be commensurate with the importance of the work.

1-5. Admixtures. An admixture is an extra component sometimes added to a concrete mixture for the purpose of creating a special property or for neutralizing a normal characteristic of the concrete or to correct some deficiency of the mixture. The number of admixtures and the variety of advantages ascribed to them by their proponents are great. There are differences of opinion among engineers, with reference to some of them, as to their reliability and worth. Some are manufactured products of consistent behavior. Others are industrial waste products of little sales value, often representing unwanted expense or a troublesome disposal problem to their owners. The latter are seldom produced purposely, and sometimes their performance as concrete ingredients is inconsistent. These occasional components of concrete can, by their behavior, be arranged in three groups. (1) Some act mechanically during the plastic life of the concrete. (2) Others react chemically with one or more of the constituents of portland cement. (3) The most significant, however, are initially mechanical in their action, although later they participate in the reaction of cement with water.

Bentonite and other clays, silty sand, talc, and other chemically inert pulverized stones, all of subsieve particle size, are of the first category. Their main effect is provision of cohesion and plasticity to poorly workable mixtures, and they act in two ways. Least significantly, they are fillers of small voids. Most importantly, they are spacers which mechanically separate aggregate particles; and as such, they are analogous to ball bearings in a raceway. They separate aggregate particles and thus reduce internal friction, and they provide room for more paste. They present great surface areas to be wet by cement paste, and because of their small size they stiffen and add skeletal structure to the paste. In this manner they reduce bleeding of water and sedimentation of cement. Furthermore, they alleviate harshness during placing, unreasonable shrinkage during the setting process, and abnormal permeability of the hardened concrete.[1] They are often effective for correcting deficiencies of poorly graded aggregate, but they usually require that *more water* be used for a prescribed consistency of concrete.

Air-entraining agents are peculiarly different admixtures of the mechanical category which usually permit a reduction of mixing water. As organic chemicals they are detrimental to strength, but this incidental behavior must be tolerated if their mechanical advantage is to be used. Microscopic bubbles of air are whipped into concrete during the mixing

---

[1] T. C. Powers, The Use of Admixtures for the Correction of Aggregate Gradation, *J. ACI*, Vol. 22, No. 1, p. 36, September, 1950.

operation, and if they are stabilized so that they cannot collapse or escape, they perform all the mechanical functions of an inert fine solid.   Moreover, they comprise innumerable discontinuous voids filled with elastic gas which cushions the strain of freezing and the expansion forces of chemical changes of the cement when instigated by harmful substances. It is reemphasized that air is the effective admixture, that it can be gotten into the concrete only by mechanical mixing, and that it can be kept there in an effective amount only by the great surface tension created by the foam-stabilizing organic agent.   The remarkable discovery of aeration[1] has immeasurably improved our expectations of concrete durability and has been one of the most important developments of cement technology.

Chemical admixtures are added to concrete for the purpose of modifying the normal plastic life of the mixture, or for influencing its rate of gaining hardness and strength.   A disadvantage of most chemical admixtures is that small changes in their amount cause great changes in their action. Furthermore, some may retard one cement and accelerate another.   In this connection it should be remembered that cements, even though they may be of successive production lots from the same mill, are not always similar in constitution.   Moreover, since chemical agents modify the normal reaction of one or more of the cement constituents, and since they are effective in extremely small amounts, their action is critically influenced by changes in proportioning of mixtures.

Calcium sulfate in small amounts is a commonly used retarder and yet, with increased amounts, its behavior changes and it becomes a powerful accelerator.   Organic materials such as gelatin, glue, sugar, other carbohydrates, and ligneous constituents of spent sulfite liquor used in papermaking, even if they are used in extremely small amounts, strongly affect the reaction of cement and water.   Some of these are used as form coatings to retard surface hardening and thus make easy the production of textured effects when forms are stripped.   Salts of metallic lead inhibit cement reaction, and paints containing lead pigments may cause minor troubles at work sites.

Calcium chloride is a chemical admixture often used for the acceleration of repairs, as well as in winter concrete construction where its use is exceedingly helpful to the accomplishment of such work.   However, its variability of action with different cements makes questionable the uniformity of behavior of the concrete at later ages.   Despite the unfortunate probability that cement may vary in constitution, this frequently used admixture can be obtained in uniform quality.[2]   When used in

---

[1] O. L. Moore, Pavement Scaling Successfully Checked, *Eng. News-Record*, Oct. 10, 1940.

[2] Specifications for Calcium Chloride, ASTM Designation: D 98.

amounts not exceeding 3 per cent by weight of the cement, it shortens the plastic life of the mixture and in this manner is helpful in reducing bleeding. If it is used during warm weather or in heated concrete, the set may be accelerated to such an extent as to impair finishing, and heat of hydration may be increased to a degree that causes cracking. It may aggravate the dry shrinkage of hardened concrete and it usually increases expansion caused by alkali-aggregate reactivity. Nevertheless, it definitely improves the early strength of concrete.[1]

Proprietary brands of other chemical admixtures are numerous, but fairness to all producers of such materials precludes discussion of individuals. Some of them are harmless while others are hazardous, and any of them should be used only with discrimination. They are sold as surface hardeners, densifiers, plasticizers, water repellents, workability promoters, mixing-water conservationists, and dispersing agents. However, the last of these reputedly beneficial actions is illogical and is not supported by scientific principles.[2] Powdered aluminum deserves mention as a chemical admixture that is added to concrete to cause evolution of hydrogen gas, and this in turn creates porosity very much as does an air-entraining agent. However, the gaseous bubbles are unstabilized and they readily escape from the concrete. Much gas is necessary to accomplish the primary purpose—production of lightweight concrete—and variations of ambient temperature as well as cement alkalinity make the process difficult to control and the end result problematical.

Admixtures of the third category exhibit characteristics of both the foregoing types. They act not only as void fillers and as thrust bearings between large particles; they later associate with and participate in the reaction of cement and water. They differ from the truly chemical admixtures in that they are usable in large amounts and their reactivity is indirect. They are all finely divided pozzolanic materials which become chemically active only after the cement has combined with water. Their essential component is amorphous silica,[3] which reacts at normal temperatures in the presence of water with calcium hydroxide created by dissolution of the C_3S constituent of cement. The combination forms a strength-producing and densifying component of the hardened cement paste.

Natural materials such as pumicite or volcanic ash, diatomaceous silica, pulverized opaline cherts and shales; as well as artificial materials like

[1] J. J. Shideler, Calcium Chloride in Concrete, *J. ACI*, Vol. 23, No. 7, p. 537, March, 1952.

[2] T. C. Powers, Should Portland Cement Be Dispersed? *J. ACI*, Vol. 17, No. 2, p. 117, November, 1945.

[3] R. E. Davis, Use of Pozzolans in Concrete, *J. ACI*, Vol. 21, No. 5, p. 377, January, 1950.

pulverized aluminous and siliceous slags and calcined clays, and fly ash[1] —recovered from the stacks of coal-burning steam boiler plants—are all of pozzolanic nature.   Some of these may be substituted for as much as 30 per cent of the cement with little loss in ultimate strength of the concrete and, furthermore, with beneficial effects on the heat of hydration, durability, and impermeability.   All these are somewhat helpful in minimizing the harmful effects of alkali-aggregate reactivity.[2]   These pozzolanic materials, during the plastic life of the concrete, also exhibit all the helpful mechanical attributes and harmful demands for *more water* as the inert materials discussed above.   Natural cement, mixed with or substituted for part of the portland cement, is within this category.

The successful use of an admixture requires the solution of three serious problems.   First, because most of them are variable among their kind and different in action with cement, they deserve precautionary study with the associated cement before they are used.   Second, their adequacy of performance is difficult to measure at a construction site during the progress of the work.   Consistency of action is not visually evident, and—without technical control—abnormality of behavior is not immediately disclosed and may not become known until some years after the structure is built.   Finally, their use is accompanied by problems of materials handling and batching.   On large projects, where installation of special batching equipment, employment of capable technicians, and cost of testing devices are justified, these problems are easily solved. However, on ordinary projects where rigid control may unreasonably increase supervisory cost, their solution is almost impossible.

A good example of the first problem is fly ash, which may be as variable in its performance as are the smokestacks from which it is collected, and yet it is admittedly a worthy beneficial admixture.   Offhand it would seem that material from a single source could be trusted, but a change of fuel or operating conditions at its place of origin, or a new shipment of cement, might cause differences of reaction.   Chemical agents in general, and air-entraining agents added on the work, are convenient examples of the second control problem.   Such additives are effective in very small amounts, and sometimes they are disastrous if used beyond certain limits.   For example, the effective amount of air-entraining agent in a cubic yard of paving concrete weighing about 2 tons approximates 2 avoirdupois ounces, whereas 4 oz may cause irreparable reduction of strength.   The effect of 2 as compared with 4 oz in 2 tons of concrete is

---

[1] R. F. Blanks, Fly Ash as a Pozzolan, *J. ACI*, Vol. 21, No. 9, p. 701, May, 1950.
[2] W. T. Moran, Use of Admixtures to Correct Alkali-aggregate Reaction, *J. ACI*, Vol. 22, No. 1, p. 43, September, 1950; and C. H. Scholer and G. M. Smith, Use of Chicago Fly Ash in Reducing Cement-aggregate Reaction, *J. ACI*, Vol. 23, No. 6, p. 457, February, 1952.

not visually evident and can be determined only by test.[1] Most workmen are contemptuous of such small components of a batch and—if unsupervised—they may give some batches a double dose, leave it out of others, and generally measure the admixture to suit their own convenience. Finally, the use of admixtures requires special storage and measuring devices seldom available at ordinary batcher plants. Moreover, the repeatedly similar batching of five, as compared with four, ingredients is more difficult to accomplish, not as 5 is greater than 4, which is 20 per cent, but instead as 32 is greater than 16, or 100 per cent.[2]

There are several dispassionate reasons why some engineers are skeptical of admixtures in general. Chemical admixtures are usually critical in the amount that can safely be used. By-products are often uncertain in their behavior. Successful use of these requires watchful observation and vigorous control, and this can be achieved only by expert technicians. Admixtures obscure the incongruous proportioning of basic ingredients, and without conscientious testing, poorly compounded mixtures are often undetected. When they are used with preeminently satisfactory aggregates, many admixtures do nothing more than increase the cost of the concrete; whereas an equivalent monetary increase of cement is usually beneficial and introduces no extra problem of control. Nevertheless, in situations where undesirable aggregates are all that can be obtained, or when a specific resistance to some special condition of service must be built into the concrete, admixtures are indispensable.

1-6. Compressive Strength of Concrete. Concrete is an artificial stone, and its excellent resistance to compression resembles the principal asset of natural stone. It is a pseudofluid for a convenient part of its early life; hence it can be easily installed to form large monolithic parts of a structure. Furthermore, its strength and other properties can be regulated to some extent during its manufacture, and as a consequence, its cost need be no greater than is necessary for the service it is required to perform. For these reasons, concrete is an ideal material for foundations and for other large or intricate parts of structures that must be rigid and resistant to compressive forces.

The quality of concrete is usually specified or discussed in terms of unit compressive strength at the age of 28 days; when no age is mentioned, this interval after mixing is usually implied. Other ages of refer-

---

[1] Symposium on Measurement of Entrained Air in Concrete, *ASTM Proc.*, Vol. 47, p. 832, 1947.

[2] Consider the measurement of five different things, any one of which may with equal probability and without regard to the others be measured either right or wrong. There are 32 possible results of such an operation and only one of the results is desirable—that all be measured right and none wrong.

ence such as 3 or 7 days are also used, especially with respect to high-early-strength concrete, and in such instances the age of reference should always be defined. The compressive strength of concrete, as it is most often determined by testing molded cylinders,[1] may range from approximately 1,000 to more than 8,000 psi. Compressive strength is also measured, in some instances, by testing cores of concrete drilled from the structure or by testing remnants of beams broken in flexure.[2] It has been demonstrated by means of elaborate and costly testing procedures that—when it is simultaneously subjected to lateral as well as axial pressures—concrete of moderate quality may support axial unit loads in excess of 70,000 lb when the lateral restraint is about one-third of this value. However, the triaxial-stress relationships of concrete have received very little study, and their effect upon local areas of concrete supporting concentrated loads is seldom considered in the design of reinforced-concrete structures.

Compressive strength, because it is a most important as well as an easily measured attribute, is often used as an index of the quality of other physical properties. It is generally assumed that—for concrete containing similar ingredients—a greater compressive strength is accompanied by greater tensile strength, greater flexural strength, greater modulus of elasticity, greater density, less permeability, and greater durability. These assumptions are substantially true of any large family of observations; yet, because of somewhat poor correlation of compressive strength with other properties, and in consideration of inherent variation of individual measurements of compressive strength, the relationships should be considered little more than indications of probable trends when only a few observations of compressive strength are involved.

Even in most favorable circumstances, when the batching of ingredients is conscientiously performed with greatest accuracy and when specimens are prepared and tested by thoroughly competent technicians, the individual results of a series of tests are scattered around their collective average value in accordance with the principles of mathematical probability. It cannot be too strongly emphasized that—even when using the most precise equipment with the greatest of personal care—the results of a series of similar observations are quite different among themselves; many are similar and some may even be identical, and yet others may be so incredibly different as to seem not to be members of the same family. Such is the quality of concrete; most of it approximates what we wish it

[1] Method of Test for Compressive Strength of Molded Concrete Cylinders, ASTM Designation: C 39.

[2] Method of Test for Compressive Strength of Concrete Using Portions of Beams Broken in Flexure, ASTM Designation: C 116.

to be, much we would rather have nearer an average, and even with the best of control, some is inexplicably poor or superlative.

A most useful measure of the scatter or dispersion of a number of observations is their standard deviation; i.e., the square root of the average of the squares of all individual deviations from the common average. This root-mean-square measure of dispersion is very helpful for estimating the trustworthiness of a few observations, estimating the over-all range of quality indicated by available observations, judging the normality among others of a seemingly incredible observation, and establishing limits beyond which needs for control are indicated.[1] The standard deviation is sometimes expressed as a percentage of the mean value of the observations with which it is associated, in which case it is called the *coefficient of variation*. This percentage measure of the dispersion of compressive tests—with respect to concrete having a reasonable degree of conscientious control—is usually somewhat more than 10 and less than 15 per cent. Allowance for this normal variation in quality is reflected in the long-established practice of proportioning controlled concrete mixtures in such a manner as to produce concrete having an average strength 15 per cent greater than is required by plans or specifications.

The ultimate strength of concrete is influenced primarily by the richness in cement of the water-cement paste used to glue the aggregate into a solid mass. Other factors such as aggregate type and quality, temperature and duration of water-cement reaction, workmanship during mixing and placing, treatment during the early hardening period, and age are of secondary but truly significant influence. These secondary factors approach primary importance as the quantity of cement in a mixture is reduced. In other words, a potentially strong mixture—containing a large amount of cement per unit of water—will tolerate a greater number and degree of unfavorable factors of secondary influence with less evident harm than will a mixture of lean cement content.

Attainment of a desired compressive strength of concrete was, until the year 1918, largely dependent upon auspicious circumstance; empirical proportions of cement and aggregate were hopefully expected to produce certain strengths. For example, volumetric proportions of $1:2\frac{1}{2}:5$ were expected to produce concrete having a compressive strength of 1,500, and others of $1:2:4$ were confidently presumed to attain 2,000 psi at the age of 28 days. What few supervisory tests were then performed were often disquieting.

Vastly significant progress in concrete technology—the result of research sponsored by the Portland Cement Association and performed

[1] ASTM Manual on Quality Control of Materials, *ASTM Special Tech. Pub.* 15-C, January, 1951.

at Lewis Institute—was made known during 1918 by Duff A. Abrams, then professor in charge of the Structural Materials Research Laboratory. The major result of these cooperative studies of cement and concrete was the disclosure of a means by which the strength of concrete could be more assuredly controlled. The fundamental concept of Abrams's water-cement-ratio theory of proportioning concrete was that, "With given concrete materials and conditions of test the quantity of mixing water used determines the strength of concrete, so long as the mix is *of a workable plasticity*."[1] This precept has since revolutionized the design and control of concrete mixtures, and its truth has been fully confirmed in practice.

This early research was performed when concrete ingredients were measured by volume. As a consequence, Abrams's empirical equations expressed the strength of concrete as a function of the volumes of water and cement comprising the paste. This water-cement ratio by volume, as it was once so extensively used, represents a somewhat illogical association of an actual volume of water with an apparent volume of cement. This peculiarity of the volumetric ratio should be recollected if reference is made to the literature of some 20 years following 1918. Use of the ratio of cement to water *by weight* was initiated by Inge Lyse in 1932, as a result of his discovery that the relationship of strength to C/W could be manipulated as an empirical equation of the first degree. This ratio, C/W by weight, is of peculiar utility for the study of experimental mixtures. It was used in much of the technical literature of the decade following 1932, and yet, because of the inveterate custom of relating all the ingredients of mixtures to cement rather than to water, it is now seldom encountered and the inverse ratio W/C by weight is of most frequent current use.

There are indications that the strength attained at 28 days by concrete made with modern cement is approximately double that reported by Abrams in his tests using material of 1914 to 1918 production.[2] So also, but to a much lesser degree, have other physical properties of concrete been benefited by industrial improvements in the composition and processing of cement. These improvements, some of great and others of little significance, have curtailed the once common practice of proportioning mixtures for optimum compressive strength with minimum

---

[1] PCA Minutes of Annual Meeting, December, 1918. Italics have been added, inasmuch as the specific limitation of the principle is often overlooked. The rule does not apply to damp, sandy, crumbly mixtures used for precast units such as pipe or block; nor does it apply to harsh and greatly undersanded concrete.

[2] H. F. Gonnerman and W. Lerch, Changes in Characteristics of Portland Cement as Exhibited by Laboratory Tests over the Period 1904 to 1950, *ASTM Special Tech. Pub.* 127, p. 22, June, 1951.

Table 1-1   **Assumed Strength of Concrete Mixtures**

Water content, U.S. gal per 94-lb sack of cement	*Assumed compressive strength at 28 days, psi*
$7\frac{1}{4}$	2,500
$6\frac{1}{2}$	3,000
$5\frac{3}{4}$	3,500
5	4,000

cement content. They have also made obsolete much of the mixture design data of the period 1920 to 1940. Now more emphasis is placed on the attainment of durability by limitation of excessive water, and rightly so, since a saving of cement is absurd if such economy is detrimental to the life of a structure.

When average materials are used and no preliminary tests of the materials are made, the extensively approved Code[1] of the American Concrete Institute limits the water in a mixture of non-air-entrained concrete—including that contained in the aggregates—and designates the assumed compressive strength of the concrete as shown in Table 1-1. However, when artificial aggregates or admixtures are to be used, the Code requires, and when the use of controlled concrete is contemplated, it permits that other water contents determined in accordance with designated ASTM testing procedures shall be used.[2] The method of determining maximum water for controlled concrete requires the establishment of a curve showing the relationship of water content to compressive strength. This curve should be based on a minimum of four observations made at each of three different water contents. In contrast, the New York City Building Code requires that such a curve for controlled concrete shall be based on four similar observations, but at each of four water contents. Both these procedures should require that the water content used on the work be that indicated by the curve as producing a strength 15 per cent greater than is called for on the plans or in the specifications.

The foregoing regulations give only minimum requirements; in many cases a greater number of observations may be desirable. For example, when only three observations of a relationship are available—and the nature of the relationship is unknown—such data are always most accurately described by a circular arc or by a straight line that might be envisioned as a special case of an arc of infinite radius. As a matter of fact, four observations are also inadequate for a clear indication of the

---

[1] Standard Building Code Requirements for Reinforced Concrete (ACI 318-63). Separate copies can be obtained at nominal cost by addressing the American Concrete Institute in Detroit, Mich.

[2] Occasionally, the use of a vacuum has been employed to reduce the water content of a concrete after it has been deposited, especially for slabs.

Table 1-2   **Compressive Strength, psi, at 28 Days, as Influenced by Various Ratios of Water to Cement by Weight**

W/C 0.37	W/C 0.46	W/C 0.55	W/C 0.64	W/C 0.73
6,390	5,230	4,000	3,450	2,720
5,910	4,390	3,450	2,700	2,560
6,280	4,500	3,650	3,200	2,300
5,220	4,680	4,350	3,010	2,650
5,950	4,700	3,860	3,090	2,560

true nature of the relationship, or for the choice of an empirical equation most suitable for describing the data. Furthermore, in so few batches, if the relative amount of fine and coarse aggregate in any single batch was poorly balanced, this might produce freakish results and a misleading curve. For these reasons, in any research where secondary factors may influence the results, a spread of observations is essential. Though it is not invariably true, nevertheless it is generally assumed that an exponential equation is a most suitable approximation of the strength vs. water ratio relationship.

Hypothetical data, suitable for demonstrating the relationship of strength to water content, are shown in Table 1-2. These data are illustrative of some of the considerations involved in establishing a curve. They give a result that is similar to contemporary information having widespread usage;[1] such curves are, of necessity, conservative in their indications. More reliable information is obtainable by making tests with the materials actually to be used. These data simulate a situation wherein five batches of concrete of suitable consistency might have been prepared, and four specimens were molded from each batch. As is customary, unit strengths are rounded off to the nearest 10 lb. It should be noted that no single observation is of any worthwhile significance alone; each must be considered with regard to all the data. Furthermore, it is unequivocally stated that—with given aggregates—paste of greater fluidity requires more sand; otherwise the concrete might be cohesionless, harsh, or susceptible to segregation during placing, and permeable to water after hardening. In contrast, an overabundance of sand for a specific fluidity of paste is detrimental to strength, requires more cement for a prescribed consistency of concrete, aggravates settlement shrinkage, and may adversely influence durability. We shall assume that proper adjustment of fine and coarse aggregate—suitable for the desired consist-

[1] "Design and Control of Concrete Mixtures," 9th ed., Portland Cement Association, 1948.

ency of concrete and for the different dilutions of paste—was made in these batches during the mixing operation.

Data from the table are shown plotted, using semilogarithmic coordinates, in Fig. 1-3. One of several methods might have been used to locate the curve, the least accurate being to plot the average values and then draw a straight line best fitting all of them. However, with a table of logarithms and a calculating machine, the constants of the curve can be computed from the average values in about the same time. Occasionally, when observations are meager in some areas of interest and prolific in others, or when they cannot be conveniently grouped to provide average values of equal weight, computation of the curve using individual values is the only reliable resort. The mechanics of this method, using average values from the table, are shown hereafter.

In principle, the general exponential equation

$$S = ab^x$$

is converted to the more easily managed linear form

$$\log S = \log a + x \log b$$

Test results are substituted there, forming separate equations for each associated pair of observed values. These are then solved simultane-

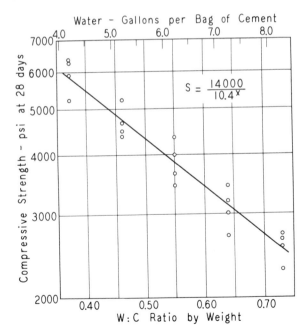

Fig. 1-3   Relation of compressive strength to water content.

Table 1-3   **Illustrative Computations**

	(1)	(2)	(3)	(4)	(5)
	W/C	Compressive strength	log S		
	x	S	y	xy	x^2
	0.37	5,950	3.7745	1.396565	0.1369
	0.46	4,700	3.6721	1.689166	0.2116
	0.55	3,860	3.5866	1.972630	0.3025
	0.64	3,090	3.4900	2.233600	0.4096
	0.73	2,560	3.4082	2.487986	0.5329
$n = \sum^5$	2.75	.....	17.9314	9.779947	1.5935
Mean (M)	0.55	.....	3.58628		

ously for the two unknown constant terms.   In practice, by using tabular methods shown below, the equations need not even be formed.   Moreover, the individual entries shown in columns 4 and 5 of Table 1-3 need not be made when a calculating machine is used, since they can be accumulated in the machine; only their summation need actually be noted.   The general form of the least-squares method of solving several simultaneous equations having two unknowns is

$$b = \frac{\Sigma xy - n(M_x M_y)}{\Sigma x^2 - n(M_x)^2}$$

and
$$a = M_y - b(M_x)$$

Substituting the tabular values in the above,

$$\log b = \frac{9.779947 - 5(0.55 \times 3.58628)}{1.5935 - 5(0.55 \times 0.55)} = -1.0163 \ldots$$
$$\log a = 3.58628 - (-1.0163 \times 0.55) = 4.1453$$

and
$$S = \frac{13,970}{10.38^x}$$

1-7.   Modulus of Elasticity of Concrete.   The modulus of elasticity of concrete in compression, E_c, is the ratio of the unit stress expressed as pounds per square inch to the deformation in inches per inch of gauge length.   A portion of a typical stress-strain diagram is illustrated in Fig. 1-4.   It should be noted that the locus is a curve and that E_c is of variable value depending upon the magnitude of the stress.   Many contradictory values may be assigned to the modulus of elasticity, as is shown in the illustration.

The 1951 ACI Code assumed that

$$E_c = 1,000 f_c'$$

where f_c' is the compressive strength of the concrete at the age of 28 days, unless otherwise specified. In this expression, no allowance is made for varying densities and unit weights of concrete. The 1963 ACI Code states that one may assume

$$E_c = w^{1.5} \times 33 \sqrt{f_c'}$$

where w is the unit weight, which may vary from 90 to 155 pcf.

Various factors influence the elastic behavior of concrete, the primary one being the relative amounts of cement and water used to bind the aggregates together. The age and condition of curing affect E_c in that both factors are of direct influence on the completeness of combination of cement and water. The density and degree of saturation of the hardened concrete are also influential factors.

In general, greater cement content, greater compressive strength, better curing conditions, greater age, and greater density of concrete tend to increase the value of the secant modulus E_c, which approaches the value of the initial tangent modulus as a limit. E_c varies slightly with the degree of water saturation of the hardened concrete mass, completely saturated concrete having a slightly higher modulus than the same concrete in a partially saturated condition. Another factor having a pronounced influence on the value of E_c is the type and quality of aggregate contained in the mass. As a general rule, the value of E_c is greater for concrete made with crushed stone aggregate consisting of prismoidal-shaped particles as compared with concrete made with gravel largely composed of rounded particles.

Figure 1-4A contains the results of one set of tests showing directly the variations in the modulus of elasticity as well as in the compressive strengths of a concrete because of changes in the water-cement ratio. In character, these results are what should be expected.

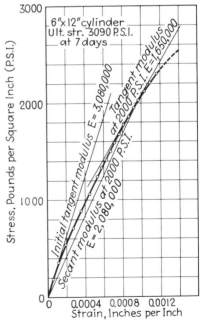

Fig. 1-4 Modulus of elasticity of concrete.

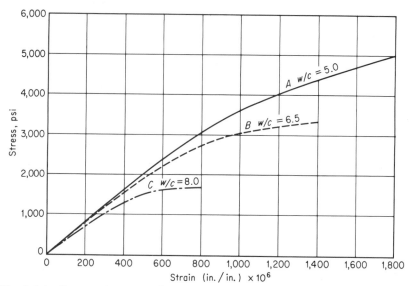

Fig. 1-4A Compressive stress of concrete vs. strain for various water-cement ratios. (*After tests made by F. Clark and N. Friets, Yale University, 1956.*)

The secant modulus of elasticity is inconvenient to determine as a routine procedure, inasmuch as the operation is time-consuming and requires facile use of optical, mechanical, or electrical strain gauges of great delicacy and precision. In contrast, with suitable instrumentation, the tangent modulus can be easily and quickly determined by sonic or vibrational methods.[1] The elastic properties of concrete are fairly well correlated with both the flexural and the compressive strengths of the material, and the elastic modulus is a useful index of the state of these other properties. It can be used to trace a progressive change in the quality of concrete with regard to time, and this can be done repeatedly and in a non-destructive manner, using only a few specimens. Observations of elastic characteristics are especially useful in studies of durability, because the deterioration of specimens can be noted as changes in elastic properties long before failure is visible or otherwise evident. Furthermore, in actual structures the elastic modulus influences the rate of conduction of vibrations through the concrete. Deterioration of structures can sometimes be disclosed as discontinuities or differences in the rate of propagation of wave fronts deliberately created within the concrete.[2]

[1] Method of Test for Fundamental Transverse Frequency of Concrete Specimens for Calculating Young's Modulus of Elasticity (Sonic Method), ASTM Designation: C 215.

[2] Johannes Andersen and Paul Nerenst, Wave Velocity in Concrete, *J. ACI*, Vol. 23, No. 8, April, 1952.

In the design of reinforced-concrete structures by the straight-line (elastic or working-stress) method, it is assumed that concrete is an elastic material. For concrete of good quality the assumption is safe and practical, but it is fundamentally incorrect. In actual fact, concrete must be recognized as an extremely rigid plastic which is subject to progressive and cumulative deformation under load. This property, termed *plastic flow*, or *creep*, is negligible in most instances, but its existence should be recognized and considered in the design of unusual structures. Figure 1-5 shows the results of tests made by The Port of New York Authority to determine the magnitude of the permanent set or creep under sustained load in connection with a study of the design of a reinforced-concrete arch crossing Riverside Drive and connecting with the George Washington Bridge.

The tests illustrated in Fig. 1-5 were made on concrete columns 10 by 10 by 48 in. containing 1 per cent reinforcement. Curves *B*, *C*, and *D* indicate the permanent deformation over a period of 1 year of specimens loaded with sustained loads of 100, 300, and 500 psi, respectively. Permanent deformation under normal working load is not great, but it is measurable and is proportional to the amount and duration of loading.

Apparently, creep is approximately proportional to the amount of cement paste[1] in the concrete and is not much affected by the water-cement ratio used, the quality of the concrete, or the age of the concrete

[1] I. Lyse, Shrinkage and Creep of Concrete, *J. ACI*, February, 1960.

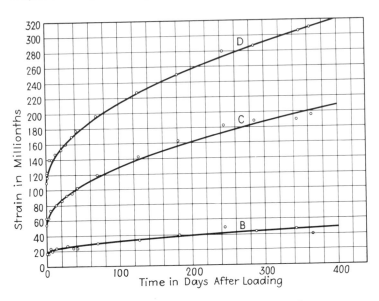

Fig. 1-5 Plastic flow due to sustained load.

at the time of application of the sustained load.   High relative humidity decreases the creep somewhat.

1-8.   Tensile Strength of Concrete.   Concrete is poorly resistant to tensile loads; consequently, its tensile strength is usually assumed to be of negligible value in the design of reinforced-concrete structures.   Tensile strength varies from about 7 to slightly more than 10 per cent of the associated compressive-strength value; it is, proportionally, least for rich mixtures and for concrete of greater age.

The data in Fig. 1-5*A* show the effect of the water-cement ratio upon both the tensile and compressive strength of concrete as determined by one set of tests.

Plain concrete is used for some structural work; e.g., small footings, retaining walls, pavements, and sidewalks.   When these are subjected to bending, their resistance to flexure depends upon their flexural strength. One method of judging the tensile strength of concrete in flexural action is the splitting test.[1]   In this test, a cylinder is laid on its side (like a roller) between the bearing blocks of a testing machine.   The compressive load is thus applied laterally as a line load on opposite ends of a vertical diameter of the cross section.   When tested to failure, the specimen tends to split longitudinally because of the tension set up in the concrete. There appears to be a definite relation between the tensile resistance in the cylinder and that which may be obtained in unreinforced concrete in

[1] See I. Narrow and E. Ullberg, Correlation between Tensile Splitting Strength and Flexural Strength of Concrete, *J. ACI*, January, 1963.

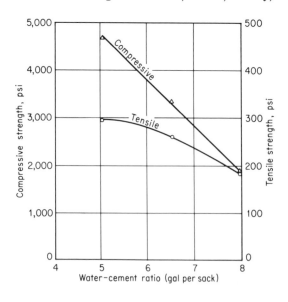

Fig. 1-5*A* Strengths of concrete vs. water-cement ratio. (*After tests made by F. Clark and N. Friets, Yale University,* 1956.)

flexural action, although the relationship may not be the same at all intensities of stress.

1-9. Shearing Strength of Concrete. The shearing strength of concrete is influenced by the ingredients and the proportions of the mixture. It can be reasonably assumed that the strength in shear approximates one-fifth, and seldom exceeds one-fourth, of the strength of the concrete in compression. Pure, or "punching," shear is the sliding of adjacent planes, one upon the other. This type of failure is resisted not only by the inherent cohesion of a material but also by internal reactions similar to friction and incidental to the granularity of the material. Concrete is of heterogeneous internal structure, and the interlocking and bridging action of strong aggregate particles distributes much of any shearing load as local concentrations of tension and compression within the mass. Shearing strength is improved by stronger paste, by greater integrity of the bond of paste to aggregate, and by stronger and more angular aggregate. This subject is discussed at greater length in Chap. 4.

1-10. Miscellaneous Properties. The weight of ordinary concrete ranges between 140 and 160 pcf[1] and is dependent upon the apparent densities and relative proportions of its ingredients, the entrained-air content, the degree of consolidation attained during its plastic life, and the amount of water physically absorbed by the hardened material. It has been customary for many years to assume the average weight of plain concrete to be 145 and that of reinforced concrete to be 150 pcf. Even for concrete containing deliberately entrained air, these assumptions are suitable for estimating purposes. With specific exception to some mixtures containing little or no coarse aggregate, which in respect to general custom should be called mortar, concrete weighing as little as 75 pcf can be made with special aggregates.

Concrete mixtures made with normal portland cement and lightweight aggregates are often disconcertingly harsh and difficult to place using ordinary construction equipment and technique. Such mixtures usually weigh between 90 and 115 pcf. A phenomenal gain in plasticity and workability, accompanied by reduction of unit weight to a range approximating 75 to 100 pcf, can be imparted to lightweight concrete mixtures by using either normal or air-entraining cement with an extra increment of air-entraining agent. About twice the amount of air-entraining agent normally used in ordinary concrete is essential for good workability and minimum weight, but this is not conducive to producing good compressive strength. Nevertheless, the best of lightweight aggregates are usually intractable in ordinary construction unless they are so plasticized, and

[1] For data on heavy concrete for radiation shielding, see Art. 14-16.

furthermore, all of them make concrete that is relatively weak in comparison with ordinary concrete. This inherent disadvantage must be provided for in design.[1]

An unfortunate practice of some proponents of lightweight concrete and of many vendors of lightweight aggregates is to discuss the unit weight of their sponsored product in terms of "oven-dry" weight, and to compare this with the water-saturated weight of ordinary concrete. Such prejudicial comparison is analogous to making measurements with an elastic yardstick. It should be remembered that concrete in a structure is almost never free of mechanically absorbed water, and that such moisture comprises approximately 10 to 20 per cent of the weight of the concrete. The propensity to absorb water is a natural attribute of hardened cement paste, and conditions of either complete or partial saturation are common to both types of concrete. Therefore, in any weight comparison of light and ordinary concrete the datum of reference should at least be the same; both might be compared saturated, both might be compared in equilibrium with atmospheric moisture, or both should be converted to the artificial condition of "oven-dried to constant weight." Inapt comparisons have caused misunderstandings among engineers, contractors, and materials producers and may have hampered progress in the art of lightweight construction. Nevertheless, the potential worthiness of such construction is unquestionable.

At the opposite end of the scale are heavy concretes which are made for such purposes as shielding to guard against radioactivity. Heavy aggregates such as iron ore, Ilmenite, or even metallic iron particles may be used. Concretes weighing approximately 235 pcf have been produced in this manner, but the cost may be in the neighborhood of $200 per yd[3]. Of course, segregation during deposition of the concrete is a serious problem.

Again, a sort of substitute concrete may be made by the Prepakt method. This is usually accomplished by placing a coarse aggregate in the forms first and then forcing a mortar or grout under pressure into the interstices to form what is hoped will be a dense and strong product. Of course, trapped air or water must be prevented if good results are to be attained.

Concrete shrinks when stored in air and expands in water. The amount of volume change is a function of the chemical composition of the cement, the water-cement ratio of the paste, the richness of the mixture, the reactivity of the aggregates, and the chemical environment of the concrete. The amount and condition of the magnesia, as well as the amounts of uncombined lime, calcium sulfate, tricalcium aluminate, tetracalcium aluminoferrite, and the so-called "free alkali" constituents of the cement

[1] See Art. 10-4.

are predominant factors influencing the permanent volume change.   However, participation of some of these in producing expansion is dependent upon external circumstances such as exposure of the concrete to waterborne chlorides or sulfates, or a casual presence in the aggregates of critical amounts of reactive silica.   Temporary and reversible changes in volume are caused by temperature changes and also by colloidal swelling and contraction of cementitious calcium silicates actuated by changes in the degree of water saturation of the concrete.   The total volumetric variation of concrete is the algebraic sum of permanent and temporary change.   It is exceedingly difficult, if not impossible, to predict or control its full amount.[1]

Seasonal and daily variation of ambient temperature of concrete is the most troublesome of the temporary changes which must be provided for in the design of large structures.   The thermal coefficient of expansion of concrete is approximately 0.000006 in./(in.)(°F).   With respect to a not uncommon seasonal change of 100°F., this represents $\frac{3}{4}$ in. change in length per 100 ft of structure.   Even though the thermal conductivity of concrete is fairly good, sudden daily or hourly variations of temperature may challenge the capability of the concrete to accommodate the change. One example of this, among many others, is the case of a pavement slab having its upper surface suddenly warmed by the sun while its lower surface is maintained cool by a damp or possibly frozen subgrade.   If thermal accommodation of the concrete is too slow, the slab will warp; and if it is prevented from warping by mechanical restraint, tensile fractures are likely to occur in its coolest surface.   Less troublesome, but still serious, problems of temporary change are created by the slow shrinkage of concrete stored in air and by the slow expansion of concrete that is constantly wet by water.   The shrinkage of concrete caused by evaporation of absorbed water ranges between 0.0002 and 0.0005 in./in.   A convenient average figure to remember is $\frac{3}{8}$ in./100 ft of structure.   Expansion of concrete in water may be as great as 0.0002 in./in. per year, and this approximates $\frac{1}{4}$ in./100 ft of structure.   These volumetric changes must be considered in the design of long members such as pavement slabs, retaining walls, and dams; and provisions must be made for sufficient and properly located expansion and contraction joints.   Means of caring for such deformations will be discussed later in greater detail.

Apparently, the trap rock found in the northeastern portion of the

---

[1] With specially selected aggregates and cements, some engineers have succeeded in minimizing the shrinkage.   In fact, there are apparently new expansive cements which, when the clinkers are mixed and ground in the right proportions with those of normal portland cement, will produce a concrete that does not shrink but may even expand to some extent.   However, these are special products which should be used as directed by the manufacturers.

United States is one of the best materials available for use in that area as aggregates when one wishes to minimize cement-aggregate reactivity. Used with certain special cements, trap rock as both fine and coarse aggregate is reported to have resulted in specimens of concrete having expansions of less than 0.01 per cent at an age of 12 months and also a contraction of less than 0.01 per cent at an age of 24 months.

The reader should be sure that he understands the difference between creep and shrinkage.   The former denotes a gradual decrease in the length of a piece of concrete because of a sort of plastic yielding under prolonged pressure.   If there is no such pressure applied, there will be no creep as such.   On the other hand, shrinkage of concrete exposed to ordinary air occurs whether there is pressure on the concrete or not, being caused partly by chemical action and partly by drying.   Creep is permanent and is not to be confused with so-called "elastic deformation" under pressure and then recovery after the load is removed.

Expansive cements[1] have been developed which, if used properly, seem to have excellent prospects for great usefulness.[2]   Of the several known types of expansive cements, one is under commercial development in the United States.[3]   It is of the type discovered and patented[4] by Alexander Klein, lecturer and research engineer at the University of California. With this cement expansions are generated by the formation of ettringite $(C_3A \cdot 3CaSO_4 \cdot 30\text{-}32H_2O)$ in freshly hardened concrete.

Earlier expansive cements based on ettringite formation provided independent sources of lime, alumina, and sulfate and depended upon solution chemistry in fresh concrete to produce the required compound. The variable composition of the portland cement constituents together with generally poorly controlled field conditions precluded adequate prediction of how much ettringite would form.   Control of the expansiveness of the concrete could not be attained to the degree necessary for commercial work.

Klein's cement differs from these earlier attempts in that anhydrous calcium sulfoaluminate $(C_4A_3\bar{S})$ is formed in a clinkering operation, and its amount in the finished cement can be closely controlled.   Controlling the amount of the anhydrous precursor of ettringite in the cement permits control of the expansion magnitude in the concrete.

The theory of expansive cements requires that the expansion occur early in the life of the concrete and that the expansion take place against

---

[1] A. Klein, T. Karby, and M. Polivka, Properties of an Expansive Cement for Chemical Prestressing, *J. ACI*, July, 1961.

[2] See Art. 13-4 for the application of this quality to prestressed concrete.

[3] Data furnished by C. W. Blakeslee & Sons, Inc., New Haven, Conn.

[4] U.S. Patent 3,155,526, issued Nov. 3, 1964.

some type of restraint, usually in the form of reinforcing steel. Unrestrained expansion produces no useful work and is a waste of the energy stored in the cement.

The expansive cement being introduced in the United States is a compensated shrinkage stress type and has been trademarked ChemComp. It is intended to compensate for drying shrinkage stresses in concrete and thereby eliminate the problem of drying shrinkage cracking.

The mechanism is simple. During the first few days of the concrete's life, while it is curing and hardening, bond to the reinforcing steel is developed. A very small expansion of the concrete places the restraining element in tension and the concrete in slight compression. When the concrete begins to dry out and shrink, this slight compression is relieved, and the concrete does not go into tension. Portland-cement concrete, upon drying, does go into tension and often cracks. ChemComp-cement concrete compensates for this by providing a means of achieving a small precompression before the start of drying shrinkage.[1]

1-11. Proportioning of Concrete. Aside from the busy and engaging field of research, the proportioning of concrete begins with some person who must define, for the understanding of others, the minimum quality and other characteristics of the material with which a structure is to be built. To decide about details, this person might consider the recommendations of some authoritative guide such as the ACI Building Code, or he could call upon his own experience or possibly make some exploratory tests of available materials. He might even be directed by local regulations as to the details of his definition. The end result would be a specification. He should forthrightly ensure that his specification properly describes concrete capable of performing without failure the work required of it and, furthermore, that the concrete shall perform its assigned task for the contemplated life of the structure. Since concrete can be gravely harmed by the acts of irresponsible or unscrupulous persons, the engineer should also give directions regarding its manufacture, conveyance, installation, and early care.

In the preparation of specifications, the attainment of adequate strength was once of primary concern; whatever durability was perhaps attained was accepted as the best that could be obtained of the particular concrete. Because cement has been remarkably improved with respect to strength, and because of our greater technical knowledge of factors influencing durability as well as the strength of concrete, it is now the habit to consider durability as being of primary importance. Adequate strength can easily and assuredly be attained with properly adjusted mixtures.

---

[1] See Art. 13-4 for the application of this expansive cement to prestressed concrete.

Recommendations of a joint committee of several engineering societies,[1] with regard to suitable water contents for various conditions of exposure, were once widely used for ensurance of durability. Now, since air entrainment is conclusively effective and special-purpose cements are available, this former means of assurance is less necessary. One such restriction used in former practice is to the effect that the water content of concrete exposed to freezing weather shall not exceed 6 gal per bag of cement. We can now be assured that a structure will endure, first by choosing a proper cement, and second by requiring that, in relationship to water content, enough of it be used; and furthermore, wherever it might also be effective, we can demand that a specific amount of air be incorporated in the concrete. As contributive ensurance of durability we can insist that plastic cohesive mixtures of suitable consistency be installed, by methods which preclude adulteration, segregation, or other harmful mistreatment of the concrete.

One of several methods can be used to define and guarantee the functional quality of the concrete. The least dependable of these is the designation of arbitrary amounts, such as 1:2:3, or 1:2:4, or sometimes even 1:3:5 parts by volume,[2] of the solid ingredients comprising the mixture. Volumetric measurement of such proportional increments, and the quality of the resulting concrete, is extremely erratic because the loosely compacted volume of damp aggregate is amazingly variable at different moisture contents. Fine aggregate is particularly susceptible to what is called *bulking*.[3] If this seeming increase of the apparent volume of aggregates is uncompensated during measurement, batches may be deficient by as much as 25 per cent of their solid components; they are usually deficient in some components and they are always variable in quality, and they are not proportioned as specified. The student is warned that, so far as the hardened product is affected, such arbitrarily selected mixtures, prepared as either arithmetic or geometric series relationships, contribute no remarkable properties to the concrete.

[1] "Report of the Joint Committee on Standard Specifications for Concrete and Reinforced Concrete," ASCE, AREA, AIA, ASTM, ACI, PCA, June, 1940.

[2] Fuller's rule for estimating the quantities of materials for concrete from known or assumed proportions given by volume can be useful on small projects of minor importance. It is as follows:

$$C = \frac{11}{c + s + g} \qquad S = C \times s \times \frac{3.8}{27} \qquad G = C \times g \times \frac{3.8}{27}$$

where $c:s:g$ represents the ratios of ingredients by volume, such as 1:2:4, and C = barrels of cement per cubic yard of concrete, S = cubic yards of fine aggregate per cubic yard of concrete, and G = cubic yards of coarse aggregate per cubic yard of concrete. See Edward E. Bauer, "Plain Concrete," 3d ed., McGraw-Hill Book Company, New York, 1949.

[3] A. T. Goldbeck, National Crushed Stone Association, *Bull.* 1, 1927.

Weighing of arbitrarily selected increments is somewhat more desirable than is volumetric measurement for small projects or for work of minor importance. More accurate compensation can be made for water carried into the batch by the aggregates. Arbitrary proportions are completely satisfactory, however, when provision is made to permit some shifting of aggregates from one part to the other if their gradation should change, and when the materials are weighed and some limitation is placed upon water. Another widely used method of ensuring adequate quality is to require that a minimum cement content—such as 5, 6, or 7 bags in each cubic yard of concrete—be used. This is not entirely effective unless the maximum allowable water content or the consistency of the concrete is also defined. A third way of establishing a quality level is to require that the concrete conform with a specific water-cement ratio; this, if it is to be effective, should also be accompanied by a definition of the consistency of the mixture. A fourth way is to specify the performance required, e.g., a statement to the effect that the concrete is to attain a compressive unit stress of 3,500 psi at an age of 28 days. The contractor is then to secure the proper materials, develop the proper proportioning, and prove by tests that the concrete meets the requirements.

The ailments and deteriorative tendencies of aged concrete are influenced greatly by the consistency at which the concrete was placed. The cohesion, consistency, plasticity, and workability of pseudofluid concrete mixtures are elusive properties. They are explainable, but such a discussion is too involved to be presented here. These properties are crudely but effectively measured by what is called the *slump* of the concrete.[1] The slump of freshly prepared concrete is influenced by the fluidity of the water-cement paste, by the relative amounts of paste and aggregates, and by the textural characteristics of the aggregates used to give the paste skeletal structure.

Concrete mixtures are suspensions of solids in water. Excessive water, producing more fluid consistencies of greater slump, encourages segregation and sedimentation of the solid components of the semifluid mixture. When concrete mixtures are placed, at first the larger particles of coarse aggregate segregate and settle faster, leaving concentrations of both water and mortar at the surface of the pour. This is especially undesirable in pavement-slab construction. Later, while the concrete is quiescent, but before it has begun to harden, the larger particles bear against each other and interlock; the smaller particles continue to settle and fall away from them, leaving voids filled with water. These water-filled voids are reservoirs susceptible to freezing. These voids can also form

---

[1] Slump Test for Consistency of Portland-cement Concrete, ASTM Designation: C 143. The slump is the subsidence, caused by gravity, of a plastic concrete specimen having an original height of 12 in.

continuous or interrupted channels for the capillary percolation of water. Excessive manipulation of the concrete during placement aggravates these troubles.   Therefore, it is desirable that the slump of concrete be the least that can be placed, with due regard to the nature of the structure, the size of the forms, and the clearances of reinforcement.

For pavements, or other construction where concrete can be discharged into the form at its final resting place, mixtures of preferably less than 2-in. and certainly less than 4-in. slump are desirable.   For more intricately shaped structures or for deeper members such as walls, piers, or columns, concrete of less than 6-in. slump should suffice in all but the most difficult of circumstances.   Such slumps are good for the potential durability of the structure, but they are not always liked by contractors because they are not conducive to effortless placement of the concrete.

The formulation and adjustment of actual mixtures almost always involves awareness of, and the deliberate choice of, the least harmful of several conflicting interests.   We must neutralize as best we can the inconvenient characteristics of cement paste;[1] and for this purpose we must, in most instances, use whatever natural materials happen to be handy in large amounts needing little processing.   Casual deposits of the finer sizes of such materials are seldom as clean or as well graded as we might wish that nature had made them.   Whatever fine aggregate we may by chance have must be pitted against coarse, and both of these against a semifluid paste of water and cement, to arrive at a mixture of suitable characteristics for placement; and, while doing this, we must keep in mind the effect of each of these components upon the integrity of the hardened concrete.

The first valid system of proportioning mixtures, based upon many experiments and free from the necromancy of favored numbers, was that proposed by Abrams,[2] in which he established the fundamental principle that both the water and the cement are the major factors determining the strength of concrete and, of almost equal importance, in which provisions were made for necessary adjustments incidental to the size, texture, and gradation of the aggregates.   Another logical proposal of great significance was the announcement in 1921, by Talbot and Richart, of their voids–cement-ratio theory of proportioning mixtures.[3]   Though this later theory resembled Abrams's theory in some respects, it differed

[1] T. C. Powers and T. L. Brownyard, Studies of the Physical Properties of Hardened Portland Cement Paste, *J. ACI*, October, 1946, to April, 1947.

[2] Duff A. Abrams, Design of Concrete Mixtures, Structural Materials Research Laboratory, Lewis Institute, *Bull.* 1, 1918.

[3] Albert N. Talbot and Frank E. Richart, The Strength of Concrete in Relation to the Cement, Aggregates and Water, *Univ. Illinois Eng. Expt. Sta. Bull.* 137, October, 1923.

specifically in its priority of consideration and emphasis upon the spatial characteristics of mixture components. This method was never widely used because of the soon realized obsolescence of volumetric batching of aggregates. Nevertheless, it had profound influence upon the direction of later research toward consideration of the absolute, rather than the apparent, proportions of mixtures.

A discovery of great practical significance was made in 1932 by Inge Lyse, then professor of engineering at Lehigh University, as a result of his study of mixtures with respect to their absolute volume relationships. Lyse found that the volume of water in a unit volume of concrete was substantially constant in mixtures of different water-cement ratio, when such mixtures were prepared with similar ingredients and were of similar workability. In other words, with little regard to cement content, the quantity of water in 1 yd³ of concrete of prescribed consistency is for most practical purposes a constant. This is not invariably true, inasmuch as very rich and very lean mixtures require somewhat more water than do those of ordinary cement content; but, when the fluidity of the pastes for the strengths involved are not too different, the rule is useful for making moderate shifts from one strength to another. Another discovery of practical utility made by Swaze and Gruenwald[1] was that, if cement is included in the computation of fineness modulus of the granular ingredients, a substantially constant value is indicated for all mixtures from lean to rich. Direct use of this principle is somewhat involved, but it can be approximated in practice with another generality to the effect that the absolute volume of the coarse aggregate, in mixtures of similar consistency made with similar ingredients but of different water-cement ratios, is substantially constant.

Three proportions of materials for concrete are given in Table 1-4, for the specific purpose of illustrating several elementary principles of pro-

[1] Myron A. Swaze and Ernst Gruenwald, Concrete Mix Design—A Modification of the Fineness Modulus Method, *J. ACI*, Vol. 18, No. 7, March, 1947.

Table 1-4   **Examples of Proportioning of Concrete**

Slump, in.	Parts by weight				Cement bags per cu yd	Ratio $C:f + c$	Ratio $f:f + c$
	Cement (C)	Fine aggregate (f)	Coarse aggregate (c)	Water (w)			
2	1.00	1.71	3.21	0.41	6.83	1:4.92	0.348:1
4	1.00	1.62	2.97	0.41	7.16	1:4.59	0.353:1
6	1.00	1.54	2.70	0.41	7.58	1:4.24	0.363:1

portioning.   It should be noted that, even though these mixtures are of different slumps, they are nevertheless of similar water-cement ratio. They should, as is indicated by Abrams's theory, be of similar strength. It is evident that their cement contents are greater with greater slumps; this is so because the amount of aggregate serving to stiffen the paste is respectively less at greater slumps, as is indicated by the ratio $C:f + c$. So also do the so-called "wetter" or more workable mixtures have greater fine-aggregate components, as is indicated by the ratio $f:f + c$.

Foregoing principles and approximations of fact are inadequate, in themselves, for arriving at dependable mixtures.   Something should also be known of the ultimate influence upon quality of each component of a mixture.   Following is a brief discussion of some of the more important considerations.

Cement paste is, with respect to durability, the weak link in the chain of ingredients.   Aggregates have lasted and will probably continue to last for ages; but hydrous cement compounds are susceptible to leaching, to volume changes induced by temperature and absorbed water, and to chemical changes caused by external influences.   Nevertheless, enough paste must be used in mixtures to coat all the aggregate, to fill interstitial voids, thus minimizing porosity, and to separate aggregate particles so that mixtures are of suitable workability for placement.   Hardened cement paste, in comparison with aggregates, is relatively soft and poorly resistant to mechanical wear.   The least viscous and more diluted pastes are most susceptible to weathering, mechanical wear, and other damage. Furthermore, excessive manipulation during placement and during finishing of mixtures rich in paste can be more harmful to durability than poorly proportioned mixtures.   Entrained air as well as some other admixtures, because they inhibit segregation and sedimentation of mixture components, are effective aids for ensuring durability.

The relative water content of cement paste is the most influential factor determining strength.   However, the shape and surface texture and the inherent strength of the aggregates are secondary factors not to be ignored.   Furthermore, the relative fullness of large interstices, with smaller particles or with paste, has an appreciable influence upon strength. Excessive amounts of fine aggregate are detrimental to strength as well as to durability; so also are excessive slumps, which may have been arrived at by adding water, rather than by withholding aggregates from the mixture.   Entrained air is detrimental to strength, but if not excessive, its effect can be neutralized by reducing the fine-aggregate component and also the water.

The volume constancy of concrete is influenced almost entirely by the behavior of the cement paste.   The paste, in comparison with aggregates, is much more reactive to thermal changes.   Even the ailment of aggregate reactivity is partly attributive to the nature of the paste. Hardened cement paste is a gelatinlike colloidal material, and it is typical of these that they are avid breathers of moisture; they quickly occlude and readily release great amounts of water in an attempt to maintain equilibrium with external conditions. In doing this, they swell when they are wet and contract when dry. Pastes richest in cement are most active in these respects. These swellings and contractions of paste are resisted by what might be envisioned as the bulk inertia of individual aggregate particles. Easily worked mixtures of great slump having a greater ratio of paste to aggregate, and especially those of great cement content, are most susceptible to expansion and contraction. Minimum film thickness of paste between aggregate particles, the paste being of lean cement content, is conducive to minimum volume change instigated by absorbed water. In contrast, the thermal expansion and contraction of concrete is influenced little by richness of paste or richness of mixture. However, thermal conductivity and rapidity of accommodation to sudden changes of temperature are improved by richer pastes and richer mixtures.

Concrete having one of its faces seemingly dry can actually be transferring great amounts of water through the paste within the mass. However, the water leaves the drier surface as vapor at a rate usually so rapid as to cause the surface to appear dry. This avid permeation of water throughout the paste cannot be avoided. It is not too alarming, inasmuch as this water is saturated with leachings during most of its travel and leaves its dissolved constituents within the concrete. Much more serious, however, is the capillary movement of water through discrete channels within the concrete, perhaps with only occasional delays required to permeate a few intervening bulkheads of hardened paste. This action can dissolve and remove from the concrete essential material necessary for continuing integrity. Such troubles are minimized by placing concrete of minimum slump made with paste rich in cement. Furthermore, the creation of percolation channels as a result of sedimentation of the finer components of mixtures is discouraged greatly by entrained air, and to some extent by other admixtures.

After good workmanship, cement is the most costly component of concrete. If a structure is to endure and perform its function for a life span commensurate with its cost, adequate amounts of this essential ingredient must be used. However, the leaner and less viscous mixtures of great slump are least costly to make and are easily placed. From a contractor's point of view, water is the cheapest component of concrete. It facilitates effortless placement; and furthermore, if $7\frac{1}{2}$ gal more than is really needed can be added to a batch of concrete, it seems when it is in the structure to be another cubic foot of concrete. It is, in these circumstances, paid for by the owner at the prevailing unit price of concrete; and later, it may again be paid for by the owner in his correction of undesirable attributes of the structure.   Again from the contractor's point of view, sand is the next most economical ingredient of concrete. It is of fine particle size and great surface area, and it holds a wet mixture together well; although it is not so effective as water, it facilitates the self-induced flow of mixtures and it makes less laborious the efforts and the number of workmen required.   One hundred pounds of sand will increase the bulk of a mixture perhaps 20 per cent more than will 100 lb of cement, and pound for pound, it costs perhaps one-tenth as much. Unfortunately, when it is used in excessive amounts, it is not beneficial to the longevity of a structure.

It is obvious that water is the most critical and least beneficial ingredient of concrete, yet we cannot make concrete without it.   Nevertheless, we can discourage the indiscriminate use of water, and we can search out and control sources of unavoidable supply.   For example, the aggregates usually carry a considerable amount of water upon their surface. Between 20 and 30 per cent of the total water in a mixture is introduced in this manner; if ignored it can seriously modify the intended characteristics of the concrete.   The amount should be determined[1] and a corresponding reduction should be made in the amount added to the batch at the mixer.   Moreover, corrections should also be made in the measurement of what seems to be aggregate but is, instead, partly aggregate and partly water.   This should be done often.   The surface moisture of fine aggregate in particular can change appreciably during a day's work and, unless corrections are conscientiously made, the concrete produced during the day can be of unwanted variable quality.

Air is a desirable component of concrete.   If it is to be effective in promoting durability, it must be present in certain minimum amounts and it must also be dispersed as globules of almost microscopic size. From 2 to 6 per cent air by volume is enough to provide adequate results; more adds little benefit and is inconveniently harmful to strength.[2]   The amount of air entrained is obviously influenced most by the quantity of frothing agent used.   However, for a specific amount of agent, lean mixtures entrain more air than do others of greater cement content, sandy

---

[1] Method of Test for Surface Moisture in Fine Aggregate, ASTM Designation: C 70. Surface moisture carried by coarse aggregate is sometimes estimated, but can be measured by drying a sample of known weight upon a hot plate, using care that the absorbed moisture is not also evaporated; in the hands of a careful operator, this method is also suitable for fine aggregate.

[2] Method of Test for Weight for Cubic Foot, Yield, and Air Content (Gravimetric) of Concrete, ASTM Designation: C 138. Also Symposium on Measurement of Entrained Air in Concrete, *ASTM Proc.*, Vol. 47, 1947.

mixtures more than those of less fine-to-coarse ratio, and finer sands more than those of coarser gradation. Higher temperatures of concrete and longer mixing time cause more entrainment of air. Concrete having an entrained-air component acquires a peculiar property. It possesses what is called a *thixotropic structure;* that is to say, it behaves like a fluid while it is agitated but when left undisturbed it acquires a false rigidity. Yet, even though it may have assumed an apparently rigid state, it resumes the fluid condition if it is remanipulated. This phenomenon of false set is the beneficial property that discourages segregation and sedimentation and thus improves durability. Such concrete can be struck off, troweled, and otherwise finished sooner than can ordinary concrete. Wet shrinkage of the mass, segregation of mortar, and bleeding of water are minimized by its falsely rigid structure.

The probable effect of modifying any single component of a properly adjusted concrete mixture is indicated in Table 1-5. This should be interpreted as indicating the probable trend that would be caused by a moderate excess of any single constituent, and we suggest that it should not be applied to changes of more than one component at a time. It is

Table 1-5   **Effect upon Quality Attributes of Concrete, Caused by an Increase of Any Single Component of the Mixture**

Attribute	Cement	Fine aggregate	Coarse aggregate	Water	Air	Mixing	Age
Slump...............	−	−	+	+	+	+	
Cohesion..............	+	+	−	−	+	+	
Workability...........	+	+	−	+	+	+	
Segregation...........	−	−	+	+	−	−	
Sedimentation.........	−	+	−	+	−	−	
Wet consolidation.......	−	+	−	+	−	−	
Bleeding..............	−	−	+	+	−	−	
Entrained air..........	−	+	−	+		+	
Durability............	+	−	+	−	+	+	−
Strength..............	+	−	+	−	−	+	+
Elastic modulus........	+	−	+	−	−	+	+
Freeze resistance.......	+	−	+	−	+	+	−
Wear resistance........	+	−	+	−	+	+	+
Chemical resistance.....	+	−	+	−	+	+	+
Permeability..........	−	+	−	+		−	−
Wet expansion.........	+	−	−	−	+	−	−
Dry shrinkage.........	+	−	−	−	+	−	−
Density..............	+	−	+	−	−	+	
Form finish..........	+	+	−	−+	+	+	

generally safe to assume that a deficiency would produce opposite effects. It is quite possible that in some cases gross rather than moderate changes might produce somewhat different effects, and this should be kept in mind. Nevertheless, if the table is used with discrimination it can be helpful as a guide to the formulation of satisfactory mixtures.

1-12. Computations. The computations incidental to concrete proportioning require the application of little more than accurate arithmetic. With the exceptions of those concerned with the statistical analysis of data, and the computation of constants of curves such as the water-cement vs. strength relationship, all can be performed with suitable accuracy using a slide rule. Volumetric batching devices are now seldom encountered, and they are notoriously inaccurate; weighing batchers are seldom more accurate than about ¾ per cent of any value that they may indicate. Often, the novice computes batch quantities to a greater apparent accuracy than is justified by the trustworthiness of field measuring equipment. Aside from the problems of research, the use of more than three significant figures can seldom be justified.[1]

Certain values frequently used in concrete computations are customarily assumed to be constants. For example, the weights of a bag and a barrel of cement are always assumed to be 94 and 376 lb, respectively, since these are requirements of cement specifications. The specific gravity of cement is so nearly similar among different lots that it can be assumed to be 3.15 in value. Water is generally assumed to weigh 62.3 pcf, and 8.33 lb/gal. One cubic foot of water is equivalent to 7.48 gal. These assumptions are sufficiently correct to be within the limit of accuracy of field measuring devices. Other values essential to the computation of proportions and other characteristics of mixtures are related to the particular aggregates and the specific mixture involved, and these must be determined by test. Among these are the specific gravities of the respective aggregates, the amount of water required to produce a prescribed slump, the entrained air content of the mixtures, and the so-called "cement" factor or the quantity of cement in 1 yd³ of concrete.

The volumetric yield of a mixture is of great interest to both the contractor and the engineer. It is easily determined by weighing a known volume of concrete—usually ½ ft³—in an accurately calibrated measure. The true volumetric yield of a batch is equal to the sum of the weights of all its solid and liquid components, divided by the actual weight of 1 ft³ of the mixture. Knowing the amount of cement incor-

[1] Consider the following as an illustration of the importance of figures in some such cases: A professor was giving out the examination papers to a class of graduate students. He said: "There are three questions. You may insert the figure 7 in the third place in each answer. Your problem is to find the first two figures."

porated in the batch, and the yield of the batch, we can then compute the cement factor; this is usually done in terms of bags, barrels, or pounds of cement per cubic yard of concrete. It should be noted that this method of determining yield and cement factor automatically takes into consideration the bulking effect of any entrained air, and the shrinkage incidental to the solution of any of the solids in water. The volume of entrained air is, for all practical purposes, the difference in the volume of the batch as compared with the sum of the volumes of its solid and liquid components.

Computation of the absolute volume, or what might also be called the equivalent solid volume, of the granular ingredients is a basic calculation that may at first be difficult to grasp. It should be recollected that the specific gravity of a material is, by definition, the relative weight of the material as compared with the weight of an equal volume of water. If, for example, an aggregate could be melted and poured without voids into a cubic foot measure, it would weigh perhaps 2.65 times as much as 1 ft^3 of water. Therefore, the weight of a solid or absolute cubic foot of any material is its specific gravity multiplied by 62.3 lb, the latter being the weight of 1 ft^3 of water. The true displacement or absolute volume of a granular material is its actual weight divided by its so-called weight per solid cubic foot.

Concrete proportions are indicated in many ways. For example, volumetric proportions much used in the past were often referred to as a one-bag batch such as 1:2:4, in which case the one part, or one bag, of cement was assumed to have an apparent volume of 1 ft^3. Weight proportions are often given in similar form, and in such instances, the ratios refer to pounds of aggregate per pound of cement. Weight proportions are sometimes given as 94:216:412, and it should be obvious that these are weights of aggregate per bag of cement. Water, when it is mentioned, is most often given as gallons per bag of cement. However, a most convenient form of designation that is rapidly coming into widespread use is 565:1,160:2,150:256, where the materials are arranged in the conventional order and represent the weights approximating 1 yd^3 of concrete, the last figure given being pounds of water per cubic yard.

1-13. Mixing Concrete. Except under very unusual circumstances, concrete is mixed in a power-driven drum equipped with blades so arranged as to agitate, stir, and interfold the plastic mass. The size and shape of the drum, its speed of rotation, the arrangement and condition of the blades, and the duration of mixing all influence the quality of concrete. Of these factors it is customary for the engineer to control the rate and duration of mixing. From experience it has been determined that a concrete mixer should revolve at an approximate peripheral speed of 200 fpm.   Speeds greater than about 225 and less than 100 fpm are usually found to be unsatisfactory.   As a general rule, the peripheral speed for greatest mixing efficiency is inversely proportional to the diameter of the drum.

The quality of concrete is improved to a certain extent by longer and more thorough mixing, but, beyond an indefinite optimum time, further mixing may produce a reduction in quality.   Experience indicates that in no instance should the duration of mechanical mixing be less than 1 min after all the materials including the water are in the drum.   Current practice on projects requiring concrete of uniformly good quality is to require a minimum mixing time of $1\frac{1}{2}$ min for a mixer of 1 yd^3 capacity and a greater time for larger mixers.   A safe rule which may generally be used to indicate the desirable duration of mixing is to require $1\frac{1}{2}$ min for the first cubic yard of mixer capacity and an additional $\frac{1}{2}$ min for each additional cubic yard thereafter.

It cannot be too strongly emphasized that the duration of mixing should be counted from the time when all the materials, including the water, have been introduced into the drum.   A batch of materials mixed with less than the designed amount of mixing water is prone to roll and ball up in the mixer and, except with continued mixing of abnormally long duration, will be nonuniform in composition when discharged from the drum.   This condition is particularly evident in large portable mixers of the truck-mounted type.

Job mixers differ mainly in details of design and arrangement of their operating controls rather than in their type or principle of mixing.   The main abuse to which they are subjected is that of overcharging.   They cannot efficiently or satisfactorily function when they are charged with a volume of material much greater than one-half the gross volume of the mixer drum.   They are designed by their manufacturers efficiently and properly to mix a specific volume of concrete, and their capacity is indicated by their catalogue numbers.   Thus a No. 54S mixer is designed to mix 54 ft^3 of concrete, whereas a No. 27E mixer is intended to mix only 27 ft^3.

The use of mixers mounted on truck bodies, for the delivery of batched materials or premixed concrete from a remote central batcher plant to the site of the work, is now commonly encountered.   Sometimes these are merely agitators of concrete already mixed before loading aboard the truck; sometimes they are used to mix the measured materials while in transit; and at other times they are restricted to the conveyance of measured materials and their mixing under supervision at the work site.   They range in capacity from 2 to 8 yd^3.   Most truck mixers are properly designed to produce uniform mixtures.   Best designs are such that the ratio of drum length to diameter is little more than 1:1, the drum being equipped with suitable mixing blades arranged to provide good interfolding and some end-to-end transfer of concrete within the drum. Tilted drums, permitting high discharge of the batch for longer reach when chuting into a form, is also a desirable feature. Most truck mixers are equipped with mechanical water pumps, because rapid transfer of water into the drum materially reduces the time required for satisfactory mixing and improves the uniformity of the concrete throughout the batch.

Portland cement is generally well proportioned and well calcined, but even now some underburned cement having an undesirable component of gamma C_2S may occasionally be produced. Material of this nature is likely to form greater than normal amounts of laitance[1] upon continued mixing. Prolonged or erratic duration of mixing also makes difficult the control of entrained air. For these reasons, the mixing of concrete for variable periods, or for intervals greater than about 30 min, should be discouraged.

1-14. Handling and Placing of Concrete. A most thorough and careful design can be completely defeated by improper practices in the handling of ingredients and the placing of concrete. It is obvious that cement deserves proper storage and protection from moisture, but it is less often observed that aggregates also require attention and care. Aggregates should be protected from contamination with earth, coal dust, plaster, scrap lumber, and similar materials encountered at work sites. They should not be intermingled in storage, because this makes impossible their proper batch measurement. Coarse aggregate is susceptible to size segregation if it is dropped from a crane bucket or other conveyor; it particularly should be stored in bins, or in stock piles having slopes less than the normal angle of repose of the material. Unrestrained dropping of coarse aggregate causes large particles to concentrate at the toe of a slope, and this should be discouraged.

Similar considerations should be observed in the handling and placing of plastic concrete, and this is especially so of less cohesive and overly wet mixtures. Unrestrained dropping, steep chuting, and horizontal flow of concrete are extremely harmful and should not be tolerated. When concrete must be deposited in a deep form or slid down a steep chute, its progress should be restrained by baffles so that it continues to be a cohesive mass of uniform composition, rather than a goblike shower of segregated components. Whenever possible, concrete should be placed in a form at its final resting place in a structure. Chuting of con-

---

[1] Laitance is often a chalky-colored collection of cement and fine aggregates which is flushed to the top of a pour of concrete having an excess of water. It is weak and nondurable. It should be removed before concrete is deposited on the area concerned.

crete into a form at an angle, and flow of concrete within a form, should be permitted only when no other method of placement is possible. Lateral flow of concrete causes the coarse aggregate and the mortar to come to rest at different places in a form, and this may result in porous, honeycombed, frost-sensitive, or other weak regions of unsuitable concrete.

In almost all situations, concrete should be deposited vertically and in horizontal layers of reasonable depth. Great lifts of a single pour encourage segregation of coarser and sedimentation of the finer constituents of mixtures and, moreover, may cause unwanted displacement of forms. The creation, during deposition of concrete, of sloping planes or surfaces should be avoided; first, because such slopes aggravate differential flow and segregation of mixture components and, second, because such pitched surfaces constitute weak planes of potential shearing of the structure.

The deposition of concrete in deep water is especially difficult with respect to the foregoing considerations. This problem is further complicated, inasmuch as turbulence of water at the plastic concrete interface is likely to wash away cement, may greatly aggravate segregation, and causes weak and sloping bearing planes for subsequent lifts. Placement of concrete in water is accomplished by means of vertical pipes called *tremies*, which must be maintained filled with concrete and must have their lower ends submerged in plastic concrete for the duration of a pour. Flow of concrete down the tremie is assisted by increasing the hydraulic head of concrete in the tube, by pounding or vibrating the tremie, and by raising the tube in its entirety so that its lower end is nearly withdrawn from the submerged plastic mass. If the tremie is withdrawn so far as to lose its charge of concrete and become filled with water, then great difficulty is sometimes experienced in reestablishing placement of concrete without harming the partially completed pour. An illustration of this method of underwater placement is shown in Fig. 1-6. Rich mixtures are always necessary for underwater placement to compensate for possible losses of cement.

Many ingenious devices are used for the conveyance of concrete to its place in the structure. Wheelbarrows and two-wheeled buggies have been long used and are reliable conveyances for small projects. When they are equipped with pneumatic tires they are quite easily manhandled while heavily loaded. Cubical hinged-bottom buckets were once used extensively, but these have been superseded almost completely by cylindrical buckets provided with a rolling gatelike bottom discharge. At one time tall towers, provided with a fast elevator much like a mine skip, and equipped with long chutes supported by booms and guy wires, were often used for placing concrete. However, excessively long and too steep chutes caused great segregation of concrete, and such equipment is now

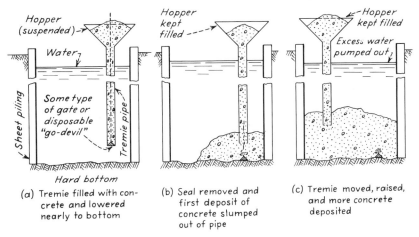

Fig. 1-6 Simplified illustration of depositing concrete by tremie.

in disfavor.   Concrete has been successfully placed at great distances
from the mixer by means of high-speed conveyor belts.   Great quantities
are placed by pumping from the mixer to the forms through several hundred feet of 6- or 8-in. pipe, with vertical lifts of as much as 200 ft.   This
is done with a ramlike pumping machine and has the trade name Pump-
crete.   Concrete is often placed pneumatically to create vertical walls of
thin section and to provide encasement of irregular objects, such as steel
beams and columns, by means of the proprietary Gunite devices.   These
comprise mixing chambers, hose lines, and spray nozzles, operating on
much the same principle as paint-spraying equipment.   In tunnel construction, concrete is often conveyed to the form in side-dumping nar-
row-gauge mine cars.   Although it was once commonly done, it is now
considered poor practice to modify the proportions of good concrete to
suit the peculiarities and limitations of conveying and placing equipment.
Now it is more often required that contractors use equipment suitable
for the kind of concrete desired in the structure.

Concrete, except when it is placed underwater, should be consolidated
by tamping, rodding, and spading.   This is essential to the elimination
of large casual voids, the complete encasement of reinforcement, and the
proper contact of concrete with form faces and embedded fixtures.
Mechanical vibration of the concrete is often used for this purpose, but
this should be considered only as an auxiliary means of consolidation
rather than as a substitute technique.   Spading, in particular, is necessary for the production of good form finishes and smooth exposed surfaces of optimum durability.   Spading should be done in such a manner
as to work coarse aggregate *away* from the form and not toward it.   This
ensures that mortar is against the form where it is needed to minimize

the possibility of honeycomb or air cells, at what might be thought of as the skin of the structure. Smoothly uniform surface textures enhance durability. Coarse aggregate worked away from a form during proper spading tends to return to the face because of semifluid pressure of other coarse particles within the plastic mass.

The *proper* use of mechanical vibration is beneficial to the compressive strength and to the bond strength between concrete and steel. It facilitates the placement of more economical mixtures having lesser slump because of greater aggregate components. The *improper* use of vibrators, or their use on the wrong type of concrete, can cause greater harm than would placement of concrete without any effort toward consolidation. The consistency of concrete to be vibrated, as measured by its slump, should not exceed 3 in., and less workable mixtures are even more desirable. In addition, mixtures placed with vibrators should definitely be undersanded to the extent of being harsh. Workmen soon observe that concrete can be guided as a stream by leading the mass with a vibrator, and this requires less effort than does the use of a shovel. They tend to use them excessively and often move the concrete horizontally with the machine, thus causing unwanted segregation. Vibrators applied directly to the reinforcing bars can displace them from their proper locations and also tend to cause subsidence of the concrete from the undersurfaces of the steel.

Air entrainment is indescribably effective in neutralizing almost all the harmful aspects of poor workmanship, unsuitable equipment, bad construction practices, and willful abuse of concrete. Its use for ensuring durability despite all the pitfalls of construction is almost indispensable. Nevertheless, all concrete should be tamped, spaded, vibrated, and otherwise manipulated as little as is consonant with suitable consolidation and satisfactory finish. Pavement slabs in particular, as well as other pours of large area, and especially those which are made with overly workable mixtures, should be protected from excessive manipulation of concrete in the forms. The surface of such pours, when the concrete is overworked, accumulates a layer of mortar comprising fine sand, water, and less reactive components of the cement. This accumulation is especially weak and susceptible to spalling when frozen. Such accumulations should not be permitted to occur. Where they may have inadvertently done so, they should be removed by chipping, wire brushing, sandblasting, or other abrasive measures and the surface should be broomed with a slurry of rich mortar.

Brief mention should be made of a proprietary aid to construction called the *vacuum process.*[1] In this procedure domelike mats are applied

[1] K. P. Billner, Applications of Vacuum Concrete, *J. ACI*, Vol. 23, No. 7, March, 1952.

to the fresh concrete immediately after strike-off, and special mats are sometimes even used as forms for a pour. These mats are evacuated with an air pump, and this helps to consolidate the concrete and removes some of the batched water from the plastic mass. Leaner mixtures and concrete of somewhat undesirable water content are sometimes placed in this manner, and their residual water content is reduced after placement in the form. Specialized methods of finishing are required when pavement slabs are so constructed, and all the usual curing practices must also be applied. This construction procedure should not be confused with vacuum devices that are sometimes used for lifting and transporting hardened precast units.

1-15. Curing. Some aspects of the water-cement reaction are not yet clearly understood. For that reason, any outline of the process must be somewhat speculative. Nevertheless, the following brief discussion is offered as being helpful to an understanding of the need for curing, even though it is admitted that much of it may be conjectural. The reaction of cement and water, if it is to satisfy our objective of producing strong and durable concrete in a reasonable time, requires for its suitable progress not only time and favorable temperatures; it also requires the close and prolonged association of suitable amounts of the two reagents. The process is a progressive reaction of a liquid phase (water) with a relatively insoluble solid phase (cement). It appears, within our present knowledge, that the reaction continues certainly for many years and may require many decades if it ever is completed. The presence of a great amount of absorbed water within the concrete is essential to its worthwhile progress.

Water can react only at the surface of a cement particle and, at the instant this occurs, the particle is covered with a layer of hydrous calcium silicate gel. The reaction progresses inwardly—in a radial manner—from the new boundary surface of the particle, and this can occur only at such a time as more water may reach the new boundary. Considering only the calcium silicates of cement, their gelatinous reaction products are peculiarly bulky substances of great porosity and surface area. After some of the available water has been chemically fixed as a nonfluid part of the silicate molecule, the gel then seizes and forcefully retains a concentrated, densified, and viscous film of closely packed water molecules upon its great surface. This film of *adsorbed* water is more intimately associated with the solid gel than it is with the remaining free-water phase, and it increases the apparent bulk of the gel. The distinctiveness of adsorption, as compared with absorption, should be apparent. The first is a surface-energy phenomenon that is common to almost all matter but is especially evident of materials of colloidal size, whereas the second is a capillary effect.

It appears that part of the batch water is solidified as a chemical appendage of the silicate molecule; some more of the batch water is arrested, viscously concentrated, and forcefully attached by forces incidental to interfacial energy to the extensive surface of the hydrophilic gel; and the remainder of the batch water permeates the porous structure of the gel. The latter is retained mainly by forces of capillarity. It seems reasonable to assume that only the capillary water is favorably disposed toward ultimate reaction with more cement to form more gel.

The porosity of the hydrous silicate gel is such that the vapor pressure of water exceeds the force of capillarity, the latter being a function of pore size and surface tension. Much of the absorbed capillary water escapes from the concrete by evaporation. Loss from within the concrete of the uncombined and unassociated capillary water can be avoided only if the concrete is maintained in an atmosphere saturated with water vapor, or in contact with liquid water. Since unassociated capillary water is all that can probably react with the remaining nuclei of uncombined cement, it should be prevented from escaping from the pores of the gel. If it should escape, it must be replenished if cement is to continue to react. It is quite likely that some of the interfacially captured water may revert, under the influences of environmental changes such as temperature, to capillary water. It is also a possibility that adsorbed water may slowly react with cement. Nevertheless, the actual behavior of concrete provides a strong implication that the capillary pore water is the more effective reagent.

Curing of concrete is the maintenance of pore saturation or, in effect, establishment of the handy presence of water capable of ultimately forming more gel. It is clear that the smallest cement particles are the first that should be completely converted to gel. This is confirmed by the fact that the same cement clinker, when it is more finely ground, produces better early strength. Present indications are that medium and coarse cement particles are not completely converted to gel even after many years. This probably accounts for the so-called "autogenous" healing of small cracks in excessively strained concrete, the effect even being evident in new cracks in old concrete. Incompletely converted cement particles are fractured by the strain, thus giving them new facial access to water. However, water must be within the concrete and must be free to make the healing reaction. Conservation of pore water is especially necessary during early life, if enough gel is to be formed to produce adequate strength and impermeability.

Curing is also a matter of thermal conservation, inasmuch as rates of chemical reaction are greater at higher temperatures. In fact, the reaction of cement and water at 35°F is almost insignificant in amount, and concrete maintained at that or lesser temperature is dangerously weak and porous. However, if free water is available and temperature of the concrete is subsequently increased, the reaction proceeds at a satisfactory rate. Since newly placed concrete, if it is cold, cannot create enough gel to give it worthwhile strength, there is great danger of damage being done to the concrete by the disruptive forces incidental to the freezing of water within the mass. The hydroxide-laden capillary water freezes at about 28°F.

Exothermic reaction of cement and water provides heat, and the presence of an actual excess required to produce plasticity of the mixture provides water, and these along with the cement are the essentials for the creation of hydrous gel. These favorable conditions can be conserved by holding forms in place and by sealing off exposed surfaces likely to evaporate water. Early covering of the freshly hardened concrete with damp earth, wet burlap, or impervious paper sheets is a practical means of minimizing evaporation. Spray-applied coatings of bitumen and other water repellents are also used for this purpose and, although they are not so effective as the foregoing, still they are much better than complete inattention to reduction of evaporation.

During winter construction of flat slabs and other thin sections, of relatively large surface area, the radiation of heat from the concrete is greater than that supplied by chemical reaction. In such circumstances, the concrete ingredients must be warmed to provide additional reserves of heat. This preheating must not be carried to extremes. Water in particular should not be overly hot because this causes too rapid reaction of the cement and expansion of the mass which, upon cooling, is almost certain to fracture extensively because of its low tensile strength. Good practice aims for temperature of the fresh concrete between 50 and 90°F, and this without using water in excess of 120°F. A better practice is to place the concrete at a temperature somewhat less than the average annual temperature of the locality of the structure. This permits the concrete to cure at nearly the average seasonal shape and size of the structure, thus minimizing high tensile strains in the hydrated cement paste during abnormally cold temperature cycles. It is also necessary during winter construction that additional heat be supplied for several days after installation of the concrete. This is accomplished by enclosing structures with tarpaulins or by other means, and heating the enclosure with fire pots called *salamanders*. Electrical resistance heating has also been of limited application.[1] Pavement slabs are maintained heated by covering them with salt hay in layers sufficiently thick to provide insulation. One danger in localized heating by salamanders and elec-

[1] Chuzo Itakura, Electric Heating of Concrete in Winter Construction, a paper presented at the annual meeting of the ACI, Feb. 27, 1952.

trical heaters is the serious evaporation near the heating units caused by too high a temperature in an attempt to keep parts further away at the proper temperature.[1]

An illustration of the lack of proper curing can often be seen in the disintegration of concrete sidewalks and curbs. Too often the contractor—or even the do-it-yourself homeowner—pours the concrete on parched ground, finishes off the surface, and then leaves the work to take care of itself. In hot weather the sponge action of the dry ground steals the water from below and the hot sunshine on top evaporates the water from the top so that, between them both, there is not enough water left to produce the needed strength and durability. The result is sometimes complete disintegration in a few years. Moistening the ground first and keeping the concrete wet for a week probably would have avoided the unfortunate results.

From the preceding, it is obvious that special care is needed when curing concrete in desert conditions, where the humidity of the atmosphere is extremely low. Sprinkling and covering with wet burlap, paper, and similar materials is generally ineffective because the evaporation is so rapid that it is practically impossible to keep the concrete moist. Sealing compounds may help considerably, but there is a question whether or not vapor caused by heat from the sun will form under the membrane and produce tiny ruptures of the seal so that water escapes unnoticed. One of the most effective ways of ensuring a fair degree of curing under arid conditions is to leave the forms in place for 2 or 3 weeks and, after the concrete has hardened sufficiently, to cover top surfaces with about 2 in. of sand which is frequently sprinkled.

Extensive discussions of the influence of good and poor curing on strength and other physical properties of concrete are available.[2] Strength is very seriously impaired by poor curing practices, and it is suggested that reference be made to the technical literature. Although prolonged and efficient curing is most desirable, it is obvious that some compromise with ideals must be accepted in practice. To facilitate speedy construction and economy in the building of forms, it is desirable that the engineer establish a reasonable schedule of curing that is adequate for the integrity of his structure. This should also be fair to the builders of structures, most of whom reason that, when the concrete has

[1] On the other hand, artificial cooling is sometimes used to reduce the heat of chemical action and the subsequent shrinkage of mass concrete. Refrigerants applied through piping and pumping may be used, and ice substituted in lieu of water can be resorted to. Retarding admixtures and pozzolanic materials may be used to slow down the rapidity of the chemical reactions, thus reducing the initial expansion and the later shrinkage.

[2] Curing of Concrete, a round-table discussion by various persons, *J. ACI*, Vol. 23, No. 9, May, 1952.

Table 1-6   **Curing and Removal of Forms**

Item	Forms held in place after concrete has set, days		Concrete kept wet after setting, days	
	Cold weather	Warm weather	Cold weather	Warm weather
Self-supporting floors and beams	14	10	7	10
Thin walls, 8 in. and less	4	3	7	7
Thick walls, and massive piers	3	2	7	7
Floors, and pavements on soil			4	7

set, it is at least hard and are hopeful that it is also sufficiently strong for them to forget it and proceed with other obligations. For ordinary concrete, not of a high-early-strength nature, the general procedures shown in Table 1-6 are suggested. It should be realized that wetting of concrete for more prolonged periods is necessary during hot summer months, and at all times in arid climates. Because continuous wetting during dry spells is troublesome and expensive it is often done fitfully and is sometimes neglected. Such bad practices are particularly detrimental to pavements, since shrinkage cracking and unreasonable wear are thus encouraged. Apparent hardness of dry surfaces should not be mistaken for good concrete inasmuch as gel formation is retarded in such circumstances.

1-16. Forms. Forms[1] are intended to define the contour and locate the position of individual members with reference to the structure as a whole. To limit satisfactorily the size, shape, and position of parts of the structure, it is necessary that forms be built to resist the forces imposed upon them. It should never be forgotten that concrete is a semifluid during its early life. It is usually assumed that it exerts a horizontal pressure equal to the hydrostatic head of a liquid weighing 145 pcf.

Concrete, when it is vibrated, acts as a fluid throughout its depth; consequently, the full hydrostatic head should be considered. In contrast, concrete placed without the aid of vibration exerts a fluid pressure for a depth depending upon the rate and temperature of placement. The consolidation and interlocking of aggregate and the initial setting of cement tend to neutralize fluidity; therefore, the pressures are actually somewhat less than are estimated for the full hydrostatic head. From literature of the Universal Form Clamp Company it is indicated that

[1] Formwork for Concrete, *J. ACI*, SP-4, 1963.

concrete having a temperature of 50°F and placed at a rate of 6 vertical ft/hr exerts a maximum horizontal thrust of 1,030 psf, at a 9-ft depth of head.   At greater depths the horizontal pressure is reported to be less until, at a head of 12 ft, it is reported to be 870 psf.   The foregoing maximum pressure is approximately 80 per cent of the theoretical pressure that might be expected at 9-ft depth of action.   Conditions such as higher ambient temperature, and others that are favorable to earlier setting of concrete, result in lesser maximum form pressures and, furthermore, the effects are evident at lesser depths.   For example, the literature referred to indicates that concrete placed at 70 rather than at 50°F exerts only 740 rather than 1,030 psf maximum pressure, and the maximum effect is evident at only 7- rather than at 9-ft depth of head.

The foregoing comments are indications of the problems involved in the design and fabrication of adequate forms.   Discussion of other aspects of the problem of forms will be found in Chap. 11.

1-17.   Reinforcing Steel.   Concrete cannot be relied upon to withstand much tensile stress.   This deficiency is overcome by embedding steel bars in those parts of a section that are subjected to tension.   The two materials act in conjunction with each other, each doing the work for which it is best suited.   The combination acting together as a unit, concrete resisting compression and steel resisting tension, is called *reinforced concrete*.   It can be made a strong, durable, and economical system of construction, and it has been proved satisfactory for a great variety of structures.

Much has been said about concrete, and little about steel.   The latter is of equal interest and importance.   However, because of greater technical knowledge of the material, and better control during manufacture, reinforcement is a standardized material of great dependability.   It is manufactured in the form of plain, deformed, or twisted bars and rods of various cross-sectional areas, ranging from the area of a $\frac{1}{4}$-in. to nearly the area of $1\frac{1}{2}$-in. round stock, or even larger.   It is also regularly available as wire, and as wire mesh in various sizes and combinations of weave.

We are experiencing a transitional improvement of the surface texture and some modification of available sizes of reinforcement.   The strength and other physical properties, excepting size and shape of surface deformations, are common to both the older and the newer styles of bars.[1]   However, the new system of numbering rather than the formerly used designation of bars by their nominal size, along with specific requirements for surface texture intended to improve mechanical bond of the bar to the

[1] Billet-steel Bars for Concrete Reinforcement, ASTM Designation: A 15; Rail-steel Bars for Concrete Reinforcement, ASTM Designation: A 16; and Axle-steel Bars for Concrete Reinforcement, ASTM Designation: A 160.

concrete, are now commonly used.[1] The numbering system is such that, beginning with 3 and including 11, each number represents the eighths of an inch diameter of an equivalent round bar having the same weight per foot as the deformed bar.[2] Data pertaining to both the older and newer bar sizes are given in Tables 1 to 3 of the Appendix. Allowable bond stresses for both styles are given in the following section in Table 1-8.

Reinforcing bars are hot-rolled from car axles, T rails of standard section, and new steel billets. Special high-strength steels are also being used. Because of greater dependability from the standpoint of favorable bending characteristics, it is desirable, in the case of important structures, to require the use of bars manufactured from new billets. Bars are produced in what are called structural, intermediate, and hard grades, the essential differences in these grades being increasingly greater ultimate strengths and higher yield point values accompanied by progressively less ductility. The new high-strength steels have been designed to obtain both the high strength and ductility. Naturally, the bars with higher yield value permit the use of greater working unit stresses, but since concrete is generally weak in tension, high stresses in the bars may cause prolific cracking of the concrete. The best balance between these properties is a matter of personal opinion, professional judgment, and the satisfaction of building code requirements.

Wire and wire mesh are ordinarily available in sizes ranging from the smaller No. 14 to the larger No. 7-0 gauges and are manufactured from cold-drawn steel wire.[3] This is fabricated as mesh by welding in a large variety and combination of weaves. The ultimate strength of cold-drawn reinforcing wire is approximately equal to that of hard grades of hot-rolled bars, but the yield point of this material is appreciably greater than that of rolled bar steel. Table 1-7 shows data about reinforcement.

The modulus of elasticity in tension of plain carbon steel varies between 28,000,000 and 31,000,000 psi. It may be assumed to be a constant having a value of 30,000,000 psi, although the Code specifies 29,000,000 psi. At normal atmospheric temperatures, steel is considered to be a truly elastic material throughout a loading range below a certain value termed its *proportional*, or *elastic*, limit. The proportional limit is not coincident with the yield point as determined by ordinary acceptance tests, but is sufficiently close for all practical purposes.

[1] Minimum Requirements for the Deformations of Deformed Steel Bars for Concrete Reinforcement, ASTM Designation: A 305.

[2] Nos. 9 to 11 depart somewhat from this rule in that they are equivalent to former 1-in., $1\frac{1}{8}$-in., and $1\frac{1}{4}$-in. square sizes.

[3] Cold-drawn Steel Wire for Concrete Reinforcement, ASTM Designation: A 82; Fabricated Steel Bar or Rod Mats for Concrete Reinforcement, ASTM Designation: A 184; Welded Steel Wire Fabric for Concrete Reinforcement, ASTM Designation: A 185.

Table 1-7   **Recommended Permissible Unit Stresses, Reinforcement**

Material		Max value, psi	Recommended value, psi
Tension:			
Structural grade	f_s	18,000	18,000
Bars $\frac{3}{8}$ in. or less in one-way slabs not more than 12-ft span	f_s	$0.5f_y$ or 30,000	
Deformed bars with $f_y = 60,000$ psi or more and sizes No. 11 or less	f_s	24,000	24,000
All other reinforcement	f_s	20,000	20,000
Column reinforcement:			
Spiral columns (vertical):	f_s	$0.4f_y$ or 30,000	
$f_y = 35,000$	f_s		14,000
$f_y = 40,000$	f_s		16,000
$f_y = 50,000$	f_s		20,000
$f_y = 60,000$	f_s		24,000
Tied columns (vertical):	f_s	$0.34f_y$ or 25,500	
$f_y = 35,000$			11,900
$f_y = 40,000$			13,600
$f_y = 50,000$			17,000
$f_y = 60,000$			20,400
Spirals (yield strength):	f_y		
Hot-rolled, intermediate grade	f_y	40,000	40,000
Hot-rolled, hard grade	f_y	50,000	50,000
Hot-rolled, A432 grade and cold-drawn wire	f_y	60,000	60,000

Steel should be well embedded within the concrete, to protect it from corrosion and from damage by fire. If the concrete is plastic during placement and is made neither so dry as to be porous or honeycombed nor so wet as to be highly permeable to water, it provides good protection against corrosion of the reinforcement. Concrete, because its absorbed water is saturated with calcium hydroxide, is highly alkaline as is indicated by a hydrogen-ion concentration of somewhat more than 9.0 pH. Such an environment makes almost impossible the oxidation of steel and, in most instances, the demolition of old structures has shown that the steel was well protected. Ordinarily, a cover of $1\frac{1}{2}$ or 2 in. is desirable for steel behind large flat surfaces exposed to air, 2 to $2\frac{1}{2}$ in. at corners in air, and 3 in. when concrete is exposed to fresh water or moist earth. Local regulations with regard to fireproofing may sometimes dictate greater coverings, but too much concrete beyond the outermost bars of steel may aggravate spalling of the concrete. Rusting of the reinforcement is invited when cover of steel is inadequately thin.

Much controversy exists regarding the proper cover that should be used over reinforcement in structures immersed in or wetted by sea water.   At least 3 in. cover is desirable and 4 in. is probably better. Even though types II and V cements are resistant to the sulfate constituents of sea water, and concrete made with these cements in company with entrained air is unquestionably resistant to such an environment, this may not afford enough protection to steel.   Some engineers are of the opinion that very rich mixtures having a cement factor of 7 to 8 bags per yd^3 are necessary for suitable protection of steel.   They believe that the extra cost of concrete is justified when interference with service or replacement would be costly.   Such rich concrete is, however, most susceptible to thermal cracking and this permits direct access of sea water to steel.   Bituminous coatings, of a coal tar rather than an asphaltic base, and certain epoxies applied to the concrete within the tidal and spray range, are helpful in providing protection in such situations, but the coatings must be renewed every few years.   An unfortunate example of the results of improper workmanship or materials is shown in Fig. 1-7.

Fig. 1-7   A situation illustrating the need for good materials and the best workmanship for concrete construction that is exposed to sea water, especially in the tidal range.

1-18.   Allowable Unit Stresses and Safety Factor.   The allowable unit stress and safety factor are interdependent and are based upon our experience with reinforced-concrete structures.   The safety factor is the proportion by which the ultimate strength of the concrete, or the elastic limit of the steel, exceeds the permissible unit stress used in the design. Values shown in Table 1-8 are the maximum permissible unit stresses to be used in the design of structures as given in the American Concrete Institute Building Code Requirements for Reinforced Concrete (ACI 318-63).   (This will be referred to herein as the "Code.")   There are other local rules and regulations, or codes, guiding both the design and the construction of reinforced-concrete structures but the authoritative nature and reliability of the ACI Code are of nationally recognized significance.

The Code formula for the modulus of elasticity of concrete, as shown for the determination of n in Table 1-8, is

$$E_c = w^{1.5}33\sqrt{f_c'}$$

For stone concrete weighing 145 pcf without reinforcement, this formula gives the following:

f_c', psi	2,500	3,000	4,000	5,000
E_c, psi	2,880,000	3,150,000	3,640,000	4,060,000

Table 1-8 shows recommended unit stresses for concrete of various compressive strengths.   In general, these are based on the ACI Code. However, the allowable punching shear and the recommended bond stresses for bars have been introduced by the author.   These will be discussed later in this text.   The stated unit stresses are for normal concrete having an assumed weight of 145 pcf.   Lightweight concretes are special products whose strength should be determined or ascertained from the experience of the producers.

Some students may reach the opinion that the ACI Code, so often referred to here, is extremely arbitrary.   It is, instead, the condensed wisdom of experienced engineers who know the many facets of the problem of designing structures.   By open discussion of their opinions and experiences, and by mutual compromises, they have agreed upon this guide that serves the useful purpose of ensuring their less experienced colleagues of satisfactory attainment of safe results.   Nevertheless, good judgment is still needed in the design of structures, inasmuch as there may be inconsistencies within any code.   If good judgment and conservatism make it seem desirable to do so, there is no reason why an engineer should not make a structure better than the minimum limits set by such specifications of safe practice.

Table 1-8 Recommended Permissible Unit Stresses, Concrete

Allowable unit stresses

Description		For any strength of concrete $n = \dfrac{29,000,000}{w^{1.533}\sqrt{f_c'}}$	Max value, psi	For strength of concrete shown below			
				$f_c' = 2,500$ psi $n = 10$	$f_c' = 3,000$ psi $n = 9$	$f_c' = 4,000$ psi $n = 8$	$f_c' = 5,000$ psi $n = 7$
Flexure f_c:							
Extreme fiber stress in compression	f_c	$0.45f_c'$		1,125	1,350	1,800	2,250
Extreme fiber stress in tension in plain concrete footings	f_c	$1.6\sqrt{f_c'}$		80	88	102	113
Shear v (as a measure of diagonal tension at a distance d from the face of the support):							
Beams with no web reinforcement	v_c	$1.1\sqrt{f_c'}$		55	60	70	76
Joists with no web reinforcement	v_c	$1.2\sqrt{f_c'}$		61	66	77	86
Beams with properly designed web reinforcement	v	$5\sqrt{f_c'}$		250	274	316	354
Slabs and footings $\left(\dfrac{d}{2}\text{ outside of periphery of concentrated load}\right)$	v	$2\sqrt{f_c'}$		100	110	126	141
Punching	v_T	$0.15f_c'$	$0.2f_c'$	375	450	600	750
Bond u (A305 bars):							
Top bars (tension)	u	$\dfrac{3.4\sqrt{f_c'}}{D}$	350				
All other bars (tension)	u	$\dfrac{4.8\sqrt{f_c'}}{D}$	500				
	u	$6.5\sqrt{f_c'}$ $\frac{1}{2}\times u$ for A305	400				
Recommended for top and bottom bars							
Size No. 3–No. 7	u			225	250	275	300
Size No. 8–No. 11	u			150	175	200	225
Compression bars							
Plain bars							
Recommended	f_c			113	125	138	150
Bearing f_c:							
On full area	f_c	$0.25f_c'$		625	750	1,000	1,250
On one-third area or less	f_c	$0.375f_c'$		938	1,125	1,500	1,875

The problems given in the text are for the purpose of demonstrating the design of reinforced concrete as it is done in practice. A wide range of unit stress is used in problems, mainly to illustrate and emphasize that the quality of the materials, the uncertainties of construction, and the nature of the work are factors that must be considered. For example, the unit stress that may permissibly be used in steel reinforcement is not the same in different building codes; the obtainable quality of concrete or of steel may vary; wartime restrictions or other necessities for economizing on steel tonnage or individual opinion or professional judgment of the engineer may influence a design. As a matter amenable to judgment the prevailing situation regarding older styles of deformed reinforcement might be considered. Table 1-8 implies that these obsolescent types should now be classified as plain bars. Nevertheless, the Joint Committee Report referred to in Art. 1-11 recommended for such bars a permissible bond unit stress of $u = 0.05f_c'$ but not to exceed 200 psi for beams and one-way footings; this was increased to $u = 0.056f_c'$ in two-way footings when the bars were hooked. Whenever the older style of bar is used, it is suggested that these former criteria of safe practice might also be used instead of A 305. The following are cited as other examples of variations in practice that are matters amenable to judgment with respect to specific conditions of service:

1. Culverts or tunnels under deep embankments, bins designed for maximum filling, foundations supporting massive superstructures, and, in general, most structures carrying relatively large determinable loads compared with any possible future increase—these may take advantage of greater allowable unit stresses that might be used in their design, provided that the accompanying cracking of concrete in regions of tension is not objectionable.

2. Bridges, crane girders, some floors, and other structures or members that could at some time be subjected to great loads, impacts, earth tremors, and other things causing forces of unknown magnitude—these should be designed conservatively. This should also be done in the case of important structures which, in the event of their failure, could cause great loss of life or of their economic value.

All should realize that the saving of steel resulting from an increase in the permissible unit stress used in a design is not proportional to that increase. In fact, in most large structures properly designed, the saving is likely to be in the neighborhood of 5 to 10 per cent of that relative increase. Obviously, this is a small part of the cost of the reinforcement which, in turn, is usually a rather small percentage of the cost of the job. However, the safety of the structure may really decrease as the permissible unit stress is raised, hence the need for good judgment in determining these matters.

Table 1-9   Allowable Unit Stresses in Materials as Specified by American Association of State Highway Officials (AASHO), psi

Concrete				
f_c'				
2,000–2,400 $n = 15$	2,500–2,900 $n = 12$	3,000–3,900 $n = 10$	4,000–4,900 $n = 8$	5,000 and over $n = 6$

Flexure:
  Extreme fiber in compression $f_c = 0.4f_c'$
  Extreme fiber in tension, plain concrete, primarily in footings $f_c = 0.03f_c'$
  Extreme fiber in tension, reinforced concrete $f_c = 0$

Shear:
  Beams without web reinforcement:
    Longitudinal bars not anchored, or plain concrete footings $v_c = 0.02f_c'$
                                                                          (max 75 psi)
    Longitudinal bars anchored $v_c = 0.03f_c'$
                                                                          (max 90 psi)
  Beams with web reinforcement $V = 0.075f_c'bjd$
  Horizontal shear in shear keys between slab and stem of
    T beams and box girders $v = 0.015f_c'$

Reinforcement (deformed)		
	Structural grade	Intermediate, hard, and rail steel grade
Tension:		
In flexural members	18,000	20,000
In web reinforcement	18,000	20,000
Compression:		
In columns	13,200	16,000
In beams if well tied	$-2n \times f_c$ alongside	(max 16,000)
Straight or hooked ends, exclusive of top bars:		
In beams, slabs, and one-way footings	$0.10f_c'$ (max 350 psi)	$0.10f_c'$ (max 350 psi)
In two-way footings	$0.08f_c'$ (max 280 psi)	$0.08f_c'$ (max 280 psi)
Top bars of beams with more than 12 in. of concrete under them	$0.06f_c'$ (max 210 psi)	$0.06f_c'$ (max 210 psi)
Bars larger than No. 11	$\frac{2}{3}$ of above values	

These comments are generally applicable to the concrete also, although the strength of the concrete is not so likely to be critical and the percentage of saving in the cost due to increasing permissible unit stresses in the concrete is usually far less than in the case of the steel.

1-19. Importance of Workmanship.   The need for honest and intelligent workmanship in the field during the building of a concrete structure has been emphasized here.   It should be emphasized again and again. The best of designs can be ruined if the intent of the plans is not carried out in the field.   Proper reinforced-concrete construction depends upon men—men who understand the action of structures, men who know the characteristics and the limitations of the material that they are handling, and men who are conscientious and determined to conduct their work with honor to themselves and with credit to their profession.

BEAMS SUBJECTED TO BENDING

2

2-1. Introduction. In general, a beam is a member which carries loads that act transversely with respect to its longitudinal axis so as to cause the member to bend. The most simple reinforced-concrete beams are those whose cross sections are rectangular in shape. In fact, an ordinary floor slab, like that shown in Fig. 2-1, may be thought of as a series of such beams of unit width b (usually 12 in.), represented by the piece $ABCD$ having a depth equal to t and a span equal to L.

The effective span of a beam should be taken equal to the clear distance between supporting members plus some allowance for end bearing but not more than the distance between the centers of the supports or the clear span plus the depth of the beam. If a beam or slab is not poured integrally with its supports, the allowance for bearing need not exceed the depth of the beam or slab. Some engineers assume the span to be the clear distance plus 4 in. for light members or plus 12 in. for heavy construction. When the end is supported by an edge beam or in a manner that restrains it only slightly, the end is assumed to be at the center of the support.

In dealing with such a beam there are two types of problems to consider. One is the analyzing or testing of existing and assumed beams of given dimensions to compute what forces they can withstand safely; the other is the designing or proportioning of beams to support certain given forces or loads. Two ways of attacking these problems will be explained. One is the *working-stress design* method (WSD), sometimes called the straightline or elastic theory, in

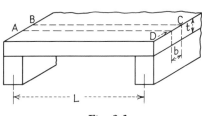

Fig. 2-1

which the computations are based upon the use of certain allowable unit stresses in the materials.   The other is the *ultimate-strength design,* or ultimate-load, theory (USD) in which the calculations are based upon the ultimate resisting capacities of the materials, with some *load factor* (or safety factor) applied to determine the safe bending moment to be permitted in practice under working conditions.   The student should be familiar with both procedures.[1]

A good training in theory is essential for any designer.   However, when one enters practical engineering work he will find that the design of reinforced-concrete structures is influenced to a large extent by general specifications, codes, and customary practices.   Such codes may change, and they should do so as the art progresses and our knowledge increases. Each engineer should have a copy of the latest codes, such as the Building Code Requirements for Reinforced Concrete (ACI 318-63) that has been prepared by the American Concrete Institute.   The designer should be careful to comply with whatever specifications, building codes, and regulations govern his work, bearing in mind that they are general rules to be followed unless he has important reasons for making the construction even better and safer.   For bridge design, the Standard Specifications for Highway Bridges, published by the American Association of State Highway Officials, dated 1962, is applicable.   It will be called "AASHO" when referred to in this text.

2-2.   Table of Symbols and Their Meanings.   Most of the symbols which are used in the Code have been adopted for this text as far as it has been practicable to do so.   Those which are used in this chapter are grouped for convenience.   They are fundamental symbols for use throughout the work.

The list is as follows:

a = depth of compression area assumed for stress block in USD, in.
A_s = area of steel in tension, in.[2]
A_s' = area of steel in compression, in.[2]
A_{sf} = area of reinforcement to develop compressive strength of overhanging flanges in I and T sections, in.[2]
b = width of rectangular beam or flange of T beam, in.
b' = width of stem of T beam, or web of I section, in.
c = distance from extreme compression fiber to neutral axis for USD, in.

[1] For the ultimate-strength theory, see Charles S. Whitney, Plastic Theory of Reinforced Concrete Design, *Proc. ASCE,* Vol. 66, December, 1940; V. P. Jensen, Ultimate Strength of Reinforced Concrete Beams as Related to the Plasticity Ratio of Concrete, *Univ. Illinois Eng. Expt. Sta. Bull.* 345; articles by Corning, Anderson, Hognestad, Siess, Reese, and Lin, *J. ACI,* June, 1952.

C, C_c = total force of compression in concrete, lb or kips.

C', C_s = total force of compression in steel, lb or kips.

d = depth from compression face of beam or slab to the center of gravity of the longitudinal tensile reinforcement, in. (called *effective depth*). In some special problems with reinforcement distributed over a considerable depth of the member, d may be measured from the compression face to the row of bars that is farthest from it.

d' = depth from compression face of beam or slab to center of gravity of the longitudinal compressive reinforcement, in.

d_1 = distance between centers of gravity of tensile and compressive reinforcement in USD, in.

D = total overall depth of a beam, in.

E_c = modulus of elasticity of concrete in compression, psi.

E_s = modulus of elasticity of steel in tension or compression, psi.

ϵ = unit strain in concrete, in./in.

ϵ' = unit strain in steel, in./in.

f_c = compressive unit stress in extreme fiber of concrete, psi.

f_c' = ultimate compressive strength of concrete, usually at age of 28 days, psi.

f_s = tensile unit stress in longitudinal reinforcement, psi.

f_s' = compressive unit stress in longitudinal reinforcement, psi.

f_y = yield point stress of tensile and compressive reinforcement, psi.

I_c = moment of inertia of transformed section in terms of concrete, in.[4]

j = ratio of distance between centroid of compression and center of gravity of tensile reinforcement or the extreme row of tensile reinforcement to the depth d.

k = ratio of distance between the compression face of the beam or slab and the neutral axis to the depth d.

k_1 = factor to be used in computing the depth a of stress block in USD.

kip = 1,000 lb (sometimes abbreviated as k).

L = span of beam or slab, usually in ft.

m = $f_y/0.85f_c'$.

M = bending moment, ft-lb or in.-lb.

M_c = internal resisting moment in terms of the strength of the concrete, in.-lb.

M_s = internal resisting moment in terms of the strength of the steel, in.-lb.

M_u = ultimate resisting moment, in.-lb.

n = ratio of modulus of elasticity of steel to that of concrete.

p = ratio of area of tensile reinforcement to the effective area of concrete in beams and slabs = A_s/bd.

$p' = $ ratio A'_s/bd.

$p_b = $ reinforcement ratio producing balanced conditions at ultimate strength.

$p_f = A_{sf}/b'd$.

$p_w = A_s/b'd$.

$q = A_s f_y/bdf'_c$.

$S_c = $ section modulus of transformed section in terms of concrete, in.[3]

$S_s = $ section modulus of transformed section in terms of steel, in.[3]

$t = $ thickness of slab or flange of T or I section, in.

$T, T_s = $ total force of tension in steel, lb or kips.

$U = $ required ultimate load capacity of section.

$bd = $ effective area of beam or slab, in.[2]

$\phi = $ capacity reduction factor for USD.

2-3.   General Action of Reinforced-concrete Beams.   A reinforced-concrete beam differs considerably in its internal action from one made of homogeneous materials such as steel.   The internal forces are generally provided by two separate partners—the concrete to resist compression and the steel to resist tension.

A beam must curve if it is subjected to a bending moment because the parts that are in compression must shorten and those which are subjected to tension must elongate.   The typical method that is used here to enable one to visualize these actions is shown in Fig. 2-2.   The short irregular lines represent cracks.   They are drawn in the regions where tension exists—the weak portions of the concrete which must be *reinforced* with steel.   No cracks are shown in the portions that are subjected to compression because the concrete there will be effective by itself and cracks will not occur where there is pressure.   Such pictures do not mean that concrete beams always crack so excessively; they are to indicate the condition that the beams may reach if they are loaded sufficiently.   This

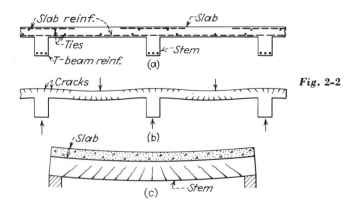

Fig. 2-2

visualization of excessively deformed members as almost a series of cracked portions that are tied together by the bars acting somewhat as a chain is helpful in understanding the action of reinforced concrete. If the bars are not parallel to the tensile force, it is satisfactory to assume that the effective area of a bar is its normal cross-sectional area times the cosine cf the angle between the longitudinal axis of the bar and the direction of the force.

If one were to imagine that he cut out a piece of the beam of Fig. 2-2(b) near where a concentrated load is shown by an arrow, he might picture the internal resisting forces as in Fig. 2-4(a) or as in Fig. 2-8(a). In each case the resisting moment consists of a pair of equal and opposite forces in the materials, with some lever arm between them.

2-4. Distribution of Stresses in a Reinforced-concrete Beam.
First of all, examine Fig. 2-3 in which (a) represents the general character of the stress-strain curve of the reinforcement and (b) pictures that of the concrete. Sketch (c) shows the cross section of a rectangular beam with reinforcing bars in the bottom. If this beam is bent so as to cause compression in the top and tension in the bottom, it will be assumed that the deformations or strains of the materials at any point will vary in propor-

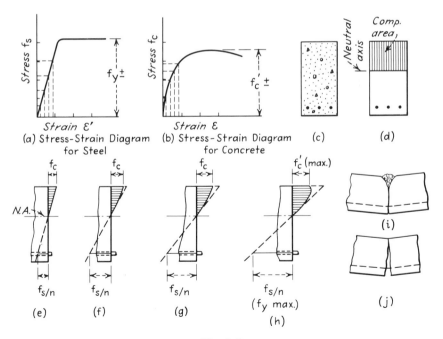

Fig. 2-3

tion to the distance of that point from a point of zero strain called the *neutral axis*. At very small loads applied for the first time, both the concrete in the bottom and the steel will resist tension, but such loads are too small for practical consideration. However, as the loads and the bending increase, the concrete near the bottom will soon reach a tensile stress that causes it to crack. Then the bars must resist practically all of the tension alone, because the beneficial effect of any tension in a small area of uncracked concrete just below the neutral axis will be negligible. Now the member may be assumed to have an effective cross section like that pictured in Fig. 2-3(*d*), where tension in the concrete below the neutral axis is neglected entirely. Furthermore, once the concrete has cracked in tension under the first application of loading it will not resist tension a second time even though the first heavy load has been removed.

As the beam bends more, it would seem that the stresses in the concrete will vary from zero at the neutral axis to a maximum at the extreme top "fiber." However, Fig. 2-3(*b*) shows that the concrete is not truly elastic. Therefore, even though the strains increase, the stresses in the concrete will not do so proportionately. The assumed general character of the distribution of compressive stresses in the concrete as the bending increases is pictured in sketches (*e*) to (*h*). As this increase of strain progresses toward a maximum, the top fibers yield plastically, or at least they continue to deform without offering proportionately increased resistance, causing the pressure diagram to be somewhat as shown in (*h*). Additional loading might then cause the beam to fail in compression somewhat as pictured in Fig. 2-3(*i*). Such failures are likely to be sudden and without warning; hence they are dangerous.

On the other hand, as the loads increase, the bars resist tension in proportion to their strain until the elastic limit is reached. Beyond this point, such large plastic deformation of the steel would occur that the beam might "pop open" and fail in tension or by local crushing of the concrete at the top of the crack, as illustrated in Fig. 2-3(*j*). Such a failure is usually accompanied by such serious cracking and deflection that trouble can be detected in time to remove the excessive loads and to avoid collapse.

During the loading, after the concrete of the tensile area has cracked, the neutral axis may not remain stationary as indicated in Fig. 2-3(*e*) to (*h*), inclusive. Certainly it shifts upward as the tensile failure of Fig. 2-3(*j*) approaches, and it may shift downward as the compression failure of sketch (*i*) occurs.

The working-stress theory assumes that the stress distribution in the concrete at working stresses of approximately $0.45f'_c$ is like that of Fig. 2-3(*e*). At ordinary working loads, this assumption seems to be reasonably satisfactory. It is probably impossible to predict accurately what the real stresses in a reinforced-concrete beam will be. However, such beams can be made to support loads safely.

At ultimate loads, the compression diagram may not be exactly as shown in Fig. 2-3(h), but it can be approximated satisfactorily. This will be explained later.

In both theories, the stress in the tensile steel is assumed to be that caused by the strain that would be in the concrete at the same distance from the neutral axis if the concrete could resist such a strain. That is why the steel stress is labeled f_s/n in Fig. 2-3, where n is the modular ratio for which values are given in Table 1-8.

To understand this, assume that a cylinder of concrete has a reinforcing bar embedded longitudinally in it with the ends of the bar flush with those of the cylinder. The cylinder is placed in a testing machine and compressed 0.005 in. Then, since $\Delta L = PL/AE$ in general for an elastic material, and since ΔL is the same for both the steel and concrete,

$$\Delta L = 0.005 \text{ in.} = \frac{P_c}{A_c} \times \frac{L}{E_c} = \frac{P_s}{A_s} \times \frac{L}{E_s}$$

$$f_c = \frac{P_c}{A_c} \quad \text{and} \quad f_s = \frac{P_s}{A_s}$$

Since L is constant or the same for both,

$$\frac{f_c}{E_c} = \frac{f_s}{E_s} \quad \text{or} \quad f_s = \frac{f_c E_s}{E_c} = n f_c$$

This is assumed to apply for both tension and compression. Another way to look at this is to say that a given area of steel is equivalent to n times that area of concrete. This is called a *transformed area* substituting for the steel, and this principle will be used later.

Experiments seem to indicate that concrete in flexural action will resist a higher unit stress in compression than it will in 6- by 12-in. cylinders. However, because of our present limited knowledge of this action, the cylinder strength will be used as the measure of the ultimate strength f_c'.

2-5. Beams with Tensile Steel Only; Working-stress Theory.

Figure 2-4(a) shows a small portion of a beam that is isolated as a free body. Neglecting the dead load and the shearing forces, let M represent the magnitude and the direction of the bending moment which is caused by the external loads. Since the beam is in equilibrium, this moment must be counteracted by a moment of equal magnitude and opposite direction. Obviously, this resisting moment must be provided by the strength of the materials composing the beam itself.

As previously explained, the tensile strength of the concrete is rather low and unreliable, and it is to be neglected.   Therefore, the tensile force T in Fig. 2-4(a) is assumed to be provided by the steel alone.   It is sufficiently accurate to assume that a reinforcing bar has an intensity of stress f_s which is equal to the unit stress theoretically at its center.   This value, multiplied by the area of the steel, gives the tensile force

$$T = A_s f_s \tag{2-1}$$

For multiple layers of bars, the total area of the steel is often assumed to be concentrated at the center of gravity of the group of bars, and the tensile stress f_s is computed at that center of gravity.   This is assumed to be an average stress on all the bars even though those farther from the neutral axis may be somewhat more highly stressed.

Figure 2-4(a) shows the assumed triangular distribution of stress in the concrete.   Therefore, the maximum compression occurs at the top fibers, and the stress decreases to zero at the line O-O, the neutral axis. The height of this area which is in compression is called kd.   Since this assumed pressure diagram of the compressive forces is a triangular wedge, the resultant force C must equal the volume of the wedge, and it must be located at the center of gravity of this imaginary solid, which is at a distance $kd/3$ from the top fibers.   It is therefore clear that

$$C = \frac{1}{2} f_c(kd)(b) = \frac{1}{2} f_c kbd \tag{2-2}$$

It is also apparent that the magnitudes of these forces C and T must be equal in order to have equilibrium, for which $\Sigma H = 0$.   Therefore,

$$\frac{1}{2} f_c kbd = A_s f_s \tag{2-3}$$

From Fig. 2-4 it is easily seen that the lever arm of the internal resisting couple with forces C and T is the distance between the points of

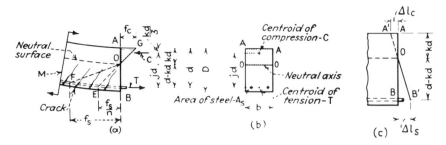

Fig. 2-4

application of these forces, jd. Thus,

$$jd = d - \frac{kd}{3} \quad \text{or} \quad j = 1 - \frac{k}{3} \tag{2-4}$$

Inasmuch as the moment of a couple equals the magnitude of either of the equal opposite forces times the lever arm between them,

$$M_c = \frac{1}{2} f_c kbd(jd) = \frac{1}{2} f_c kjbd^2 \tag{2-5}$$

and

$$M_s = A_s f_s jd \tag{2-6}$$

Formulas (2-5) and (2-6) are fundamental in the analysis and design of rectangular beams with tensile reinforcement only, and with no tension in the concrete, when WSD is employed. Both formulas are needed because reinforced concrete is not a homogeneous material. In practical design, at ordinary allowable stresses, the steel is usually the critical part of most reinforced-concrete beams.

If the beam of Fig. 2-4 is progressively loaded to failure, the concrete may yield under compression before the steel gives way under tension. If so, the beam is said to be *overreinforced* because it has more than the necessary amount of steel. In case the reverse is true, and the steel fails first, the beam is *underreinforced*. From the standpoint of the efficient use of materials, the best design is one that results in a beam in which the maximum safe working strength of the concrete and that of the steel are reached simultaneously—a *balanced design*. However, cost and other practical matters affect one's designs. Sometimes it is wise to use more than the theoretical amount of reinforcement in order to simplify the details by making the bars in many different beams alike; sometimes, it is advisable to use more concrete than needed for strength alone in order to have the thickness or general dimensions desired and to reduce cracking and deflection; often it is best to avoid an excessive variety of sizes which increases the formwork because the cost of a reinforced-concrete structure does not vary directly with the volume of concrete used in it.

The economy of any particular construction depends upon the relative costs of concrete and steel in place instead of upon whether or not it is a balanced design. Sometimes it is important to save steel even though more concrete is necessary; at other times, especially in order to minimize the weight of the structure, thin sections may be the best and most economical even though they are heavily reinforced.

By definition,

$$p = \frac{A_s}{bd} \quad \text{or} \quad A_s = pbd$$

Substituting this value of A_s in Eq. (2-6) gives

$$M_s = pf_s j bd^2 = Kbd^2 \qquad (2\text{-}7)$$

where K is computed as a coefficient which can be tabulated for balanced designs as shown in Tables 5 and 6 in the Appendix.   Equations (2-5) and (2-7) may be restated as

$$bd^2 = \frac{2M_c}{f_c kj} = \frac{M_c}{K} \qquad (2\text{-}8)$$

and

$$bd^2 = \frac{M_s}{pf_s j} = \frac{M_s}{K} \qquad (2\text{-}9)$$

Another convenient form for Eq. (2-6) is

$$A_s = \frac{M_s}{f_s j d} \qquad (2\text{-}10)$$

Equations (2-5) and (2-6) are in convenient form for analyzing beams; whereas Eqs. (2-8), (2-9), and (2-10) are handier for use in designing them on the assumption that they are to be balanced designs.

Let line AB of Fig. 2-4(c) represent a plane cross section through a beam before the external loads are applied.   If this section is considered to remain a plane after bending has taken place, it will move relatively to $A'B'$.   The shortening due to compression at the top can be represented by Δl_c; the elongation of the bars caused by tension can be represented by Δl_s.

For any elastic material that is not stressed beyond its elastic limit, the modulus of elasticity equals the stress per square inch divided by the corresponding deformation in inches per inch of length, or $E = f/\delta l$. Therefore, although the magnitude of E_c may be somewhat uncertain, as shown in Fig. 1-4, assume that

$$E_c = \frac{f_c}{\delta l_c} \qquad \text{and} \qquad E_s = \frac{f_s}{\delta l_s}$$

where δl_c and δl_s represent the strains per unit of length.   From Fig. 2-4(c), by the use of similar triangles, it is found that

$$\frac{\Delta l_c}{\Delta l_s} = \frac{kd}{d - kd} = \frac{k}{1 - k} \qquad (2\text{-}11)$$

The ratio of the modulus of elasticity of steel to that of concrete might be called n where, as shown in the preceding article,

$$n = \frac{E_s}{E_c} \qquad (2\text{-}12)$$

Then

$$n = \frac{f_s/\delta l_s}{f_c/\delta l_c} = \frac{f_s(\delta l_c)}{f_c(\delta l_s)} \qquad (2\text{-}13)$$

Based upon Eq. (2-13), it seems that, if the deformations of the steel and of the concrete are equal, as when side by side at the same distance from the neutral axis of the beam, then one can say that, as previously shown for a cylinder,

$$n = \frac{f_s}{f_c} \quad \text{or} \quad f_s = nf_c \tag{2-14}$$

These equations neglect the effect of creep and shrinkage.

Equation (2-14) explains again why EB of Fig. 2-4(a) is labeled f_s/n. To some scale, AG and EB represent stresses in the concrete for a straight-line variation of stress and for truly elastic action. The stress in the steel would then have to be $n(EB)$, which is plotted as FB.

As shown in Table 1-8, the Code states that, for design purposes, the value of n to be used is

$$n = \frac{29,000,000}{w^{1.5}33 \sqrt{f_c'}}$$

Some assigned values for n are given in Table 1-8. The value of n is to be taken as the nearest whole number but not less than 6. For light-weight concrete, n is assumed to be the same as for normal concrete of the same strength except for the calculation of deflection, even though w in the preceding formula represents the unit weight of concrete varying from 90 to 155 pcf. For computing dead loads, the weight of normal reinforced concrete is often taken as 150 pcf.

Another formula that is useful for the analysis of beams by WSD is

$$k = \sqrt{2pn + (pn)^2} - pn \tag{2-15}$$

This may be derived as follows:

Referring to Fig. 2-4(b), imagine that A_s is replaced by an area of concrete equal to nA_s at the same distance from the neutral axis, as illustrated in Fig. 2-12(b). Then the neutral axis of the cross section will be at its center of gravity. Taking moments of both sides about line O-O, with kd as the unknown,

$$\frac{b(kd)^2}{2} = nA_s(d - kd)$$

$$bk^2d^2 = 2nA_sd - 2nA_skd$$

$$k^2bd = 2nA_s - 2nA_sk$$

$$k^2 - 2pnk = 2pn \quad \text{and} \quad k^2 - 2pnk + (pn)^2 = 2pn + (pn)^2$$

$$k = \sqrt{2pn + (pn)^2} - pn \tag{2-15}$$

Notice that this does not take into account any effect of creep and shrinkage.

Notice also that p is whatever the area of the tensile reinforcement and the effective area of the beam cause it to be, whether it is a balanced design or not.   Furthermore, this shows that k and the area of concrete in compression are larger for a heavily reinforced beam than for one that is lightly reinforced, and that k is also larger for weaker concrete having a larger value of n.   Notice also from the fact that $p = A_s/bd$ that the effective depth does not include the concrete cover below the tensile bars, this being for protection of the steel and therefore not useful in resisting bending.   Of course, concrete added on the compression side is another matter since it does increase the effective depth of the beam.

The assumptions upon which these formulas for WSD are based should be studied carefully.   They assume that the concrete cracks so that it cannot withstand tension.   This is equivalent to saying that before the member will fail, these conditions will occur and the beam will be safe in spite of them, but that any resistance to tension which the concrete may provide will merely add to the safety of the structure.   Then, finally, the value of n for concrete is considered to be the same for tension as it is for compression.

Now apply the preceding principles to specific examples.   The problems may be solved with sufficient accuracy by the use of a slide rule. Inasmuch as different materials may be used and various specifications may apply to any specific situation, the problems used herein will employ a wide range of unit stresses in order to illustrate what the student may encounter in practice.   First, analyze given members.

Example 2-1.   If the beam shown in Fig. 2-5(a) is subjected to a bending moment of 300,000 in.-lb, and if $n = 10$, compute f_s and f_c.

Incidentally, the designer is not so much interested in the magnitudes of f_s and f_c for any given case for their own sakes as he is in comparing them with the allowable unit stresses to see whether or not the member is safe on the one hand and economical on the other.

$A_s = 3 \times 0.44 = 1.32$ in.^{2} (see Table 3, Appendix)

$$p = \frac{As}{bd} = \frac{1.32}{10 \times 14} = 0.0094$$

$$k = \sqrt{2pn + (pn)^2} - pn = \sqrt{2 \times 0.0094 \times 10 + (0.0094 \times 10)^2}$$
$$- 0.0094 \times 10 = 0.35 \qquad \text{From Eq. (2-15)}$$

This value of k may be roughly checked by the use of Fig. 10 of the Appendix, or its magnitude may frequently be determined with sufficient accuracy directly from this diagram.

$$j = 1 - \frac{k}{3} = 1 - \frac{0.35}{3} = 0.883$$

$$f_c = \frac{2M}{kjbd^2} = \frac{2 \times 300,000}{0.35 \times 0.883 \times 10 \times 14^2} = 990 \text{ psi}$$

$$f_s = \frac{M}{A_s jd} = \frac{300,000}{1.32 \times 0.883 \times 14} = 18,400 \text{ psi}$$

Example 2-2. Compute the safe resisting moment of the beam shown in Fig. 2-5(*b*) in in.-k if the allowable $f_s = 20,000$ psi, $f_c = 1,350$ psi, and $n = 9$.

$$A_s = 4 \times 0.79 = 3.16 \text{ in.}^2$$

$$p = \frac{A_s}{bd} = \frac{3.16}{15 \times 28} = 0.0075$$

$$k = \sqrt{2pn + (pn)^2} - pn = \sqrt{2 \times 0.0075 \times 9 + (0.0075 \times 9)^2}$$
$$- 0.0075 \times 9 = 0.306$$

$$j = 1 - \frac{k}{3} = 1 - \frac{0.306}{3} = 0.898$$

$$M_s = A_s f_s jd = 3.16 \times 20 \times 0.898 \times 28 = 1,590 \text{ in.-k}$$

$$M_c = \tfrac{1}{2} f_c kjbd^2 = \tfrac{1}{2} \times 1.35 \times 0.306 \times 0.898 \times 15 \times 28^2 = 2,180 \text{ in.-k}$$

From the fact that the foregoing figures show the safe value of M_s to be considerably less than M_c, it is apparent that this beam is under-reinforced. Using the magnitude of M_s and solving for the simultaneous value of f_c for purposes of illustration gives

$$1,590,000 = \tfrac{1}{2} \times f_c \times 0.306 \times 0.898 \times 15 \times 28^2$$
$$f_c = 984 \text{ psi}$$

It can also be said that f_c varies as the magnitude of the resisting moment, or

$$f_c : 1,350 :: 1,590 : 2,180$$

Example 2-3. Assume that Fig. 2-6(*a*) shows the cross section of a simply supported 9-in. concrete slab in the second floor of a multistory

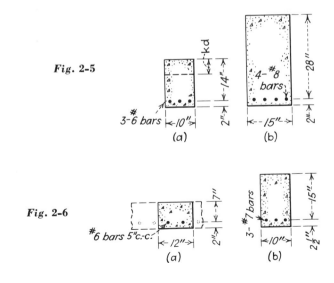

Fig. 2-5

Fig. 2-6

warehouse. It spans over a vault. The owner wants to know how much uniform live load he can put on this slab safely if $L = 12$ ft, $f_s = 20,000$ psi, $f_c = 1,200$ psi, and $n = 10$. Is this an efficient design? What is the stress in the steel at the safe load? Notice that the cover over the bars is 2 in.

First, imagine a slice 12 in. wide to be cut out of the slab from one support to the other, parallel to the main reinforcement. Each such piece will be equivalent to a rectangular beam 12 in. wide. It will contain the equivalent of 12/5 No. 6 bars.[1] Then

$$A_s = 12 \times \frac{0.44}{5} = 1.06 \text{ in.}^2 \qquad d = 7 \text{ in.} \qquad b = 12 \text{ in.}$$

$$p = \frac{A_s}{bd} = \frac{1.06}{12 \times 7} = 0.0126$$

$$k = \sqrt{2pn + (pn)^2} - pn = \sqrt{2 \times 0.0126 \times 10 + (0.0126 \times 10)^2}$$
$$- 0.0126 \times 10 = 0.392$$

$$j = 1 - \frac{k}{3} = 1 - \frac{0.392}{3} = 0.869$$

$$M_s = A_s f_s jd = 1.06 \times 20,000 \times 0.869 \times 7 = 129,000 \text{ in.-lb}$$
$$M_c = \tfrac{1}{2}f_c kjbd^2 = \tfrac{1}{2} \times 1,200 \times 0.392 \times 0.869 \times 12 \times 7^2 = 120,000 \text{ in.-lb}$$

The strength of the concrete, in this case, limits the safe resisting moment. The slab is only slightly overreinforced, and the design is therefore fairly efficient. The simultaneous computed magnitude of f_s is

$$f_s = \frac{20,000 \times 120,000}{129,000} = 18,600 \text{ psi}$$

Considering that the weight of reinforced concrete is usually assumed to be 150 pcf, the dead load of the slab is $150 \times \frac{9}{12} = 112$ psf of horizontal area. Assuming that $M = wL^2/8$ for a simply supported beam and that the safe live load is the reserve supporting capacity of the slab over and above that required for the dead load, it is apparent that

$$M = \frac{120,000}{12} = \frac{w \times 12^2}{8} \qquad \text{or} \qquad w = 556 \text{ psf}$$

Therefore, the live load permissible in this case is $556 - 112 = 444$ psf.

Next, apply these principles to the design of members to serve specific purposes.

Example 2-4. Design a beam to resist a bending moment of 400,000 in.-lb if $n = 10$, $f_s = 18,000$ psi, and $f_c = 1,100$ psi.

Ordinarily there are many beams of varying proportions which can safely support a required load or resist a specified bending moment.

[1] See Table 2 in the Appendix.

However, it is generally reasonable and economical to proportion a rectangular beam so that its depth equals about twice its width unless these dimensions are controlled by other conditions. Sufficient lateral stiffness,[1] economy and efficiency in the use of materials, strength in shear as well as in bending, space for proper placing of reinforcing bars—these are some of the practical reasons for using such proportions.

From Table 5 of the Appendix, with $f_s = 18{,}000$ psi, $f_c = $ a bit less than 1,125 psi, and $n = 10$, find $k = $ approximately 0.38 for a theoretically balanced design. Then

$$j = 1 - \frac{k}{3} = 1 - \frac{0.38}{3} = 0.873$$

Using Eq. (2-8)

$$bd^2 = \frac{2 \times M_c}{f_c k j} = \frac{2 \times 400{,}000}{1{,}100 \times 0.38 \times 0.873} = 2{,}190$$

It would be satisfactory to use $K = 189$ from Table 5 of the Appendix instead of $f_c k j / 2$, as shown in Eq. (2-8).

The problem now resolves itself into a case of "cut and try." There are a multitude of possible values for the width and depth of the beam. However, one way is to assume d and test for b, changing the assumptions until proper and reasonable dimensions are found. Assuming $d = 18$ in. gives

$$b = \frac{2{,}190}{18^2} = 6.75 \text{ in.}$$

Experience will soon show that this value of b is so small that it will be difficult or impossible to place the reinforcing bars properly. However, assuming $d = 15$ in. gives

$$b = \frac{2{,}190}{15^2} = 9.75 \text{ in., or, say, } 10 \text{ in.}$$

Taking this value of $d = 15$ in. and substituting it in Eq. (2-10) yields

$$A_s = \frac{M_s}{f_s j d} = \frac{400{,}000}{18{,}000 \times 0.873 \times 15} = 1.7 \text{ in.}^2$$

If three No. 7 bars are used, $A_s = 3 \times 0.6 = 1.8$ in.². Placing these bars in the assumed beam gives a section as pictured in Fig. 2-6(b). The $2\frac{1}{2}$ in. of concrete below the steel provides slightly more than the minimum required cover of 2 in. This allows for stirrups.

[1] The Code specifies that the clear distance between lateral supports of a beam shall not exceed 50 times the least width of the compression flange. This is because of the possibility of failure by lateral buckling. Personally, the author prefers to be more conservative than this, using a limit of $L/b = 20$.

Ordinarily it would be unnecessary to test this beam further because the width of the member and the area of the steel used are slightly greater than the minimum required by the calculations. However, for illustration, the values of f_s and f_c will be computed by analyzing the beam as follows:

$$p = \frac{A_s}{bd} = \frac{1.8}{10 \times 15} = 0.012$$

$$k = \sqrt{2pn + (pn)^2} - pn = \sqrt{2 \times 0.012 \times 10 + (0.012 \times 10)^2}$$
$$- 0.012 \times 10 = 0.383$$

$$j = 1 - \frac{k}{3} = 1 - \frac{0.383}{3} = 0.872$$

$$f_s = \frac{M_s}{A_s jd} = \frac{400,000}{1.8 \times 0.872 \times 15} = 17,000 \text{ psi}$$

$$f_c = \frac{2M_c}{kjbd^2} = \frac{2 \times 400,000}{0.383 \times 0.872 \times 10 \times 15^2} = 1,066 \text{ psi}$$

The beam is slightly overreinforced, a fact which is shown by the relative magnitudes of f_s and f_c given above.

In assuming depths of beams for designing ordinary structures, the following may be of some service as a general guide or starting point, d being in inches and the span L in feet:

1. For slabs for roofs and floors, assume $d = L/3$ to $L/2$ in. (if $L = 8$ ft, $d = 2.6$ to 4 in.). The larger depth is preferable for stiffness.

2. For light beams and heavy slabs, assume $d = 0.8L$ in.

3. For heavy beams, and headers or girders supporting crossbeams, assume $d = 1.0L$ to $1.25L$ in., depending upon the intensity of loads and lengths of spans.

4. For ordinary continuous beams and girders, assume d somewhat less than given above.

The planning of the arrangement and spacing of reinforcement is far more important than one might at first expect. The necessary cover over the bars should be secured; this means real cover of concrete beyond the surface of the steel, not a dimension to the center of a bar. Not only should the spacing of the main reinforcement be ample to permit the aggregate to pass between adjacent bars, as shown in Tables 8 and 8A of the Appendix, but this should also be adequate at splices, at overlaps, where bent bars adjoin others, and where main bars are beside dowels.

Here is one case as an example. Some continuous T beams in a wharf were 16 in. wide with No. 5 stirrups and four No. 11 main bars about $3\frac{1}{4}$ in. c.c. Near the supports three of these bars were bent up from each side and continued across the top as shown by bars a and b in Fig. 2-7 to resist tension in the top of the beam.   The fourth bar from each side continued across and overlapped the bent ones, as shown by c and d. All were detailed to lie in the same plane at the top and bottom of the beam.   Bars a and b formed a screen at the top where they passed each other.   Bars c and d did likewise where they lapped over a and b, which were already close together.   After a few years it was discovered that the latter had caused honeycombing in the concrete cover under the bars.   The sea water and salty air entered and caused rusting of the steel.   This, with probable aid from freezing of water in the voids, produced longitudinal cracks in the bottoms of the beams, even loosening the concrete cover in several places.   Extensive and costly repairing of the beams was necessary, whereas some deeper girders that were immersed more but were made properly showed no disintegration.

Incidentally, a bitumastic or epoxy coating sprayed or brushed on such exposed beams may help a little in protecting the members.

Many times it is desirable to use approximate formulas when making a first try at the design of a beam in order to minimize time and labor. This tentative design can be checked later by more theoretically correct methods.   A casual inspection of Tables 5 and 6 in the Appendix shows that the value of k generally lies somewhere between 0.3 and 0.45; hence j is somewhere from 0.9 to 0.85.   Therefore, assume $k = 0.38$ and

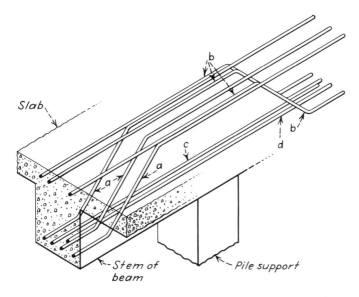

Fig. 2-7 Example showing how bars too close together may cause honeycombing of the concrete.

$j = 0.88$, as average values.   Then Eq. (2-5) gives

$$M = \frac{1}{2} f_c \times 0.38 \times 0.88 b d^2 = \frac{1}{6} f_c b d^2 \qquad \text{or} \qquad f_c = \frac{6M}{bd^2} \qquad (2\text{-}5a)$$

and Eq. (2-6) becomes

$$M = 0.88 A_s f_s d \qquad \text{or} \qquad A_s = \frac{M}{0.88 f_s d} \qquad (2\text{-}6a)$$

These formulas will be labeled as shown to denote that they are approximations of the original ones.   They are easily remembered and are useful if one wishes to get a scale on sizes required, especially when he has no books to which he can refer.

The values of K shown in Tables 5 and 6 in the Appendix are also very useful in expediting the design of rectangular beams.   To illustrate this, assume the following problem:

Design a beam along the edge of a large hatch in a floor.   It is simply supported and has a span of 22 ft.   It carries a uniformly distributed live load of 1,500 plf, as well as its own weight.   An intersecting beam at its center delivers to it a reaction of 20,000 lb.   Assume $n = 8$ and the allowable f_c and $f_s = 1,200$ and 20,000 psi, respectively.

This is a girder with a very heavy load.[1]   From the ratios of depth s to spans previously given, assume $d = 1.2 \times 22 = 26.4$ in., or let $d = 27$ in., $D = 30$ in., and $b = 18$ in.   The dead load of the beam is

$$w = \left(\frac{18 \times 30}{144}\right) 150 = 560 \text{ plf}$$

$$M = (1,500 + 560) \frac{22^2}{8} + 20,000 \times \frac{22}{4} = 235,000 \text{ ft-lb}$$

From Table 5 of the Appendix, $K = 173$, then

$$bd^2 = \frac{M}{K}$$

or, using the assumed d,

$$b = \frac{235,000 \times 12}{27^2 \times 173} = 22.4 \text{ in.}$$

This is greater than the assumed b of 18 in.   It seems wise to deepen the beam; hence assume $d = 30$ in., $D = 34$ in., and $b = 20$ in.   The new dead load is 710 plf, and $M = 244,000$ ft-lb.   Testing for b again,

$$b = \frac{M}{d^2 K} = \frac{244,000 \times 12}{30^2 \times 173} = 18.8 \text{ in.}$$

[1] Although the value of $n = 8$ applies to 4,000-lb concrete, the engineer has deliberately set the allowable unit stresses on a conservative basis in order to reduce live-load deflections.

This shows that b might be 19 in., but the 20-in. dimension is conservative, satisfactory, and simple for formwork.

Technically, k and j should be computed for these conditions, but the magnitude of j would change only slightly, being a little more than the 0.892 given in Table 5. Therefore, since the loads are approximations to start with, it will be close enough for practical purposes to use $j = 0.892$.

$$A_s = \frac{M}{f_s j d} = \frac{244,000 \times 12}{20,000 \times 0.892 \times 30} = 5.47 \text{ in.}^2$$

Table 3 in the Appendix shows that nine No. 7, seven No. 8, six No. 9, five No. 10, or four No. 11 bars may be used. Next, considering Table 8 in the Appendix, and assuming $\frac{3}{4}$-in. aggregate, seven No. 8 bars with five in the bottom row and two in a second row 3 in. above it gives a satisfactory spacing and arrangement of the bars, with $A_s = 5.53 \text{ in.}^2$.

Checking this beam, if considered necessary, and assuming that the bars are concentrated at their center of gravity ($\frac{7}{8}$ in. above the bottom row), the following are found:

The minimum depth D equals 30 in. for d plus $\frac{7}{8}$ in. for the distance from the center of gravity to the bottom row of bars plus $\frac{1}{2}$ in. for stirrups (to be discussed later) plus $2\frac{1}{2}$ in. for cover, giving a total of $33\frac{7}{8}$ in. The assumed D of 34 in. is therefore satisfactory.

$$p = \frac{5.53}{20 \times 30} = 0.0092$$
$$k = \sqrt{2 \times 0.0092 \times 8 + (0.0092 \times 8)^2} - 0.0092 \times 8 = 0.316$$

(Check this with Fig. 10 of the Appendix.)

$$j = 1 - 0.316/3 = 0.895$$
$$A_s = \frac{244,000 \times 12}{20,000 \times 0.895 \times 30} = 5.45 \text{ in.}^2 \text{ required}$$
$$f_c = \frac{2 \times 244,000 \times 12}{0.316 \times 0.895 \times 20 \times 30^2} = 1,150 \text{ psi}$$

2-6. Beams with Tensile Steel Only; Ultimate-strength Theory.[1] Assume a portion of a beam that is bent and cracked as pictured in Fig. 2-8(a). It is assumed to be loaded to the point of imminent failure. If it is a balanced design under this condition, the concrete will be stressed to what will be called its ultimate compressive strength, and the steel will be stressed to its yield point in tension. If the beam is underreinforced, the steel will reach its yield point before the concrete is stressed

[1] AASHO specifies WSD only.

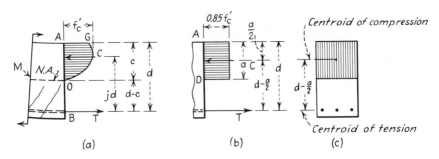

Fig. 2-8

to its ultimate value; if overreinforced, the concrete will reach its maximum resistance first.

Figure 2-8(a) shows an assumed shape for the pressure diagram for the concrete. It is obviously far different from the diagram in Fig. 2-4(a). The curved diagram for the compressive stress means that, when the compression edge is stressed highly, plastic redistribution of stress occurs, but the total compressive resistance and the resisting moment continue to increase to some maximum as the beam is loaded further so that the beam really can support more bending moment than it could if the compressive stresses were assumed to vary as in Fig. 2-4(a), even with the same safety factor for the stresses at the extreme compression edge.

The strains in the materials at any given points are still assumed to vary directly as the distances of those points from the neutral axis. The maximum limit for the strain in the concrete at the compression edge should be $\epsilon = 0.003$ in./in. The maximum strain in the reinforcement is to be taken as $\epsilon' = f_y/29{,}000{,}000$, the strain at the yield stress. At larger strains (overstressing), the assumed stress in the steel is to be limited to f_y, and it is to be assumed to be independent of further strains, as indicated by the straight line in Fig. 2-3(a) beyond the yield point. Of course, tensile stresses in the concrete are to be neglected.

Instead of the curved diagram AGO in Fig. 2-8(a), Whitney[1] proposed the substitution of a fictitious rectangular "stress block" like that shown in sketch (b). This is chosen so as to result in approximately the same compressive force C and to have its center of gravity correspond closely with that of the diagram in sketch (a). This substitution simplifies the computations somewhat. It is not intended to mean that the actual compressive stresses are constant from the compression edge to (or close to) the neutral axis, but it is a convenience in making computations. Actually, any moderate variation of the stresses near the neutral axis

[1] Charles S. Whitney, Plastic Theory of Reinforced Concrete Design, *Trans. ASCE*, Vol. 68, 1942.

will not cause much effect upon the resisting moment because of their short lever arms from this axis. The uniform ordinate for the pressure diagram is assumed to be $0.85f_c'$, where f_c' is determined by test cylinders. This is seemingly a conservative value to use as the assumed maximum compressive stress. Other shapes and other values have been proposed for the compressive stress block but Whitney's method will be used here.

The ultimate-strength theory seems to produce results that agree fairly well with those of many tests. In fact, certain empirical values of factors have been proposed to make this so.

The depth AD of Fig. 2-8(b) is called a to emphasize the fact that point D is not at the real neutral axis. The lever arm of the resisting couple will be called $d - a/2$. Then, as shown in Fig. 2-9(a), the maximum compressive force $C = 0.85f_c'ab$, where b is the width of the assumed rectangular beam. The maximum tensile force $T = A_sf_y$. It should be remembered that the total compression C must be equal and opposite to the total tension T. Hence the maximum values shown in Fig. 2-9(a) will occur simultaneously only when the member is a balanced design. Otherwise, either C or T may control the magnitude of the ultimate resisting moment M_u.

Now refer to Fig. 2-9, sketches (b) to (e). Let AB represent the

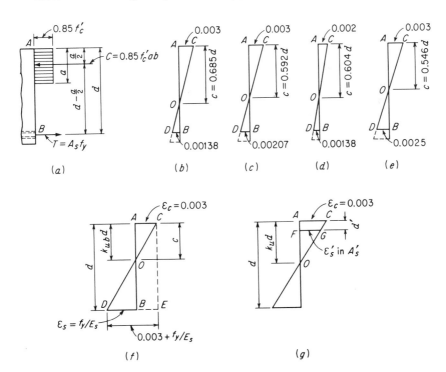

Fig. 2-9

effective depth of a rectangular beam from the compression edge to the center of gravity of the reinforcement, this being kept a constant dimension. In all these sketches, $ACDB$ represents the strain diagram in the beam due to the bending moment, AC being compression deformation. In (b), let AC = the allowable strain of 0.003 in./in. in the concrete and DB = the strain in the steel if f_y = 40,000 psi. Then $c = 0.685d$. In (c), the steel strain is for f_y = 60,000 psi, and $c = 0.592d$. In (d), the beam is assumed to be underreinforced so that the strain in the concrete is less than the maximum, but the bars are stressed to their yield point, 40,000 psi, so that $c = 0.604d$. Finally, in (e), the concrete is assumed to be strained to the maximum allowable, whereas the steel is overstressed. In this case, $c = 0.546d$. All of the preceding is for the purpose of showing the reader that the position of the neutral axis and the height of any stress diagram will vary as the strain conditions change.

The Code specifies that the dimension a for the height of the assumed stress block in Fig. 2-8(b) is to be the distance c from the compression edge to the neutral axis multiplied by a reduction factor k_1 as follows: $k_1 = 0.85$ for compressive strengths f'_c = 4,000 psi or less, but this coefficient is to be reduced by 0.05 for each 1,000 psi excess of f'_c over 4,000 psi.

Let Fig. 2-9(a) be used as a reference. If $C = T$, then $0.85f'_c ab = A_s f_y$ and

$$a = \frac{A_s f_y}{0.85 f'_c b} \tag{2-16}$$

Also, by definition from Art. 2-2, $q = A_s f_y / (bd f'_c)$ and $p = A_s / bd$. Then

$$q = \frac{A_s f_y}{bd f'_c} = \frac{p f_y}{f'_c} \tag{2-17}$$

Refer again to Fig. 2-9(a). The internal resisting moment is $T(d - a/2)$ or $C(d - a/2)$. In terms of the strength of the concrete, use the latter. Then $M_u = 0.85 f'_c ab(d - a/2)$. Using Eq. (2-16) for a,

$$M_u = 0.85 f'_c b \frac{A_s f_y}{0.85 f'_c b} \left(d - \frac{A_s f_y}{2 \times 0.85 f'_c b} \right)$$

Multiplying and dividing by d, find

$$M_u = 0.85 f'_c b \frac{A_s f_y d}{0.85 f'_c bd} \left(d - \frac{A_s f_y d}{1.7 f'_c bd} \right)$$

Then, using Eq. (2-17),

$$M_u = bd^2 f'_c q(1 - 0.59q)$$

Similarly, in terms of steel,

$$M_u = A_s f_y \left(d - \frac{a}{2} \right)$$

However, the Code requires that a capacity reduction factor ϕ is to be included in these moment equations for USD. This factor, as shown in Table 16 of the Appendix, varies for different types of structural members and action. It is a safety provision based upon good engineering judgment and experience. Its reducing effect is made greater for members having larger uncertainties in material action, design, and construction probabilities. It is also more severe when the seriousness of failure is worse—as for columns compared to beams. For flexural members, ϕ is to have a value of 0.9. Therefore,

$$M_u = \phi[bd^2 f'_c q(1 - 0.59q)] = \phi\left[A_s f_y\left(d - \frac{a}{2}\right)\right] \qquad (2\text{-}18)$$

The Code indicates that USD (and prestressed concrete) requires better concrete than that allowed for WSD. For the latter, not more than 20 per cent of the strength tests of laboratory-cured cylinders is allowed to show values less than the specified f'_c; for the former, not more than 10 per cent shall be less than f'_c.

To derive an expression for the percentage of reinforcement for balanced design, refer to Fig. 2-9(f) in which ϵ_c and ϵ_s are the strains for balanced conditions at ultimate strength. Call $AO = k_{ub}d$. By similar triangles,

$$\frac{k_{ub}d}{d} = \frac{0.003}{0.003 + f_y/E_s}$$

With $E_s = 29,000,000$,

$$k_{ub} = \frac{29,000,000 \times 0.003}{29,000,000 \times 0.003 + f_y} = \frac{87,000}{87,000 + f_y}$$

The force in the concrete from Fig. 2-9(a) is $C = 0.85f'_c ab$, but by definition, $a = k_1 c$, where c is the distance from the compression edge to the neutral axis, as shown in sketch (f). The force in the steel is $A_s f_y$. Then, using these forces and assuming balanced conditions,

$$A_s f_y = 0.85f'_c ab = 0.85f'_c k_1(k_{ub}d)b$$

Substituting the value of k_{ub} previously found and $p_b = A_s/bd$,

$$p_b = \frac{0.85f'_c k_1}{f_y} \times \frac{87,000}{87,000 + f_y} \qquad (2\text{-}19)$$

However, the actual percentage p must not exceed $0.75p_b$. Otherwise, the large amount of reinforcement required is likely to be impracticable. Also, the factor ϕ is not included in Eq. (2-19). The factor k_1 in Eq. (2-19), used later in Eq. (2-28), should be assumed to be 0.85 for f'_c up to 4,000 psi, and reduced by 0.05 for each 1,000 psi by which f'_c exceeds 4,000 psi.

In using USD, the safety factor may be applied through what is called the *load factor*. Since the computations give the ultimate load or resisting moment for the beam, the working load must be smaller. Much good judgment is required in determining the proper magnitude for this load factor in various circumstances.

Since the dead load is usually determinable with considerable accuracy, the load factor need only provide for the necessary reserve strength which one wishes to have in the materials, and its magnitude can therefore be moderate. In the case of live loads, however, the situation is different because the future use of a structure and the loads that may come upon it are so uncertain. One should therefore be more conservative in making allowances for future live loads. When impact loads are probable, he should be even more conservative. If the repetition of loads will be great, and if repeated reversal of stress is likely, there may be danger of fatigue. If so, the member may be about 60 to 70 per cent as strong as its computed strength indicates.

Remember that, with USD, one is not interested in the actual stresses in a beam at working loads. He is, however, very much interested in what reserve strength is available between the working conditions and failure of the beam.

The Code specifies that the design capacity shall be the following:

1. For cases where wind and earthquake forces may be neglected in the design,

$$U = 1.5 \text{ DL} + 1.8 \text{ LL} \tag{2-20}$$

where U equals the ultimate capacity, DL equals the actual computed dead load, and LL equals the live load specified by whatever codes or specifications are applicable. The coefficients are intended to make allowances for possible excesses of loads above the assumptions, various approximations and assumptions in the theories used and in the calculations, and the effects of field operations that are not ideal.

2. For cases where wind loads W must be included in the design,

$$U = 1.25(\text{DL} + \text{LL} + W) \tag{2-21}$$

or $$U = 0.9\text{DL} + 1.1W \tag{2-22}$$

whichever is the larger, but no member shall have an ultimate strength less than that required by Eq. (2-20). Earthquake loads E may be substituted for wind loads in Eqs. (2-21) and (2-22) but both wind and earthquake loads need not be assumed to act simultaneously. Of course, partial live loading must be considered in computing the critical effects to be allowed for at any given point.

The Code also states that the effects of creep, elastic deformation, shrinkage, and temperature should be taken into account in structures for which it is normal practice to do so. This sounds very simple but, according to the author's observations, such effects are often overlooked or outright neglected. Furthermore, since the failure of a structure is completely inexcusable, and since the protection of life and property is so important, the author prefers to design beams with load factors which seem to be appropriate but not less than those specified by the Code. Of course, the factor ϕ in Eq. (2-18) is an indirect way of increasing the load factors specified previously. However, here are a few reasons which may influence one to be conservative:

1. Stiffness is often an essential element of structures if they are to be satisfactory to the user.

2. Cracking under large stresses may be objectionable even if not unsafe.

3. Field work under competitive conditions is likely to result in failure to attain the perfection desired by the designer.

4. If trouble develops with a structure, the engineer, as well as the contractor, may be sued for damages—and thereby lose his shirt.

5. Property may be repaired, but lives cannot be replaced.

As a comparison, compute the ultimate moment M_u for the member shown in Fig. 2-5(b), which is the same beam as was analyzed in Example 2-2, using the Code formula. Assume $b = 15$ in., $d = 28$ in., $A_s = 3.16$ in.^{2}, $p = 0.0075$, $f'_c = 3,000$ psi, and $f_y = 40,000$ psi. Let $a = 3.31$ in.

From Eq. (2-18), with the steel as the weak partner,

$$M_u = 0.9 \left[3.16 \times 40,000 \left(28 - \frac{3.31}{2} \right) \right] = 3,000,000 \text{ in.-lb}$$

If the safety factor is to be 2, then the allowable

$$M = \frac{3,000,000}{2} = 1,500,000 \text{ in.-lb}$$

Of course, the coefficient ϕ is an added safety factor. Without this, $M_u = 3,000,000/0.9 = 3,330,000$ in.-lb and $M = 1,650,000$ in.-lb. This shows that, with the same safety factors and the steel controlling the moment, the results of both theories are comparable. However, when the concrete controls, the results may differ considerably, the beams designed by the ultimate-strength methods probably being somewhat smaller, but the greater deflection may be undesirable.

Perhaps the reader thinks that conservatism in design is outdated. The author received a shock when he saw pages 14 and 15 of *Engineering News-Record* for October 29, 1964. Here there are the following articles:

1. Timber Dome Spanned 240 ft Until Guying Cables Parted

2. Engineer on Trial for Dam Collapse

3. Owners Question Dam's Concrete

4. Cairo Building Collapse Kills 31

5. Wall Splits Bowling Alley Building

Some of the above difficulties arose from the design of parts used in construction, some from foundations, some from materials, and some from apparently wrong design of structures.

Example 2-5.   Design a simply supported slab to support a uniform live load of 300 psf, using USD. Assume the following: $L = 12$ ft, $f'_c = 3,000$ psi, $f_y = 40,000$ psi, the load factor for dead load (DL) = 1.5, and the load factor for live load (LL) = 2. $M = wL^2/8$. Use No. 5 bars.

This is a problem of design rather than one of analysis. Equation (2-18) cannot be used directly. It is necessary to get trial values for d and A_s or p. Since the live load is large and stiffness is desired, assume d to be greater than $L/3$ in. = 4 in. Try $d = 5.5$ in. and $t = 7$ in. Then, with $b = 12$ in. for a typical slice, DL = 88 psf. For design,

DL: $\qquad$ $88 \times 1.5 = 132$
LL: $\qquad$ $300 \times 2 = 600$ $\qquad$ $732\,\text{psf}$ = design load w

$$M = \frac{732 \times 12^2 \times 12}{8} = 158,000 \text{ in.-lb/ft of width}$$

Whitney found that, for a balanced design, it seemed to be a good approximation to assume $a = 0.537d$ and $d - a/2 = 0.732d$. Then, approximately, the resisting moment in terms of concrete strength would be

$$M_u = \phi \left(\frac{f'_c b d^2}{3} \right) \tag{2-23}$$

However, having assumed d, the design may or may not be balanced.

Next, select a value for A_s or p directly. Generally, if p exceeds 0.025 to 0.03 for a beam with tensile steel only, the member will be very heavily reinforced and relatively flexible. Since stiffness is desired here, try $p = 0.02$. Then $A_s = pbd = 0.02 \times 12 \times 5.5 = 1.32$ in.^{2}/ft of width. From Eq. (2-16),

$$a = \frac{A_s f_y}{0.85 f'_c b} = \frac{1.32 \times 40,000}{0.85 \times 3,000 \times 12} = 1.72 \text{ in.}$$

Using the second expression in Eq. (2-18) in terms of steel, with $\phi = 0.9$,

$$M_u = \phi \left[A_s f_y \left(d - \frac{a}{2} \right) \right] = 0.9 \left[1.32 \times 40,000 \left(5.5 - \frac{1.72}{2} \right) \right]$$
$$= 220,000 \text{ in.-lb}$$

This is considerably larger than the 158,000 in.-lb needed.

Keeping the depth because of the desired stiffness, try $p = 0.014$. Then $A_s = 0.014 \times 12 \times 5.5 = 0.92$ in.^{2}/ft.

$$a = \frac{0.92 \times 40,000}{0.85 \times 3,000 \times 12} = 1.2 \text{ in.}$$
$$M_u = 0.9[0.92 \times 40,000(5.5 - 1.2/2)] = 162,000 \text{ in.-lb}$$

This is satisfactory. Therefore, from Table 2 in the Appendix, No. 5 bars at 4 in. c.c. give $A_s = 0.93$ in.^{2}/ft.

Just for illustration, see what the resistance of the concrete would be from the left-hand part of Eq. (2-18). From Eq. (2-17),

$$q = \frac{pf_y}{f_c'} = \frac{0.014 \times 40,000}{3,000} = 0.187$$
$$M_u = \phi[bd^2f_c'q(1 - 0.59q)] = 0.9[12 \times 5.5^2 \times 3,000$$
$$\times 0.187(1 - 0.59 \times 0.187)] = 163,000 \text{ in.-lb}$$

This is as it should be. However, using Eq. (2-23) to find the maximum resistance which the concrete could offer with a balanced design,

$$M_u = \phi \frac{f_c'bd^2}{3} = 0.9 \frac{3,000 \times 12 \times 5.5^2}{3} = 327,000 \text{ in.-lb}$$

This is not applicable here. It merely shows what the slab might withstand if enough steel were piled into it.

2-7. Beams with Tensile and Compressive Reinforcement. In the case of continuous beams, such as those shown in Fig. 2-2(b), the maximum bending moment is usually at internal supports. These are points where compression in the concrete at the bottom of the beam may be critical. Therefore, it is necessary to reinforce the top of the beam strongly for tension at these points, as indicated in Fig. 2-2(b). However, it is general practice and desirable to extend some of the bottom bars on through the bottom of the beam at such a point, generally lapping them from both sides over the column or other support. This is partly because moving and partial live loads may cause the points of inflection in such a beam to move closer to the supports under some conditions so that some tensile resistance by means of bars is needed even near the supports. Of course, the bottom bars will also assist the concrete in resisting compression. There are also likely to be members having compression in the top but having reinforcement there as shown in Fig. 2-10(a) and (b).

In cases like these, it is necessary to realize that both the steel and the concrete in the compression area of the beam will act together in withstanding compression. The resistance to the applied moment M

is now made up of two parts. Instead of the concrete having to resist
the compression all alone, it has a helper in the form of the reinforcement
represented by A_s' in Fig. 2-10. If M were to remain unchanged while
A_s' is added to the member, the concrete would resist enough to cause
some force C_c, whereas the steel would resist so as to produce a force C_s.
The resultant of these two compressive forces and the tensile force T,
with some lever arm between them, will constitute the internal resisting
couple. The effect on T in this case will be only the effect of the change
made in the lever arm of the internal resisting couple, so that no harm
is done, but the concrete may have less pressure in it. However, if
A_s' is added to assist the concrete in resisting compression and to keep
it from being overstressed, resulting in an increase of the total compressive forces, the tension T must be made equally strong. Therefore, more
steel will be needed in A_s.

The action of a rectangular beam is basically the same whether the
stress conditions are represented on the basis of ultimate strengths as in
Fig. 2-10(c) or on the basis of the working stresses as in (d). In either
case, it is possible to look upon the resisting moment as made up of two
parts. Then, from (c), for USD,

$$M_u = \phi \left[C_c \left(d - \frac{a}{2} \right) + C_s d_1 \right] = \phi \left[0.85 f_c' ab \left(d - \frac{a}{2} \right) + A_s' f_y d_1 \right] \quad (2\text{-}24)$$

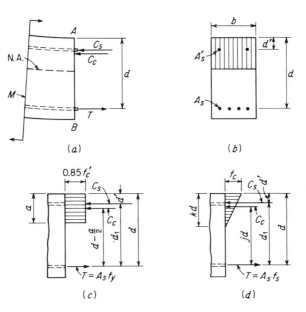

Fig. 2-10

From sketch (d), for WSD,

$$M = C_c jd + C_s d_1 = \tfrac{1}{2} f_c kdb(jd) + A_s' f_s' d_1 \qquad (2\text{-}25)$$

Equation (2-24) assumes that the compressive reinforcement A_s' is stressed to the yield point, which may not be true, whereas f_s' in Eq. (2-25) may be whatever value is necessary for equilibrium. Also, the safety factors are automatically included in the values used for f_c and f_s' for design in Eq. (2-25).

Since the total tension $A_s f_y$ must equal the total compression in the concrete and in A_s', the force resisted by the concrete for ultimate strength can be called $(A_s - A_s')f_y$. With this substituted in Eq. (2-24) for $0.85f_c' ab$, and with $d_1 = d - d'$, find

$$M_u = \phi \left[(A_s - A_s')f_y \left(d - \frac{a}{2} \right) + A_s' f_y (d - d') \right] \qquad (2\text{-}26)$$

which is the Code formula and where

$$a = \frac{(A_s - A_s')f_y}{0.85f_c' b} \qquad (2\text{-}27)$$

Of course, Eq. (2-26) applies only when the compression steel reaches the yield stress f_y.

To find an expression for the percentage of steel for balanced conditions at ultimate strength, let Fig. 2-9(g) represent the strain conditions when $\epsilon_c = 0.003$ and the strain FG in the compression steel is sufficient to produce the yield stress f_y in A_s'. By similar triangles,

$$\frac{\epsilon_s'}{k_u d - d'} = \frac{0.003}{k_u d}$$

where

$$\epsilon_s' = \frac{f_y}{E_s} = \frac{f_y}{29{,}000{,}000}$$

Then

$$k_u = \frac{87{,}000}{87{,}000 - f_y} \times \frac{d'}{d}$$

The equation for equilibrium for balanced conditions is

$$A_s f_y = 0.85f_c' ab + A_s' f_y$$

With $a = k_1 c = k_1 k_u d$,

$$A_s f_y - A_s' f_y = 0.85 f_c' (k_1 k_u d) b$$

Substituting the value of k_u and dividing by bd gives

$$p - p' = 0.85k_1 \times \frac{f_c' d'}{f_y d} \times \frac{87{,}000}{87{,}000 - f_y} \qquad (2\text{-}28)$$

When $p - p'$ is considerably less than the value given by Eq. (2-28), A_s' will not be stressed to f_y. The compression reinforcing may then

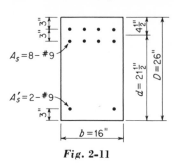

$A_s = 8 - \#9$

$A'_s = 2 - \#9$

$b = 16''$

Fig. 2-11

be neglected and M_u computed by the use of Eq. (2-18), which is conservative, or an analysis may be made on the basis of the strains and their accompanying stresses. In any case, it is essential to be sure that $p - p'$ does not exceed the value of p_b as given by Eq. (2-19). The same values apply for the factor k_1 as given for Eq. (2-19).

Using the WSD concepts shown in Fig. 2-10(d), kd may be affected by the presence of A'_s, and jd will therefore change slightly also. The value of f'_s may be approximated by assuming that it equals $2n$ times the stress in the concrete beside the compression bars. This makes some allowance for creep and shrinkage of the concrete as they affect the compression reinforcement, whereas WSD in general makes no such allowance. Then, on the basis of lever arms from the neutral axis,

$$f'_s = 2nf = \frac{2nf_c(kd - d')}{kd} \qquad (2\text{-}29)$$

Compression reinforcement in beams and girders should be held in by ties or stirrups in order to prevent the bars from buckling and thereby spalling the concrete. This matter will be discussed in Chaps. 4 and 6.

Example 2-6. Compute the ultimate resisting moment M_u for the beam shown in Fig. 2-11. This is assumed to be a section of a girder at an interior column where the concrete in compression at the bottom may be critical. Assume $f_y = 40,000$ psi and $f'_c = 3,000$ psi.

From Eq. (2-28), with $k_1 = 0.85$,

$$p - p' = \frac{0.85 \times 0.85 \times 3,000 \times 3}{40,000 \times 21.5} \times \frac{87,000}{87,000 - 40,000} = 0.0178$$

However, actually

$$p - p' = \frac{8 - 2}{16 \times 21.5} = 0.0174$$

This is close enough so that Eq. (2-26) will be used, and it is much less than $p_b = 0.0371 \times 0.75$, found from Eq. (2-19), with the 0.75 limiting factor applied.

From Eq. (2-27),

$$a = \frac{(A_s - A'_s)f_y}{0.85f'_c b} = \frac{(8 - 2)40,000}{0.85 \times 3,000 \times 16} = 5.87 \text{ in.}$$

Then, from Eq. (2-26), with $\phi = 0.9$ and $a = 5.87$ in.,

$$M_u = \phi\left[(A_s - A_s')f_y\left(d - \frac{a}{2}\right) + A_s'f_y(d - d')\right]$$
$$= 0.9[(8 - 2)40{,}000(26 - 5.87/2) + 2 \times 40{,}000(21.5 - 3)]$$
$$= 6{,}320{,}000 \text{ in.-lb}$$

Example 2-7. Compute the safe resisting moment M of the beam shown in Fig. 2-11, using WSD. Assume $f_s = 20{,}000$ psi, $f_c = 1{,}350$ psi, $n = 9$, and $f_s' = 2n$ times the concrete stress at the center of A_s'.

Use the approximate method shown in Eq. (2-25). With

$$p = \frac{A_s}{bd} = \frac{8}{16 \times 21.5} = 0.0233$$

and $n = 9$, Fig. 10 of the Appendix gives $k = 0.47$ for a beam with tensile bars only, but this will serve as an approximation here. Then

$$j = 1 - \frac{0.47}{3} = 0.84$$

Let $kd = 0.47 \times 21.5 = 10.1$ in. With $f_c = 1{,}350$ psi at the bottom edge, Eq. (2-29) gives

$$f_s' = \frac{2 \times 9 \times 1{,}350(10.1 - 3)}{10.1} = 17{,}100 \text{ psi}$$

Then Eq. (2-25) gives

$$M = \tfrac{1}{2}f_c kjbd^2 + A_s'f_s'd_1$$
$$= 0.5 \times 1{,}350 \times 0.47 \times 0.84 \times 16 \times 21.5^2 + 2 \times 17{,}100$$
$$\times 18.5 = 2{,}600{,}000 \text{ in.-lb}$$

However, using Eq. (2-6), the allowable resisting moment as controlled by the tensile steel is

$$M = A_s f_s jd = 8 \times 20{,}000 \times 0.84 \times 21.5 = 2{,}880{,}000 \text{ in.-lb}$$

Therefore, the compression controls the resistance of the member.

2-8. The Transformed-section Method. The transformed-section method is a method of analysis of reinforced-concrete beams. It is based upon the elasticity of the materials and upon the working-stress theory.

The methods of analysis given in the preceding articles for WSD are not well adapted to the computation of resisting moments of beams of irregular shape or those having the bars distributed over the general cross section of the member. Formulas might be developed to cover such cases, but they are generally too cumbersome for satisfactory use. However, for much ordinary design work, it is sufficient to use the previously illustrated formulas to approximate the area of tensile reinforcement and to make preliminary tests of the strength of the reinforcement.

Therefore, it is desirable to have a general, flexible, and standard procedure for use in design offices that is applicable to all problems, and to have it so made that any man can check the computations and obtain the same results as any other man, without resorting to assumptions and approximations where personal judgment may be involved. Furthermore, in important work, the computations are to be a matter of record and open to inspection. In case of any difficulty in the future, they may even become evidence in court. On this account, each designer should be able to use the transformed-section method of analysis when it is wise to do so even though much of his work in practice may be done by more easily performed methods.

It has been stated previously that, if concrete and tensile steel are deformed equally, the unit stress in the steel at working loads may be assumed to be practically n times as great as that in the concrete alongside the steel, provided the concrete could take the theoretically computed stress. This relationship can be utilized to good advantage because 1 in.² of steel may be considered equivalent to n in.² of concrete, as far as its resistance to deformation is concerned. Therefore, if all the tensile bars in a beam are assumed to be replaced in the cross section by the equivalent square inches of concrete in the same location with regard to the neutral axis, an imaginary beam is obtained in which the steel is said to be "transformed" into concrete. Thus, for any beam of given shape, dimensions, and make-up with tensile steel only, there is a definite substitute beam of homogeneous material which can be used in its stead for the purpose of calculation. Such a substitute beam will have a definite location for its neutral axis, which is the center of gravity of the section; it will also have a definite moment of inertia. Therefore, these values can be substituted in the equation $M = sI/c$ in order to find the resisting moment of the beam.

To illustrate the method clearly, it will be applied first to the ordinary rectangular beam shown in Fig. 2-12(a). The substitute, or transformed, beam is pictured in Fig. 2-12(b). The neutral axis of this "transformed-concrete" beam is represented by the line O-O.

The location of this axis is found by utilizing the fact that, for a beam of homogeneous material but of any shape, the static moment of the area of the cross section about the neutral axis is zero. Then the area above the neutral axis times the lever arm to its own center of gravity must equal the area below the neutral axis times the lever arm to its particular center of gravity.

Furthermore, to find the moment of inertia of the transformed-concrete beam about this neutral axis (called I_c) it is merely necessary to make the usual calculation for $\Sigma x^2(\Delta A)$ about the line O-O. Then the section modulus of this transformed beam will naturally have two values, namely, I_c/kd for the top or compression side and $I_c/(d - kd)$ for the bottom or tension side. With these values, the compressive stress in the concrete is simply

$$f_c = \frac{M}{I_c/kd} \tag{2-30}$$

whereas the tensile stress in the steel is

$$f_s = n\,\frac{M}{I_c/(d - kd)} \tag{2-31}$$

Although the method of finding f_c and f_s that is indicated above is satisfactory, it is sometimes hard to visualize what is going on and to avoid errors resulting from improper use of n or of the distances to the extreme fibers. On this account it is an advantage to have the section modulus for use in determining the compressive stress in the concrete differentiated from that utilized in finding the tensile stress in the steel. Therefore, call the former S_c and the latter S_s. Then

$$S_c = \frac{I_c}{kd} \tag{2-32}$$

and
$$S_s = \frac{I_c}{n(d - kd)} \tag{2-33}$$

Equation (2-30) is then

$$f_c = \frac{M}{S_c} \tag{2.34}$$

and Eq. (2-31) becomes

$$f_s = \frac{M}{S_s} \tag{2-35}$$

Therefore, by computing S_c and S_s immediately after calculating I_c, the designer can realize thereafter that the one with the subscript c goes with

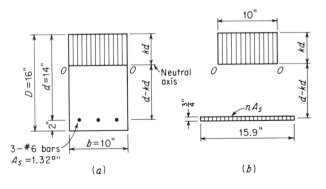

Fig. **2-12**

the concrete and the one having the subscript s is for the steel-stress calculations.

Curves for use in practical design are given in Figs. 4 and 5 of the Appendix. When working to any given specifications as to f'_c and therefore n, tables or curves giving section moduli can be prepared for a variety of beams so that these properties can be used in designing without repeated computations.

Example 2-8. Let Fig. 2-12(a) represent the cross section of a beam with the dimensions and make-up shown. Assume that this beam carries a bending moment of 300,000 in.-lb and that the materials are such that n equals 9. Compute f_c and f_s.

The area of the bars is $3 \times 0.44 = 1.32$ in.^{2}. The bars carry a total tensile force of $A_s f_s$, which, with the compressive forces, is needed to hold the beam in equilibrium. Since the steel is n times as effective as the same area of concrete, the bars are replaced by an area of concrete that will be $n A_s = 9 \times 1.32 = 11.9$ in.^{2}. If this area is arbitrarily assumed to have a depth equal to the diameter of the bars, its length will be $11.9/0.75 = 15.9$ in., as shown in Fig. 2-12(b), which pictures the transformed section of the beam of Fig. 2-12(a) in terms of concrete.

The magnitude of the unknown distance kd is found by taking moments of the equivalent areas of Fig. 2-12(b) about the neutral axis O-O, where the sum of these moments is zero. Then solving for kd gives

$$\frac{bkd(kd)}{2} = nA_s(d - kd) \tag{2-36}$$

$$\frac{10(kd)^2}{2} = 9 \times 1.32(14 - kd)$$

$$kd = 4.7 \text{ in.} \quad \text{and} \quad d - kd = 14 - 4.7 = 9.3 \text{ in.}$$

$$I_c = \frac{b(kd)^3}{3} + nA_s(d - kd)^2 = \frac{10 \times 4.7^3}{3} + 9 \times 1.32(14 - 4.7)^2$$

$$= 1,373 \text{ in.}^4$$

Notice that I of the transformed section of the bars about their own axes is neglected since it is very small.

When there are two or more rows of reinforcement in tension or compression, it may be theoretically desirable to consider each row separately in the computations, using the lever arms from the neutral axis for each row individually. The effective depth d should then be measured from the compression edge of the beam to the row of bars that is farthest from that edge. However, for much ordinary construction, it is sufficiently accurate to assume that the tensile steel is concentrated at its center of gravity, and to measure d from the compression edge to that center of gravity. Furthermore, multiple rows of compressive reinforcement are seldom used.

From Eq. (2-32),

$$S_c = \frac{1,373}{4.7} = 292 \text{ in.}^3$$

Then
$$f_c = \frac{M}{S_c} = \frac{300,000}{292} = 1,030 \text{ psi}$$

Similarly, from Eq. (2-33),

$$S_s = \frac{I_c}{n(d - kd)} = \frac{1,373}{9(14 - 4.7)} = 16.4 \text{ in.}^3$$

Therefore,
$$f_s = \frac{M}{S_s} = \frac{300,000}{16.4} = 18,300 \text{ psi}$$

It is important to notice that, if a structure is designed upon the basis of a concrete of a certain strength (and hence with a certain value of n) and if a stronger concrete is used later on in the real building of the structure, the computed stresses in the concrete will exceed the previously calculated values of f_c. However, this excess will not be greater than the proportional increase in the ultimate strength of the concrete. On the other hand, the steel will have less than its calculated stress. No structural harm will result. If the design is made upon the basis of high-strength concrete, but poor workmanship or materials cause the concrete to be much weaker, the result is on the side of danger.

Now consider a beam with compressive reinforcement in the top. Let Fig. 2-13(a) represent a side view of a very short portion of a simply supported beam having compressive reinforcement as shown. The beam is subjected to a bending moment M as indicated. Sketch (b) is a cross section of the member. Let A_s' equal the area of the compressive reinforcement in in.² and f_s' equal its unit stress. It is apparent that the internal resisting couple is made up of the tensile force T and the resultant of the two compressive forces C in the concrete and C' in the steel, with the proper lever arm. Figure 2-13(c) shows an imaginary transformed cross section of this beam with the tensile steel replaced by its equivalent amount of concrete on the basis that the stress in the bars is n times that in the concrete alongside the bars. For the compressive reinforcement, assume that the transformed area is $2nA_s'$. The area of the concrete that is displaced by the compression bars should be deducted from the cross section of the concrete added in the imaginary "wings." This is equivalent to saying that the total area of concrete which is added as a substitute for the compressive reinforcement is $(2n - 1)A_s'$, as shown in the sketch.

The Code states that the stress in the compression bars should not exceed the allowable tensile stress in any case.[1]

---

[1] AASHO limits f_s' to 16,000 psi.

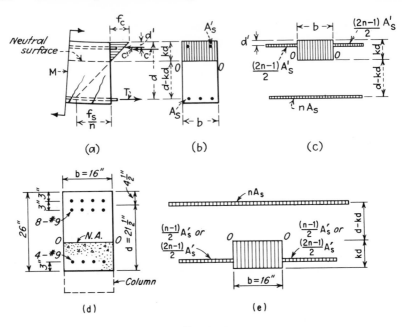

Fig. 2-13

Some engineers prefer to be more conservative and to adhere to the use of $(n - 1)A'_s$ as the imaginary area of concrete to be substituted for the compressive reinforcement. On this account, the latter has been used in the preparation of Figs. 6 to 9 in the Appendix. These figures are for preliminary design and checking purposes.

One reason why the author suggests that $(n - 1)A'_s$ might be used for the transformed area of the compression steel is the fact that a constant moment of inertia is generally assumed in the bending-moment formulas usually employed for the computation of the negative moments at interior supports of continuous beams and slabs. The presence of A'_s and the use of heavier tensile reinforcement at these points of maximum bending produce a larger moment of inertia than that at mid-span. Therefore, this added stiffness tends to increase the actual bending moment at the supports over that computed. It is obvious then that any possible or probable increase in the effectiveness of the compression steel might well be reserved to offset this neglected increase in M. For example, a fixed-end beam with a span of 20 ft, and having a concentrated load of 10 kips at mid-span with I for the end quarters of the span equal to 1.33 times that for the central half was analyzed by methods[1] used for indeterminate

[1] C. W. Dunham, "Advanced Reinforced Concrete," McGraw-Hill Book Company, New York, 1964.

structures.　The end moment was found to be 7 per cent larger than that for a uniform moment of inertia.[1]

Example 2-9.　Assume that Fig. 2-13(d) shows the cross section of a girder at an interior column.　The top and bottom are reinforced heavily, as shown, in order to keep the girder shallow because of clearance needed below it.　Let $n = 9$, $f_s = 20,000$ psi, and $f_c = 1,350$ psi.　Compute the safe resisting moment of the member by using the transformed-section method and $(n - 1)A'_s$ for the compressive reinforcement first and then $(2n - 1)A'_s$ in the second solution.　Compute f'_s in both cases.　The transformed section is shown in principle in Fig. 2-13(e).　Assume the tensile steel to be concentrated at its center of gravity.

First solution, with $(n - 1)A'_s$:

$$A_s = 8 \text{ in.}^2 \qquad nA_s = 9 \times 8 = 72 \text{ in.}^2 \qquad A'_s = 4 \text{ in.}^2$$
$$(n - 1)A'_s = (9 - 1)4 = 32 \text{ in.}^2$$

To find the location of the neutral axis, take moments of the areas about line *O-O* with kd as the variable.　Then, since the moments must balance about the neutral axis,

$$\frac{16(kd)^2}{2} + 32(kd - 3) = 72(21.5 - kd)$$

$$(kd)^2 + 4kd - 12 = 193.5 - 9kd$$

$$kd = 9.2 \text{ in.} \qquad \text{and} \qquad d - kd = 21.5 - 9.2 = 12.3 \text{ in.}$$

Compute I_c about the neutral axis, neglecting I of the bars about their own center lines.

$$I_c = \frac{16 \times 9.2^3}{3} + 32 \times 6.2^2 + 72 \times 12.3^2 = 16,280 \text{ in.}^4$$

The section modulus in terms of concrete is

$$S_c = \frac{I_c}{kd} = \frac{16,280}{9.2} = 1,770 \text{ in.}^3$$

The section modulus in terms of steel is

$$S_s = \frac{I_c}{n(d - kd)} = \frac{16,280}{9 \times 12.3} = 147 \text{ in.}^3$$

M_c in terms of concrete strength $= 1,770 \times 1.35 = 2,390$ in.-k
M_s in terms of the steel strength $= 147 \times 20 = 2,940$ in.-k

Therefore, the strength of the concrete controls.　In this case,

$$f'_s = n(\text{stress in concrete beside the steel}) = \frac{9 \times 1,350 \times 6.2}{9.2} = 8,200 \text{ psi}$$

---

[1] This is to show the qualitative effect of greater stiffness near the ends.

Second solution, with $(2n - 1)A_s'$:

$$A_s = 8 \text{ in.}^2 \qquad nA_s = 72 \text{ in.}^2 \qquad A_s' = 4 \text{ in.}^2 \qquad \begin{aligned}(2n - 1)A_s' \\ = (18 - 1)4 = 68 \text{ in.}^2\end{aligned}$$

$$\frac{16(kd)^2}{2} + 68(kd - 3) = 72(21.5 - kd)$$

$$kd = 8.43 \text{ in.} \qquad d - kd = 21.5 - 8.43 = 13.07 \text{ in.}$$

$$I_c = \frac{16 \times 8.43^3}{3} + 68 \times 5.43^2 + 72 \times 13.07^2 = 17,500 \text{ in.}^4$$

$$S_c = \frac{17,500}{8.43} = 2,080 \text{ in.}^3 \qquad S_s = \frac{17,500}{9 \times 13.07} = 149 \text{ in.}^3$$

$$M_c = 2,080 \times 1.35 = 2,810 \text{ in.-k} \qquad M_s = 149 \times 20 = 2,980 \text{ in.-k}$$

$$f_s' = 2n(\text{stress in concrete beside the bars}) = \frac{18 \times 1,350 \times 5.43}{8.43}$$

$$= 15,650 \text{ psi}$$

In the second case, the concrete strength still controls the magnitude of the resisting moment, but the member is almost a balanced design theoretically. In terms of the strength of the steel, there is little difference in the two cases. However, the theoretical value of f_s' in the second solution is much larger than in the first, but this is not important as long as f_s' is less than the 20,000 psi allowed in tension.

2-9. Design. The preceding problems show the analysis of an existing or assumed beam. In practical design work it is important to be able to determine tentative sections by approximate methods which yield results that will serve as fairly good trial members. Obviously, it is tedious to have to make a guess and then to have the analysis of the beam show that the guess was not even approximate. Therefore, the following procedures are recommended, being based upon fundamental principles:

1. *The Problem.* Assume that a continuous T beam has spans of 24 ft. The beams are 10 ft c.c. The slab is to be 7 in. thick. The live load is 300 psf. Let $f_y = 40,000$ psi, $f_c' = 3,000$ psi, $f_s = 20,000$ psi, $f_c = 1,350$ psi, $n = 9$, M at an interior column $= wL^2/12$, and the total depth is limited to 24 in. Make a trial design by both USD and WSD for the section at a column.

2. *Ultimate-strength Design*

a. Loads and moments. This is a heavily loaded beam. Ordinarily, one would assume d equal to about 1 in./ft of span or 24 in. in this case. However, if 4 in. is allowed for the distance from the top of the slab to the center of gravity of the tensile steel, this limits d to 20 in. Try a width of 14 in. which matches the size of the assumed columns.

DL: Slab = $10 \times 150 \times 7/12 =$ 8 7 5
             Beam stem = $14 \times 17 \times 150/144 =$ 2 4 8
                                              Total = $\overline{1,1\ 2\ 3}$ plf
LL: $10 \times 300 = 3,0\ 0\ 0$ plf

Then, from Eq. (2-20),

$$U = 1.5 \times 1,123 + 1.8 \times 3,000 = 7,080 \text{ plf}$$

Therefore, M_u should equal $7,080 \times 24^2/12$ or 340,000 ft-lb

b. Trial A_s. In general, a from Eq. (2-27) may be about $0.2d$ to $0.3d$. Try $a = 0.24d = 4.8$ in. A trial value for the lever arm

$$(d - a/2) = (d - 0.12d) = 0.88d$$

However, the factor ϕ in Eq. (2-26) has to be included. One could then assume an adjusted lever arm of $0.88 \times 0.9d = 0.79d$ (approx). Then try $0.79 \times 20 = 15.8$ in. Therefore,

$$A_s = \frac{M_u}{15.8 f_y} = \frac{340,000 \times 12}{15.8 \times 40,000} = 6.45 \text{ in.}^2$$

Try six No. 10 bars ($A_s = 7.62$ in.^{2}), with four in the top row and two in a row 3 in. below them, as in Fig. 2-25(d). From Table 8A in the Appendix, four No. 10 bars can fit in the 14-in. width, but they will be rather close.

c. Test of concrete. From Eq. (2-16), for tensile steel only,

$$a = \frac{A_s f_y}{0.85 f'_c b} = \frac{7.62 \times 40,000}{0.85 \times 3,000 \times 14} = 8.55 \text{ in.}$$

This is larger than the 4.8 in. originally assumed. However, from Eq. (2-17),

$$q = \frac{A_s f_y}{b d f'_c} = \frac{7.62 \times 40,000}{14 \times 20 \times 3,000} = 0.363$$

From the first term of Eq. (2-18),

$$M_u = \phi[bd^2 f'_c q(1 - 0.59q)] = 0.9[14 \times 20^2 \times 3,000 \times 0.363(1 - 0.59 \\ \times 0.363)] = 4,330,000 \text{ in.-lb or } 360,000 \text{ ft-lb}$$

This is more than the 340,000 ft-lb required, so that no compression reinforcement is really needed although two bars should be extended the full length of the bottom in any case. These will automatically help the concrete even when not relied upon. The tensile steel will also be satisfactory. To test it, use the second form of Eq. (2-18).

$$M_u = \phi\left[A_s f_y\left(d - \frac{a}{2}\right)\right] = 0.9\left[7.62 \times 40,000\left(20 - \frac{8.55}{2}\right)\right]$$
$$= 4,330,000 \text{ in.-lb}$$

as it should.

3. *Working-stress Design*

a. Loads and moments. The loads are as before. The bending moment due to DL + LL is

$$M = \frac{(1,123 + 3,000)24^2}{12} = 198,000 \text{ ft-lb}$$

b. Trial A_s. In general, k will be between 0.33 and 0.40 for such beams. Therefore, try $j = 0.88$. As before, $d = 20$ in. Then, from Eq. (2-6), a trial value for A_s is

$$A_s = \frac{M}{f_s j d} = \frac{198,000 \times 12}{20,000 \times 0.88 \times 20} = 6.75 \text{ in.}^2$$

Try six No. 10 bars ($A_s = 7.62$ in.^{2}) arranged as before.

c. Test of concrete. With $k = 0.38$ and $j = 0.88$ as a trial, the safe resisting moment of the concrete alone is, from Eq. (2-5),

$$M_c = \tfrac{1}{2}f_c k j b d^2 = \frac{1,350}{2} \times 0.38 \times 0.88 \times 14 \times 20^2 = 1,260,000 \text{ in.-lb}$$
$$\text{or } 105,000 \text{ ft-lb}$$

d. Trial A_s'. The moment to be resisted by the compressive reinforcement is $198,000 - 105,000 = 93,000$ ft-lb. If

$$kd = 0.38 \times 20 = 7.6 \text{ in.}$$

and $d' = 3$ in., then, by proportion, the concrete stress alongside the compressive steel will be

$$f = \frac{f_c(kd - 3)}{kd} = \frac{1,350 \times 4.6}{7.6} = 818 \text{ psi}$$

Then, if $f_s' = 2nf$, $f_s' = 2 \times 9 \times 818 = 14,720$ psi. Using this, the needed A_s' with a lever arm of $d - d'$, is

$$A_s' = \frac{93,000 \times 12}{14,720 \times 17} = 4.46 \text{ in.}^2$$

This requires four No. 10 bars ($A_s' = 5.08$ in.^{2}).

e. Final analysis. Using this trial section, which looks somewhat like Fig. 2-25(c), find the following by use of the transformed-section method, with $n = 9$ and $(2n - 1)A'_s$ as the substitute for the compressive reinforcement:

$$kd = 7.67 \text{ in.} \qquad d - kd = 12.33 \text{ in.} \qquad I_c = 14,400 \text{ in.}^4$$
$$S_c = 1,880 \text{ in.}^3 \qquad \text{and} \qquad S_s = 130 \text{ in.}^3$$

Then $M_c = 1,350 \times 1,880 = 2,540,000$ in.-lb or 212,000 ft-lb and $M_s = 20,000 \times 130 = 2,600,000$ in.-lb or 217,000 ft-lb. The value of M_c is the smaller but is safe since it exceeds the 198,000 ft-lb needed.

4. *Comparison.* In the USD calculations under 2, it was found that six No. 10 bars in the top with four in the top row and two located 3 in. below them were satisfactory and that no compressive steel was theoretically required, whereas, in the WSD, six No. 10 were needed in the top and four No. 10 in the bottom. This difference is primarily due to the difference in the safety factors or load factors used in the two cases. In WSD, the safety factor SF for the concrete is

$$3,000/1,350 = 2.22$$

The uniform load for USD is 7,080 plf. Then $7,080/4,123 = 1.72$. If this is adjusted for ϕ, $1.72/0.9 = 1.91$, then the approximate difference in SF is $2.22/1.91 = 1.16$. Thus the WSD is 16 per cent stronger in this case.

2-10. T Beams. The use of simple concrete slabs of moderate depth and weight is generally limited to spans of 10 to 15 ft. Where it is desired to use concrete for long spans without excessive weight and material, a common type of construction is that shown in Fig. 2-2(a). It consists of a relatively thin slab with deep haunched portions or stems at intervals. Figure 2-2(b) gives an exaggerated picture of the action of the slab under vertical loads, whereas (c) shows the action of the stem if it is simply supported. Instead of considering the stem to be a rectangular beam which carries the load by itself, it is better and more important to realize that all parts of the structure must act simultaneously. In general, the stem and the slab near it can be assumed to act as a unit, forming a "T beam" as shown in Fig. 2-14. The slab and the stem are to be poured monolithically, or they are to be bonded securely together.

In this type of construction there are theoretically two general cases to consider. In the first, the neutral axis is located in the slab or flange section as shown in Fig. 2-14(a), which is the usual condition. The problem then is the same as for a rectangular beam of the size shown by the dotted lines, since direct tensile stress in the concrete below the neutral axis is neglected anyway. The second case is one in which the neutral axis lies in the stem as pictured in Fig. 2-14(b). The diagram

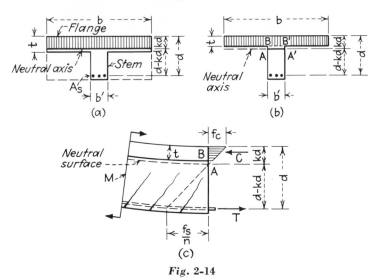

Fig. 2-14

representing the compressive unit stresses in the flange according to WSD is a trapezoidal wedge instead of a triangular one.   This is shown in Fig. 2-14(c).   The pressure on the small portion of the top of the stem $ABB'A'$, which is subjected to compression, is generally of little importance.   When USD is used, the triangular or trapezoidal diagram in Fig. 2-14(c) will be replaced by an imaginary rectangular one.   The basic action for a simply supported T beam or any case with compression in the top is the same in both theories because the resisting couple will be the forces C and T with some lever arm between them.

The Code states that the maximum effective width of the flange on each side of the stem of a symmetrical T beam which is part of a monolithic floor can be assumed to equal eight[1] times the thickness of the flange or one-half the clear distance between adjacent stems or webs. The total flange width must not exceed one-fourth the span of the beam. For beams with a flange on one side only, the assumed effective width of flange beyond the stem shall not exceed six times the thickness of the slab or more than one-half the clear span of the slab to the next beam or more than one-twelfth the span of the beam.

Of course, when T beams are continuous over a series of supports, the compression resistance C is in the bottom.   The beam is then considered to be rectangular with the width b' because the concrete of the flange is not relied upon to withstand any tension in the top.   Some common remedies to help resist the large compression in the bottoms

---

[1] AASHO limits this to six.

of T beams are sideward flaring of the stem, deepening of the stem
locally, addition of compressive reinforcement, or proportioning of the
stem as required at the column and use of this section throughout. The
first two cause expensive formwork.

The compressive stress in the flange of most ordinary T beams is
relatively small and seldom needs to be investigated. The bottom
reinforcement is the controlling feature. The chief difference in the
results of the analysis of the steel by WSD and USD methods is in the
lever arm between C and T. From Fig. 11 of the Appendix it is seen
that k for ordinary T beams is very small, and that j varies from about
0.9 to 0.95. It is therefore conservative to compute M_s, f_s, and A_s for T
beams by using $j = 0.9$ in Eq. (2-6) when using WSD. Theoretically,
USD will result in a slightly larger lever arm. The Code specifies that,
when the flange thickness equals or is greater than the depth to the neutral
axis ($1.18qd/k_1$), Eq. (2-18) shall be used, with a and q computed as for a
rectangular beam of width b. When the flange thickness is less than
$1.18qd/k_1$, M_u is to be computed from the following:

$$M_u = \phi\left[(A_s - A_{sf})f_y\left(d - \frac{a}{2}\right) + A_{sf}f_y(d - 0.5t)\right] \qquad (2\text{-}37)$$

where A_{sf} is the steel area needed to develop the ultimate compressive
strength of the overhanging flanges beyond the web and is assumed to be

$$A_{sf} = \frac{0.85(b - b')tf_c'}{f_y} \qquad (2\text{-}38)$$

This is equivalent to saying that the portion of the tensile steel needed
to develop the web as a rectangular beam can be treated in the same
way as in the second form of Eq. (2-18). Similarly, the portion of the
tensile steel required to develop the compression in the flange can be
assumed to have a force of $A_{sf}f_y$ and a lever arm from its centroid to
the middle of the flange thickness.

Also, the Code states that

$$a = \frac{(A_s - A_{sf})f_y}{0.85f_c'b'} \qquad (2\text{-}39)$$

For a rectangular beam, Eq. (2-16) gave $a = A_sf_y/0.85f_c'b$. For the
T beam, A_{sf} represents the area of the tensile steel needed to develop the
compression in the flanges. Thus one might assume that each foot of
the effective width of the flange requires the same amount of tensile steel
to counterbalance the compression in the flange. Then $A_s - A_{sf} =$ the
steel for the width b', so that Eq. (2-39) is really similar to Eq. (2-16)
for a width b'.

With $p_f = A_{sf}/b'd$ and $p_w = A_s/b'd$, $(p_w - p_f)$ should not exceed $0.75p_b$, with p_b computed from Eq. (2-19).

Actually, most T beams are very much underreinforced, and properly so. With so much concrete available for resisting compression, an attempt to pack bars in the bottom so as to approach a balanced design would be extremely unwise. Common sense and consideration for the workmen in the field should be used. The author has seen some cases where the men tried to pack the bars in as called for by the drawings but simply had to let things go as best they could.

For WSD it is usually sufficient to assume that

$$M = A_s f_s(0.9d) \text{ to } A_s f_s(0.92d)$$

and for USD that

$$M_u = \phi[A_s f_y(0.9d)] = A_s f_y(0.8d)$$

If a flange is used to stiffen an isolated beam laterally, the thickness t of the flange should not be less than $0.5b'$, and the total width should not exceed $4b'$.

Example 2-10. Design a simply supported T beam to span 25 ft and to carry a uniformly distributed live load of 300 psf plus the dead load. Assume that the slab is 6 in. thick and that the stems are 8 ft cc. Use WSD with $f_s = 20,000$ psi.

This problem is one which requires the making of assumptions to get a trial member. The procedure is as follows:

1. Assume d in inches $= 1.2L$, where L is in feet; $d = 1.2 \times 25 = 30$ in.
2. Assume $b' = 16$ in., a little over $d/2$.
3. Assume two rows of bars, giving a cover of about 4 in. from the center of gravity of the reinforcement.
4. Compute the trial dead load:

$$\text{Slab} = \quad 8 \times \tfrac{6}{12} \times 150 = 600$$
$$\text{Stem} = \frac{16}{12} \frac{(30 + 4 - 6)}{12} 150 = 470 \qquad 1{,}070 \text{ plf} = \text{total DL}$$

5. Compute the live load:

$$8 \times 300 = 2{,}400 \text{ plf}$$

6. Compute the total bending moment:

$$M = \frac{wL^2}{8} = \frac{(1{,}070 + 2{,}400)25^2}{8} = 270{,}000 \text{ ft-lb}$$

7. Assume $j = 0.9$, and solve for A_s:

$$A_s = \frac{M}{f_s j d} = \frac{270{,}000 \times 12}{20{,}000 \times 0.9 \times 30} = 6 \text{ in.}^2$$

Try six No. 9 bars with four in the bottom row and two in an upper row 3 in. above them. From Table 8 in the Appendix, assuming $\frac{3}{4}$-in. aggregate, it is seen that four bars can be used in the 16-in. width. With the center of gravity 1 in. above the bottom row, the 4 in. added in item 3 will give adequate cover for the reinforcement.

2-11. Irregular Beams. By using the transformed-section method, it is generally possible to analyze and design irregular, unusual, and unsymmetrical shapes as beams for WSD only. Practice in the solution of such problems is the best way in which to fix the methods in one's mind. The problem which follows is used for that purpose.

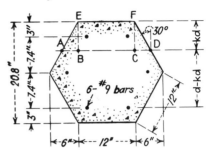

Example 2-11. Compute M_c and M_s for the hexagonal section shown in Fig. 2-15 if $n = 9$, $f_s = 22{,}000$ psi, and $f_c = 1{,}350$ psi.

The neutral axis lies somewhere in the upper half of the beam. The

Fig. 2-15

rectangle $EFCB$ has a width of 12 in. and a height of kd. The triangles ABE and FCD have altitudes that are equal to kd, but the bases AB and $CD = kd \times \tan 30°$ or $0.577kd$. Therefore, the equation for kd can be expressed as follows:

$$\frac{12(kd)^2}{2} + \frac{2kd(0.577kd)kd}{2 \times 3} + (2n - 1)A_s'(kd - 3) = nA_s(10.4 - kd)$$
$$+ nA_s(17.8 - kd)$$

where $nA_s = 9 \times 1 \times 2 = 18$ in.^{2}
and $(2n - 1)A_s' = 17 \times 1 \times 2 = 34$ in.^{2}

Solving this equation gives

$kd = 5.6$ in.   and $d - kd = 12.2$ in.

$$I_c = \frac{12 \times 5.6^3}{3} + \frac{2(0.577 \times 5.6)5.6^3}{12} + 34 \times 2.6^2 + 18 \times 4.8^2$$
$$+ 18 \times 12.2^2 = 4{,}030 \text{ in.}^4$$

$$S_c = \frac{I_c}{kd} = \frac{4{,}030}{5.6} = 720 \text{ in.}^3$$

$$S_s = \frac{I_c}{n(d - kd)} = \frac{4{,}030}{9 \times 12.2} = 37 \text{ in.}^3$$

$M_c = f_c S_c = 1{,}350 \times 720 = 970{,}000$ in.-lb
$M_s = f_s S_s = 22{,}000 \times 37 = 810{,}000$ in.-lb

PRACTICE PROBLEMS

Bending-moment formulas for simply supported beams are assumed to be known by the reader. The bending-moment diagrams for some beams of uniform section with one or both ends fixed and for various conditions of loading are shown in Figs. 1 and 2 of the Appendix. These may be substituted for continuous beams in the problems for the calculation of bending moments. In all cases, assume that the weight of concrete plus reinforcement is 150 pcf.

A. Analysis.

In all the following problems assume $f'_c = 3,000$ psi, $f_y = 40,000$ psi, $n = 9$, allowable $f_c = 1,350$ psi, and allowable $f_s = 20,000$ psi unless stated otherwise in the problem. The reader is to use whatever method he believes is appropriate for each problem unless the method is specified.

2-1 Assume a beam of the cross section shown in Fig. 2-16(a). It is subjected to a bending moment of 2,000,000 in.-lb. Compute f_s and f_c.

Ans. $f_s = 21,700$ psi; $f_c = 1,230$ psi.

2-2 The beam shown in Fig. 2-16(b) is simply supported and has a span of 14 ft. Compute the uniformly distributed live load that this beam will safely support in excess of its own dead load.

2-3 Compute the safe bending moment for the beam shown in Fig. 2-16(c) if $n = 8$, $f_s = 22,000$ psi, and $f_c = 1,800$ psi.

Ans. $M_s = 91,000$ ft-lb; $M_c = 130,000$ ft-lb.

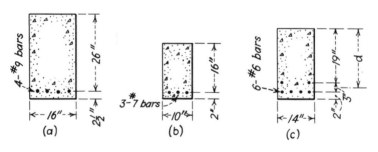

Fig. 2-16

2-4 Assume the slab shown in Fig. 2-17. Compute the safe M_s and M_c, also M_u. Using the load factors given in Eq. (2-20), is the beam safe for the loading shown?

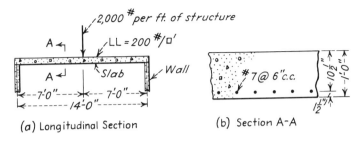

(a) Longitudinal Section   (b) Section A-A

Fig. 2-17

2-5 The beam shown in Fig. 2-18 is subjected to a bending moment of 80,000 ft-lb. Is it safe if the dead load is one-third of the total load?

2-6 Compute the resisting moments M_s, M_c, and M_u for the beam shown in Fig. 2-19 if $d = 16\frac{1}{2}$ in., $f_s = 18,000$ psi, and $f_y = 36,000$ psi. *Ans.* $M_s = 918,000$ in.-lb; $M_c = 815,000$ in.-lb; $M_u = 1,700,000$ in.-lb.

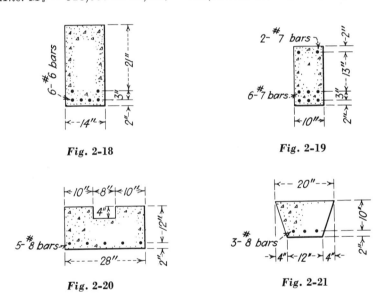

Fig. 2-18   **Fig. 2-19**

Fig. 2-20   **Fig. 2-21**

2-7 If the beam shown in Fig. 2-20 carries a bending moment of 400,000 in.-lb, what are f_c and f_s?

2-8 By the transformed-section method, compute the safe resisting moment of the beam shown in Fig. 2-21.

Discussion. The trapezoidal area in compression may be considered as a rectangle minus two triangles having altitudes of kd and bases equal to $kd/3$.

2-9   Figure 2-22 shows a bracket for a viaduct.   Compute the total bending moment about the center line of the column.   If the dead load is one-half of the live load, is the member safe, using the load factors as in Eq. (2-20)?   Also, compute M_s and M_c.

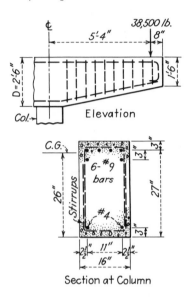

Fig. **2-22**

2-10   By the transformed-section method, compute the safe resisting moment of the box section shown in Fig. 2-23 if the allowable $f_c = 1,200$ psi.

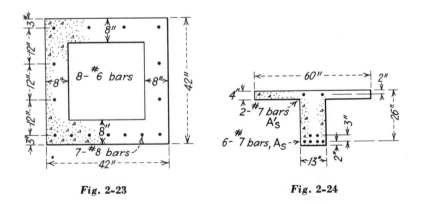

Fig. **2-23**   *Fig.* **2-24**

2-11   Compute M_s and M_u for the T beam shown in Fig. 2-24.

2-12 A designer has proposed the construction shown in Fig. 2-25(a) for a simply supported slab over a long passageway. It is to support a uniform live load of 300 psf. Is this construction satisfactory?

2-13 Compute M_u for the continuous T beam shown in Fig. 2-25(c) where the beam crosses a column and has tension in the top.

Ans. $M_u = 920,000$ ft-lb.

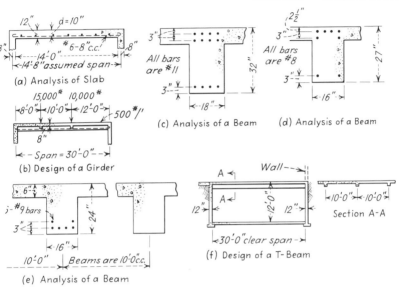

(a) Analysis of Slab

(b) Design of a Girder

(c) Analysis of a Beam

(d) Analysis of a Beam

(e) Analysis of a Beam

(f) Design of a T-Beam

Fig. **2-25** Studies of beams and slabs.

2-14 Figure 2-25(d) shows the section of a continuous T beam at an interior column. The slab is 6 in. thick and the beams are 9 ft c.c. If the span is 24 ft, compute the safe uniformly distributed live load that the beam can support, assuming it to have fixed ends.

2-15 Figure 2-25(e) shows the construction proposed for the floor over a narrow basement in an industrial plant. The T beams are simply supported and have a span of 25 ft. The live load is 150 psf. Is the design safe? *Ans.* $M_u = 3,680,000$ in.-lb vs. $4,000,000$ in.-lb needed.

$M_s = 2,210,000$ in.-lb vs. $2,390,000$ in.-lb needed if $j = 0.92$.

B. Design.

2-16 Design a simply supported T beam girder to support the loads shown in Fig. 2-25(b) plus its own weight.

2-17 A structure has a small boiler room in a basement as pictured in Fig. 2-25(f). Design the floor slab and beams over this basement, with simply supported T beams as shown. The live load = 300 psf. Make sketches to show the dimensions and reinforcement to be used for the typical construction.

2-18 Beam AB is simply supported and carries the framing shown in Fig. 2-26(a) and a brick wall weighing 800 plf. The reactions at C and D are shown in sketch (b). Estimate the size and weight of the beam; then compute the bending moment at its center. Design the beam as a rectangular one with reinforcement in the bottom only.

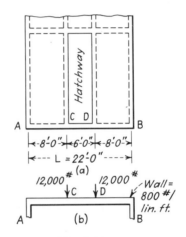

Fig. 2-26

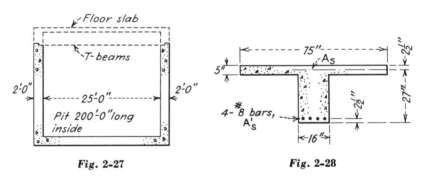

Fig. 2-27   **Fig. 2-28**

2-19 Design a simply supported slab- and T-beam floor for the top of the long pit shown in Fig. 2-27. The live load is to be 200 psf. Neglect stairways and hatches.

2-20 Assume that the T beam shown in Fig. 2-28 is continuous and that the negative bending moment at an interior column is 2,000,000 in.-lb. Determine the bars needed to withstand the tension in the top if they (or the top row) are $2\frac{1}{2}$ in. from the top of the slab. A second row will be 3 in. below the upper one if needed. All bars are to be No. 8.

BOND

3

3-1. Nature and Magnitude of Bond Stress. Some symbols to be used in this chapter are as follows:

D = diameter of bar, in.

L_s = embedded length of bar; anchorage length, in.

ΣO or Σo = sum of perimeters of bars at a section times 1 in., in.^{2}

u = bond unit stress for working-stress design, psi.

u_u = bond unit stress for ultimate-strength design, psi.

v = shearing unit stress in concrete as a measure of diagonal tension, psi.

v_c = shearing unit stress in concrete without web reinforcement, psi.

v_L = longitudinal shearing unit stress in concrete (numerically equal to v), psi.

V = shearing force at a section, lb or kips.

V_u = ultimate shearing force at a section, lb or kips.

When a reinforcing bar is embedded in concrete, the concrete adheres to its surface, resisting any force that tends to pull or push out the bar. This is called *bond* between the concrete and the steel. The intensity of this adhesive force is called the *bond stress*, or *bond unit stress*. In reality, this bond stress is a resistance to shearing between the surfaces of the steel and the concrete. The action is that of resistance to forces that try to break the concrete away from the surface of the steel in a direction parallel to that surface and lengthwise of the bar.

The function of bond in a reinforced-concrete member is somewhat analogous to that of bolts, rivets, and welds in structural steelwork. It is the force that holds the two materials together so as to develop their simultaneous and mutually helpful action. If the bars have no change of stress—and therefore no change of length—as the result of the application of a load on the member, then there will be no bond stress set up by it, but as soon as flexural action causes the steel to stretch or to compress, the bond stresses must come into action in order to cause these changes.

When the bar in Fig. 3-1(a) is stretched, the elongation in the length L_s is greatest at the point where the steel enters the bottom of the concrete block. It then decreases to zero at or somewhere below the top end of the bar. A little reflection will show that the intensities of the bond stresses along the bar must vary somewhat in proportion to the stretching of the bar inside the concrete block unless the bond is broken. Probably the bond stresses are very high near the point at which the bar enters the concrete. It is also likely that they are very high (probably at the point of local failure) at the cracks pictured in Fig. 3-7(b). The distribution of the bond stresses is very uncertain; yet, for analysis and design, it is generally considered to be uniform over some length that is necessary to develop the force of compression or tension in the bar. This length times the average bond stress acting along the surface of the bar parallel to the axis of the bar may be said to *anchor* or *develop* the bar. However, one should realize that the bond will develop the bar as quickly as possible so that a part of the bar that is a long way (15 to 30 diameters) from the point of entry of the bar in Fig. 3-1(a) may have no stress at all; in other words, anchorage far from the point where the bar is needed may not be brought into action if the bond can develop the required resistance before tensile stresses can reach the anchor.[1]

Referring to Fig. 3-1(a), let o = the perimeter of the cross section of this bar, D = its diameter, L_s = the length of embedment, and u = the average bond unit stress. It is clear that the total strength of the bond per inch of bar equals ou. It is also apparent that the embedded length

[1] R. M. Mains, Measurement of the Distribution of Tensile and Bond Stresses along Reinforcing Bars, *J. ACI*, November, 1951.

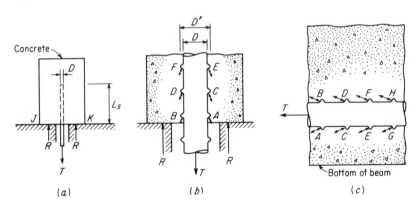

Fig. **3-1**   Illustration of how lugs may cause splitting.

of the bar should be great enough to develop the required tensile (or compressive) strength of the bar or, at least, the required force in order to avoid having it pull out of the concrete. Then, for a round bar that is subjected to a tensile force T,

$$T = \frac{\pi D^2 f_s}{4} = \pi D u L_s \qquad \text{or} \qquad L_s = \frac{D f_s}{4u} \qquad (3\text{-}1)$$

This shows that the length of embedment that is needed to develop the strength of the bar in tension (or in compression) increases with the tensile stress in the steel and decreases with an increase in the magnitude of the permissible bond stress. Notice that Eq. (3-1) gives the average stress and does not take into account the local and severe increases in the bond stresses at and near the cracks like those pictured in Fig. 3-7(b).

Theoretically, many small bars are better than a few large ones, as far as bond is concerned. This is evident if one studies Eq. (3-1), bearing in mind that the cross-sectional area of a bar increases as the square of D, whereas the surface of the bar per inch of length varies as the first power of D. However, there must be adequate space between adjacent bars—also between the bars and the forms—so that the concrete will completely fill the forms and thoroughly encase the steel if the bond is to be developed fully and if the bars are to be protected. See Table 8 in the Appendix for recommended spacing of bars in beams.

The American Concrete Institute Code (ACI 318-63) specifies the formulas shown in Table 3-1 for the determination of the allowable bond unit stresses for various sizes and types of bars for both the working-stress and ultimate-strength methods. In connection with beams, the reader should notice that there is a difference between the formulas for top and bottom bars. The former refer to horizontal bars that have more than 12 in. of concrete cast below them. This is apparently due to the tendency of wet concrete to settle, or at least for the coarse aggregate to do so. This may cause voids under horizontal bars that are supported in fixed positions when the depth of concrete under them is considerable, thus weakening the bond on the under side of the bars. Vibration may or may not increase this action. Tables 13 to 15 in the Appendix show the numerical values of u for various sizes of bars and strengths of concrete.

Notice that the allowable magnitude of the bond stress u for A305 bars and for plain bars varies inversely according to their diameters. This may be logical because, as the bar sizes increase, the needed force of bond resistance becomes more critical since large bars have relatively less bond strength per inch of length to develop their strength. However, this variability of the allowable value of u is a nuisance from the designer's standpoint because he has to use a different value for each size. Further-

Table 3-1   Formulas for Bond Unit Stresses, psi

Type of bar	Working-stress design			Ultimate-strength design*		
	Tension		Compression	Tension		Compression
	Top	Bottom		Top	Bottom	
A305	$\dfrac{3.4\sqrt{f'_c}}{D}$ 350 max	$\dfrac{4.8\sqrt{f'_c}}{D}$ 500 max	$6.5\sqrt{f'_c}$ 400 max	$\dfrac{6.7\sqrt{f'_c}}{D}$ 560 max	$\dfrac{9.5\sqrt{f'_c}}{D}$ 800 max	$13\sqrt{f'_c}$ 800 max
A408	$2.1\sqrt{f'_c}$	$3\sqrt{f'_c}$	$6.5\sqrt{f'_c}$ 400 max	$4.2\sqrt{f'_c}$	$6\sqrt{f'_c}$	$13\sqrt{f'_c}$ 800 max
Plain	50 per cent of above values 160 max			50 per cent of above values 250 max		

* See Table 16 in Appendix for safety provisions.

more, the use of one set of values for "top" bars and another for "bottom" bars complicates the situation still further.

It seems to the author that the proper functioning of reinforced concrete depends so largely upon the action of bond in developing the steel that one should be very conservative in this feature of his designs. The author therefore prefers to use the specified values for top bars as applicable to most tensile steel so that, no matter where the bars are located in beams, walls, slabs, etc., the bond strength is sure to be adequate. A double standard may also be confusing and incorrectly applied. Of course, in design work one should follow whatever specifications govern a given job. On the other hand, one does not violate the spirit of a specification when he uses limiting values that are more conservative than the maximum limit permitted. If the loss caused by a failure is not likely to be serious, and if the lower limit seems to be an unnecessary handicap, the higher stress may be considered where applicable. Also, if the live load is relatively small and the dead load is accurately determinable, there may be more justification in raising the bond limit. If the live load is relatively large, and especially if impact is severe, one will be wise to adhere to a conservative limit in so vital a matter.

Regarding this matter of allowable bond stresses, it will be recalled that the ACI Code of 1951 set a limit of 210 psi for the top bars and 300 psi for the bottom bars for 3,000-lb concrete at working stresses regardless of bar size. The 1963 Code now sets an upper limit of 350 psi for the top bars and 500 psi for bottom ones for the same working-stress conditions.

However, these limits apply primarily to the small sizes. Using No. 6, for instance, the maximum bond stress on the top bars as shown in Table 13 of the Appendix can now be 250 psi and that on the bottom bars can be 350 psi. For the larger bars like No. 9, the new Code actually specifies a limit that is lower than formerly required.[1]

Table 3-1 and Tables 1 and 15 of the Appendix give data regarding A408 bars, which are very large sizes for special heavy construction. It will be seen that the bond stresses permitted on these are relatively low, as they should be. Another point to notice in Table 3-1 is the fact that the allowable bond stress on reinforcement in compression is made independent of the bar sizes but is limited to 400 psi, corresponding to the value of about $f_c' = 4,000$ psi.

When it is practicable, the longitudinal bars in a beam should be "anchored" in regions where the concrete is under compression. They may be bent into the compression side as in Fig. 3-9(a) or extended to the support like bar a in Fig. 3-11(c). In such places as the latter, the pressure itself may help to hold the bars to some extent. When they must be anchored in the portion of the beam that is subjected to tensile stresses, the bars should extend across the possible cracks and not be even approximately parallel to the lines along which the tensile cracks in the concrete might occur; in other words, the bars must not depend upon the tensile strength of the concrete to anchor them. In the case of a beam, the bond stresses along the bars develop the tensile forces T by transferring the longitudinal shear in the concrete to the bars.

All bars must be clean and free from dirt, grease, scale, and loose rust. Anything that destroys the ability of the concrete to stick to the steel may prove to be serious because it may prevent the stress in the latter from being fully developed, and therefore it may keep the steel from performing its function properly.

The permissible bond stress along the reinforcement can be increased by making the surface of a bar rough or irregular. Such bars are called *deformed* bars, and they are used generally. Some types of bars are pictured in Fig. 3-2. The lugs, or corrugations, produce a mechanical bond which helps to lock the concrete and the steel together. Although the lugs do not seem to cause much increase in the bond stress at which a bar will have its initial slip[2] when tested, the wedging or locking action of the projections has a considerable effect in raising the ultimate strength of the bond provided the cover of concrete over the steel is sufficient to prevent stripping or spalling. Any type of deformation that is equivalent

---

[1] Deformed bars are not made in sizes smaller than No. 3 except by special order.
[2] H. J. Gilkey, S. J. Chamberlin, and R. W. Beal, The Bond between Concrete and Steel, *J. ACI*, September, 1938.

to a raised rib that extends directly or diagonally around the bar appears
to be better than narrow ridges running lengthwise along the bar.

Refer to Fig. 3-1(*a*) again where the sketch shows diagrammatically
an arrangement for a pull-out test of a bar.   Sketch (*b*) shows the bar of
diameter D and its corrugations of diameter D' to exaggerated scale.
Under the action of the pull T, the corrugations serve as lugs to bear
against the concrete.   There are resultant actions and reactions somewhat
as shown by the double arrows.   Thus there is a tendency to split the
concrete block as the bar is pulled out.   In sketch (*a*) where a large block
of concrete is shown, the strength of the concrete itself may be helped by
frictional resistance to bursting along the bearing surface JK.   Therefore,
pull-out tests may show too favorable results.   In a beam the situation
is more serious.   If the enlarged bar in (*b*) is rotated as in (*c*) and is
assumed to be in the bottom of a simply supported beam, it can be seen
that the action of the lugs at A, C, E, and G can tend to spall off the thin
cover of concrete below them, whereas the forces at B, D, F, and H act
against the main body of the beam where the concrete is deep and strong.
Furthermore, the action of bending and shearing introduces more complica-

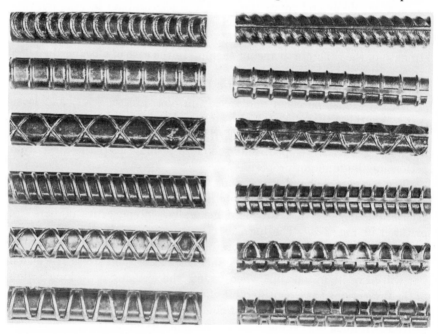

Fig. 3-2   These reinforcing bars are manufactured to meet the requirements of ASTM
Tentative Specifications for Minimum Requirements for the Deformations of Deformed
Steel Bars for Concrete Reinforcement A 305-50T.   (*Courtesy of the American Iron and
Steel Institute.*)

tions which will be discussed later. However, in general, deformed bars tend to distribute the deformation in the concrete in a series of hair cracks closely spaced instead of in a few localized but large cracks.

When there are two or more rows of reinforcement in the bottom of a beam, as shown in Fig. 3-5, the splitting tendency previously referred to is accentuated. The stirrups discussed in the next chapter can be very helpful in resisting this action.

The magnitude of the bond stress probably does not vary directly with the ultimate compressive strength of the concrete in actual structures. The report of pull-out tests by Arthur P. Clark[1] gives bond-slip curves that show considerably higher values of bond for a given slippage in the case of 6,000-lb concrete than for 2,000-lb concrete at slippages of 0.001 in., but less difference at very small slippages. Concrete of 3,500-lb strength gave bond-slip curves that do not differ greatly from those for 2,000-lb concrete. However, the specifications have to be tied to something definite.[2]

It is obvious that high-strength bars that are fully developed will require more bond strength to accomplish this development than would be necessary for bars of the same size but having less intensity of longitudinal stress.[3]

In short members like those often tested, especially those with loads at or near the third point of the span, a relatively large portion of the bar is affected by the compression at and near the supports. This may be a helpful feature in obtaining a relatively high value for the bond resistance in simply supported beams.

In light floors, pavements, and other construction where welded wire mesh can be used to advantage, bond is no problem. This is partly because of the large surface area of the wires in comparison with their cross-sectional area, but mainly because of the effective anchorage provided by the welded junctions of the cross wires. Woven mesh and the ordinary wiring of crossed bars do not provide effective anchorages.

3-2. Hooks. In many cases it is not feasible or possible to extend straight bars far enough to develop their strength sufficiently by bond alone. A common way to remedy this trouble is to bend or hook the bars so as to obtain additional length for their development through

---

[1] Arthur P. Clark, Bond on Concrete Reinforcing Bars, *J. ACI*, November, 1949.

[2] See Test Procedures to Determine Relative Bond Value of Reinforcing Bars (ACI 208-58), *J. ACI*, July, 1958.

[3] See R. G. Mathey and D. Watstein, Investigation of Bond in Beam and Pull-out Specimens with High-yield-strength Deformed Bars, *J. ACI*, March, 1961. Also, see P. M. Ferguson and J. N. Thompson, Development Length of High Strength Reinforcing Bars in Bond, *J. ACI*, July, 1962.

bond.   When a hooked bar is pulled, it tries to slip or slide around the curve.   The hook provides a certain amount of mechanical locking of the steel into the concrete, but this is too indefinite to be relied upon.

Considerable publicity has been given to the idea that the use of A305-type bars will completely eliminate the need for hooks to anchor the reinforcement.   The use of judgment is still desirable.   If a bar has to resist large tension very near its end, and if failure of the structure would be a serious matter—as it usually is—an engineer is unwise in failing to provide some means for ensuring suitable anchorage for the end of the bar.

To illustrate this idea in just one example, assume that the ground floor of a large warehouse is to be constructed as shown in Fig. 3-3(a). This design is used so that the floor can be poured after the walls and roof are completed.   Straight bars a project over the 3-in. seat but end there. Heavy loads on the floor will tend to cause cracks as shown because of both bending and shear.   It would be much better if the ends were bent up as shown by the dotted lines.   Another and better remedy is shown in Fig. 3-3(b) where small dowels b lap over bars a and tie the floor to the foundation wall as well as serving as an extension of a.   If the foundation is strong enough to restrain the end of the floor slab, bars c shown in sketch (c) will resist the tension in the top, thereby transferring to the right the point of inflection and therefore the place where bar a is really needed.   Still another scheme is shown in sketch (d) where bars d and the structural floor extend across the foundation wall and a topping is later placed over the slab.   This may be inconvenient because it requires construction of the floor before the walls can be built.

Strong development of the tensile steel is especially important at the fixed ends of cantilevered beams, at the ends of simply supported beams, and at the tops of continuous or restrained beams where they pass over their supports.   This action will be discussed more fully in Arts. 3-7 and 3-8.

Hooks are generally necessary for plain bars in tension.   One exception is when such bars terminate at the interior supports of continuous beams because they are then in a region of compressive stresses.

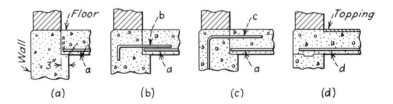

Fig. 3-3   A floor slab seated on a foundation wall.

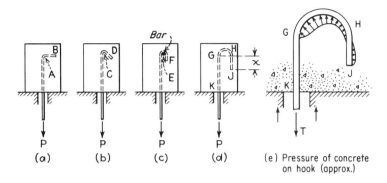

(e) Pressure of concrete
on hook (approx.)

Fig. 3-4

Certain principles should be followed in designing hooks.   Figure 3-4 shows some types of anchorage that are often specified in designs.   Sketch (*a*) is a sharp right-angular bend that may be used when a little mechanical anchorage is desired but full development of the bar is unnecessary. When the bar is pulled downward, the bent portion produces a compressive stress in the concrete.   However, this arm usually has insufficient strength as a cantilever to spread the load over the entire length *AB*. It therefore tends to crush the concrete locally at *A*.   It is clear also that a downward pull on the bar cannot produce a horizontal motion of the portion *AB* and therefore cannot develop its bond resistance until the bar begins to pull out below *A* and until it crushes a fillet in the concrete.   Consequently, when a large tension is to be developed in the bar, this type of anchorage should be made with a reasonably large radius at *A*—usually at least 3*D* for the inside of the curved bar for minor anchorages, but larger for heavy longitudinal bars, such as those at the knees of rigid-frame bridges and the bottoms of retaining walls.

The anchorage shown in Fig. 3-4(*b*) is a modification of that pictured in (*a*).   It is not an improvement, because the acute angle tends to make the compression at point *C* even worse than that at *A* in Fig. 3-4(*a*). Also, the concrete may not fill the triangular space at *C* completely, so that the portion *CD* may be of little use until the bond below *C* is broken. In fact, the larger the bar the worse the condition becomes.

Figure 3-4(*c*) shows another kind of bend in which the bar is hooked back upon itself with a small radius at the bend.   This is not much better than the two previous types unless bent around a bar as shown; it may then make a good anchorage for bars of small diameter.   However, such sharp bends cannot be made with hard steels.   It is much better to go a little farther and to bend the bar as pictured in Fig. 3-4(*d*).   This gives sufficient concrete area inside the bent portion of the bar to withstand the compression which is caused by the tension in the steel.   It is also desirable to provide a straight portion beyond the bend as an additional anchorage.

The minimum inside diameters of such bends as specified by the Code are given in Table 10 of the Appendix.   This shows the minimum dimensions for standard hooks and 90° bends.   For bars No. 14S and 18S, special fabrication procedures are needed to make the bends shown in Table 10 if f_y is 50,000 psi or over.   Notice also that the straight portion at the free end beyond the 90° bends is to be 12 diameters.   This is for full development of tension in the bars.   If only a little anchorage is desired just to avoid the need for a sudden development of tension through bond, a short 90° bend like that shown dotted in Fig. 3-3(a) and bar b in (b) will be helpful.

If a bend is made in a tension member at a point of high stress, the radius of bend must be large enough to prevent crushing of the concrete.

When a hook passes around a longitudinal bar, as shown in Fig. 3-4(c), the pull of the former is transferred at least partially into the latter as a small beam.   This is a sort of mechanical connection.   In most cases, however, bond must be the chief means of developing the hooked bar.

A hook should not be depended upon to anchor a bar subjected to compression, because the bend merely accentuates the tendency of the bar to buckle.

Of course, a hook may have some mechanical strength due to the stiffness of the bar, but this is not relied upon when determining the anchorage needed to develop working stress or ultimate strength of a bar. The bond is considered to act along the surface of the bar whether the latter is straight or bent into a hook.   In fact, one of the principal purposes of a hook or of a 90° bend is to obtain sufficient length of bar when the end of the member is so close that the bar cannot be developed to the required amount if it is straight, since there simply is not enough room. Furthermore, since bond failure is likely to be progressive along the length of a long bar, a hook will be ineffective if it is located beyond the point where bond can develop the working stress or ultimate strength of the bar anyway.   It is probably advisable to assume that the value of a standard hook is not to be trusted to develop more than 50 per cent of f_s or f_y in the steel used regardless of the computed bond resistance beyond the point at which the curve starts.   In fact, the Code states that standard hooks[1] may be assumed to develop 10,000 psi in the bars for WSD and 19,000 psi for USD, or they may be regarded as extensions of the bars at appropriate bond stresses.

Example 3-1.   Analyze the anchorage of a No. 4 bar of the type shown in Fig. 3-4(d) if it is a standard hook having the dimensions given

[1] For hooks and other reinforcing details, see Manual of Standard Practice for Detailing Reinforced Concrete Structures (ACI 315-65).

in Table 10 of the Appendix.   Assume that the bar has a tensile stress of 20,000 psi and the allowable bond stress is 350 psi.

From Eq. (3-1), the length of bar required to develop the tensile strength by bond is

$$L_s = \frac{Df_s}{4u} = \frac{0.5 \times 20,000}{4 \times 350} = 7.14 \text{ in.}$$

The length HJ is $2\frac{1}{2}$ in. and GH is $4\frac{3}{4}$ in., giving a total of $7\frac{1}{4}$ in., which is more than the 7.14 in. required for L_s.

Now investigate the pressure on the concrete at the inside of the hook, which is likely to be greatest at or near point G [Fig. 3-4(e)].   The actual stress condition here is uncertain, but an arbitrary and theoretical maximum value for this pressure may be found by assuming that the action is similar to that of a hoop or pipe which is subjected to normal pressure from the inside, for which $T = pr$, where T is the tension in the bar, p is the normal pressure in pounds per linear inch of the bar, and r is the radius of the curve in inches—in this case r is assumed to be the radius of the inside of the bar.   Then

$$T = A_s f_s = pr \qquad \text{or} \qquad p = \frac{0.2 \times 20,000}{1.25} = 3,200 \text{ pli}$$

Then the unit compressive stress in the concrete is

$$f_c = \frac{p}{D} = \frac{3,200}{0.5} = 6,400 \text{ psi}$$

This is a very large compressive stress even though it is localized and may soon be dissipated through the body of the concrete.   However, this example shows that one should not expect such a hook to develop the full strength of a tensile bar all by itself.   The hook will be a great help but there should be some additional length GK for bond to act on the straight bar, or else the hook should be made with a much larger diameter.

Just to see how the specified length works for a 90° bend, test this same bar if the bend is like that in Fig. 3-4(a) with the dimensions specified in Table 10 in the Appendix.   Dimension $E = 2\frac{3}{8}$ in. and $C = 6$ in. This gives a total length of $8\frac{3}{8}$ in., which is more than the length of the standard hook.

From the preceding figures, the reader can see that one should not concentrate too much tension at a hook regardless of the theoretical computations.   Of course, any mechanical device which will develop the strength of the bar without harming the concrete will serve the purpose of a hook but such devices are not often used in reinforced concrete. Naturally, hooks are not of any use in developing compressive stress in the reinforcement.   It is generally advisable to hook any plain bar if it has to develop tension quickly—or not to use a plain bar for that particular purpose.

3-3. Bond of Multiple Layers of Bars.   Let Fig. 3-5(a) represent a beam with two layers of bars having diameters equal to D.   Let the cover of concrete on the sides be s, and let the clear distance between bars be m.   When the beam is loaded it bends, and the portion below the neutral axis *O-O* elongates, thus stretching the steel.   The increment of tension must be transfered to the bars by the concrete.   If the bond strength of the bottom half of the lower bars is to be developed, it must be done by shear in the concrete across the section AB.   The bond on the top half of these bars is developed directly.   Therefore, let v_L = the allowable unit stress in the concrete (longitudinal shear); $T'_1 - T_1$, the increment of tension in the bottom bars; and $T'_2 - T_2$, the increment of tension in the upper bars per inch of length.   Then

$$0.5(T'_1 - T_1) = 3\left(\frac{\pi D}{2}\,u\right) = (2s + 2m)v_L \tag{3-2}$$

The section CD must develop the entire lower set of bars and half the stress in the upper set, or

$$(T'_1 - T_1) + 0.5(T'_2 - T_2) = 3\left(\pi D + \frac{\pi D}{2}\right)u = (2s + 2m)v_L \tag{3-3}$$

It must be noticed that u in the foregoing cases may not be the full allowable value of the bond stress, but it will be as much as is required to develop the increment of tension.   However, the concrete must be strong enough in shear to act in the manner shown.   This is a possible weakness in short beams with heavy reinforcement in two or more rows.   In such cases, it may not be desirable to use the close spacing shown in Tables 8 and 8A of the Appendix for inner rows of bars.

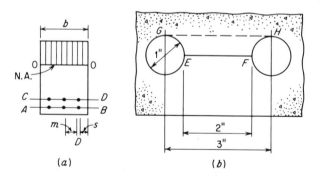

Fig. **3-5**   Critical longitudinal sections along bars.

Actually, the weakest section might be above the plane CD of Fig. 3-5(a). If sketch (b) represents a section at a pair of No. 8 bars spaced 3 in. c.c., the concrete at section EF between them, neglecting the corrugations, is 2 in.^{2} of concrete per inch of length. The surface area of the bottom portions of the bars affecting EF is $\Sigma o = \pi D/2 = 1.57$ in.^{2}. Now, if the section at GH is examined, the concrete area is 3 in.^{2}/in. of length of beam, but the new surface area carrying bond stress is $\Sigma o' = \pi D = 3.14$ in.^{2}. Therefore,

$$GH = 1.5EF \qquad \text{but} \qquad \Sigma o' = 2\Sigma o$$

This shows that the critical section in sketch (a) might actually be the concrete above the upper layer of bars, depending upon the relative magnitudes of the stresses involved.

A further explanation of the bond stresses resulting from beam action will be given in Art. 3-6 and in the next chapter.

3-4. Splices. The ordinary method of splicing reinforcement is by the lapping of the bars past each other so that the bond stresses will transfer the load out of one bar into the concrete and thence into the other bar. Tensile bars might be hooked at splices but it is not usually practicable or desirable to do so. The length of lap should be at least that given for L_s in Eq. (3-1). However, such splices should not be made at points of maximum bending unless they are in the portions of the beam in which compressive stresses exist and in which the steel is not the principal stress-carrying part of the section. In general, it is best to locate splices near points of contraflexure. Many times they can be staggered so that all splices do not come at the same point or even close together.

If splices have to be at points of maximum tension, lapped splices should develop the full strength of the steel as required by M or M_u/ϕ at not over 75 per cent of the values given for u in Tables 13 and 14 in the Appendix. The Code states that such lapped splices should not be used for deformed bars larger than No. 11. In general, one should provide a length of lap not less than $24D$ for $f_y = 40{,}000$ psi, $30D$ for $f_y = 50{,}000$ psi, $36D$ for $f_y = 60{,}000$ psi, but not less than 12 in. in any case. Of course, plain bars need twice as much lap. The lap should also be increased 20 per cent if the splices at such maximum bending points are located within $12D$ laterally, $6D$ or 6 in. from the edge of the beam, or if over one-half of the splices are within $40D$ lengthwise; and the area surrounding these splices should be enclosed by closely spaced stirrups, since lapped splices tend to produce a certain amount of twist.

Splices of reinforcement made by means of the bond strength of the concrete may theoretically be somewhat more effective if the bars are staggered as shown in Fig. 3-6(a). This is satisfactory in a wide slab

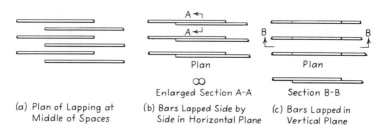

Fig. 3-6   Arrangements for splicing reinforcement by means of bond.

or mat where the spacing of the bars is rather large, but it is undesirable in a beam or other member having several closely spaced bars because the overlapped section acts as a screen to interfere with proper encasement of the bars and filling of forms.   In the case of a horizontal beam, the bars may be lapped in a horizontal plane as shown in sketch (*b*) or one above the other as illustrated in (*c*).   These two facilitate wiring of the bars to hold them in position during concreting.   The wiring does not add appreciably to the strength of the splices.   The arrangement in sketch (*b*) is likely to cause air pockets and poor bond in the space just below the junction between bars.   Encasement in (*c*) is better, but the top bars do not fit the stirrups properly, and they cause the beam to have a smaller effective depth at one place than at another one.   In actual practice, bars in such a position are likely to be knocked down into the position shown in (*b*); hence the latter is the more practicable arrangement if the spacing provides proper clearance for the passage of aggregate.

A lawsuit in which the author was a witness illustrates how bad results can be caused by "a little knowledge."   A man engaged a contractor friend to build a commercial garage for him, letting the contractor design the structure.   The second floor was to have a series of three 20-ft bays made of T beams.   The bars were ordered 30 ft long.   The contractor laid them in the bottoms of the beams.   At the center of the middle span, a point of maximum bending, he butted them end to end instead of lapping them, thus making the total just the right length to reach across the structure.   Of course, when he tried to remove the forms, the beams opened up in the middle and came down with them.   Since it was the contractor's design, the court compelled him to tear down the damaged side spans also and to replace the entire floor at his own expense.

Splices in welded wire mesh should not be made at points of maximum bending either.   If the tension in the wires is over 50 per cent of the allowable stress, the lap should be a minimum of one full space between end cross wires plus at least 2 in.; if the stress is lower, the end cross wires may be lapped a minimum of 2 in., but it is best to lap them as for more highly stressed regions in all cases.

Sometimes it is desirable (but expensive) to splice large bars by welding.[1] Butt splices or lapped splices may be used; the latter being made by having fillet welds along the sides at their junction. Such splices may be conservatively designed on the assumption that 1 lin in. of fillet weld has a strength of 800 to 1,000 lb for each $\frac{1}{8}$ in. of thickness of the weld itself. Ordinary fillet or bead welds are $\frac{1}{4}$ to $\frac{1}{2}$ in. thick. The welds should develop the full strength of the bar, not depend upon the help of the bond. Special patented devices for welding may also be used.

3-5. Development of Longitudinal Reinforcement. The longitudinal reinforcement in a beam should be fully developed if the design is to be safe (unless the amount of steel is excessive) and if the bars are to be effective. Of course, the required number of bars is to be determined by the bending moment to be resisted. Not all of them need be extended for the full length of the beam. They can be discontinued as fast as the decrease in the bending moment will permit—somewhat in the same manner as the cover plates on steel girders. This will be illustrated in Arts. 3-7 and 3-8. However, the shifting of the point of inflection in continuous beams due to live loads must be taken into account; at least one-fourth of the positive reinforcement in the bottom must be carried along the bottom and at least 6 in. into the support; not less than one-third of the negative reinforcement in the top must be carried beyond the inflection point for $\frac{1}{16}$ of the span or the effective depth of the member, whichever is greater. In simply supported beams, at least one-third of the bars should be extended 6 in. or more into the support.

If a simply supported beam has a concentrated load at its center, the bending-moment diagram is of the character shown in Fig. 3-7(*a*). Assume that four No. 6 bars are required at the center *C*. Then it is customary to assume that each bar resists an equal share of the maximum bending moment *DC*. It is therefore theoretically possible to stop one bar at each of points *E*, *F*, and *G* without overloading the remaining ones. However, it is also customary and advisable to extend the bars as shown

---

[1] Splices in compression members will be discussed in Chap. 6.

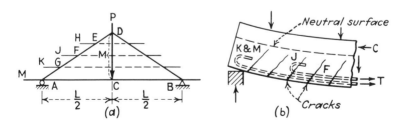

Fig. **3-7**

by *EH*, *FJ*, and *GK* so as to develop them by bond as given by Eq. (3-1) or at least to partially develop them before they reach the point at which they begin to be needed to help in supporting the load.   These bars may be either straight as shown for *c* in Fig. 3-9(*a*) or bent up as shown for *a* and *b*.   It is customary to stop or bend the bars in pairs for symmetry unless the number is odd or unless the conditions are unusual.   The Code requires that any reinforcing bar shall extend beyond the point at which it is not needed for resisting flexural stress for a distance equal to the effective depth of the beam or 12 bar diameters, whichever is greater. Furthermore, main bars should not terminate in a region of tensile stress unless one of the following is satisfied: the shear is not over 50 per cent of that normally permitted, extra stirrups are provided for a distance of three-fourths of the effective depth of the member each side of the cutoff point, the remaining area of the main bars is double the area required for moment, or the perimeter of the remaining bars is double that required for flexural bond.

To illustrate the anchorage of longitudinal bars, let Fig. 3-7(*b*) represent the anchorage of the bar which could terminate at point *F* in Fig. 3-7(*a*).   If the bar is hooked, then the hook *FJ* causes a localized pressure on the concrete of the lower portion of the beam, which is in tension and probably cracked.   The longitudinal shearing strength of the beam must resist this force.   Thus it is easy to see that the hook is of little advantage or possibly harmful because of the concentration of the pull at an inadvisable point in the beam.   It is therefore better to straighten out the bar and extend it an adequate distance beyond *F* or to bend it up and hook or anchor it in the compression side of the beam.

The bars in Fig. 3-7(*a*) which extend to *K* and *M* should preferably be hooked as shown, made with a 90° bend, or properly extended for anchorage by bond, as stated previously.   The author likes to have about 8 or 12 in. of a straight bar extend over the support, depending upon the size of the bar.   In any case, be sure that the anchorage is strong enough to keep the bars from being pulled out.   Any hooks should have adequate cover on the sides because, otherwise, the concrete may spall off, owing to its inadequate strength in tension due to the wedging effect of the hook.

A report of tests[1] made to ascertain the bond strength along horizontal bars gives, in part, the following tentative but interesting conclusions:

1. Straight horizontal bars fixed in position near the tops of beams may be weak in bond because the wet shrinkage or settlement of the concrete tends to cause voids under the bars.   Vibration of the forms

---

[1] R. C. Robin, P. E. Olsen, and R. F. Kinnane, Bond Strength of Reinforcing Bars Embedded Horizontally in Concrete, *J. Inst. Engrs., Australia*, Vol. 14, No. 9, September, 1942; and *Highway Res. Abstr.*, February, 1943.

may help to remedy the situation. This feature may be a source of weakness in continuous beams.

2. Straight horizontal bars released so as to "float" in the concrete near the tops of beams develop good bond strength.

3. Horizontal bars with hooked ends, but which are rigidly held vertically, may develop harmfully large slip before the anchorage will be effective.

4. Straight horizontal bars in the bottoms of beams (and vertical ones) develop good bond.

5. End anchorage beyond the points of inflection of continuous beams is likely to permit large deflections of the beams before these anchorages become effective.

6. Completely watertight forms tend to improve the bond along the top bars.

The data in item 5 are especially worthy of thought in connection with such beams as those pictured in Figs. 3-9 to 3-11, inclusive. They seem to show the desirability of bending the bars so that they can, before failure, serve somewhat the same function as the cables in a suspension bridge.

A few additional items to bear in mind to enable the reinforcement in a beam to be adequately developed by bond are the following:

1. Bars in the same layer should not be closer than one nominal diameter, $1\frac{1}{3}$ times the maximum size of the aggregate, or 1 in. Preferably, the spacing should be as shown in Table 8 of the Appendix.

2. Multiple layers of bars should not be closer than 1 in., but preferably as shown in Table 8. The bars in the upper layer should be vertically over those in the bottom layer to avoid screening effect.

3. The clear space between bars at splices should not be less than specified in items 1 and 2.

4. Actual cover of concrete over the bars (meaning stirrups, also) should not be less than $1\frac{1}{2}$ in. for beams and girders, $\frac{3}{4}$ in. for slabs, 2 in. for exposed formed surfaces, and 3 in. where the concrete is poured directly on soil.

Chamberlin[1] reported that curves of ultimate load-carrying capacity of beams increased with wider simulated spacing of the bars up to a stress of f_y.

It was stated previously that the bond on top bars would be better if the bars could float in the concrete. Figure 3-8 is a picture of some highway viaduct construction. One can see at a glance that floating bars cannot be used in heavy construction where they might be most desirable from the standpoint of bond strength. The bars have to be placed, sup-

[1] S. J. Chamberlin, Spacing of Reinforcement in Beams, *J. ACI*, July, 1956.

Fig. 3-8   Reinforcement in the deck of some viaduct construction.

ported off the forms, and wired together properly prior to concreting. Furthermore, the top reinforcement in continuous structures must be held securely if the effective depth is to be maintained.   Of course, such things as highway pavements are specialties in which it is common practice to place some of the concrete, put reinforcing mats on top of this concrete, and then to deposit the rest of the concrete thereon.

3-6.   Bond Stresses on Longitudinal Bars in Beams.   It is important for a designer of concrete structures to be able to visualize clearly the action of bond on the reinforcement in a beam.   As an example to aid in the understanding of this action, assume the two simply supported beams shown in Fig. 3-9(a) and (d).   The respective shear and bending-moment diagrams are pictured in sketches (b), (c), (e), and (f).   The discussion will be based on the viewpoint of working-stress design, but the ideas will hold equally well for ultimate-strength design.

One way to investigate the bond stresses is to determine the change in the bending moment M in a short length of the beam.   This change in M produces a corresponding change in the tension in the reinforcement, and this change of tension is a measure of the bond stress required to develop it in the given distance.   Therefore, for a distance Δx,

$$\Delta M = \Delta T j d \qquad \text{but} \qquad \Delta T = u(\Sigma o)\Delta x$$

Therefore,
$$u = \frac{\Delta M}{(\Sigma o) j d (\Delta x)} \qquad (3\text{-}4)$$

For example, using Fig. 3-9(c), the bond from A to B is

$$u = \frac{125,000 - 120,000}{(\Sigma o)jd} = \frac{5,000}{(\Sigma o)jd} \quad \text{psi}$$

Since the change in ΔM is constant, u remains constant for one half of the beam if Σo and jd are constant also.

However, there is a more customary and convenient method for computing u. Let Fig. 3-9(g) represent to exaggerated scale a small piece of the beam of sketch (a). Treating this as a free body in equilibrium, and balancing moments about C_B at the right-hand face,

$$(T_B - T_A)jd = V(\Delta x) = u_{AB}(\Sigma o)x(jd)$$

or, in general terms, and for tensile reinforcement, with $\Delta x = x = 1$ in.,

$$u = \frac{V}{(\Sigma o)jd} \tag{3-5}$$

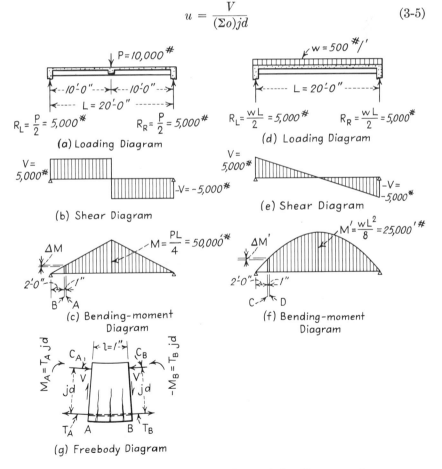

Fig. 3-9 Relation of bond to changes in bending moment.

where V equals the total transverse shear in the beam at the cross section being considered and Σo equals the total surface area of the bars per inch of length of beam at that location—numerically equal to the total perimeter of the bars.   From Eq. (3-5), it is clear that the bond stresses are greatest where the shear is the largest, where the total surface area of the bars is the least, and where jd is the smallest.

Referring to Fig. 3-9(b), Eq. (3-5) gives

$$u = \frac{5,000}{(\Sigma o)jd} \quad \text{psi}$$

This agrees with the previous calculation, as it should, and the shear diagram itself shows that the magnitude of u remains constant for one half of this beam if Σo and jd do not change.

Figure 3-9(e) and (f) produces similar results except that ΔM and V are changing from a maximum rate at the end to zero at the center.

To reiterate, regardless of how the computations are made, a difference in the stress in the reinforcement at two neighboring points along a particular bar requires that there be bond stresses acting upon the bars between these points in order to develop the change in stress.   This is true whether the stresses are tensile or compressive, and even whether one is of the first type and the other of the second.   The only danger to guard against is that of a required change of stress that is more rapid than the bond can provide.   When the bond strength is exceeded the bar may slip, and then the member will no longer act in the manner intended by its designer even if it does not fall down.

Assume a simply supported beam in which the bars are bent up as pictured in Fig. 3-10(a), this being done for purposes that will be explained in Chap. 4.   In the portion near A of sketch (a), the two bars marked c are the only ones that can be fully relied upon to resist the tension and the effect of the shear V_1 on the bond—and to develop the change of bending moment—although a and b help some in the region near their bends.   In the portion from B toward C it is safest to assume that the four bars b and c resist the tension and the effect of V_2.   To the right of

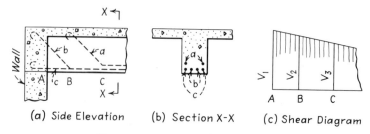

(a) Side Elevation   (b) Section X-X   (c) Shear Diagram

Fig. 3-10   Beam with bent-up bars.

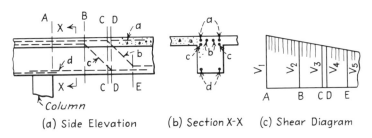

(a) Side Elevation   (b) Section X-X   (c) Shear Diagram

Fig. 3-11   Continuous beam.

C, the six bars a, b, and c are available for V_3. It is therefore obvious that the bond on bars c near A will be critical, and it may be serious. If the latter, it may be undesirable to bend bars b as intended, but they should be extended and bonded or hooked as are c. ·

Again, assume a continuous beam with bars bent up as pictured in Fig. 3-11(a). The tension is in the top near the column. Six bars a, b, and c are therefore available to resist the bond stresses in the tensile reinforcement caused by V_1 in the portion AB. It is advisable to rely upon only four bars a and b to resist the effect of V_2 in BC. Similarly, the two bars a are available and trustworthy to resist V_3 and any tensile stresses at the top of the beam to the right of C.

In the bottom of the beam of Fig. 3-11(a), bars d are in compression. The bond stress on them may be tested by computing the stress in the concrete alongside them near A, then similarly 1 ft to the right of A. Then n (or perhaps $2n$) times these computed stresses will represent sufficiently well the compressive stresses in the bars, and their difference is the amount of stress to be developed in the steel by bond in that 1-ft distance. This is seldom if ever critical.

In the vicinity of D and E of the beam of Fig. 3-11(a), or even at the left of D, tensile stresses may occur in the bottom of the beam for certain conditions of loading. If so, the two bars d are the only ones to be relied upon in the portion from A to near D; four bars c and d in DE; and six bars b, c, and d at E and to the right of E. The worst case for any particular portion of the beam may have to be found by determination of the loading conditions, bending moments, and shears that produce the highest bond stresses at that place. This may require some trial computations. At any rate, satisfactory bond stresses at one point may not mean that they are equally satisfactory at all other points. Furthermore, it should be remembered that bars d in the compression zone do not help bars a, b, and c perform their duties in the tensile region.

In order to illustrate the computation of bond unit stresses in a beam, assume the case shown in Fig. 3-12, which pictures half of a heavy continuous beam 24 ft long. Using Eq. (3-5) and the values in Fig. 3-12, but with d measured to the center of gravity of the group of bars, and with no reduction of the shear due to the width of the column, the bond stress on the eight No. 9 bars at A is

$$u_A = \frac{V}{(\Sigma o)jd} = \frac{59,600}{28.4 \times 0.872 \times 28.5} = 85 \text{ psi}$$

Also, approximately,

$$u_C = \frac{53,200}{14.2 \times 0.88 \times 30} = 142 \text{ psi just left of } C$$

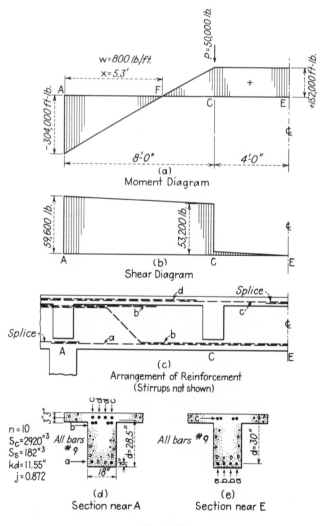

Moment Diagram
(a)

Shear Diagram
(b)

Arrangement of Reinforcement
(Stirrups not shown)
(c)

(d)
Section near A

(e)
Section near E

Fig. 3-12

There is another way of looking at this question of bond unit stresses. Figure 3-12(a) shows that a point of contraflexure (zero moment) is at F, the distance x from A to this point scaling 5.3 ft. Obviously, the tensile stress in the bars at F should be zero. Using the moment at A and the properties of the section given in Fig. 3-12(d) based on the transformed-section method, the unit stress in these bars at A is

$$f_s = \frac{304{,}000 \times 12}{182} = 20{,}000 \text{ psi}$$

Therefore, this stress must be imparted to the steel through bond in the distance x. Since the shear is nearly constant from A to F, assume that this "pickup" of tension is at a constant rate. Therefore, on this basis, the bond stress on one top bar is

$$u = \frac{f_s A_s}{x(\Sigma o)} = \frac{20{,}000 \times 1}{5.3 \times 12 \times 3.54} = 89 \text{ psi}$$

This is only slightly different from the result given by Eq. (3-5). Exact agreement of these figures is not to be expected because the bending-moment diagram in Fig. 3-12(a) should be curved slightly.

The foregoing method is particularly useful in computing bond stresses in members subjected to longitudinal thrusts or tensions combined with bending. The basic idea is to compute the stress in the steel at a given point, then calculate it a foot (or two) from the first point. Obviously, the bond stress must be

$$u = \frac{\text{difference in unit stress} \times \text{area of bar}}{\text{distance between points in inches} \times \text{perimeter of bar}} \quad (3\text{-}6)$$

Unless the direct forces are large compared with the bending, their effects upon the bond stresses are not important.

One must not forget that bond stresses also act upon bars in compression. However, the lower unit stresses generally existent in such bars seldom cause trouble with the bond. For instance, Fig. 3-12(d) shows two No. 9 bars near the bottom where they are subjected to a compressive unit stress of about 9,250 psi if based on nf. From Eq. (3-6), and considering the distance x in Fig. 3-12(a),

$$u = \frac{9{,}250 \times 1}{5.3 \times 12 \times 3.54} = 41 \text{ psi}$$

The designer is not so much interested in learning what the magnitudes of the bond stresses are as he is in making sure that they are not excessive. Bond generally becomes critical only in the case of short, heavily loaded members such as footings, beams carrying offset columns, and other cases where the shears are relatively large compared with the bending moments.

When bond stresses are too high, or when the designer does not want
to depend upon bond alone, it is often feasible mechanically to fasten
bars to something that anchors them thoroughly.   An example of this
in industrial construction is shown in Fig. 3-13.   The roof of this feeder
compartment must support a possible depth of ore of about 70 ft; hence,
it must be very strong.   In order to ensure against bond failure, some
of the critical splices are welded so as to make the bars continuous from
one side to the other.   Short diagonal bars might also be welded to the
main bars to tie the structure together.   In fact, this idea has frequently
been employed in the pedestals for turbogenerators in power plants where
vibration and possible fatigue have to be considered.

For ultimate-strength design, the specified load factors are applied to
the dead load and the live load first, then both the ultimate shear V_u and
the ultimate moment M_u are computed by customary applicable methods.
Then Eq. (3-5) is changed to

$$u_u = \frac{V_u}{\phi(\Sigma o)jd} \tag{3-7}$$

where u_u = ultimate bond stress

V_u = total ultimate shearing force

ϕ = safety provision for bond as given in Table 16 of the Appendix
In computing Σo, one may include bent-up bars that are not over $d/3$
above the main longitudinal tensile reinforcement, but it is safer to disregard them since they are sloping away from the main steel toward the
compression side of the beam.   Of course, the computed tension or compression in any bar at any cross section must be developed on each side
of that section.   Anchorage or development of the bond stress u_u must

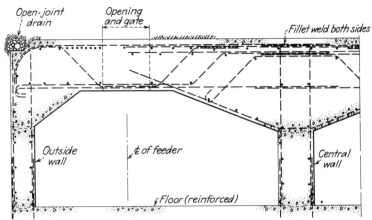

Fig. 3-13   Roof of feeder compartment under 8,000-ton ore bin.   (*Cananea Consolidated Copper Company, Cananea, Sonora, Mexico.*)

be on the basis that the stress in the bar is found from M_u/ϕ divided by $(\Sigma o)jdL_s$.

As an example, assume a simply supported beam. Let $V_u = 50,000$ lb, the reinforcement $=$ five No. 7 bars, $f'_c = 3,000$ psi, $jd = 16$ in., and $\phi = 0.85$ from Table 16 in the Appendix. Then

$$u_u = \frac{V_u}{\phi(\Sigma o)jd} = \frac{50,000}{0.85 \times 13.8 \times 16} = 266 \text{ psi}$$

This is far below the allowable 595 psi shown in Table 14 of the Appendix.

Example 3-2. Assume a continuous T beam similar to that shown in Fig. 2-25(d), where the section is taken at the edge of an interior column. Assume six No. 11 bars in the top as pictured, $d = 23.5$ in., $j = 0.88$, V for live load $= 26,000$ lb, $V = 30,000$ lb for dead load, and $f'_c = 3,500$ psi. Compute the bond stress (1) by working-stress methods and (2) by ultimate-strength methods, using Eqs. (3-5) and (3-7), also $U = 1.5DL + 1.8LL$. Find the relative safety factors compared to the allowable values for bond given in Tables 13 and 14 in the Appendix.

(1) $$u = \frac{V}{(\Sigma o)jd} = \frac{56,000}{26.6 \times 0.88 \times 23.5} = 102 \text{ psi}$$

$$SF = 140/102 = 1.37$$

(2) $$V_u = 1.5 \times 30,000 + 1.8 \times 26,000 = 92,000 \text{ lb}$$

$$u_u = \frac{V_u}{\phi(\Sigma o)jd} = \frac{92,000}{0.85 \times 26.6 \times 0.88 \times 23.5} = 198 \text{ psi}$$

$$SF = 280/198 = 1.41$$

The above results are reasonably comparable.

3-7. Distribution of Bond Stresses in Simply Supported Beams.

In the preceding article the bond stresses in beams have been discussed without regard to the cracked condition of the members when they are highly stressed. The situation, from a practical standpoint, deserves a further examination.

Assume the simply supported beam AB of Fig. 3-14. Bars a are shown extending the full length and terminating over the supporting columns. When a heavy load is applied at the center C, the beam deflects and may crack somewhat as pictured to exaggerated scale in (b). The length of the neutral axis remains unchanged, but the bottom corners and the slender columns each move outward some distance Δ. Bars a, being held by bond, are stretched until they offer sufficient resistance to produce equilibrium.

Now cut out an imaginary cracked piece of this beam as pictured in Fig. 3-14(c). Let the bending moment be computed about an axis in the vertical plane X-X through the uncracked portion EO above the point where the crack FO or its projection will intersect the neutral axis of the beam. Then, neglecting the weight of the beam, the external bending moment at E is

$$M_E = R_A x = Vx$$

since $R_A = V$ in this instance. This causes a clockwise rotation that is counteracted by the internal forces represented by C' and T'. Therefore, taking moments about C' in the face EO, and assuming that no stresses cross the crack except the tension in the steel,

$$M_E = T'jd \qquad \text{or} \qquad T' = \frac{M_E}{jd}$$

On the other hand, notice that this tension must be physically developed by bond on the bars from the end A to F. Similarly, the tension at crack OG is

$$T = \frac{M_D}{jd}$$

and T must be developed by bond between A and G. The difference $T' - T$ must be developed in the bars by bond in the distance GF. The action of the bond is similar for all portions from A to the center C.

Now notice the offset between the axis of moments and the point where the bars "emerge" from the assumed free body in Fig. 3-14(c). As assumed pieces closer and closer to the end of the beam are investigated, such as the piece from A to JOH, it is seen that, if the beam does crack close to the support, the bond length AH becomes very short whereas the

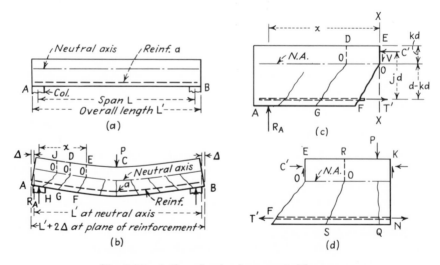

Fig. 3-14   Action of a simply supported beam.

computed tensile force needed for equilibrium is based upon a considerably longer distance from the reaction to the center used in computing the moment. Therefore, it seems that the bond strength needed near the end of a simply supported beam is greater than that indicated by the bending-moment and shear diagrams at that particular spot. This is a qualitative statement. The exact magnitude of the bond stress will be difficult to ascertain.

There may be some question regarding the effect of the portions of the beam at the right of the crack FO of Fig. 3-14(b). A triangular portion of this piece is on the left side of a plane through EO.

The remainder of the beam from EOF to the center is pictured in sketch (d) as a free body. One may imagine that, as the beam bends, the cracked part $OFSO$ tends to thrust toward the left so that the bond adds to the tension in the bars, causing their stress to increase from F to S. There is a compressive reaction in the upper part $OERO$, but this compression is in a region that is at the right of section EO in the picture. Because of these things it seems to be justifiable to consider that the free body pictured in Fig. 3-14(c) is satisfactory and that bond stresses in portion FS of sketch (d) do not affect conditions at the left of EOF.

If the cracks near the end are at 45° with respect to a vertical plane, the uncracked concrete that grips the reinforcement is offset a distance $d - kd$ toward the support from the center of moments. Actually, a crack may or may not exist close to the support. However, for design purposes, it seems to be advisable to make the beam so that it cannot fail even if such a crack does occur.

One might look upon Fig. 3-14(b) as though the compressive forces in the upper part of the beam tend to thrust downward and toward the ends, producing something that approaches the action of a tied arch. This concept also indicates that there is probably a tendency to cause large bond stresses near the ends of tensile bars in simply supported beams. The author therefore recommends the following procedure for simply supported ends of beams:

1. Assume that a crack crosses the tensile reinforcement a little beyond the edge of the support, as at H in Fig. 3-14(b).

2. Assume that the crack has a slope approximately 45° from the vertical.

3. Assume a center of moments, as in OJ of sketch (b), that is located $\frac{2}{3}d$ beyond where the crack crosses the tensile reinforcement.

4. Compute the tension in the steel as required by the bending moment computed at this assumed center.

5. From the end of the beam to the assumed crack, provide enough length of reinforcement to develop the computed tensile force, using the full allowable bond stress u or u_u permitted by the specifications.

6. Elsewhere, compute the bond stress by use of Eq. (3-5) or Eq. (3-7).

7. Let this procedure be assumed to apply to concentrated or uniform loads or to any combination of loads on the beam, including its own weight.

8. Since the Code limitations for the allowable bond unit stress are based upon tests that have been interpreted by means of Eq. (3-5) or Eq. (3-7), it is desirable to retain them as controlling values even when one uses the extra allowances for lengths suggested here.

The preceding discussion indicates that, when the width of the support is small, some form of hook may be desirable in order to make sure that the ends of the reinforcing bars cannot slip. Of course, if the first actual crack is considerably farther from the end than assumed here, the intensity of the bond stress required is reduced.

Adequate hooks or other anchorages of bars at simply supported ends of beams are an insurance against failure by pulling out of the bars. Even though the bars may slip elsewhere and cause bad local cracking, the anchorages will enable the bars to hang on. The failure of the beam will then be likely to occur in flexure or in shear.

By similar reasoning, it seems that the required bond stress near the center of the beam of Fig. 3-14(b) may be less than that computed by the use of the shear diagram and Eq. (3-5) or Eq. (3-7). However, this has no important significance.

On the other hand, if the beam acts as the cracked member of Fig. 3-14(b) indicates, it is advisable to be conservative in terminating any tensile reinforcement that does not extend for the full length of the member. If four bars are needed at the center of the beam of Fig. 3-14(b), draw the bending-moment diagram to scale, as in Fig. 3-15; then divide the middle ordinate of the diagram into four equal parts. Draw lines parallel to AB until they intersect the bending-moment diagram at K, J, P, Q, T, and U. Scale the lengths of KJ, etc.; then add a distance equal to about two-thirds of the effective depth of the beam to each end to allow for the offsetting effect. This is a minimum length for a particular bar. However, it is good policy to extend each bar a moderate distance beyond the point where it is supposedly needed in order to have a little

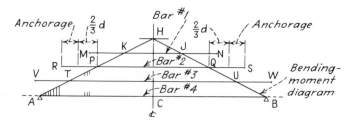

Fig. 3-15   Cutoff of longitudinal reinforcement.

chance to develop some of its strength through bond.   Also, this reserve length will allow for some unusual loading that may increase the bending moment.   This second addition to the length is to be determined by judgment for any given case.   The author prefers to make this extra length 12 in. or sufficient to develop 50 per cent of the allowable tensile stress in the reinforcement, using the maximum allowable bond stress. In the case illustrated by Fig. 3-15, it is apparent that at least two of the bars should extend the full length of the beam.

The Code requires that, except at supports, every reinforcing bar should be extended beyond the point where it is no longer needed for flexural resistance.   The minimum distance for this is to be equal to the effective depth of the beam or 12 diameters, whichever is greater.

Now, with the preceding ideas in mind, investigate the bond stresses in a particular beam in order to get some scale on what the offsetting may mean.   Figure 3-16(a) shows a beam like that of Fig. 3-14(a).   It is loaded by applying a concentration of 10,000 lb at the center, and the weight of the beam is neglected in order to simplify the computations. Assume that $b = 10$ in., $d = 12$ in., $kd = 4.8$ in., $d - kd = 7.2$ in., $L = 10$ ft, $j = 0.87$, and the reinforcement is two No. 8 bars.   Assumed cracks are shown.   The purpose is to get qualitative data.   The cracks may occur at various positions and at various slopes.   Those shown seem to be possible and reasonable, but there should probably be more of them near the center.

The bending moments are computed at sections 1, 2, etc.   From these, the tensile forces are computed, but they are assumed to be offset to points a, b, etc., as pictured in Fig. 3-16(a).   The triangle ADC in sketch (b) shows the diagram of tensile forces T as plotted for points 1, 2, 3, etc., and they are naturally proportional to the triangular bending-moment diagram.

To investigate the bond on the bars of the beam in Fig. 3-16(a), assume that the shear diagram is as pictured by the rectangle $ACFG$ in sketch (c).   According to Eq. (3-5), the bond stress will be uniform along the bars from A to C.   Therefore,

$$u = \frac{V}{(\Sigma o)jd} = \frac{5{,}000}{6.28 \times 0.87 \times 12} = 76 \text{ psi}$$

However, if the tension at a is to be developed by the bond on the 8-in. length of bars shown at the left of this crack, the average bond stress along this length must be, from Table 3-2,

$$u = \frac{T_a}{(\Sigma o)L_s} = \frac{6{,}300}{6.28 \times 8} = 125 \text{ psi}$$

Table 3-2 shows that the computed bond, or rate of pickup of tension,

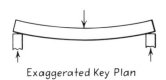

Exaggerated Key Plan

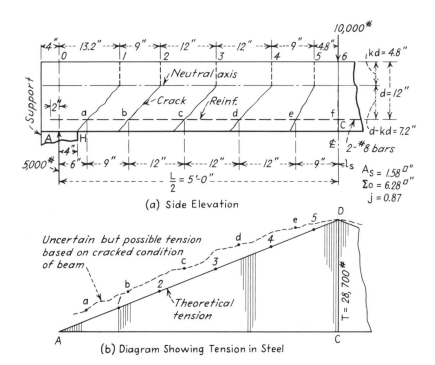

(a) Side Elevation

(b) Diagram Showing Tension in Steel

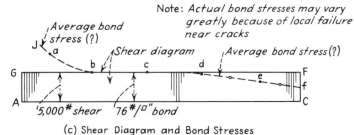

Note: *Actual bond stresses may vary greatly because of local failure near cracks*

(c) Shear Diagram and Bond Stresses

Fig. 3-16   Study of a simply supported beam with a concentrated load at the center.

Table 3-2   **Study of Bond in Beam of Fig. 3-16**

Point	$M,$ in.-lb	$T = M/jd,$ lb	$\Delta T,$ lb	l_s in.	$u = \Delta T/(\Sigma o)l_s,$ psi
1	66,000	6,300	6,300		
2	111,000	10,600	4,300		
3	171,000	16,400	5,800		
4	231,000	22,100	5,700		
5	276,000	26,400	4,300		
6	300,000	28,700	2,300		
To a			6,300	8	125
a-b			4,300	9	76
b-c			5,700	12	76
c-d			5,800	12	77
d-e			4,300	12	57
e-f			2,300	9	41

is substantially the same as given by Eq. (3-5) for points b, c, and d. Then it declines somewhat at e and f, as would be expected from the magnitudes of the tensile forces. These values for the average bond unit stresses between cracks are plotted in the position of the centers of the lengths l_s between cracks, as shown by the dotted lines in Fig. 3-16(c), and they indicate that a critical situation may exist near the ends of the beam considered. Close to the cracks themselves, the bond stresses must be considerably larger, with probable failure locally. This example shows the general tendency for higher bond stresses near the end, whether for concentrated or uniform loads, and whether the analysis is made on the basis of working stresses or ultimate strength. The purpose here is to show the advisability of extending the bars or anchoring them at least as much as required by the Code for simply supported ends.

3-8.   Distribution of Bond Stresses in Continuous Beams. When a horizontal, continuous, reinforced-concrete beam is subjected to a small load, the tension in the top over a support and in the bottom near mid-span may be sufficient to cause only a few hair cracks in these regions. However, as the load is increased, the cracking intensifies and spreads until the beam may be cracked somewhat as shown to exaggerated scale in Fig. 4-16 when the customarily allowable unit stresses exist in the longitudinal bars. This condition is one to be investigated in order to see what the bond stresses may be and to determine the proper locations of cutoff or bend points of the bars. Qualitative data are desired; exact quantitative information is probably unattainable.

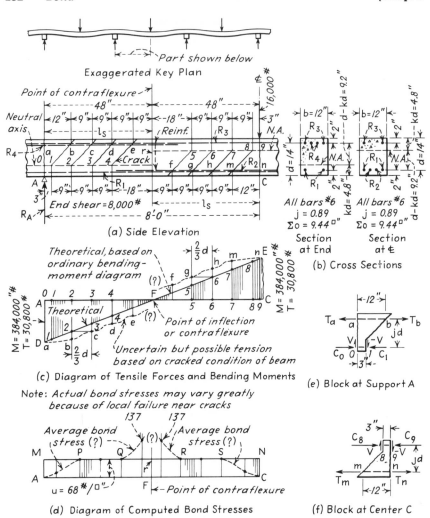

Fig. 3-17   Study of a continuous beam with a concentrated load at the center.

As an illustration, assume a continuous rectangular beam with a concentrated load of 16,000 lb at the center, as shown in Fig. 3-17(*a*) and (*b*). Neglect the dead load of the beam in order to simplify the computations. Imagine that the beam has the cracks shown, with approximately 45° slopes, except at the center and over the support at *A*. Bars R_1 and R_3 extend the full length of the beam, but R_2 and R_4 are assumed to terminate as shown. Assuming balanced loads that cause the equivalent of a fixed-end condition for the beam, the bending moments at A and C are

$$M = \frac{WL}{8} \times 12 = \frac{16{,}000 \times 16 \times 12}{8} = 384{,}000 \text{ in.-lb}$$

The bending-moment diagram is shown by the triangles ADF and FEC of sketch (c).

Analyze the beam as though it were made of a series of blocks, with each one in equilibrium. Each block is bounded on the sides by a crack and a vertical plane through the intersection of the crack or its projection and the neutral axis. The end and central blocks are pictured in Figs. 3-17(e) and (f), respectively.

In sketch (e), the left side is assumed to be cut just off the theoretical support so that V acts as shown. The tension T_a is

$$T_a = \frac{M}{jd} = \frac{384{,}000}{0.89 \times 14} = 30{,}800 \text{ lb}$$

If the moments of the forces shown in (e) are taken about the center of compression C_1, in the right-hand face,

$$-T_a jd + V \times 3 = T_b jd$$
$$-30{,}800 \times 0.89 \times 14 + 8{,}000 \times 3 = T_b \times 0.89 \times 14$$
$$T_b = 28{,}900 \text{ lb}$$

The change in tension $T_a - T_b = 1{,}900$ lb. It must be produced by bond stresses on the embedded length ab of the tensile reinforcement in Fig. 3-17(e). This offsetting of T_b means that there is more distance in which bond can develop the change in tension near the support than the distance from the latter to the center of moments. Therefore, the magnitude of the required bond stress here is less than the bending-moment diagram indicates. However, any straight bars in the top should be lengthened so as to have their ends far enough beyond the theoretical crack lines to ensure proper anchorage.

Similarly, the tensions at c, d, and e are offset *toward* the point of inflection with reference to their respective centers of moment, points 2, 3, and 4. By corresponding analyses the tensions at f, g, h, and m are seen to be offset *toward* the point of inflection with reference to their particular centers of moment, points 5, 6, 7, and 8.

This offsetting due to cracks on both sides of the inflection point at r in Fig. 3-17(a) and in a direction toward this point indicates that the required bond stresses in the vicinity of the inflection point must be larger than required by the bending-moment diagram in order to develop the tensions fast enough. Of course, the completeness, location, and direction of the cracks in the vicinity of the point of contraflexure are problematical. Table 3-3 merely shows the moments, tensions, and bond stresses computed upon the basis of the stated assumptions for this problem.

Figure 3-17(*d*) has been prepared to show the results of this problem and to picture principles graphically.   The bond stress on the tensile reinforcement over the supports of continuous beams should be computed by the use of Eqs. (3-5) or (3-7) since this is conservative, but special attention should be given to the provision of adequate lengths for bond near the points of inflection, following the Code as a minimum.   Therefore, considering the bending-moment diagram of Fig. 3-17(*c*), the tensile reinforcement should be bent or terminated about $\frac{2}{3}d$ beyond the requirements of the bending-moment diagram.   This is indicated in Fig. 3-18. The actual conditions and requirements in the vicinity of the point of inflection are unknown, but some extension of the bars here seems to be desirable for the top as well as for the bottom.   Of course, when the bars are terminated, an extra anchorage allowance L_s of 12 in. or enough to develop 50 per cent of the allowable tensile stress is desirable for conservatism.

Conclusions of similar character will be found if one analyzes a continuous beam with a uniformly distributed load upon it or one with both concentrated and uniform loads.

When moving loads are applied, the bending-moment diagrams should be those which show the maximum bending over the support for various critical positions of loads as one extreme and the maximum bending in the central portion of the span as the opposite extreme.   Bars should then be bent or cut off as determined by these maxima individually with the $\frac{2}{3}d$ added for offsetting, plus any extra needed for anchorage before

Table 3-3   **Study of Bond in Beam of Fig. 3-17**

Point	M, in.-lb	$T = M/jd$, lb	ΔT, lb	L_s between cracks, in. Section	L_s between cracks, in. Length l_s	$u = \Delta T/(\Sigma o)l_s$ psi
0	$-384,000$	30,800	1,900	*a-b*	12	17
1	$-360,000$	28,900	5,800	*b-c*	9	68
2	$-288,000$	23,100	5,800	*c-d*	9	68
3	$-216,000$	17,300	5,700	*d-e*	9	67
4	$-144,000$	11,600	11,600	*e-r*	9	137
r	0	0				
5	$+144,000$	11,600	11,600	*r-f*	9	137
6	$+216,000$	17,300	5,700	*f-g*	9	67
7	$+288,000$	23,100	5,800	*g-h*	9	68
8	$+360,000$	28,900	5,800	*h-m*	9	68
9	$+384,000$	30,800	1,900	*m-n*	12	17

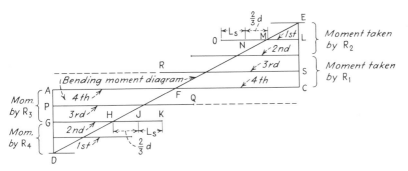

Fig. 3-18 Diagram showing cutoff or bend points of longitudinal bars in a continuous beam with a concentrated load at the center.

cutoff. Since such loading causes a shifting of the point of inflection, these critical diagrams are needed to make sure that none of the bars is terminated too soon.

The statements made for the end portion of continuous beams—from the support to the point of inflection—apply equally for cantilevered beams, except that the point of zero bending is at the outer end or at the outer load. Good anchorage, and probably hooks, are desirable at the outer ends of the longitudinal tensile reinforcement of such beams.

3-9. Bond under Impact and Reversal of Stress. Apparently the resistance of the bond of concrete to reinforcement under static or gradually applied loads is one thing; under shocks from suddenly applied loads it seems to be something else—and far less reliable. More data are needed on this subject, but experiences with reinforced-concrete structures under bombing appear to show that the concrete is likely to disintegrate and fly to pieces, leaving the bars behind as a grille of steel. This action may be due to the brittleness of the concrete and to the action of severe and rapidly traveling deformations of the steel, causing progressive local bond failures and cracking which "run" along the bars from point to point as the "wave" of sudden tensile elongation progresses through the steel. At any rate, structures designed to resist any such loads should be knitted together with a sort of steel cage having the junctions of main bars mechanically fastened in all directions—preferably by welding. In other words, the concrete should be a sort of filler between the bars, but the bond strength should be depended upon only slightly.

Rapid increases in the stresses in reinforcement, and quick reversal of these stresses, may also cause breaking of the bond. This may occur in continuous girders carrying heavy trucks, railroad trains, and cranes. Pending more reliable information than is now available, the designer should be unusually careful in checking into the bond situation. He should be conservative, basing his decisions upon the nature and importance of each special case.

3-10. Bond of Concrete to Structural Plates and Shapes.   When a large I beam is covered with concrete, a strange combination results. The former is a strong ductile member, whereas the latter is relatively stiff and brittle.   Obviously, the two materials do not willingly act in unison.   The bond of the concrete to the steel must be great enough to compel the latter to act as a piece of reinforcement in the former, or else the two will break apart.

There is much uncertainty regarding the magnitude of the bond stress that can be developed when such heavy steel members are encased in this way.   The cross section of the steel may be large compared with its surface area; the surface of the steel is generally flat and smooth; there may be dirt, rust, or grease on the steel when it is embedded; the details of the concrete section may be such that it tends to break into isolated parts or chunks so that it becomes merely a filler.   Therefore, the steel member must be properly anchored to the concrete.   The greatest permissible value for the bond stress in such a case is the same as for plain bars.

The methods to use in designing such members are described in Chap. 5.

PRACTICE PROBLEMS

3-1   Compute the necessary length through which two No. 8 bars must be lapped in order to splice them fully if $f_s = 18,000$ psi and the allowable bond stress $u = 170$ psi.

3-2   Two No. 10 bars are lapped 20 in.   If the allowable bond stress $= 160$ psi, what stress can be developed safely in the bars by such a lapped splice?

3-3   Two No. 11 bars in a beam extend 10 in. over the support, as pictured for bar d of Fig. 3-3(d).   If the allowable bond stress is 175 psi, what unit stress can be developed in the bars?   Assume $f'_c = 3,000$ psi for one case; then compare the result if $f'_c = 5,000$ psi, and $u = 240$ psi.

$Ans.$ $f_s = 5,000$ psi; f_s for part 2 $= 6,900$ psi.

3-4   Two No. 7 bars are to be spliced by lapping.   If $f_s = 20,000$ psi, $f'_c = 3,000$ psi, and $u = 215$ psi, what length of lap is needed?

3-5   Two No. 5 bars are lapped 24 in.   The bars are stressed to 20,000 psi.   What is the intensity of the bond stress that must be developed at the splice?

3-6 A No. 6 bar has a hooked end similar to that of Fig. 3-4(d). If $HJ = 3$ in., the allowable $u = 150$ psi, and the radius of the inside of the bend $= 4D$, locate the point at which the bond alone can be said to have developed 18,000 psi in the bar. *Ans.* $L_s = 8.9$ in. from tangent point.

3-7 Compute the maximum crushing stress on the concrete at the beginning of the hook of the bar in Prob. 3-6, point G of Fig. 3-4(d), using the principles illustrated in Example 3-1.

3-8 Assume the beam shown in Fig. 2-18. If the maximum shear $= 46,000$ lb and $f_c' = 3,000$ psi, is the bond stress satisfactory? Assume $d = 23$ in. and $j = 0.88$. *Ans.* $u = 160$ psi and less than 350 allowed.

3-9 Assume a continuous T beam similar to Fig. 2-25(d). The top row of steel is four No. 8 bars; two other No. 8 bars are located 3 in. below these. Is the bond stress satisfactory if $f_c' = 2,500$ psi, allowable $u = 170$ psi, $j = 0.88$, $d = 26$ in., and $V = 42,000$ lb?
Ans. Yes, since $u = 98$ psi.

3-10 Using the beam of Prob. 3-9, assume that the bending moment at the support is 2,000,000 in.-lb and that it is 1,500,000 in.-lb 12 in. away from the support. If $n = 10$, $kd = 8.86$ in., $d - kd = 18.14$ in., $S_c = 2,130$ in.^{3}, $S_s = 104$ in.^{3}, and the top bars are as given in Prob. 3-9, compute the bond stress on one of the top bars, using the idea of pickup of bending moment. *Ans.* $u = 100$ psi.

3-11 Assume a beam like that of Fig. 3-10(a). If the shear V near $A = 30,000$ lb, bars c are two No. 9, $d = 22$ in., $j = 0.9$, and the allowable $u = 190$ psi, is the bond safe? *Ans.* No. $u = 214$ vs. 190 allowed.

3-12 Assume that bars a of Fig. 3-3(a) are No. 6 at 8 in. c.c., $d = 8$ in., $j = 0.9$, $f_s = 20,000$ psi, and allowable $u = 250$ psi. The engineer in charge says that the bars are to be 50 per cent developed by the time they pass the edge of the 3-in. shelf. Design a means of doing this.

3-13 Assume the same slab and conditions given in Prob. 3-12, except that the engineer wants these bars to be 50 per cent developed by the means shown in Fig. 3-3(b). The shelf is 3 in.; the wall beyond it is 13 in. Design and dimension the bars b.

3-14 Assume a simply supported T beam having a span of 24 ft. It is to support a dead load of 1,200 plf and a live load of 2,000 plf. Assume No. 8 bars, $f_c' = 3,000$ psi, $f_s = 20,000$ psi, $f_y = 40,000$ psi, $j = 0.92$, and $U = 1.5DL + 1.8LL$. Design the beam for bending and bond for both working-stress and ultimate-strength conditions, and make a sketch showing the details of the reinforcement.

3-15 Assume the continuous beam shown in Fig. 3-12(d) and (e). The span is 30 ft; the load is 3,000 plf DL and 1,200 plf LL uniformly distributed; the ends are assumed to be fixed and the bending-moment diagram is assumed to be as given in Fig. 1 of the Appendix. Determine the bend points, cutoffs, and splices of reinforcement, also the maximum u and u_u if $\phi = 0.85$.

3-16   A continuous T beam has a reaction of 30,000 lb DL and 15,000 lb LL.   The top reinforcement consists of six No. 10 bars, $f'_c = 3,500$ psi, $d = 20$ in., $j = 0.88$, $\phi = 0.85$, and $U = 1.5$DL $+ 1.8$LL.   Is the bond stress safe for ultimate-strength conditions?

Ans. $u_u = 202$ psi, but 310 psi is allowed.

SHEAR AND WEB REINFORCEMENT IN BEAMS

4

4-1. Introduction. The determination of the magnitudes and the distributions of the shearing stresses in reinforced-concrete beams is a very troublesome problem. Many tests have been made, and these have determined the ultimate shearing strengths of certain beams. Their results can be accepted as being correct for those particular cases only. It is clear that the total safe shear for any given member must be that which the tests show it to be. However, the question to settle is that of the probable action of the beam that one is analyzing or designing and the best method of finding the shearing stresses in it or of predicting the shearing resistance that it can develop safely.

Reinforced-concrete beams are not homogeneous. When they are subjected to shearing forces, they therefore seem to behave in a manner that is peculiar to themselves. It should be admitted that this behavior is not well understood. The subsequent portions of this chapter are parts of a general attempt to clarify what may be the nature of this behavior, to present a reasonable and understandable theory to be used in estimating the shearing strength of such beams, and to determine how to proportion the reinforcement for them. The explanations given here are designed to make it possible to visualize more clearly what may be the action of each element of the member—or at least to understand what that action may become before the beam will fail.

It seems that the basic action of a beam in resisting shearing forces probably will be similar whether one is looking at the matter from the viewpoint of working stresses or ultimate strength, but the magnitudes of the forces and resistances will vary in one's calculations. Obviously, a beam will behave differently after it has cracked.

4-2. Determination of Shearing Stresses. Assume a reinforced-concrete beam that has transverse loads that cause shearing and bending stresses in it. Under this action, the beam will curve as shown in Fig. 4-1(a), producing compression in the top and tension in the bottom part of the member. The portion of this beam above the neutral axis shown will not crack open prior to real failure. However, as pictured for the lower region, cracks will form when the concrete is unable to elongate sufficiently to equal the deformation of the bars.

It seems that there are two cases to consider when studying the behavior of reinforced-concrete beams, viz., the shearing stresses in the uncracked beam and those in the member after the cracks have formed. The latter condition is the one of practical importance. As stated in the

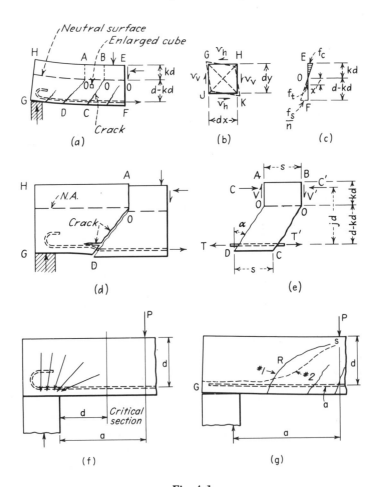

Fig. **4-1**

preceding chapter, there must be transverse shearing forces to be resisted, and there are longitudinal shearing stresses that accompany them.

First, investigate the *uncracked* condition, neglecting the reinforcement. Let Fig. 4-1(*b*) represent a small, square prism of concrete having a length equal to *b*, the width of the beam. Assume a vertical or transverse shearing unit stress v_v acting upon the face *HK* and an equal but opposite stress acting on *GJ*. These two forces constitute a couple having a moment equal to $v_v b(dy)(dx)$. Since the conditions for equilibrium require that there be no rotation of this prism, there must be another couple which counteracts the first one. The latter must be composed of another set of shearing forces which act longitudinally on faces *GH* and *JK*. Assuming the intensity of the latter to be v_h, then

$$v_v b(dy)(dx) = v_h b(dx)(dy) \qquad \text{or} \qquad v_v = v_h \qquad (4\text{-}1)$$

Equation (4-1) indicates that at any point in the uncracked beam there must be vertical and longitudinal shearing stresses of equal intensity. These tend to distort the material as shown in Fig. 4-1(*b*), causing the diagonal *GK* to lengthen and the other diagonal *JH* to shorten. Therefore, compressive stresses must exist on a plane passed through *GK* perpendicular to the paper, and tensile stresses must act on such a plane through *JH*. The latter is called *diagonal tension*, and its critical direction is assumed to be inclined at an angle of 45° from the beam's axis unless affected by other conditions. The intensities of these diagonal compressions and tensions on planes through *GK* and *JH* are each equal to v_h. Before any cracking of the concrete occurs, these shearing stresses may be distributed over the entire depth of the member somewhat the same as in homogeneous beams.

Now, consider the *cracked* condition of the beam. Referring to Fig. 4-1(*a*), it is clear that, when the beam bends sufficiently, the concrete will fail in tension somewhere below the neutral axis. To illustrate this, let $n = 10$, the allowable $f_s = 20,000$ psi, and the ultimate tensile strength of the concrete $f_t = 400$ psi. Then, from similar triangles in Fig. 4-1(*c*),

$$f_t : \frac{f_s}{n} :: x : (d - kd)$$

$$400 : \frac{20,000}{10} :: x : (d - kd) \qquad \text{or} \qquad x = \frac{d - kd}{5}$$

It is obvious that the concrete will crack below this limit *x*.

There are two kinds of tensile cracks to consider. The first ones are those which occur in regions of large bending moments but where the shear is very small or is zero. They are due to the elongation of the bars and to the concrete's inability to stretch equally. These cracks are usu-

ally normal to the axis of the beam, or nearly so.   Figure 4-5[1] shows
some of these cracks between the two loads on the beams.   The second
kind of crack consists of those which are caused primarily by the combined
action of the longitudinal tension and the transverse shearing forces
applied to the beam.   Henceforth, this combined action will be referred
to as *diagonal tension.*

When a crack has formed, it seems that the shearing and diagonal
tensile stresses cannot be transmitted across the opening by the concrete
alone, even though the opening is only a hair crack.   Therefore, the solid
part of the beam above the crack must be the principal thing which
prevents the transverse failure of the member.   The shearing resistance
of the tensile reinforcement acting as dowels in beams without web
reinforcement may be helpful but cannot be relied upon or, at best, is
relatively small and probably should be neglected.[2]   The resistance to the
transverse shearing forces will therefore be assumed to be confined mostly
to the uncracked portion when the only reinforcement in the beam is
longitudinal tensile bars, as in Fig. 4-1(*a*).

Figure 4-1(*d*) is an attempt to illustrate this.   The right-hand portion
of the beam of Fig. 4-1(*a*) is pictured as though it had moved downward
slightly with respect to the part beyond section *AO* and the assumed
crack *OD*.   The cracked surfaces will separate under this movement so
that it is difficult to see how they will offer resistance to it.   The dowel
action of the bars will tend to spall off the concrete from *D* toward *G*,
but this resistance is not trustworthy.   The surfaces at *AO* would have
to slide past each other to cause such a failure.   The forces which might
cause this movement would have to overcome the shearing resistance of
section *AO*.   Even frictional resistance caused by the compression acting
on *AO* will try to prevent such transverse movement.

According to these assumptions, if v_T is the average transverse shearing
stress in the portion above the neutral axis of Fig. 4-1(*d*), and if V is the
total transverse shear at section AO,

$$v_T = \frac{V}{bkd} \tag{4-2}$$

However, this is seldom critical because the resistance of the concrete to
such shearing action seems to be very large.   In footings and thin-slab
floors across the tops of columns where the shearing forces are large, this
feature may deserve investigation.   This transverse shearing action is
sometimes called *punching shear.*

---

[1] Frank E. Richart, An Investigation of Web Stresses in Reinforced Concrete
Beams, *Univ. Illinois Eng. Expt. Sta. Bull.* 166.

[2] See Art. 4-4 for a discussion of dowel action.

The allowable intensity for v_T is somewhat uncertain[1] and may be greatly affected by the amount of compression on the uncracked section. A tensile force—or negligible compression on this area—may cause shearing failure, even under small transverse forces.   Perhaps the following is satisfactory as a guide for WSD when a substantial compressive force acts on the uncracked area of a beam:

$$\max v_T = 0.2f_c' \qquad (4\text{-}3)$$

Ordinarily, it is advisable to limit v_T to $0.15f_c'$.

One unusual but illustrative case of failure of a slab by punching or diagonal tension is illustrated in Fig. 4-2.   This structure is a tunnel under a coal pile at an industrial plant.   The original design of the floor as made by the engineer was somewhat as shown in sketch (*a*).   Notice that the pile reaction at the center of the floor was supposed to be spread through a thick cap; then the floor slab itself was to have nominal bending, and compression from pressure on the side walls.   Somehow or other the contractor omitted the thick pile cap, he deepened the floor somewhat, and he placed the same reinforcement near the bottom, as shown in sketch (*b*).   The piles in some places punched conical holes through the floor, somewhat as pictured by the indicated cracks.

Next, consider the longitudinal forces that are set up in the member when it is loaded.   As the beam of Fig. 4-1(*a*) bends, it must develop the tension in the bars.   In Chap. 3 it was stated that this change in tension in the reinforcement is produced by the bond stresses acting on the steel. To study this action further, assume that the portion of this beam marked

[1] See B. Bresler and K. S. Pister, Strength of Concrete under Combined Stresses, *J. ACI*, September, 1958.

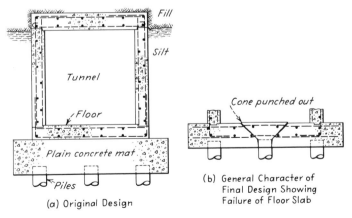

(a) Original Design

(b) General Character of Final Design Showing Failure of Floor Slab

Fig. 4-2   Unusual example of failure by diagonal tension or punching.

$ABCD$ is cut out as pictured in Fig. 4-1(e). Assume that the length s is short. The cracks are at some angle α, the maximum value of which is assumed to be 45° near the support. The section above the neutral axis O-O is cut by imaginary vertical planes. It seems to be obvious that, in horizontal planes above the bars, there must be a horizontal shearing force that equals the change in tension $T' - T$. If the intensity of this horizontal or longitudinal shear is called v_L, the total shear just above the steel for a length s and a width b is

$$T' - T = v_L bs \tag{4-4}$$

However, for the same horizontal area, the total bond strength is

$$T' - T = u(\Sigma o)s \tag{4-5}$$

or $$v_L bs = u(\Sigma o)s \tag{4-6}$$

Then, substituting $u = V/(\Sigma o)jd$ from Eq. (3-5),

$$v_L bs = \frac{V(\Sigma o)s}{(\Sigma o)jd}$$

or $$v_L = \frac{V}{bjd} \quad \text{or} \quad v = \frac{V}{bjd} \tag{4-7}$$

This is the former conventional and general formula for the magnitude of the shearing stresses in a reinforced-concrete beam in terms of the transverse shear V and the dimensions of the section, remembering that b is the width of the stem of a T or I beam or the full width of a rectangular beam. It means that one may then design or analyze a reinforced-concrete beam for practical purposes as though there were a vertical shearing stress, which will be called v, that acts upon the area bd times j.[1] However, v is merely a measure of the diagonal tension.[2] Therefore, when one uses values of v that have been found to be satisfactory by tests and experience, Eq. (4-7) for WSD may be reduced to

$$v = \frac{V}{bd} \tag{4-8}$$

as given in the Code, and, in terms of USD,

$$v_u = \frac{V_u}{bd} \tag{4-9}$$

For T and I beams, b is replaced by b'.

In Eqs. (4-8) and (4-9), V and V_u are supposed to be the total transverse shearing force at the point where the diagonal tension is being inves-

---

[1] This formula is used in AASHO.

[2] See P. M. Ferguson, Some Implications of Recent Diagonal Tension Tests, *J. ACI*, August, 1956.

tigated. Naturally, these forces are a maximum at the edge of a support. However, the critical section for diagonal tension is apparently some distance away from a reaction point. The Code states that this section is to be assumed to be at the distance d from the face of the support except for brackets and other short cantilevered members. The magnitude of v in the portion of the beam from the critical section to tne support may decrease toward the support, but it is conservative to assume that v remains constant in this region. The justification for the assumption that the critical section is some distance from the support may also be illustrated by an examination of possible failure cracks shown in Fig. 4-1(g). Figure 4-5 shows the results of tests. It is obvious that the maximum diagonal tension must have been where the cracks are located, and these are not at the extreme ends except where stripping of the bars occurred.

The effects of torsion and of inclined flexural compression in members having variable depths are to be considered wherever applicable. These may modify and increase the effect of transverse loads. Torsion will be considered in Chap. 14.

Working-stress values of v_c for beams without web reinforcement are given in Table 1-8, but the computed shearing stress to be resisted by the concrete alone is not to exceed $1.1\sqrt{f'_c}$ at the distance d from the face of the support for WSD. For USD, the maximum v_c is to be limited to $2\phi\sqrt{f'_c}$, where the reduction factor $\phi = 0.85$ from Table 16 in the Appendix.

Another formula for the maximum shearing stress in an unreinforced web of an I or T beam for WSD, based largely on test results,[1] is

$$v_c = \sqrt{f'_c} + \frac{1,300p_w V d}{M} \qquad (4\text{-}10)$$

where $p_w = A_s/b'd$, but v_c should not exceed $1.75\sqrt{f'_c}$. For USD, this formula becomes

$$v_c = \phi\left(1.9\sqrt{f'_c} + \frac{2,500p_w V d}{M}\right) \qquad (4\text{-}11)$$

but the maximum should not exceed $3.5\phi\sqrt{f'_c}$. Equation (4-10) is merely Eq. (4-11) divided by a safety factor of 1.9 if ϕ is neglected. In both of these equations, V and M are the shearing force and the bending moment at the section being considered, but M should not be less than Vd or V_ud, whichever applies.[2] These formulas are to be used at the option of the designer in lieu of Eqs. (4-8) and (4-9).

---

[1] Report of ACI-ASCE Committee 326, Shear and Diagonal Tension, *J. ACI*, February, 1962, pp. 277–333.

[2] In AASHO, $v_c = 0.03f'_c$ (90 psi max).

The location of a heavy load with respect to the face of the support is important. This distance is labeled a in Fig. 4-1(f) and (g), and it may be called the *shear span*. The ratio of a/d greatly affects the magnitude of the diagonal tension. When a is short, as in (f), v is generally less serious than when a is large, as in (g), even though the transverse shearing force is the same in both cases.[1]

Consider again the isolated piece of a cracked beam shown in Fig. 4-1(e), acting as a free body in space. The fact that a reinforced-concrete beam, with longitudinal bars only, does not always fail as soon as one of these cracks occurs may be due largely to the strength of the portions like $OCDO$ acting as short cantilevered blocks below the neutral surface O-O (locus of neutral axes), each block resisting the bending caused by the corresponding increment of tension $T' - T$. If the beam is loaded further and the cracking increases, it can be seen that the strength of the member is severely limited. It may be that a crack caused by diagonal tension will start at the vicinity of point R of Fig. 4-1(g) and progress with further loading until it is something like crack No. 1. This crack may progress until the concrete crushes near S—a shear-compression failure—provided the bars do not fail and provided that they can hang on at the end G. On the other hand, the crack may be more like No. 2, where the bond has failed near G and the bars have slipped, as shown for beam 221.1 in Fig. 4-5, causing a sudden failure. If the bars are hooked or otherwise well anchored at G, the beam may crack progressively until it resembles a sort of flat tied arch. This action seems to be what occurred with beam 222.1 in Fig. 4-5. Such action will be accompanied by severe cracking and large deflection because of the stretching of the bars, but this will give one warning of danger rather than cause a sudden collapse.

Briefly reviewing the matter, notice that the real weakness of concrete in beams seems to be its lack of ductility. Its strength in true punching shear is very great. However, when the steel elongates and the cracks open, the latter seem to extend farther and farther toward the compression side of the beam until it fails, unless the bars fail in tension first. This increasing lengthening of the cracks really accompanies a shifting of the neutral axis, and it may cause a decrease of the real area of the concrete which can resist the transverse shear and compression.[2]

When lightweight concrete is used, the Code specifies that the limiting

[1] G. N. J. Kani, The Riddle of Shear Failure and Its Solution, *J. ACI*, April, 1964. Also J. Taub and A. M. Neville, Resistance to Shear of Reinforced Concrete Beams: Part 1, Beams without Web Reinforcement, *J. ACI*, August, 1960; and Shear Strength of Reinforced Concrete Beams without Web Reinforcement: Part 2, Factors Affecting Diagonal Cracking, *J. ACI*, November, 1962.

[2] C. S. Whitney, Ultimate Shear Strength of Reinforced Concrete Flat Slabs, Footings, Beams, and Frame Members without Shear Reinforcement, *J. ACI*, October, 1957.

value of v_c for a beam without web reinforcement should be $0.17F_{sp}\sqrt{f_c'}$ for WSD and $0.3\phi F_{sp}\sqrt{f_c'}$ for USD where F_{sp} is the ratio of splitting tensile strength to the square root of the compressive strength as determined by special testing procedures, but may be taken as 4.0 unless actual tests show otherwise. Also, Eq. (4-10) is to become

$$v_c = 0.15F_{sp}\sqrt{f_c'} + 1{,}300\frac{p_w V d}{M} \tag{4-10a}$$

and Eq. (4-11) is to be replaced by

$$v_c = \phi\left(0.28F_{sp}\sqrt{f_c'} + 2{,}500\frac{p_w V d}{M}\right) \tag{4-11a}$$

The preceding discussion applies to beams and girders. The calculation of shearing stresses in footings and in large slabs will be discussed in Chaps. 9 and 10.

Example 4-1. Assume a simply supported rectangular beam similar to that in Fig. 4-1(a). The shear at the point d from the face of the support is 10,000 lb DL and 12,000 lb LL. Compute the value of v and v_u if $f_c' = 3{,}000$ psi, $b = 14$ in., $d = 20$ in., $\phi = 0.85$, and $U = 1.5\,\text{DL} + 1.8\,\text{LL}$. Is the member safe? See Tables 1-8 and 4-1 for allowable working stresses.

$$v = \frac{V}{bd} = \frac{22{,}000}{14 \times 20} = 79 \text{ psi}$$

Table 4-1   **Limiting Values of Shearing Unit Stresses, psi**

			Working-stress design			
f_c'	$1.1\sqrt{f_c'}$	$1.75\sqrt{f_c'}$	$2\sqrt{f_c'}$	$3\sqrt{f_c'}$	$4\sqrt{f_c'}$	$5\sqrt{f_c'}$
2,500	55	88	100	150	200	250
3,000	60	95	110	164	219	274
3,500	65	104	118	178	237	296
4,000	70	111	126	190	253	316
5,000	78	124	141	212	282	353

		Ultimate-strength design		
f_c'	$2\phi\sqrt{f_c'}$	$3.5\phi\sqrt{f_c'}$	$4\phi\sqrt{f_c'}$	$6\phi\sqrt{f_c'}$	$10\phi\sqrt{f_c'}$
2,500	85	149	170	255	425
3,000	93	163	186	279	466
3,500	100	176	200	301	502
4,000	107	188	215	322	537
5,000	120	210	240	360	601

This compares with 60 psi allowed, and the beam is therefore overstressed.

$$U = 1.5 \times 10,000 + 1.8 \times 12,000 = 36,600 \text{ lb}$$

$$v_u = \frac{V_u}{bd} = \frac{36,600}{14 \times 20} = 131 \text{ psi}$$

The maximum allowable $v_c = 2\phi \sqrt{f_c'} = 2 \times 0.85 \sqrt{3,000} = 93$ psi. Again, the beam is shown to be overstressed. It will need web reinforcement.

4-3. Vertical Stirrups.   If the loads acting upon a beam are great enough to cause shearing stresses that exceed the safe values for the concrete alone when used with longitudinal reinforcement only, then it becomes necessary to strengthen the beam.   Although an increase of the tensile bars will reduce their unit stress and elongation, and thereby reduce the cracking, this is not sufficient to remedy the situation.   The strengthening which should be done is the addition of "web reinforcement" which will prevent the cracks from spreading and from causing failure of the structure.

Web reinforcement may consist of stirrups at 90° to the longitudinal bars, stirrups inclined at 45° or more with respect to the longitudinal reinforcement, longitudinal bars bent up 30° or more with respect to the longitudinal reinforcement, or a combination of vertical or inclined stirrups and bent-up bars.   In any case, the stirrups or bent-up bars must be properly anchored at both ends.   However, the stress in any web reinforcement shall not exceed the stress allowed in tension for the grade of steel used.   Furthermore, even with web reinforcement, the computed shear v in the concrete at the critical point distant d from the face of the support shall not exceed $5 \sqrt{f_c'}$ for working-stress design, and v_u for ultimate-strength design shall not exceed $10\phi \sqrt{f_c'}$.   For the convenience of the reader, these and other limiting values are given in Table 4-1.

Figure 4-3(*a*) and (*b*) represents a rectangular simply supported beam with vertical stirrups placed in it as illustrated by *EF*.   These stirrups are carried under the longitudinal reinforcement at *E* and are anchored into the concrete near *F*.   The bends at the outside of the tensile reinforcement are important not only to serve as anchors but also to help prevent spalling of the corners and sides of the beam because of the wedging action of the corrugations previously discussed in Chap. 3.   The presence of these stirrups makes considerable difference in the shearing strength of the beam.

Consider further the probable action of deformed bars in a beam, using Fig. 4-3*A* for reference.   Sketch (*a*) shows a deformed bar to exaggerated scale.   Imagine it to be in the bottom of a simply supported beam—such as the bars above *DC* in Fig. 4-3(*c*).   Let *A* be an assumed fixed point for reference, and assume that the force T stretches the bar from the original length shown by the solid lines to the positions pictured by the dotted ones. The lugs at B and C are pulled to the right, or one can imagine that the concrete tries to push to the left with respect to the bars. Anyway, the longitudinal shearing forces H in the concrete[1] are the ones that, through bond, cause the change in the tension in the bars. There may be a tendency not only to crack the concrete in tension but to shear the concrete off along some area like the plane through DEF or below it, but real shearing resistance of the concrete is large. It seems to be more probable that there is a mechanical action produced by the lugs which causes them to tend to "jump" out of the upper grooves in the concrete because the latter is the main portion of the member and is the "immovable body," so that the weak part of the system is the tensile resistance of the concrete in a horizontal plane—perhaps through ABC. The only thing that the bar can then do is to try to move downward as it is stretched toward the right in (a), carrying the concrete cover with it and causing a longitudinal crack along the centers of the bars or somewhat above that

[1] In many cases herein, the word "shear" will be used in connection with bond and web reinforcement in beams. It is to be understood that the transverse shear is used as a measure of diagonal tension, the horizontal component of which is what affects the bond and the development of tension (or compression) in the steel due to flexural action.

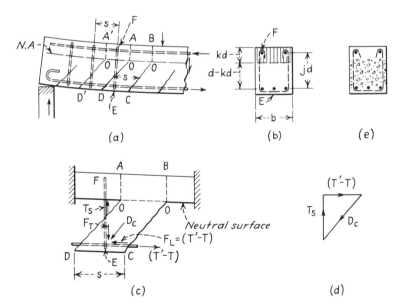

Fig. 4-3 Possible action of vertical stirrups.

level.   Such action may explain the crack patterns shown in beams 221.1, 222.1, and 2210.1 in Fig. 4-5.

A secondary useful function of stirrups is therefore illustrated in Fig. 4-3A (*b*), where the arrows show the imagined directions of the tendencies for spalling of the concrete.   If the bars K and L are too far from the bends in the stirrups at G and J, the center of the concrete cover might spall off.   That is why the double stirrup arrangement is desirable for very wide beams.

It seems that the preceding speculation about the action of the lugs on the bars may help the reader to see why it is desirable to make sure that the main longitudinal bars are well tied in.   It also shows one reason why the spacing s in Fig. 4-3(a) should not be too large.   However, this action of stirrups is a secondary one.   Their chief function is to resist diagonal tension.   To describe the action of vertical stirrups (or of any web reinforcement placed perpendicular to the longitudinal reinforcement of a beam), a horizontal, simply supported beam will be assumed.

Examine what happens as the beam of Fig. 4-3(a) bends under increased loading.   Before the concrete cracks, the stirrups may be subjected to a stress of only n times that which is in the concrete beside, and parallel to, them.   They have very little effect because of their small area compared with that of the concrete.   In this case, the latter resists a large part of the longitudinal tension which is caused by bending.   It also resists most of the diagonal tension which is caused by the shearing and bending action.   However, as previously explained, the deformations accompanying the usual tensile stresses in the steel are so great that they compel the concrete to open up in "hair" cracks.   Then the stirrups come into action prominently, and they resist the spreading of the cracks.

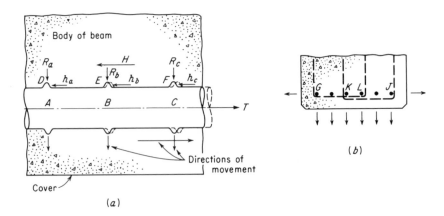

Fig. 4-3A

Next, analyze the action of these stirrups. To do so, assume that Fig. 4-3(c) represents the piece $ABCD$ of the beam of Fig. 4-3(a). Imagine that it is cut out along with the top portion of the adjacent parts of the beam and that it is placed between fixed supports. Then assume that the increment of tension $T' - T$ is applied as a horizontal pull on the longitudinal bars. This pull tends to break off the portion $OCDO$ about the neutral surface O-O, or about the faces AO and BO, an action that the concrete alone cannot resist well. Then, if the stirrup EF is in place, it will serve as an anchor or tie to stop this rotation. Thus it becomes clear that the force $T' - T$ sets up a diagonal compression D_c in the concrete and a tension T_s in the stirrup. The horizontal component of the diagonal compression will be called F_L, whereas its vertical component is F_T. Therefore, the stirrup EF, if acting without help from the concrete, must resist a pull T_s equal to F_T. In other words, $T' - T$, D_c, and T_s are three forces that are in equilibrium and that meet at a point. Therefore, their magnitudes must be proportional to the sides of a triangle, as shown in Fig. 4-3(d).

From all this, referring to Fig. 4-3(a), it is apparent that each stirrup acts as a tie to prevent the portion or portions that it occupies from breaking off. Each stirrup must withstand a force equal to the sum of the stresses acting upon the area that affects it, or $A_v f_v = F_T = F_L$ when D_c is inclined at 45°. Therefore, assuming that the diagonal tension acts at 45° with respect to the longitudinal bars, and with $v = V/bd$,

$$A_v f_v = vbs = \frac{Vs}{d} \tag{4-12}$$

for working loads. Similarly,

$$A_v f_y = \frac{V_u s}{\phi d} \tag{4-13}$$

for ultimate-strength design. A_v is the area of the bars composing one complete stirrup; f_v is the allowable unit stress in it for WSD; and f_y is the maximum for USD. This assumes that the concrete does not help the stirrup. However, this matter will be discussed in the next article.

The details of one type of stirrup can be seen by examining Fig. 4-3(b). The bottom is looped under the longitudinal reinforcement so as to get a mechanical grip around it and thus obtain a better chance to develop the stirrup. The top ends are hooked but are not overlapped, thus permitting the placing of the main bars after the stirrups are set in the forms. The ability of the concrete to develop the necessary bond along the stirrup below the neutral axis is decreased because of the cracks. The stirrups should be developed by bond in and near the region of compression. It seems to be safe to assume that the maximum depth of this portion that can be relied upon for bond is $0.5d$. Since this distance is relatively short, the hooks are needed in order to provide the necessary length of bar, which may be found from the formula

$$T_s = A_v f_v = (\Sigma o)uL_s \tag{4-13a}$$

where L_s is the length of embedment which is needed to develop the allowable tensile unit stress in the stirrup. Generally, the requirements for bond strength compel the use of rather small bars for stirrups so as to have a large ratio of surface area to cross-sectional area. The close spacing needed anyway will seldom require the use of large bars as stirrups. The small ones also are easier to bend, and the hooks are not so large. It is also advantageous to use longitudinal tie bars as pictured in Fig. 4-3(b) so as to help develop the stirrups and to hold them in position during the placing of the concrete. If such bars are used, the stirrups may be bent as shown in Fig. 4-3(e), a method that may help to hold them in better line at the top.

Vertical stirrups are so simple, they can be arranged so readily in cases of varying shear, and they can be set in the forms with the other bars so easily that they constitute one of the most practical systems of web reinforcement.

Figure 4-3(a) and (c) suggests that the beam may act somewhat like a truss. This truss action of stirrups is likely to modify the transverse shearing stresses in the uncracked portion of the beam, but this change is difficult to compute with any certainty. Therefore, v_T will be assumed to be confined entirely to the uncracked concrete.

Sometimes, specifications refer to the percentage of web reinforcement in beams. This simply means the cross-sectional area of the whole stirrup divided by the area (in plan) of the portion of the beam that it reinforces. In the case of vertical stirrups, $p_v = A_v/bs$. The Code requires that this area of the stirrups shall not be less than 0.15 per cent of this area, or $A_v = 0.0015bs$.

4-4. Spacing of Vertical Stirrups. The Code limitations for the maximum shearing stress v_c in unreinforced webs of beams have already been stated. It therefore is necessary that some type of web reinforcement should resist any excess shear above what the concrete can support by itself. Vertical stirrups can be used for this purpose. On this basis, with V' designating the shear to be carried by the stirrups, Eq. (4-12) becomes

$$A_v = \frac{V's}{f_v d} \tag{4-14}$$

for working-stress design[1] and Eq. (4-13) becomes

$$A_v = \frac{V_u's}{\phi f_y d} \tag{4-15}$$

for ultimate-strength design, where both are rearranged for convenience. For T and I beams, b is replaced by b'.

Tests reported by Richart[2] seem to indicate that the stirrups do not become stressed so highly as Eq. (4-12) would indicate. This may be caused by the fact that the concrete continues to resist shearing forces and the resultant diagonal tension as much as it is able to do, acting somewhat as it would if no stirrups were present. However, this combined action is uncertain. It seems that the concrete may become less effective in its resistance as the cracking increases under the action of larger and larger loads. Eventually, the stirrups may have to hold the member together against diagonal tension forces primarily by themselves as a sort of trussing, somewhat as indicated in Fig. 4-4.

These tests seem to indicate also that the stirrups near a concentrated load, as pictured in Fig. 4-4, have smaller stresses than those farther away. Probably this is due to local compression under and near the load and to the decrease of the bond stresses and the pickup of tension at this position as explained in Art. 3-7. Furthermore, when the ends of a simply supported beam are approached, as shown in Fig. 4-4, there seems to be another decrease in stirrup stresses—an action already stated in connection with shearing stresses. The left-hand portion of Fig. 4-4 is an attempt to picture the lines of probable diagonal compression in the concrete and the tension in the stirrups, which can be looked upon as forming the web system of a sort of Howe truss. The shaded area of the right-hand part of this beam shows the region where the stirrup stresses may be the greatest.

Therefore, the position of a concentrated load with respect to the support may not have the expected effect on the shearing stresses between itself and the end of the beam. This is sometimes referred to as the effect

[1] The AASHO formula is $A_v f_v = V's/jd$.
[2] Frank E. Richart, An Investigation of Web Stresses in Reinforced Concrete Beams, *Univ. Illinois Eng. Expt. Sta. Bull.* 166.

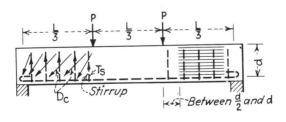

Fig. 4-4

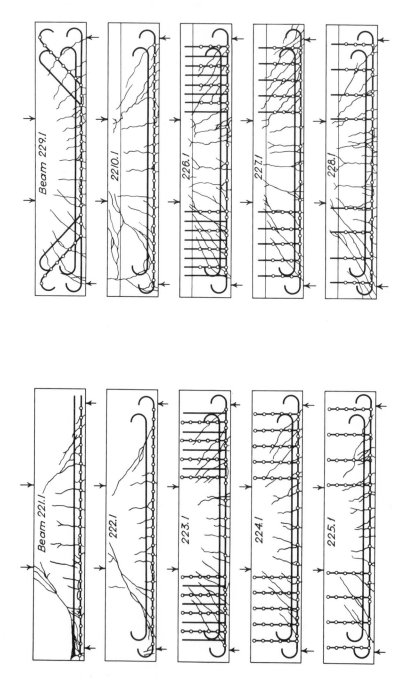

Fig. 4–5 Sketches of beams that were tested at the Engineering Experiment Station, University of Illinois.

174

of the shear span.[1]   When the first big load is near the center of the beam, it is probable that the beam will be controlled by the bending moment rather than by the shear.   When the load is close to the support, the shear may become the critical feature.

The results of tests made by Arthur P. Clark[2] on beams with and without web reinforcement to show the relation between stirrup stress and the computed shearing stress are shown in Fig. 4-6.   At first, f_v increases only moderately, as the loads and v increase, but it jumps rapidly after v reaches about 200 psi.   This seems to indicate that the concrete is helpful at first but that the shearing resistance of the beam depends more and more upon the stirrups as the beam approaches failure.

Some engineers outside of the United States believe that, once the shearing stresses reach a magnitude which is not safe for the concrete alone, the web reinforcement should be designed to resist the entire diagonal tension.   In certain cases, the author agrees with this idea but suggests that a little larger unit stress be used than that generally allowed in the grade of steel used for WSD.   However, it is customary practice in the United States to allocate to the concrete the resistance which the concrete alone might offer if there were no web reinforcement.

[1] A. F. Al-Alusi, Diagonal Tension Strength of Reinforced Concrete T-beams with Varying Shear Span, *J. ACI*, May, 1957.

[2] A. P. Clark, Diagonal Tension in Reinforced Concrete Beams, *J. ACI*, October, 1951.

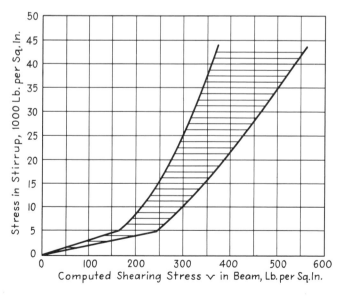

Fig. 4-6   Range of relationship of stirrup stress and shearing stress in beams as reported by Arthur P. Clark.

To illustrate the procedure when part of the shearing resistance is allotted to the concrete, let Fig. 4-7(a) be a diagram that shows the intensity of the shear or diagonal tension in one-half of a beam which carries a uniform load. The value of the maximum ordinate AE is assumed to be constant to point F, which is the critical point distant d from the support. Then AE is $v = V/bd$. The distance AB represents v_c taken by the concrete alone. The stirrups must therefore resist the remaining shear or diagonal tension, which is represented by $BEFC$. Notice that the shear from A to G is pictured as constant, even though it may increase near the support, because the stirrups from the end to G will be designed on this basis. Then from H to D no stirrups will be needed theoretically, but the Code properly requires that they be used for a distance d beyond point C, where they might terminate. This is a good procedure because the exact location of the point C can vary with unexpected live loads, and it is also soundly conservative.

Since the area $BEFC$ of Fig. 4-7(a) has different heights at various points, it means that, theoretically, one should use stirrups of different strengths at uniform spacing throughout this area or else use stirrups of one given size at varying spacing. The former method is not generally advisable. To apply the latter plan, assume a size of bar for the stirrups and find $A_v f_v$ for it, remembering that a stirrup like that in Fig. 4-3(b) consists of two bars crossing a given section. Then Eqs. (4-14) and (4-15)

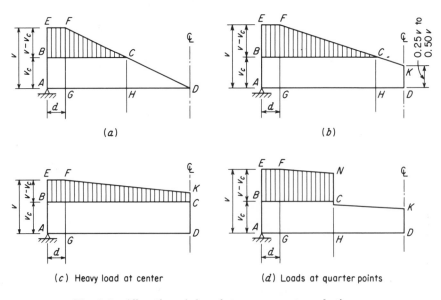

(a)

(b)

(c) Heavy load at center

(d) Loads at quarter points

Fig. 4-7   Allocation of shear between concrete and stirrups.

can be used to find the spacing s for various values of V', or they can be changed to the following:

$$A_v = \frac{(v - v_c)bs}{f_v} \qquad (4\text{-}16)$$

for WSD, and

$$A_v = \frac{(v_u - v_c)bs}{\phi f_y} \qquad (4\text{-}17)$$

for USD, where v_c is the appropriate value for the concrete alone (not the same numerical value in both cases). The spacing therefore can be uniform from A to G [Fig. 4-7(a)] then gradually increased to H, and finally carried at the maximum spacing for a distance d beyond H.

This varying spacing is often impractical for field work; therefore it is customary to use a few spaces at one dimension, then change to a larger one farther from the support, etc. However, as a study of Fig. 4-5 will show, the spacing should not be too great, otherwise a crack can form between stirrups and cause the member to fail. Some guides for the design and spacing of web reinforcement are as follows:

1. The stirrups should be spaced so that every 45° line, which might represent a diagonal crack extending from mid-depth ($d/2$) of the beam to the longitudinal tension reinforcement, shall be crossed by at least one line of web reinforcement.

2. When v exceeds $3\sqrt{f'_c}$, every 45° line in item 1 must be crossed by at least two lines of web reinforcement.

3. When v_u exceeds $6\phi\sqrt{f'_c}$, every 45° line in item 1 must also be crossed by at least two lines of web reinforcement.

4. Small sizes of bars are desirable for stirrups because they are usually strong enough when used with the required close spacing and because they can be bent more easily and sharply to such forms as those shown in Figs. 4-3 and 4-17 to provide proper anchorage.

5. In wide, heavy beams it is generally desirable to use double, overlapped stirrups like those in Figs. 4-3A and 4-8(d) both for strength and to hold in the central longitudinal bars so as to avoid spalling the concrete off the bottom.

6. No matter how slight the excess of v or v_u may be, the stirrups to be provided near a point like C [Fig. 4-7(a)] cannot be more than $d/2$ c.c.

7. From the face of the support to the critical point distant d therefrom, use the same size and spacing for all of the web reinforcement.

8. Double U-shaped stirrups like those in Fig. 4-17(d) are advantageous for erection and especially for continuous beams because they hold in the compressive reinforcement as well as the tensile bars.

9. As previously stated, A_v should not be less than $0.0015bs$.

10. At the end of a beam, it is usually satisfactory to place the first stirrup at $0.5s$ from the effective face of the support unless the support is another beam or girder, in which case $0.25s$ is more desirable.[1]

When the loads are large and concentrated, the diagrams for v and v_u or V and V_u will look more like those shown in Fig. 4-7(c) and (d). The areas where web reinforcement is required are shown hatched. When the uniform load is relatively small, it is obvious that the practicable thing to do is to use a uniform size and spacing of stirrups throughout the distance where they are required by stress plus the extra distance d beyond the theoretical end H.

Some engineers assume that the minimum shearing stress in a beam is from 25 to 50 per cent of the maximum stress. In such cases, the shear diagram would be a trapezoid, as shown in Fig. 4-7(b), even for uniform loads. The distance in which stirrups would be required would therefore be extended.

In any case, the engineer should remember that an adequate system of web reinforcement will give a beam considerable "toughness." It may crack and deflect badly, but it will not be likely to fail suddenly in diagonal tension as plain concrete beams may do. Close spacing of stirrups may also enable heavy longitudinal bars to have considerable strength as dowels in resisting shear.[2] This seems to be probable, since the bars will tend to bridge across from one stirrup to the next one nearer the reaction point. At least, this action should help before the beam will really fail in shear.

Another matter questioned by Rensaa is the difference between the effect of a seat for bearing, as in Fig. 4-1(f) and (g), on the shear-span or tied-arch effect and what occurs when a beam frames into a column, a girder, or a wall. In these last cases, there is not the same effect from the reaction, since these members supposedly have the end shear distributed (at least in part) across vertical planes at their ends. The effectiveness of the shear-span and arch action may be much less than for seated connections. The stirrups should then be extended to, or into, the edge of the support; or one or more hooked, inverted V-shaped ties should be used to strengthen the extreme end of the beam.

Example 4-2. By working-stress methods, design vertical U-shaped stirrups for the simply supported T beam shown in Fig. 4-8(a), using No. 3 deformed bars. The beam is 20 ft long, and it has a uniformly dis-

---

[1] AASHO specifies $s = 0.5d$ maximum for stress, $0.75d$ where not required for stress, and $0.5s$ at the ends. Stirrup anchorage is counted from mid-height to the center of the top hook plus the hook itself.

[2] G. J. Kani, The Mechanism of the So-called Shear Failure, *J. Eng. Inst. Can.*, 1963; and The Riddle of Shear Failure and Its Solution, *J. ACI*, April, 1964. Also, see a discussion of this paper by E. M. Rensaa and others, *J. ACI*, December, 1964.

tributed total load of 3,000 plf. Assume $b' = 12$ in., $d = 20$ in., $f'_c = 2,500$ psi, and $f_v = 18,000$ psi.

From Table 1-8, v_c is limited to 55 psi. From Table 13 in the Appendix, call the allowable u equal to 350 psi.

$$\text{Critical } V = \frac{wL}{2} - w(d + 0.5) = 3,000(10 - 2.17) = 23,500 \text{ psi}$$

$$\text{Max } v = \frac{V}{b'd} = \frac{23,500}{12 \times 20} = 98 \text{ psi}$$

The shear to be carried by the stirrups near the end is

$$v - v_c = 98 - 55 = 43 \text{ psi}$$

To find the point where the stirrups are not needed, in Fig. 4-8(b) find

$$AJ = \frac{R_A}{b'd} = \frac{3,000 \times 10}{12 \times 20} = 125 \text{ psi}$$

Then $$BJ = AJ - AB = 125 - 55 = 70 \text{ psi}$$

By similar triangles from BJC and AJD,

$$BC:AD::BJ:AJ \qquad \text{or} \qquad BC = \frac{120 \times 70}{125} = 67.2 \text{ in.} \quad \text{Say 68 in.}$$

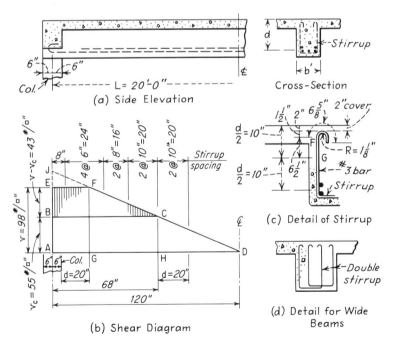

(a) Side Elevation

(b) Shear Diagram

(c) Detail of Stirrup

(d) Detail for Wide Beams

Fig. 4-8   Problem in spacing of stirrups.

From Eq. (4-16),

$$A_v = \frac{(v - v_c)b's}{f_v} \quad \text{or} \quad s = \frac{A_v f_v}{(v - v_c)b'} = \frac{2 \times 0.11 \times 18,000}{43 \times 12} = 7.68 \text{ in.}$$

Call it 6 in. to be conservative.   This applies for the distance AG.

Now, where should the stirrups start?   It is customary to place the first stirrup 2 in. or one-half s from the face of the support.   The former is a good system because the reaction is not concentrated at the edge of the column.   Therefore, start the first stirrup at 8 in. from the center line as shown in Fig. 4-8(b).   Then three spaces at 6 in. will place a stirrup at G, but it is not essential to have one exactly there.   From G to H the spacing can increase, but $d/2 = 10$ in. is the maximum allowed.   Therefore, use one space at 6 in., two at 8 in., and two at 10 in., which happens to place a stirrup at H.   Then extend them for two more spaces at 10 in. as shown in (b).

Figure 12 in the Appendix can be used to advantage in finding the allowable spacing of stirrups for working-stress design.   Having $v - v_c$ at a given point, as obtained from Fig. 4-8(b), multiply it by b or b'. Then, using the size of stirrup intended, the spacing is quickly found from the diagram.   If a given spacing is desired, the required size of stirrup can be determined similarly.

The dimensions of the stirrups shown in Fig. 4-8(c) are slightly more than the minimum for the hook.   However, a longitudinal tie bar (perhaps No. 4) should be used inside the top hooks on both sides in order to wire the stirrups to them and prevent displacement during concreting.

The length of bar needed to develop the stirrup for stress is, from Eq. (3-1),

$$L_s = \frac{Df_v}{4u} = \frac{0.38 \times 18,000}{4 \times 350} = 4.9 \text{ in.}$$

Sketch (c) shows that far more than this is available.

Notice that the area of a U-shaped stirrup is $2A_s$ because both parts

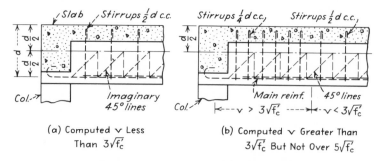

(a) Computed v Less
Than $3\sqrt{f_c'}$

(b) Computed v Greater Than
$3\sqrt{f_c'}$ But Not Over $5\sqrt{f_c'}$

Fig. **4-9**   Maximum spacing of stirrups.

are stressed simultaneously. Sometimes, as in wide and heavy beams, it is desirable to obtain more area without having the spacing too small or the bars too large. One arrangement for such beams is the use of two U-shaped stirrups placed in the same general cross-sectional plane with the two inner sides overlapping by several inches, as shown in Fig. 4-8(d), thus providing four effective bars for A_v. This is helpful since it is not wise to use stirrups larger than No. 5 or 6 because they are difficult to bend, the hooks become large, and the anchorage may become difficult to secure in one-half of the depth of the beam. Furthermore, it is not advisable to use single U-shaped stirrups that are too wide. There seems to be no specified limit for the distance between the upstanding legs of a stirrup, but it seems to be desirable to use double ones when the spread will exceed 15 to 18 in. If the standard hooks on double stirrups interfere with the placing of the main reinforcement, another set of inverted U-shaped ones may be added as shown for single stirrups in Fig. 4-17(d).

Figure 4-9 pictures graphically the Code requirements regarding the spacing of stirrups if the first one is one-half space from the face of the support. The engineer is reminded that close spacing[1] is desirable. Just because a specification might let one use stirrups at $d/2$ spacing, one may be wise to use less spacing than the maximum permissible. He may also prefer to extend the close spacing farther from the support. The total cost of a few more stirrups is small compared with the total cost of the entire structure, but the benefits may be large.

4-5. Inclined Stirrups. Another type of web reinforcement is shown in Fig. 4-10(a), which shows a reinforced-concrete beam with the stirrups placed in an inclined position. Let Fig. 4-10(b) represent the portion $ABCD$ when it is "cut out" and fixed in position as shown. The stirrups serve the same function as the vertical ones previously explained. The increment of tension in the longitudinal reinforcement is again equal to $T' - T$. This tensile force, the diagonal compression, and the tension in the stirrup constitute a system of three forces which are in equilibrium and which meet at a point. Their magnitudes must be proportional to the sides of a triangle, as shown in Fig. 4-10(c). For the illustration, the cracking is assumed to be at 45° with respect to the axis of the beam, and the horizontal shear is taken as equal to the vertical shear.

The triangle of forces in this case, assuming no help from the concrete, can be constructed by laying off the longitudinal force $T' - T = vbs$, drawing a line at 45° from one end parallel to the assumed diagonal compression, and then drawing the closing line from the other end, making it parallel to the stirrups.

[1] E. M. Rensaa, Shear, Diagonal Tension, and Anchorage in Beams, *J. ACI*, December, 1958.

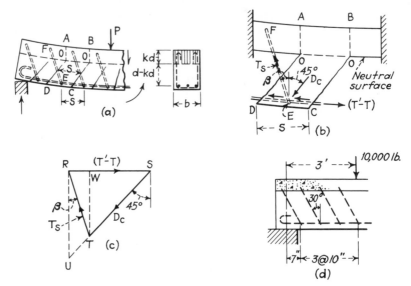

Fig. 4-10   Possible action of inclined stirrups.

An examination of Fig. 4-10(b) and (c) discloses some interesting facts. The stirrup EF withstands part of the horizontal force $T' - T$ as shown by RW. If RU represents the stress in a vertical stirrup, then the stress in the inclined stirrup is RT, which is less than RU, when the diagonal compression is assumed to be at 45°. If the stirrups are placed so that the angle β equals 45°, then the diagonal compression TS is $0.5US$, or one-half of what it would be with vertical stirrups. The tension in the stirrup RT is $0.7RU$, which means that the stress is seven-tenths of what it would be if the stirrups were vertical. In general, with WSD for a series of inclined stirrups,

$$\frac{RT}{RS} = \frac{T_s}{T' - T} = \frac{\sin 45°}{\sin (45° + \beta)}$$

$$T_s = A_i f_i = \frac{0.7(T' - T)}{\sin (45° + \beta)} = \frac{0.7vbs}{\sin (45° + \beta)} = \frac{0.7Vs}{d \sin (45° + \beta)} \qquad (4\text{-}18)$$

where A_i is the area of inclined stirrup and f_i is the unit stress in that stirrup.

Allowing the concrete to resist v_c, as for a beam without web reinforcement, Eq. (4-18) can be modified to[1]

$$A_i f_i = \frac{0.7(v - v_c)bs}{\sin (45° + \beta)} = \frac{0.7V's}{d \sin (45° + \beta)} \qquad (4\text{-}19)$$

[1] The AASHO formula for both a series of inclined stirrups and bent-up bars is $A_v = V's/f_vjd(\sin \alpha + \cos \alpha)$.

The corresponding formula given by the Code for a series of inclined stirrups[1] for WSD is

$$A_v = \frac{V's}{f_v d(\sin \alpha + \cos \alpha)} \qquad (4\text{-}20)$$

where A_v = area of the stirrup
   f_v = stress in the stirrup
   α = angle between the stirrup and the longitudinal bars
The numerical results from Eqs. (4-19) and (4-20) agree in the usual range of $\alpha = 30$ to $45°$. For USD, Eq. (4-20) is replaced by

$$A_v = \frac{V'_u s}{\phi f_y d(\sin \alpha + \cos \alpha)} \qquad (4\text{-}21)$$

There are practical disadvantages in the use of inclined stirrups. They must be held in place firmly to prevent displacement during the pouring of the concrete. Furthermore, near the ends of the beam, special short stirrups may have to be provided in order to get them in place at all. It would also be helpful if the stirrups were mechanically fastened to the longitudinal bars by welding, but this is costly and impractical, even though it would prevent slipping of the stirrups along the main steel.

4-6. Spacing of Inclined Stirrups. The spacing of inclined stirrups may be determined in the same general way as for vertical ones. Advantage may be taken of the increased length of embedment due to the slope of the stirrups when considering the length that is required to develop them through bond. The central three-fourths of an inclined stirrup is the part that is considered to be effective. The spacing used should be limited in the same manner as for vertical stirrups.

Example 4-3. By the ultimate-strength method, design stirrups inclined at 45° for a simply supported beam having a shear diagram similar to that of Fig. 4-7(d). Assume $b = 16$ in., $d = 24$ in., $L = 24$ ft, $v_c = 94$ psi, and $f_y = 40,000$ psi. The reaction at A is 41,000 lb; at the left of H the shear is 38,000 lb; at the right of H the shear is 13,000 lb; and the uniform load w is 500 lb/ft. Assume that one-half of the shearing forces represents dead load, and use $U = 1.5DL + 1.8LL$.

The allowable shear with no web reinforcement is

$$V = v_c bd = 94 \times 16 \times 24 = 36,000 \text{ lb}$$

Therefore, no stirrups are needed for stress from H to D [Fig. 4-7(d)].

---

[1] For a *single isolated* but inclined bar or stirrup to support the same load as a vertical stirrup in a beam area bs, the inclined bar should have an area $A_v = V'/(f_v \sin \alpha)$. This is because the shear is vertical, whereas the stirrup is inclined so that only the vertical component of its stress resists the shear.

At G, the total shear is $41,000 - 2 \times 500 = 40,000$ lb. Then this is critical.

$$U = 1.5 \times 20,000 + 1.8 \times 20,000 = 66,000 \text{ lb maximum at } A \text{ to } G$$
$$V'_u = 66,000 - 36,000 = 30,000 \text{ lb}$$

Using No. 4 U-shaped stirrups, $A_v = 0.4$ in.^{2} From Eq. (4-21), with $\sin \alpha + \cos \alpha = 1.414$, solving for s gives

$$s = \frac{A_v \phi f_y d(\sin \alpha + \cos \alpha)}{V'_u} = \frac{0.4 \times 0.85 \times 40,000 \times 24 \times 1.414}{30,000}$$
$$= 15.4 \text{ in.}$$

These could be used at a spacing of $d/2 = 12$ in. throughout AH and for two spaces into HD.   If No. 3 stirrups were used, s would be

$$\frac{15.4 \times 0.11}{0.2} = 8.46 \text{ in.}$$

These could be used at 8 in. c.c. for the entire distance required if desired.

4-7.   Bent-up Bars.   In Fig. 4-8(a) the upper pair of longitudinal bars is shown cut off before reaching the support.   Since the size and number of the tensile bars are determined by the maximum bending moment, it is obvious that not all the bars are needed where the bending moment is smaller.   However, such a discontinuance of part of the steel causes a rapid and localized transfer of tension from the cut bars to the continuous ones.[1]   If the excess bars are bent up into the compression zone as shown in Fig. 4-11(a), they can serve as web reinforcement, similar to inclined stirrups.   Here they are bent up two at a time for symmetry.   Sketch (c) shows that, if a piece is imagined to be cut out, the bent bars will resist the diagonal tension, but they probably will have in them some residual tension from bending action, represented by T_L.   The combined action might be as pictured in sketch (d), from which

$$T_s = A_i f_i = \frac{0.7(T' - T + T_L)}{\sin (45° + \beta)} = \frac{0.7(vbs + T_L)}{\sin (45° + \beta)} \tag{4-22}$$

if all of the shear is taken by the web reinforcement.   Here T_L is the relative amount of the total longitudinal tension at the bend point which is carried by the bent bars, and may be assumed to be $M'N_b/jdN_t$, where M' is the moment at the bend point, N_b is the number of bars bent up at that point, and N_t is the total number of bars at the bend point, including those bent up.

---

[1] P. M. Ferguson and F. N. Matloob, Effect of Bar Cutoff on Bond and Shear Strength of Reinforced Concrete Beams, *J. ACI*, July, 1959.

If the concrete is relied upon to resist an amount of shear equal to v_c, Eq. (4-22) becomes

$$A_i f_i = \frac{0.7[(v - v_c)bs + T_L]}{\sin (45° + \beta)} \tag{4-23}$$

However, such longitudinal bars are generally so large that the unit stress caused by stirrup action is relatively small.

Figure 4-11(e) and (f) is based upon suggestions by Rensaa[1] regarding the stress conditions in and around bent-up bars. These sketches are intended to indicate to the reader how the bond, longitudinal tension, and pressures at bends may act inside of a beam when large bars are bent

[1] E. M. Rensaa, Rensaa and Minsos, 9130 Jasper Ave., Edmonton, Alberta, Canada.

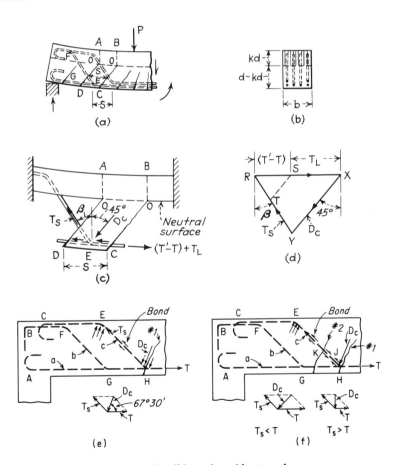

Fig. 4-11 Possible action of bent-up bars.

up.   In (*e*), *EH* does not help against crack No. 1, the bond helps to reduce T_s near *E*, and there may be large pressures at the bends.   In (*f*), a second crack No. 2 may cause a reversal of bond stresses close to the crack.   The anchorage hooks at *A* and *C*, and the bend at *B*, are large and troublesome. Many engineers prefer to avoid the use of bent-up bars in most cases.

Equations (4-20) and (4-21), used for the design of inclined stirrups, can also be used for a series of bent-up bars.   These neglect the effect of any longitudinal tension which may already be in the bars.   For a single bent-up bar and for a group of bars all bent up at the same distance from the support, the required area of the bent-up bar or bars is

$$A_v = \frac{V'}{f_v \sin \alpha} \text{ for WSD} \quad \text{and} \quad A_v = \frac{V'_u}{\phi f_y \sin \alpha} \text{ for USD} \quad (4\text{-}24)$$

Two advantages of bent-up bars for web reinforcement are the anchoring of the longitudinal bars themselves and the shifting of the bars from the bottom of the beam to the top, where they may be needed to resist negative bending moments, as for continuous beams and frames.   Bars *b* and *c* in Fig. 4-12(*c*) may also be looked upon as cradled between anchorages near both ends of the beam.

Ordinarily, it is good practice to bend the bars at an angle of 45°. Occasionally, one may bend up a pair of bars at the end of a heavy beam just to knit the member together and to give it extra toughness.   Notice that, referring to Fig. 4-3*A*, bent-up bars do not tie in the reinforcement as stirrups do.   Of course, an individual bent-up bar is pressed hard against the main body of the concrete at the bend point so that it will not cause spalling of the cover.   However, this action is local and does not seem to do much to help the other bars.   Certainly the bending of an inner layer of reinforcement in a beam does not help the outer bars and prevent their tendency to cause longitudinal cracking.

4-8.   Spacing of Bent-up Bars.   As stated in Arts. 3-7 and 3-8, the bending-moment diagram is the first thing to investigate in order to determine points at which longitudinal bars in a beam may be bent up to act as web reinforcement.   For instance, let Fig. 4-12(*a*) represent the bending-moment diagram for a simply supported beam when it has a concentrated load applied at *D*; let sketch (*d*) be a similar diagram for uniform loading.   Assume that the maximum moments are equal in both cases and that there are six bars in the bottom of the beam, as shown at the left of sketch (*c*).   Because of the cracking of the beams along lines sloping toward the supports, assume that the permissible bend points are determined by the dotted lines located 2/3*d* to the left of the boundaries of the respective bending-moment diagrams.

The shear diagram is the next thing to investigate.   Therefore, compute the intensity of the shearing stress (the measure of the diagonal tension) from Eqs. (4-8) and (4-9), and plot the diagrams for both loading conditions as in sketches (b) and (e).   If the concrete alone is relied upon to withstand a stress of v_c, the remainder of the shearing stresses must be resisted by the bent-up bars.

The rest of the procedure is as follows:

1. Assume that the bars are to be bent up at 45° and that these bars are to be bent up in pairs, for symmetry.

2. Assume that each bar withstands the same proportion of the bending moment, as shown by the horizontal dotted lines in (a) and (d).

3. Bend the first two bars c up a little to the left of the point C, where they are no longer needed to resist the bending moment.

4. Project from C downward to the other figure to locate the permissible point for the first bends.

5. Project down from B in a similar manner to find the point for the bending of the second pair of bars.

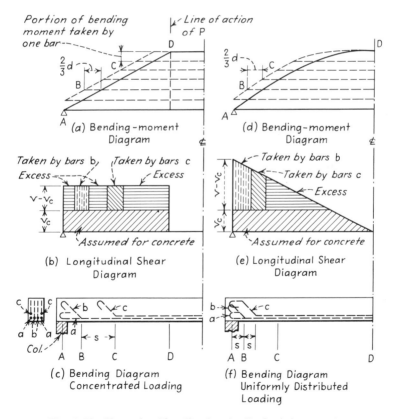

Fig. 4-12   Example of bending longitudinal reinforcement.

In general, the bend points of bent-up bars used as web reinforcement should be spaced according to the same limitations as those for stirrups. The center three-fourths of the inclined portion of such bars is the part that is considered to be effective.[1]

Some of the difficulties of arranging bent-up bars in simply supported beams are shown in Fig. 4-12. The parts of the diagram showing the intensities of the shearing stresses that are labeled "excess" have no bent-up bars in this case to reinforce them. In sketch (c) the bars are bent up as soon as the bending-moment diagram plus the extra $\frac{2}{3}d$ will permit. Since the spacing of bend points should not exceed $\frac{1}{2}d$ generally, depending upon the magnitude of v, the distance s in this case is too large. Hence, stirrups will be needed in these unreinforced areas. On the other hand, sketch (f) shows the bars bent up close to the end of the beam. Bars b are bent at $\frac{1}{2}s$ from the edge of the column or s from the column's center; then bars c are bent at the allowable distance s farther on. Again, the excess must be provided for by other means.

In sketch (f), notice that the hooks are overlapping and congested, and that bent-up bars serve as web reinforcement for only a short distance. It is possible to bend the bars one at a time, in which case they could reinforce more of the beam's length. However, this is not customary in such a case because of the resultant lack of symmetry, except for centrally located bars as shown in Fig. 4-13(f). Even if the bars are bent up one at a time, there will generally be the need for some stirrups to cover the weak spots.

Example 4-4. Using working-stress methods, bend up at 45°, for web reinforcement, the bars of the beam shown in Fig. 4-13(f) if the condition of loading is that which is pictured in sketch (a). Assume $f_v = 20,000$ psi, $f'_c = 3,000$ psi, $v_c = 60$ psi, and $d = 20.5$ in. The column at A is 14 in. wide.

The shear diagram in (b) shows that there is little change from J to K. Therefore, one could use AJ as the critical shear but, since this problem is open for inspection, use $AE = 24,400$ lb (approx) as the maximum. Then

$$v = \frac{V}{b'd} = \frac{24,400}{12 \times 20.5} = 99 \text{ psi}$$

Therefore, $v - v_c = 99 - 60 = 39$ psi, which is shown in (d) as the excess to be taken by the web reinforcement.

---

[1] AASHO requires that the spacing be measured at the neutral axis and in the direction of the longitudinal axis of the beam. The spacing s should not exceed $0.75d$. The first bar at the end should cross mid-depth of the beam at not over $s = 0.5d$ from the face of the support. In any web reinforcing, the maximum allowable stresses are $f_s = 18,000$ psi for structural grade bars and $20,000$ psi for intermediate and hard grade.

In (c), the bending moment is assumed to be resisted equally by all six bars, although the three upper ones are less effective than the lower row.

In planning this work, examine Fig. 4-13(e) and (f). Assume that bar a will be bent up first, then b, and finally both bars c. Bars d should extend the full length of the beam as shown. The maximum spacing of the bend points is $\frac{1}{2}d = 10.25$ in., or call it 10 in. These points will be spaced as shown in sketch (e). The bends are 45°.

Figure 4-13(c) shows, by the dotted lines, that the full-length bars d can be assumed to take care of the bending represented by the two bottom spaces between the horizontal lines. It also shows that bars c are needed

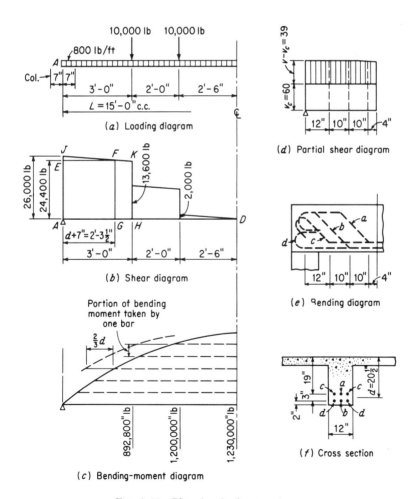

Fig. **4-13**   Planning for bent-up bars.

by the bending moment until they are close to the end of the member. When $\frac{2}{3}d$ is added beyond the limits of the diagram, the bend point will be approximately 6 in. from the center of the support.   However, in order to bend the bars at all and have them do any good as web reinforcement, they will be bent 12 in. from the center of the column, even though this may be a little too soon.   With the spacing shown in sketch (e), bars a and b will be found to be bent beyond the point where they are actually needed for main tensile reinforcement in the bottom of the beam.

Now check these bars to see if they are adequate.   Using Eq. (4-20), $s = 10$ in., and $V' = 39b'd = 39 \times 12 \times 20.5 = 9{,}600$ lb,

$$A_v = \frac{V's}{f_v d(\sin \alpha + \cos \alpha)} = \frac{9{,}600 \times 10}{20{,}000 \times 20.5 \times 1.414} = 0.165 \text{ in.}^2$$

Since No. 8 bars have an area of 0.79 in.^{2}, they are very safe for the use intended here, but be sure that the bond on d is safe.

This beam is very short and heavily loaded, with a large load near the end.   The distance to be reinforced by the bent-up bars is therefore much shorter than in most cases.   That is why this one works out fairly well.

Figure 4-14 has been prepared to show more clearly the details of the

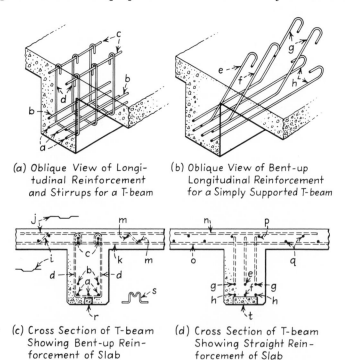

(a) Oblique View of Longitudinal Reinforcement
and Stirrups for a T-beam

(b) Oblique View of Bent-up
Longitudinal Reinforcement
for a Simply Supported T-beam

(c) Cross Section of T-beam
Showing Bent-up Rein-
forcement of Slab

(d) Cross Section of T-beam
Showing Straight Rein-
forcement of Slab

***Fig.* 4-14**   Studies of arrangements of and supports for beam-and-slab reinforcement.

web reinforcement near the end of a simply supported T beam and to compare stirrups vs. bent-up bars. Sketches (a) and (c) show three bottom bars a, the outer two of which will extend the full length of the beam. The other bar a and the two bars b will be stopped when they are not needed. Bars c at the top are small ones used as spacers to which stirrups d are wired. In sketch (b) bars h extend the entire length of the beam. In this case, as shown in (d), there are two layers of three bars each. Top bar e is bent up first; bottom bar f, next; and two top bars g, last.

In Fig. 4-14(c), the reinforcement is shown supported by small precast blocks r, although wire chairs s can be substituted. The slab reinforcement is arranged so that alternate bars k are straight. Bars i and j are bent as indicated and then lapped past each other over the girder. This scheme provides the same amount of reinforcement at the bottom of the slab at mid-span and at the top over the beam. Bars m are ties.

In Fig. 4-14(d), the bottom bars are shown supported upon long, narrow precast blocks t. Wire chairs are better for this purpose, and additional chairs are needed to hold the upper set of main bars. In the slab, top bars n are short and straight, extending each side at least one-fourth of the clear span of the slab. Bars o are straight and may be full length or, for economy, arranged as alternate long ones for the entire length and short ones that stop before reaching the beam. Bars p and q are ties and transverse reinforcement (or temperature steel) for the slab. This arrangement of slab reinforcement is much more simple than the use of bars with such small bends as those in sketch (c), although the latter do not need chairs to support the top reinforcement.

4-9. Web Reinforcement at Supports of Continuous Beams. A large portion of reinforced-concrete construction utilizes the advantage of continuity or restraint of members. This is partly in order to decrease the maximum bending moments but primarily because it is easier and more practicable to build concrete structures that way. The shears—both transverse and longitudinal—are affected by continuity to a much smaller extent than are the bending moments.

As an illustration of the effect of continuity and restraint, let Fig. 4-15(a) picture a simply supported beam with a concentrated load at its center. The bending-moment and shear diagrams are self-explanatory. Then let the ends of the beam be fixed as in Fig. 4-15(b). The bending-moment and shear diagrams for this new condition are again easily understood. The maximum bending moment in the second case is only one-half that of the first one; also, the distribution of the bending moment is different. The rate of increase of bending moment in both cases is $(PL/4)/(L/2)$. This is a constant. This fact is shown in the shear

diagrams, which are also identical.   Therefore, the required strength of
the beam against shearing forces remains unchanged, although the ends
are fixed in the second case.   This means that the relative importance of
the shear is greater than it is in a simply supported beam.

Now, examine the conditions at a point of support where the bending
moment is negative.   Figure 4-16(*a*) is an exaggerated picture of such a
case, with vertical stirrups.   The elongation of the longitudinal reinforcement causes cracking of the concrete in the same manner as for
simply supported beams except that, in this case, the top fibers are
elongated.   If the piece $ABCD$ is cut out and gripped in rigid supports,
as shown in Fig. 4-16(*b*), a brief examination of it shows that it is the
same as Fig. 4-10(*b*) if the latter is rotated 180°.   The tension in the top
longitudinal reinforcement increases from D to the support S.   The
increment of stress which is picked up by the piece $OCDO$ must therefore
act toward the left.   After $T' - T = vbs$ is computed, the diagonal
compression and the stirrup tension can be found as before by constructing the force triangle of Fig. 4-16(*c*).

Vertical stirrups, inclined stirrups, and bent-up bars serve the same
purpose whether the beams are simply supported, continuous, fixed-end,
or cantilevered.   All of them can be designed by the same general
method because the fundamental force to be considered is the diagonal
tension in the concrete.   The diagrams showing the intensity of the shear
and the bending moment should be constructed; the excess of the shear
above that which is permissible in concrete beams without web reinforcement should be found; then the size, spacing, and details of the stirrups

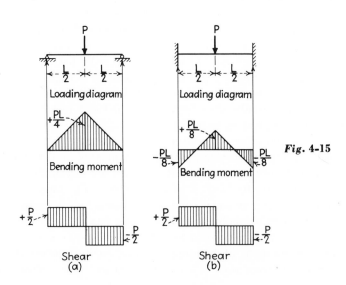

Fig. 4-15

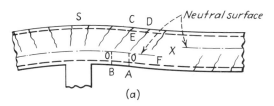

(a)

Fig. 4-16 Stirrups near support of continuous beam.

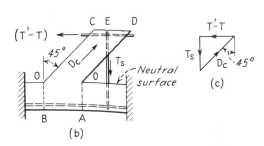

(b)

(c)

and the bent-up bars should be determined just as they were in the previous cases.

It is important to study the right-hand portion of Fig. 4-16(a). As described in Art. 3-8, the neutral axis is usually below the middle of the depth of the beam at the support S because the tension is at the top. At the center of the span, the neutral axis is above the middle of the depth. Somewhere in the region of the point of contraflexure, X, the neutral axis shifts from the one position to the other. If this portion of the beam is uncracked, the location of the neutral axis has little importance anyway. It is not difficult, however, to see how uncertain the magnitudes of the shearing stresses in continuous beams may be, especially when large live loads are applied and the point of inflection shifts with varying positions of these loads. However, because of the variation of the pickup of tension in the top bars, as described in connection with Fig. 3-16, it is desirable to extend web reinforcement to and beyond the vicinity of the point of contraflexure, even though the shear does not seem to require it to go so far. In general, this point of inflection is about 0.2L to 0.25L from the supports.

Consider still further the shearing conditions near a point of contraflexure. The loading will cause very little bending moment in this vicinity and, consequently, very little compression which pinches adjacent vertical sections together. It is possible that shrinkage, longitudinal tension caused by a drop in temperature when the ends of the member cannot move freely, or some longitudinal live load can produce a tension on the section. Then little dependence can be placed upon the ability of the concrete to withstand its supposed stress v_c. The dowel action of longitudinal bars and the interlocking of aggregates are not very reliable means for resisting transverse shear or diagonal tension failure. Thus, at what may seem to be the strongest points of the beam from the standpoint of bending, there may be weakness against failure by diagonal tension. Therefore, stirrups able to hold the shear at such points, or bent-up bars able to do so, are likely to be excellent insurance against failure.

Difficulties are encountered in arranging the details[1] of stirrups at points of negative bending moments in rectangular beams. A stirrup like that of Fig. 4-17(a) is not anchored thoroughly across the tension side of the beam; when it is inverted, it is difficult to get the hooks under the longitudinal bars if the stirrups are erected last or to place the main reinforcement if the stirrups are in position first. Sometimes two U-shaped bars without hooks are arranged as shown in Fig. 4-17(d), but the laps must be long enough to develop the stirrups, and it is advisable to have as much of the lap as possible in the compression zone of the beam, but if long enough, the lap can provide sufficient bond even in the tensile region. Still another type is that which is pictured in sketch (e). The hooks are parallel to the main reinforcement, permitting the stirrups to be slid down

[1] J. E. Bower and I. M. Viest, Shear Strength of Restrained Concrete Beams without Web Reinforcement, *J. ACI*, July, 1960. R. H. Bryant, A. C. Bianchini, J. J. Rodriguez, and C. E. Kesler, Shear Strength of Two-span Continuous Reinforced Concrete Beams with Multiple Point Loading, *J. ACI*, September, 1962.

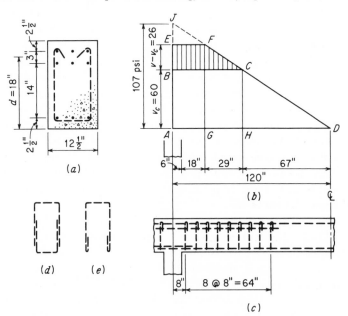

Fig. 4-17

after the longitudinal bars are in place. However, the anchorage of these stirrups is bad because the entire hook is near the surface of the beam, a fact that may cause spalling of the concrete; the compressive reinforcement is not tied in as it should be.

The stirrups in the stems of T beams may be detailed as shown in Fig. 4-17(a) because the concrete of the slab and its reinforcement will support the top adequately.

In a continuous beam having very large shearing stresses, it may be advisable to use one or two extra bent bars which slope down on both sides of the column and serve as inclined stirrups rather than as real negative reinforcement. This scheme is especially useful in deep beams with very heavy loads. These extra bars are to be in addition to regular vertical stirrups.

Example 4-5. The beam shown in Fig. 4-17(c) is continuous; it has a span of 20 ft; it carries a uniformly distributed load of 2,400 plf; its cross section is shown in (a); $f'_c = 3,000$ psi; $f_v = 18,000$ psi; and $v_c = 60$ psi. By working-stress methods, design vertical U-shaped stirrups for this beam.

$$\text{At } A, v = \frac{V}{bd} = \frac{2,400 \times 10}{12.5 \times 18} = 107 \text{ psi}$$

$$\text{At the critical point } G, v = \frac{2,400 \times 8}{12.5 \times 18} = 86 \text{ psi}$$

Then, for design at the critical point, $v - v_c = 86 - 60 = 26$ psi. Also, at G, $V' = (v - v_c)bd = 26 \times 12.5 \times 18 = 5,850$ lb. The distance in which stirrups will be needed is BC in sketch (b). Then

$$BC = \frac{120 \times (107 - 60)}{107} = 52.7 \text{ in.} \quad \text{Call it 53 in.}$$

Assume No. 3 bars for the stirrups. Then, from Eq. (4-14),

$$A_v = \frac{V's}{f_v d} \quad \text{or} \quad s = \frac{2 \times 0.11 \times 18,000 \times 18}{5,850}$$
$$= 12.2 \text{ in.} \quad \text{Call it 12 in.}$$

However, the maximum spacing is $d/2 = 9$ in. Using 8-in. spacing, and starting the first stirrup 2 in. from the face of the column, locate the stirrups as shown in Fig. 4-17(c). This covers a little more than the distance d beyond point H.

4-10. Combination of Stirrups and Bent-up Bars. As stated in Art. 4-8, and as shown in Fig. 4-12, it is often difficult to arrange the details of bent-up bars so that they will reinforce properly all parts of the web. The deficiency can be made up by adding stirrups in the otherwise unreinforced portions. It is not necessary to explain this in detail because both types of web reinforcement act in accordance with the same general principles.   However, good judgment must be exercised in combining them.

Reinforcement pictured on a drawing may seem to be simple, but placing large bars in the forms in the field may be a different matter.   Figure 4-18 shows the junction of a continuous beam and a girder.   Three bars

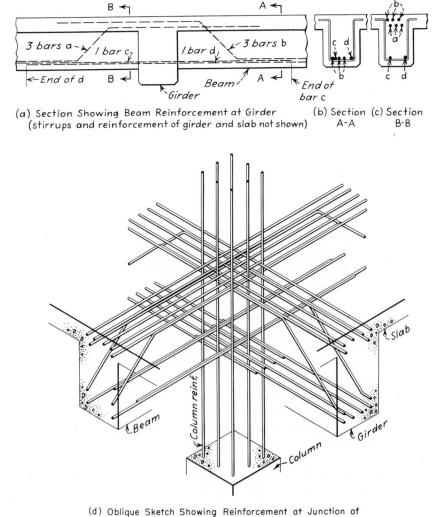

(a) Section Showing Beam Reinforcement at Girder   (b) Section (c) Section
(stirrups and reinforcement of girder and slab not shown)   A-A   B-B

(d) Oblique Sketch Showing Reinforcement at Junction of
Continuous Beam and Girder, Omitting Stirrups,
Column Ties, and Bars in Floor Slab

Fig. 4-18   Illustration of difficulties in arranging reinforcement.

a and *b* are bent up from the bottom at each side to the top to resist
negative bending. Obviously, they cannot pass through each other.
If all six are in the same plane, they will form a screen where they overlap,
and it will be difficult to place the concrete through and around them.
If they are offset to pass each other, then placed in two layers as pictured
in sketch (*c*), they still form an obstacle in the pouring of the concrete.
Bar *c* in sketch (*a*) extends across the span at the left and laps with the
others at the right; bar *d* does the reverse. If they and the bent bars
are in one horizontal plane, the overlap causes the screen effect pictured
in (*b*). Generally, if the design is not carefully planned, as indicated in
sketch (*d*), the bars will have to be rearranged somewhat in the field,
tilted a bit, raised or lowered to lap over and under each other, and
otherwise made to fit as best they can. Doing this adjusting in the field
may be a very troublesome job with stirrups, heavy longitudinal bars in
the beams, similar reinforcement in the girder, and all the slab bars that
have to fit together in the same general region.

Perhaps this shows the advantages of simplicity and easy field work
when straight bars and stirrups are used without bent bars. On the other
hand, a nominal number of bent-up bars will often be desirable in tying
together long and important structures. Whether or not to use them is
largely a matter of judgment on the part of the individual engineer.

Theoretically, it seems that inclined stirrups are to be preferred when
the number of bent-up bars is large, in order to have the tensile forces
in the web reinforcement acting in the same direction. However, when
used in combination with bent-up bars, they will cause very difficult
work in the field if they are to be properly fastened to the longitudinal
reinforcement. Therefore, vertical stirrups are more practical, but they
will generally overlap some of the inclined portions of the bent bars.
Also, when only a few bars are bent up, it seems to be advisable to reinforce the web with vertical stirrups alone, neglecting the bent-up bars and
letting their strength add to the factor of safety of the beam. Another
advantage of this last arrangement is that, in continuous beams, the
bars can be bent to meet the requirements of the bending moments alone,
rather than those of the longitudinal shear and the bending moments
together.

Example 4-6.   Assume a T beam that is continuous over a central
column as shown in Fig. 4-19(*a*). By WSD, design the web reinforcement for this beam, using bent-up bars and No. 4 U-shaped stirrups as
necessary. Assume the beam to be fixed at *B*. Let $f_v = 18,000$ psi and
$v_c = 70$ psi.

The shear and bending-moment diagrams are shown in sketches (*b*)
and (*c*) in Fig. 4-19; the cross sections, at *B* and *C* in Fig. 4-20. The
shearing-stress diagram is shown in Fig. 4-21(*a*) for the critical points *a*
distance *d* from the faces of the columns; the bending-moment diagram

is redrawn in sketch (*b*), and it is divided by the dotted lines to represent the portions of *M* resisted by the individual bars.   The curved dotted lines labeled 1, 2, and 3 show the assumed extensions required to reach the permissible bend and cutoff points for the main bars, as described in Art. 3-8.   Let $\frac{2}{3}d = 18$ in.   The reinforcing is pictured in Fig. 4-21(*c*). The bars can be identified by referring to Fig. 4-20.

Figure 4-21(*a*) shows that, since v_c for the concrete alone is 70 psi, no web reinforcement is needed at *A*, but the bars *a* will be hooked for anchorage.   In sketch (*c*), bars *b* and *c* are bent down as soon as the bending moments will permit.   There is a space of 42 in. from *B* to the top bend in bar *b*.   Stirrups are needed here.   From $A_v f_v = (v - v_c)bs$, the spacing *s*, if *v* = 112 psi is used, can be

$$s = \frac{2 \times 0.2 \times 18,000}{42 \times 16} = 10.7 \text{ in.}\quad \text{Call it 10 in.}$$

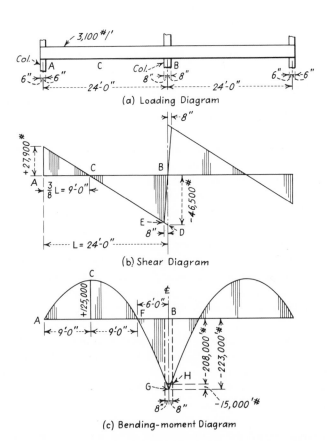

(a) Loading Diagram

(b) Shear Diagram

(c) Bending-moment Diagram

Fig. 4-19   A heavy beam continuous across a central column.

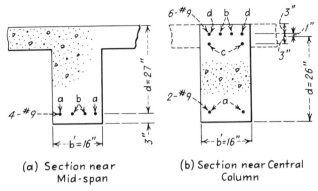

(a) Section near   (b) Section near Central
   Mid-span   Column

Fig. 4-20

Then three of these stirrups are used as shown in (c). With

$$v = v_c \doteq 21 \text{ psi}$$

s could be greater. The top bend in bar c is 60 in. from B, whereas ON of (a) is 68 in. to the point where web reinforcement is not needed. Since $d/2 = 13$ in. and the excess shear is small, add two stirrups at 12-in. spacing beyond the top bend in bar c.

The main bars are bent down where the shear is small so that they obviously have more strength than needed for stirrup action.

The arrangement shown for all of this reinforcement in Fig. 4-21(c) will be called satisfactory. However, it is obvious, without counting upon the bent-down bars, that two or three more stirrups would be a more simple design and that it would be satisfactory and suitable.

If the shears are large, the Code states that, for a combined system of web reinforcement, no one system should carry more than $\frac{2}{3}V'$. The case shown here is one in which the stirrups hold the loads in some places and the bent bars do so in others, each being able to act by itself. If the combined system is needed, it is probable that the bars cannot be bent so as to provide for all of the bending and, at the same time, be bent where they are needed to take care of all of the diagonal tension. Obviously, since neither system is complete, a design based upon vertical stirrups alone will be the best.

Example 4-7.   Figure 4-22(a) shows an end span of some continuous viaduct construction of the New Jersey approach to the Lincoln Tunnel at Weehawken. The cross section of the girder is shown in sketch (b), the shear diagram, in (c). By WSD, design vertical U-shaped stirrups for this girder, assuming $f_c' = 3,000$ psi, $v_c = 70$ psi to be conservative,[1] and $f_v = 18,000$ psi.

---

[1] AASHO allows a maximum $v_c = 0.03f_c'$.

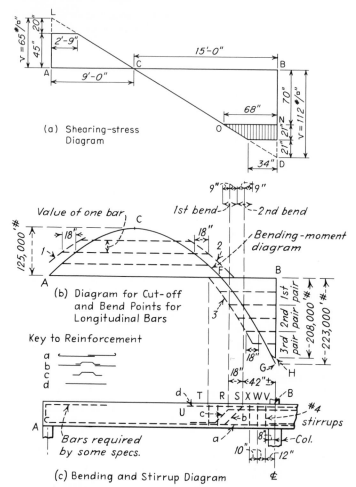

(a) Shearing-stress Diagram

(b) Diagram for Cut-off and Bend Points for Longitudinal Bars

Key to Reinforcement

(c) Bending and Stirrup Diagram

Fig. 4-21   Combined bent-up bars and stirrups.

Since this is a highway structure which is subjected to very large moving and repeated live loads, and since the major part of the shear is due to the concentrated loads brought in by the cross beams, use the total shears as the values of V, but otherwise follow AAHSO specifications. Also, be very conservative in the design.

1. *Section AB.* The critical part of this section seems to be at B, where $d = 72$ in. Assume $j = 0.9$.

$$v = \frac{V}{jb'd} = \frac{111,000}{0.9 \times 27 \times 72} = 64 \text{ psi}$$

This is less than v_c, but use a few stirrups for reasons to be explained later.

2. *Section BD.* The shear is considerably less than section AB. Use nominal stirrups only.

3. *Section DE.* The greatest intensity of shear is at D, where

$$v = \frac{V}{jb'd} = \frac{214,000}{0.9 \times 27 \times 72} = 122 \text{ psi}$$

Since the girder is a very deep one, assume No. 5 stirrups in order to have them strong enough as columns to support the top longitudinal bars and to obtain a reasonable spacing. Because of the negative moments at the supports, use inverted U-shaped stirrups as shown in Fig. 4-22(b). Then

$$s = \frac{A_v f_v}{(v - v_c)b'} = \frac{2 \times 0.31 \times 18,000}{(122 - 70)27} = 8 \text{ in. (approx.)}$$

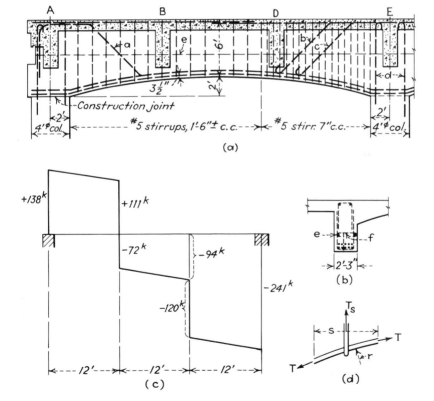

Fig. **4-22**   Girder used in New Jersey approach to Lincoln Tunnel, New York City.

Since the live loads are large, and since the full allowance for v_c in a deep member like this may be questionable, especially because the member is subjected to impact loads, be more conservative and reduce the spacing to 7 in.

With $u = 350$ psi[1] from Table 13 in the Appendix, the top and bottom portions of the stirrups should be overlapped a distance

$$s = \frac{A_v f_v}{(\Sigma o)u} = \frac{0.31 \times 18{,}000}{1.96 \times 350} = 8.2 \text{ in.}$$

However, increase this to $24D = 24 \times 0.625 = 15$ in.

4. *Practical Details.* The following details should be noticed in Fig. 4-22:

a. As shown in (d), the bottom tensile bars will tend to straighten out and to spall off the concrete below them. Therefore, stirrups must hold them back. Assuming No. 5 stirrups at 18 in. c.c., the stress in them may be approximated by assuming that $T = pr$ and that the radius of the curve is $r = c^2/8m = 32^2/(8 \times 2) = 64$ ft. The maximum tension in the bottom bars is 330 kips. Then

$$p = \frac{T}{r} = \frac{330{,}000}{64} = 5{,}150 \text{ plf}$$
$$T_s \text{ in Fig. 4-22}(d) = ps = 5{,}150 \times 1.5 = 7{,}700 \text{ lb}$$
$$f_v = \frac{T_s}{A_v} = \frac{7{,}700}{0.62} = 12{,}400 \text{ psi}$$

Therefore, these will be used from A to D.

b. Bars a are bent up and curved down at the end so as to reinforce the top corner.

c. Bars b and c are bent down to anchor them and to serve as extra web reinforcement.

d. The splicing of the bottom bars is made at the column at E; a few of the top bars are spliced near D and extended to A.

e. Bars d represent the column reinforcement which is extended up into the girder in order to reinforce the joint for continuous frame action.

f. The longitudinal bars e and the hooked ones f of sketch (b) are used to tie all the stirrups together in both directions.

g. Because of the width of the member as shown in (b), and because of the tension in the curved bottom bars from A to D, it is desirable to hook some No. 3 vertical bars under the center of the bottoms of the stirrups and over the cross ties f in this region in order to hold the radial pull of the main steel.

---

[1] AASHO requires that the maximum u for top bars be $0.06f_c'$ or 210 psi; for bottom and other bars, $0.10f_c'$ or 350 psi.

4-11. Web Reinforcement in Beams Carrying Moving Loads.
The floors of bridges and warehouses, longitudinal girders under railroad tracks,[1] crane girders, and similar structures usually carry moving live loads of large magnitude. Therefore, these structures must be designed to withstand the maximum possible combination of these loads. Such conditions often cause severe stresses in the web reinforcement. The bending moment and the rate of increase of the bending moment—and therefore the shearing stresses—differ with various positions of the loads. This means that, for the design of each section, the critical positions of the loads must be ascertained and the resultant stresses must be determined to see if the structure is safe.

The use of influence lines will greatly facilitate the design of members that carry a series of moving live loads. A few diagrams are given in Chap. 15.

When more than one moving concentrated load is used, the maximum shear can be found at enough points to enable one to plot a curve showing the maximum shear at all points in one-half of the beam. The excess of these values over the assumed allowable stress v_c for the unreinforced web of the beam can then be used in the design of the web reinforcement.

Because of the rapid changes in the magnitudes of the web stresses as the live loads pass over such beams, it is conservative to have web reinforcement throughout the length of the member, using nominal sizes in the central portions. Vertical stirrups are generally most suitable for this purpose. It also seems advisable to proportion the stirrups to carry all the shearing stresses but to use a theoretical unit stress of about 25,000 psi in them. However, the stirrups should never be weaker than those required by the use of Eq. (4-14). It is difficult to predict the shearing strengths of the portions of such beams that have tension in their tops under one position of the loads, whereas there is tension in their bottoms for other positions of the same loads almost immediately thereafter, and vice versa. At such points, bent-up bars (if well anchored) are especially beneficial in preventing disintegration of the beams and in providing steel "hangers" to knit them together.

This shifting of the point of contraflexure in continuous beams with moving live loads is of great practical importance. It spreads the region of possible large bond stresses described in Art. 3-8; it shifts the theoretical bend and cutoff points; it casts doubt upon the resistance of this area to transverse shearing; it tends to disintegrate the beam if the changes are sufficiently violent; and it requires that one be very conservative in his design when planning web reinforcement and discontinuance of the longitudinal bars. He should provide against the most extreme conditions that will affect either the top or the bottom reinforcement.

[1] AASHO requires stirrups for the full length of T beams and box girders.

Example 4-8.   Figure 4-23 shows one of the T beams in a small highway bridge[1] which is simply supported.   The loading, including impact, is shown in (a); the maximum shear diagram for one-half is shown in (c), for various positions of the moving loads.   Use AASHO specifications, which are based on WSD.   Design vertical U-shaped stirrups for this member for two cases: (1) for a design with $v_c = 75$ psi and $f_v = 18,000$ psi; (2) for the full value of v and with $f_v = 24,000$ psi. Assume $f'_c = 2,500$ psi, $v_c = 0.03f'_c$ for anchored main bars, $u = 0.10f'_c$ for bottom bars, $j = 0.92$, and stirrups the entire length.

Case (1):

If $x = 0$ in Fig. 4-23(a), the shear at A equals $39.38 + 12.6 = 51.98$ kips, including the dead load of 0.9 klf.   Also, if $x = 14$ ft, the shear at the left of C is 17.5 kips.   With $j = 0.92$,

$$v \text{ at } A = \frac{V}{b'jd} = \frac{51,980}{14 \times 0.92 \times 30} = 135 \text{ psi}$$

$$v \text{ at } C = \frac{17,500}{14 \times 0.92 \times 30} = 45 \text{ psi}$$

These values are used to construct the shear diagram in Fig. 4-23(c). Since DE is a straight line, the hatched area showing the portion needing web reinforcement is found by proportion to be 9 ft 4 in. long when $FA = v_c = 75$ psi.

From the AASHO formula, with No. 4 stirrups,

$$f_v = \frac{V's}{A_v jd}$$

$$s = \frac{A_v f_v jd}{V'} = \frac{A_v f_v}{(v - v_c)b'}$$

Therefore, at end A, with No. 4 stirrups,

$$s = \frac{(2 \times 0.2)18,000}{60 \times 14} = 8.6 \text{ in.}$$

Let $s = 8$ in. for the left part of FG.   At point G, no stirrups are needed for stress, but some are required throughout by the specifications.   Therefore, use the arrangement shown in Fig. 4-23(d), with $d/2 = 15$ in. as a maximum.

---

[1] See J. R. Verna and T. E. Stelson, Repeated Loading Effect on Ultimate Static Strength of Concrete Beams, *J. ACI*, June, 1963.

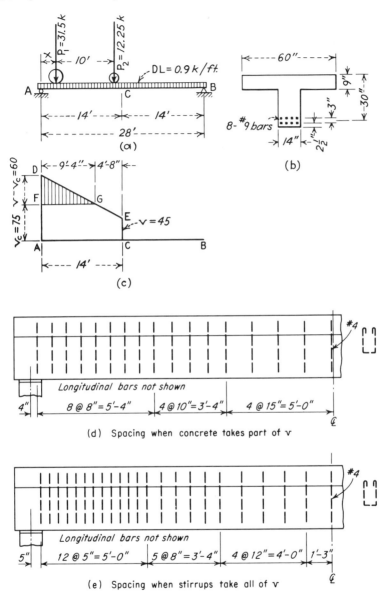

Fig. 4-23 Planning stirrups for beam with moving loads.

Case (2):

Now check the spacing if the stirrups are to withstand all of the diagonal tension at 24,000 psi.

At A:
$$s = \frac{A_v f_v}{vb'} = \frac{0.4 \times 24,000}{135 \times 14} = 5.1 \text{ in.}$$

At C:
$$s = \frac{0.4 \times 24,000}{45 \times 14} = 15 \text{ in.}$$

Therefore, use the arrangement shown in Fig. 4-23(e).

4-12. Miscellaneous Examples

Example 4-9. Figure 4-24 shows a case in which an access tunnel to a large underground parking garage was allowed to encroach within the street line below grade but the superstructure had to remain behind the building line. Sketch (a) shows the situation, but simplified somewhat. The column at A was supported by the heavy beam BC. Actually, this type of framing is more suitable for steel than for concrete construction, but reinforced concrete was used in order to conform to all of the rest of the structure. Design the web reinforcement for the end of this beam.

The end reaction at the outer wall is 398 kips; the bending moment at A, 1,552 ft-k; the effective depth, 37 in.; and the width of the T beam, 36 in. Let $f_c' = 4,000$ psi, $f_s = 20,000$ psi, and $j = 0.9$. Use WSD.

$$A_s \text{ at } A = \frac{1,552 \times 12}{20 \times 0.9 \times 37} = 28 \text{ in.}^2$$

Use eighteen No. 11 bars arranged as shown in sketch (c).

A possible diagonal-tension crack is pictured as DE in sketch (b). Some system of web reinforcement must be used to prevent this tendency from causing failure.

The computed shearing stress v in the beam should not exceed $5\sqrt{f_c'} = 5\sqrt{4,000} = 316$ psi, even with adequate web reinforcement. In this case, it is undesirable to compute the shear V on the basis of a distance d from the center of the bearing on the wall at F; much less so from the edge at E. Furthermore, the concentrated load causes the predominant part of the shearing stress from CF to AG anyway. Therefore, use the entire shear at the support for V. Although the short shear span may be beneficial, the seriousness of any failure is so great that this feature should not be depended upon. Then

$$v = \frac{V}{b'd} = \frac{398,000}{36 \times 37} = 298 \text{ psi}$$

which is just a little under the $5\sqrt{f_c'} = 316$ psi maximum allowed.

Since the direction of the possible cracking directly under and close to the column may be nearly vertical, it seems to be desirable not to depend upon vertical stirrups alone for web reinforcement, but to bend up some of the bars also. Therefore, assume that vertical stirrups are to withstand two-thirds and the bent-up bars one-third of the diagonal tension.

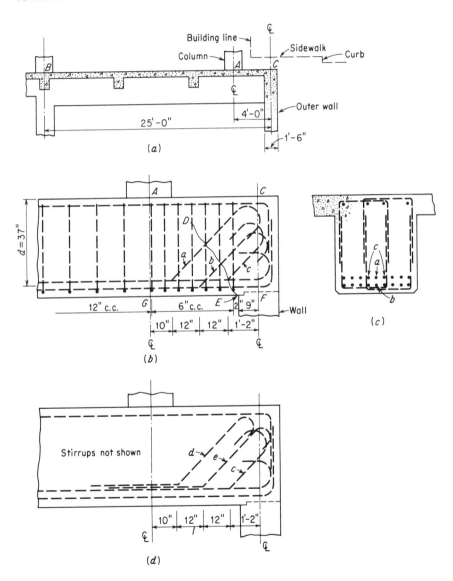

Fig. 4-24   Combined stirrups and bent-up bars.

In this case, the shear and diagonal tension are so severe that the web reinforcement will be designed on the basis that all of the shearing forces will be withstood by the web reinforcement with a maximum stress of 27,000 psi.

Vertical stirrups:

$$V = 0.667 \times 398 = 265 \text{ kips}$$

Try 6-in. spacing.

$$A_v = \frac{265,000 \times 6}{27,000 \times 37} = 1.59 \text{ in.}^2$$

Use two sets of double U-shaped stirrups overlapped as indicated in sketch (*c*). Then 1.59/4 = 0.398 in.^{2} needed per bar. Therefore, No. 6 bars (0.44 in.^{2}) will be satisfactory. These will be spaced as shown in sketch (*b*).

Bent-up bars:

First check to see how many bars must be extended to the end of the beam to provide sufficient bond strength. Table 13 in the Appendix shows that the allowed bond unit stress on No. 11 bars with $f_c' = 4,000$ psi is 150 psi for the top bars and 215 psi for the bottom bars. The latter value applies here. Therefore,

$$\Sigma o = \frac{V}{ujd} = \frac{398,000}{215 \times 0.9 \times 37} = 55.5 \text{ in.}^2 \text{ needed}$$

This requires thirteen No. 11 bars (57.6 in.^{2}), leaving only five bars which can be bent up. If bar *a* of Fig. 4-24(*c*) is bent first, then *b*, and finally two bars *c*, they might be spaced as shown in (*b*). Using 45° slopes, the bars bend away from the tensile region so quickly that their effectiveness in resisting longitudinal tension is questionable. Also, notice the congestion produced by their hooks and those of the other bottom bars. Nevertheless, these hooks are very important in this situation.

However, for the purposes of illustration, assume that the spacing of the bend points is 12 in. as shown in (*b*). Then, from Eq. (4-20),

$$A_v = \frac{Vs}{f_v d(\sin \alpha + \cos \alpha)} = \frac{133,000 \times 12}{27,000 \times 37 \times 1.414} = 1.13 \text{ in.}^2$$

This is less than the 1.56 in.^{2} of one No. 11 bar. However, the longitudinal tension in these bars has been neglected in Eq. (4-20). Therefore, it seems to be best not to bend up any of the main bars except *c* in Fig. 4-24(*d*) but to use two separate No. 8 bars for both *d* and *e* as shown in (*d*) to serve as web reinforcement. Notice that these bars are extended along the upper layer of main tensile reinforcement and are welded thereto. The welds hold these bars in place and provide a reliable anchorage for *d* and *e* as inclined stirrups. This arrangement requires more steel but it simplifies the construction and is good insurance against failure.

Notice also the arrangement of the hooks and bends at the ends of all of the bars shown in sketch (*d*). The hooks on such heavy bars are so large that they are likely to interfere with each other unless the work is carefully planned in advance.

Example 4-10. Assume a continuous T beam having a uniform load $w = 3$ klf and spans $L = 25$ ft. Let M at the support $= 156$ ft-k and $V = 37.5$ kips at this point. Assume $d = 21$ in., $b' = 15$ in., $f_v = 18,000$ psi, $f_s = 20,000$ psi, and $f'_c = 3,000$ psi. Neglect the compressive reinforcement. Design vertical U-shaped stirrups at 6 in. c.c. for the end portions of this beam. Use Eq. (4-10) in this case.

$$\text{Approx } A_s = \frac{156 \times 12}{20 \times 0.9 \times 21} = 5 \text{ in.}^2$$

Use five No. 9 bars.

The shear to use is that at a distance d from the support. V at the end $= 3 \times 12.5 = 37.5$ kips. At 21 in. from the end,

$$V_d = 37,500 - 3,000 \times 1.75 = 32,300 \text{ lb}$$

$$M_d = -156,000 + 37,500 \times 1.75 - \frac{3,000 \times 1.75^2}{2} = -95,100 \text{ ft-lb}$$

with tension in the top. Then Eq. (4-10) gives

$$v_c = \sqrt{f'_c} + \frac{1,300 p_w V d}{M}$$

where

$$p_w = \frac{A_s}{b'd} = \frac{5}{15 \times 21} = 0.016$$

$$v_c = \sqrt{3,000} + \frac{1,300 \times 0.016 \times 32,300 \times 21}{95,100 \times 12} = 67 \text{ psi}$$

This is greater than the 60 psi allowed in Table 1-8 for 3,000-lb concrete, but it will be used here.

Equation (4-8) gives

$$v_d = \frac{V_d}{b'd} = \frac{32,300}{15 \times 21} = 103 \text{ psi}$$

Therefore, for $s = 6$ in., Eq. (4-16) gives

$$A_v = \frac{(v - v_c)b's}{f_v} = \frac{(103 - 67)15 \times 6}{18,000} = 0.18 \text{ in.}^2$$

Using No. 3 stirrups gives $A_v = 0.22$ in.^{2}, which is satisfactory.

Example 4-11. Assume the same beam as in Example 4-10. Are stirrups necessary at the point of inflection, which will be assumed to be 5.25 ft from the support? Call this point E.

$$V \text{ at } E = 37,500 - 3,000 \times 5.25 = 21,800 \text{ lb} \qquad \text{and} \qquad M = 0$$

Then Eq. (4-10) cannot apply.

From Eq. (4-8),

$$v_e = \frac{21,800}{15 \times 21} = 69 \text{ psi}$$

Then Eq. (4-16), with $A_v = 0.22$ in.^{2} and $v_c = 60$ psi, gives

$$s = \frac{0.22 \times 18,000}{(69 - 60)15} = 29.3 \text{ in.}$$

However, the maximum spacing here should be restricted to $d/2 = 10.5$ in. Let $s = 10$ in. Therefore, use No. 3 stirrups at 6 in. c.c. for at least 4 ft from each end, starting the first ones 2 in. from the face of the support. Then use these stirrups at 10 in. c.c. throughout the next 3 or 4 ft.

PRACTICE PROBLEMS

In the following problems, assume $f'_c = 3,000$ psi, $f_v = 18,000$ psi, and $v_c = 60$ psi for working-stress design; and $f_y = 36,000$ psi, $\phi = 0.85$, $v_c = 93$ psi, and $U = 1.5\text{DL} + 1.8\text{LL}$ for ultimate-strength design. Solve all problems for both working-stress and ultimate-strength design unless stated otherwise in the problem.

4-1 Assume a rectangular beam for which $b = 14$ in., $d = 24$ in., and $V = 20,000$ lb DL $+ 15,000$ lb LL at the critical section 2 ft from the end support. Using No. 3 vertical U-shaped stirrups, determine the maximum allowable spacing at the end.

4-2 Design and locate No. 3 vertical U-shaped stirrups for a simply supported T beam having $b' = 10$ in. and $d = 16$ in. if the shear diagram is as shown in Fig. 4-25(a) and the dead load equals the live load.

4-3 The shear diagram for a simply supported T beam is shown in Fig. 4-25(b). If $b' = 14$ in., $d = 24$ in., and the dead load equals one-half of the total load, design No. 3 vertical U-shaped stirrups for this beam.

4-4 A simply supported T beam has a span of 20 ft, a uniform dead load of 1,200 plf, a uniform live load of 2,000 plf, an end support 12 in. wide, $b' = 14$ in., and $d = 20$ in. Design vertical U-shaped stirrups for this beam, selecting an appropriate size.

4-5 Design No. 3 inclined stirrups for the beam of Prob. 4-4.

4-6 Design vertical U-shaped stirrups to reinforce the web of a simply supported T beam if $b' = 16$ in., $d = 26$ in., and the shear diagram is as shown in Fig. 4-25(c).

4-7 If the beam of Prob. 4-6 has eight No. 9 bars in two rows of four each in the bottom, sketch how and where to bend them up as web reinforcement. Add vertical stirrups if necessary.

4-8 A simply supported T beam highway bridge member has a span of 20 ft. If $b' = 16$ in., $d = 22$ in., the reinforcement is six No. 9 bars, and

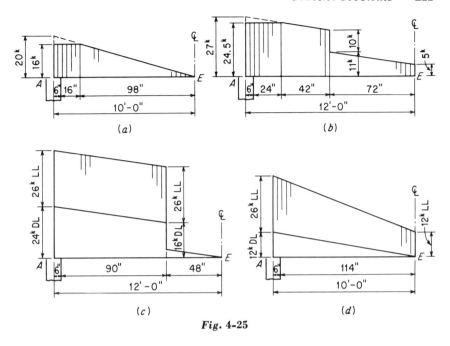

Fig. 4-25

the shear diagram is as shown in Fig. 4-25(*d*), design the web reinforcement, using bent-up bars and stirrups.

4-9 Design vertical U-shaped stirrups for the beam of Prob. 4-8 for working-stress design with stirrups for the full length to withstand all of the shear at $f_v = 25,000$ psi.

4-10 Assume that a continuous T beam has spans of 20 ft. If $b' = 16$ in., $d = 20$ in., the shear diagram is as shown in Fig. 4-25(*d*), the bottom bars at *E* are four No. 8, and the top bars at *A* are six No. 8, design the web reinforcement, using bent-up bars and vertical stirrups.

4-11 Figure 4-26 shows a simply supported T beam with a span of 20 ft.

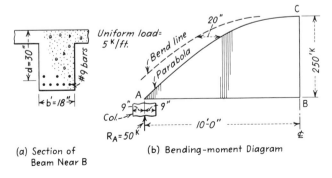

(a) Section of
Beam Near B

(b) Bending-moment Diagram

Fig. 4-26

Using a scale of $\frac{1}{2}$ in. = 1 ft 0 in., draw a side elevation for bending up the longitudinal reinforcement shown in (*a*). Add No. 4 vertical stirrups if needed.

4-12   Figure 4-27(*a*) shows the over-all dimensions assumed for a continuous T beam with a span of 20 ft. It supports a total uniform load of 4,000 plf plus a concentrated load of 30,000 lb at its center. Assume the bending-moment diagram shown in (*b*). Design the longitudinal reinforcement at *A* and *B*. Show a scheme for bending the bars for the purpose of resisting bending moment only. Design vertical stirrups for web reinforcement.

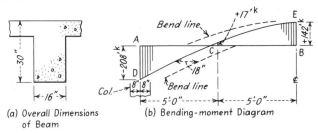

(a) Overall Dimensions   (b) Bending-moment Diagram
of Beam

Fig. **4-27**

4-13   Assume that a continuous T beam is to have the bending moment and the cross sections shown in Fig. 4-28. The span is 24 ft. The end shear is 60 kips, and the shear will be assumed to be constant from *A* to *C*. Design the web reinforcement as stirrups only, either inclined or vertical. Draw a side elevation, showing where the top and bottom bars may be stopped as straight bars. Show what the spacing of stirrups is to be.

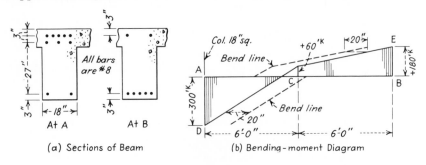

(a) Sections of Beam   (b) Bending-moment Diagram

Fig. **4-28**

4-14   Redesign the beam of Prob. 4-13, using combined bent-up bars and vertical U-shaped stirrups as web reinforcement.

4-15   Redesign the beam of Prob. 4-13, using working-stress design and vertical stirrups to take all of the shear at $f_v = 24,000$ psi.

COMPOSITE BEAMS

5-1. Introduction. It is often advantageous to use steel I beams or girders that are encased in concrete, which is used to protect the structural steel against fire and corrosion. Another type of good construction is bare steel beams with concrete slabs attached to their top flanges, designing both materials to act together as load-carrying elements of the member. The properties of the steel sections can be found in suitable handbooks.

The term *composite beam* is used to denote the foregoing cases; i.e., where the two materials are to act together in resisting the bending moments applied by superimposed loads. Such a beam is differentiated from an ordinary reinforced-concrete one primarily because the steel is a large rolled or fabricated unit which usually has great strength in itself. The concrete adds to the strength and stiffness of the structural steel, but it is the weaker of the two materials.

Another type of composite beam is one with a precast- or prestressed-concrete beam which has a concrete slab poured on top of it and attached to the top flange or portion. These two parts are also designed to act as a unit. Two such members will be discussed in detail in Chap. 13.

5-2. I-beam and Thin-slab Construction. The construction shown in Fig. 5-1(a) is a common type for use in large fireproof buildings. Ordinarily, the rather thin concrete slab is poured monolithically with the encasement of the steel I beam. When a load is placed upon such a floor, it is obvious that both the concrete and the steel will be affected.

Notice that, when a structural steel beam is encased in concrete, a material having considerable ductility is covered by another one having very little ability to stretch but great compressive strength. Therefore, when a load is applied so as to bend the member, the two materials try to act in accordance with their own particular characteristics. The top

213

flange of the I beam in Fig. 5-1(a) tends to compress inside the concrete; the bottom flange tries to elongate; both flanges endeavor to deform about a neutral axis which is at the center of the web. Simultaneously, the concrete tries to act as a T beam with its neutral axis close to the bottom of the slab. These two different actions must be in conflict because of the bond between the steel and the concrete. Unless this bond is broken, the steel cannot deform one way while the concrete deforms another way. Therefore, the action of the member is that of a composite unit.[1]

Composite construction is advantageous for obtaining stiffness as well as added strength.

It is often desirable to support the forms for the concrete from the steel beam so as to save the cost of shoring. In this case, the steel has to resist considerable bending moment by itself, so that the engineer should make sure that the top flange cannot buckle before the concrete has hardened. The steel therefore acts first as a bare steel beam, then later as part of a composite member. If shores are used to support the steel and forms until the concrete has cured sufficiently, the steel beam is not stressed appreciably until the shores are removed and the steel is obliged to act as a part of the composite beam only. Both cases will be investigated.

When composite beams frame into steel girders or columns, special care should be taken to avoid cracking along the top flanges of cross beams or girders because of the deflection of the composite beams. Where feasible, the steel beams should be made continuous with tension plates and bottom thrust angles; otherwise use heavy bars in the slab to carry tension across the girder or past the column. The whole framework should be looked over to find the places where deflections under live loads will be likely to cause cracks; then, if it is not practicable to make the steelwork continuous, it may be advisable to make definite joints or cuts in the concrete so that the cracking will occur at predetermined locations— often along the center of the flange of a cross beam.

Example 5-1. Assume the simply supported composite beam shown in Fig. 5-1(a). Using the transformed-section method and WSD, compute the maximum safe bending moment which the composite section can support if shores are used to support the beam and forms when the concrete is placed. Also, compute the longitudinal shearing stress between the concrete of the slab and the remainder of the member. Assume $f'_c = 2,500$ psi, $f_c = 1,125$ psi, $f_s = 20,000$ psi, and $n = 10$. Consider the entire width b of the slab to be effective in this case, but disregard the effect of the longitudinal bars in the slab since there is so much concrete in the compression flange anyway. General dimensions and data are given in Fig. 5-1(a). Assume that the bond does not fail.

[1] For design and analysis, either WSD or USD methods may be used.

Since very little of the steel beam is in the compression region, use nA_s as the transformed area of the I beam because, even though its area is several square inches and theoretically should be multiplied by $n - 1$ or $2n - 1$, it is too small compared with the area of the concrete to seriously affect the calculations.

$$nA_s = 10 \times 14.71 = 147.1 \text{ in.}^2$$

This may be considered to be concentrated at the center of the web of the 18-in. beam.

Compute the static moment of the transformed section about the neutral axis of the composite member, neglecting any of the stem that may be in compression unless it is found to be appreciable.

$$bt\left(kd - \frac{t}{2}\right) = nA_s(11.5 - kd)$$
$$72 \times 5(kd - 2.5) = 147.1(11.5 - kd)$$
$$kd = 5.1 \text{ in.} \qquad d - kd = 15.4 \text{ in. (near enough)}$$

Assume that the moment of inertia of the transformed section of the I beam about its own center-of-gravity axis is nI.

$$nI = 10 \times 800.6 = 8{,}006 \text{ in.}^4 \qquad \text{(see any steel handbook)}$$
$$I_c = \frac{bt^3}{12} + bt\left(kd - \frac{t}{2}\right)^2 + nI + nA_s(11.5 - kd)^2$$
$$= \frac{72 \times 5^3}{12} + 72 \times 5 \times 2.6^2 + 8{,}006 + 147.1 \times 6.4^2 = 17{,}200 \text{ in.}^4$$
$$S_c = \frac{17{,}200}{5.1} = 3{,}370 \text{ in.}^3$$
$$S_s = \frac{17{,}200}{10 \times 15.4} = 112 \text{ in.}^3$$

Then
$$\max M_c = f_c S_c = 1{,}125 \times 3{,}370 = 3{,}790{,}000 \text{ in.-lb}$$
$$\max M_s = f_s S_s = 20{,}000 \times 112 = 2{,}240{,}000 \text{ in.-lb}$$

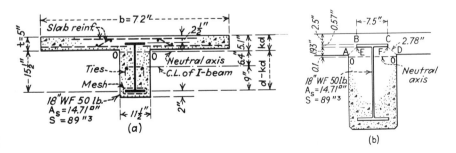

Fig. 5-1   Fully encased composite beam.

Therefore, the strength of the steel controls the resisting moment, as should be expected when so much concrete is available.

The transformed section in terms of concrete has been shown purposely because the reader may expect it to be used. However, in the case of composite sections like these, it is customary and advisable to use the transformed areas in terms of *steel*. To illustrate this here, start by keeping the depth of the slab as it is, but reduce its width to an imaginary $b/n = \frac{72}{10} = 7.2$ in. Then, assuming the neutral axis to be in the slab,

$$\frac{bt}{n}\left(kd - \frac{t}{2}\right) = A_s(11.5 - kd)$$

This will naturally give $kd = 5.1$ in. as before, as it should. Similarly,

$$I = \frac{7.2 \times 5^3}{12} + 7.2 \times 5 \times 2.6^2 + 800.6 + 14.71 \times 6.4^2 = 1{,}720 \text{ in.}^4$$

$$S_c = \frac{1{,}720 \times n}{5.1} = \frac{1{,}720 \times 10}{5.1} = 3{,}370 \text{ in.}^3$$

$$S_s = \frac{1{,}720}{15.4} = 112 \text{ in.}^3$$

Considering the I beam alone with $f_s = 20{,}000$ psi, its safe resisting moment, when laterally supported, is

$$M = Sf_s = 89 \times 20{,}000 = 1{,}780{,}000 \text{ in.-lb}$$

Therefore, the permissible increase in the safe bending moment for the composite beam over that for the plain steel beam is

$$\frac{100(M_s - M)}{M} = \frac{100(2{,}240{,}000 - 1{,}780{,}000)}{1{,}780{,}000} = 25.8\%$$

The transverse shearing forces are not important in this case because they are resisted by both the web of the I beam and the concrete. The former is usually capable of carrying the entire load.

Also, the longitudinal shear (diagonal tension) cannot cause failure of the same character as that which was discussed in the preceding chapter. However, it may produce excessive bond stresses or high local shearing stresses in the concrete.

To investigate this problem, let Fig. 5-1(b) represent an enlargement of a portion of Fig. 5-1(a). The beam will be assumed to have a span of 25 ft and a uniform load of 2,200 plf, including the dead load. The longitudinal shearing stress upon any plane of any section of the beam per unit length may be assumed as

$$S_L = \frac{VQ}{I_c}$$

where S_L equals the total longitudinal shear in pounds per linear inch of the beam, V equals the transverse shear at the given section in pounds, Q is the static moment of the part beyond the plane being considered (in this case *above AEBCFD* in the figure) computed about the neutral axis of the composite section, and I_c is the moment of inertia of the composite beam about its own center of gravity (or neutral axis). Q and I_c must be expressed in terms of inch units. They must also be computed upon the basis of the transformed section in terms of concrete. Therefore, at the end of the beam,

$$S_L = \frac{(12.5 \times 2{,}200)(72 \times 5 \times 2.6 - 7.5 \times 2.5 \times 1.35 - 1.93 \times 2 \times 0.74)}{17{,}200}$$

$$= 1{,}450 \text{ pli}$$

$$v_L = \frac{S_L}{AEBCFD} = \frac{1{,}450}{2 \times 2.78 + 2 \times 0.57 + 7.5} = \frac{1{,}450}{14.2} = 102 \text{ psi}$$

The stress on the surfaces AE and FD is shear in concrete, whereas that on $EBCF$ is bond on steel. The magnitude of the latter appears to be too large to be relied upon for such a wide flat surface of bare steel. Furthermore, since the bond stress is really a shearing stress, it is not reasonable to assume that the shear on the sections of concrete at AE and FD can have a value that is much different from the bond stress on the steel because all of the material must deform together until the bond fails or the concrete shears off. It may therefore be advisable to use some kind of anchorage between the steel and the concrete, as illustrated later.[1] Since the top flange is so close to the top of the slab, some short pieces of No. 8 bars welded crosswise of the beam flange might be suitable.

The two sides of the encasement of the steel beam should be tied together by U bars through holes in the web as shown in Fig. 5-1(*a*), by means of reinforcing on both sides for full depth, or by means of some welded anchors like Nelson studs. The encasement of the bottom flange also has to be held on by wire mesh "beam wrapper" or the equivalent to keep the bottom from spalling.

5-3. Shear Connectors. When a structural steel beam and concrete are used to constitute a composite beam like those in Figs. 5-1(*a*), 5-2(*a*), and 5-4(*a*), it is common practice to weld some kind of steel device on the top flange (1) to serve as a mechanical means for transmitting longitudinal shear between the concrete and the steel and (2) to tie the concrete to the steel beam so that they cannot separate vertically. The latter is especially important in bridges where vibration might tend to loosen the slab.

When a precast-concrete beam is combined with a poured-in-place slab to form a composite beam, ties or stirrups extending into the slab

[1] The AISC specifications do not require shear connectors in such cases.

serve both as shear connectors and vertical ties, but it is desirable to aid
the shearing resistance by bond and by roughening, serrating, or keying
the top of the precast member so as to get a positive and effective shear
resistance.[1]   This article is concerned with the steel beam and concrete
slab combination.

It is sometimes necessary in building construction, and frequently in
bridges having superelevated roadways, to have the steel beam below the
slab as shown in Fig. 5-2(a).   It is easy to see that a crack is likely to
form across the section $ABCD$.   One type of shear connector for such a
case is welded Nelson studs as shown in (b).   These are placed at whatever spacing is necessary.   Short pieces of channels, angles, or tees can
be welded to the beam to serve the same purpose.

The most common type of composite beam using structural steel is
indicated in Fig. 5-2(c).   The bottom of the slab is made flush with the

[1] J. C. Saemann and G. W. Washa, Horizontal Shear Connections between Precast
Beams and Cast-in-place Slabs, *J. ACI*, November, 1964.

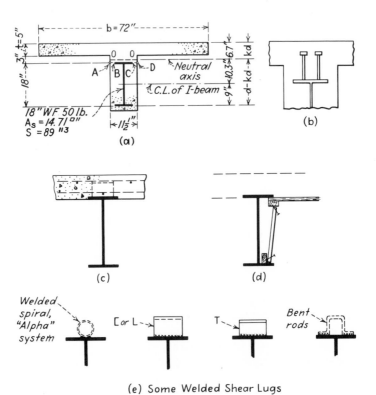

(e) Some Welded Shear Lugs

Fig. 5-2

Table 5-1   Ultimate Strength of Shear Connectors, psi
(for Stone Concrete Only)

Connector	Concrete strength f_c', psi		
	3,000	3,500	4,000
$\frac{1}{2}$-in. diameter × 2-in. hooked or headed stud	5,100	5,500	5,900
$\frac{5}{8}$-in. diameter × $2\frac{1}{2}$-in. hooked or headed stud	8,000	8,600	9,200
$\frac{3}{4}$-in. diameter × 3-in. hooked or headed stud	11,500	12,500	13,300
$\frac{7}{8}$-in. diameter × $3\frac{1}{2}$-in. hooked or headed stud	15,600	16,800	18,000
3-in. channel, 4.1 lb—for 1 in. of length	4,300	4,700	5,000
4-in. channel, 5.4 lb—for 1 in. of length	4,600	5,000	5,300
5-in. channel, 6.7 lb—for 1 in. of length	4,900	5,300	5,600
$\frac{1}{2}$-in. diameter spiral bar—for one pitch	11,900	12,400	12,800
$\frac{5}{8}$-in. diameter spiral bar—for one pitch	14,800	15,400	15,900
$\frac{3}{4}$-in. diameter spiral bar—for one pitch	17,800	18,500	19,100

Note: Generally use safety factor = 4.
Source: From American Institute of Steel Construction.

lower face of the top flange so that the concrete supports the steel later-ally. Also, as shown in (*d*), this arrangement facilitates the formwork, especially when the weight of the wet concrete and the forms is to be carried by the steel beam alone. Welded spirals, as shown in sketch (*e*), were frequently used in the past, but the modern tendency seems to be primarily toward the use of Nelson studs or some type of structural-steel shear connectors. The welded bent bars shown in Fig. 5-2(*e*) are gen-erally more expensive.

The part of a shear connector that is most effective for resisting longitudinal shear is that which is close to the flange of the I beam. The upper part is primarily to lock into the concrete to resist vertical separation.

Some types of corrugated forms for building construction are also equipped with lugs so that composite action can be obtained when the forms are welded to the beam flanges.

Recommended ultimate loads for some types of shear connectors are shown in Table 5-1. These are the same as AISC data and have been furnished by Nelson Stud Welding, a division of Gregory Industries, Inc., Lorain, Ohio. Their recommended formulas for computing these data, as determined by tests, are the following:

1. For structural steel channels:

$$Q_{uc} = 180(h + \tfrac{1}{2}t)w \sqrt{f_c'} \qquad (5\text{-}1)$$

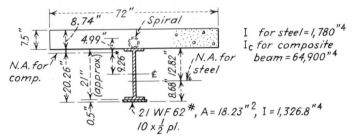

(a) Section for Positive Moments

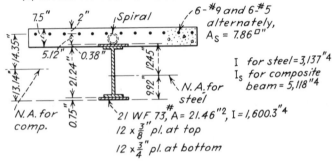

(b) Section for Negative Moments

Fig. 5-3   Spiral shear connectors used in a two-span continuous bridge.

2. For welded studs having ratios $H/d \geq 4.2$:

$$Q_{uc} = 330d^2 \sqrt{f_c'} \tag{5-2}$$

For welded studs having ratios $H/d < 4.2$:

$$Q_{uc} = 80Hd \sqrt{f_c'} \tag{5-3}$$

3. Spirals (helical bars):

$$Q_{uc} = 3,840d\sqrt[4]{f_c'} \tag{5-4}$$

where Q_{uc} = critical load capacity of one shear connector or one pitch of spiral bar

h = maximum thickness of channel flange, in., measured at the *face* of the web

t = thickness of the web of a channel, in.

w = length of channel, in., measured transversely of flange of beam

d = diameter of stud or spiral bar, in.

H = height of stud, in.

The recommended safety factor (SF) to apply to the values given in Table 5-1 is variable, as shown in Eq. (5-5).

$$\text{SF} = \frac{A(1 + C_{mc} + C_{mi}C_s) - (C_{mc} + C_{mi}) + C_V}{1 + C_V} \tag{5-5}$$

where $C_{mc} = \dfrac{\text{max DL moment on composite section}}{\text{max LL moment}}$

$C_{mi} = \dfrac{\text{max DL moment on steel beam alone}}{\text{max LL moment}}$

$C_s = \dfrac{\text{section modulus of composite beam for extreme tension fiber}}{\text{section modulus of steel beam alone for extreme tension fiber}}$

$C_V = \dfrac{\text{vertical DL shear on composite section}}{\text{vertical LL shear}}$

A = a numerical factor = $1.8 \pm$ if composite action is required to the yielding of the steel beam (structural grade) or 2.7 if composite action is required for all loading conditions.   The 2.7 value for A is preferred, with a maximum SF = 4.   A few values as given by Gregory are shown in Table 5-2.   It is good policy to use a safety factor of 3.5 to 4 in most cases.

Theoretically, it would seem that the spacing of shear connectors should vary inversely with the magnitude of the transverse shear V. However, the concrete is usually neglected in the consideration of trans-

Table 5-2   **Safety Factors from Eq. (5-5) for $C_{mc} = C_V = 0$**

C_{mi}	C_s					
	1.0	1.2	1.4	1.6	1.8	2.0
	$A = 1.8$					
0	1.80	1.80	1.80	1.80	1.80	1.80
0.2	1.96	2.03	2.10	2.18	2.25	2.32
0.4	2.12	2.26	2.41	2.55	2.70	2.84
0.6	2.28	2.50	2.71	2.93	3.14	
0.8	2.44	2.72	3.02	3.27		
1.0	2.60	2.96	3.32			
	$A = 2.7$					
0	2.70	2.70	2.70	2.70	2.70	2.70
0.2	3.04	3.15	3.26	3.36	3.47	3.58
0.4	3.38	3.60	3.81			
0.6	3.72					

Source: Nelson Stud Welding, Division of Gregory Industries, Lorain, Ohio.

verse shear and diagonal tension because the I beam is almost always strong enough by itself.   However, tests seem to indicate that the use of a uniform spacing throughout the length of the member is satisfactory. This is advantageous in the case of spirals, but uniform spacing is not very important when welding on other types of connectors.   Of course, one should be sure to have enough anchors to meet the extreme probable conditions.   For example, refer to Fig. 5-3A, showing the end of a simply supported member.   As the beam deflects under load, the slab tries to slide along the top flange of the I beam towards each end.   This action will cause the concrete, which is very strong in compression, to try to shove the shear connectors to the left in sketch (a).   If the steel beam is long and relatively flexible, this action is accentuated.   It may even cause a little yielding of the shear lugs or local crushing of the concrete. When the load is removed, the beam tries to spring back to its normal position.   If the shear connectors have yielded slightly under the deflected condition, it is probable that the tensile strength of the concrete will be unable to force the connectors back to their former position.   This will be likely to cause transverse cracking of the slab across the bridge, with resultant loss of future composite action at smaller loads when the cracks are not closed by pressure.   Composite action is a good thing under many conditions, but it should not be carried too far.   Remember that shallow depths mean more deflection, more force on the shear connectors, and the likelihood of the transverse cracking referred to previously.

The shear connectors should be designed to resist *all* of the longitudinal shear, not relying upon bond to the steel I beam for any resistance.   If the steel beam carried the dead load, the shear connectors of the composite member need be designed to resist only live load and impact.   If the I beam and forms are to be shored until the concrete has attained its proper strength, the connectors must be designed to resist DL + LL + I.   In addition, shrinkage of the concrete will tend to produce compression in the top and tension in the bottom of the beam, with shearing forces applied to

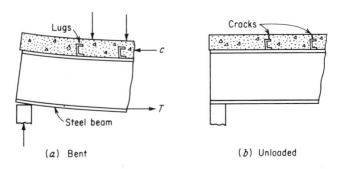

(*a*) Bent   (*b*) Unloaded

Fig. 5-3A

the connectors. However, in a simply supported composite member, if the shrinkage is symmetrical about the middle of the span, shortening of the slab will tend to shove the connectors toward the center whereas the action of the loads tends to push them the other way, as indicated in Fig. 5-3A(a). Therefore, shrinkage may add bending in the beam,[1] but it should not hurt the connectors unless harmful cracking occurs. Creep due to pressure in the concrete will usually act similarly to shrinkage. Temperature changes might have some effect, especially on a bridge deck that is exposed to the sun, but this is usually assumed to be taken care of by the safety factor. However, a warm morning sun after a cool night might cause the deck to expand enough to produce considerable stress in the connectors near the end of the span. For these reasons, and from his own experience, the author prefers to decrease the theoretically permissible spacing of the shear connectors in the end quarters of simply supported composite beams for bridges.

Most of the force delivered to the shear connectors by the slab comes through high local compression near the bottom of each connector. The strength of the concrete does not have a proportionate effect on the value of a connector, as shown in Table 5-1. On the other hand, the resistance is directly dependent upon the strength of the welds, and the strength of these should be equal to or greater than that required.

In continuous construction, the concrete is not relied upon for tensile resistance, so that shear connectors in regions of negative bending moments are used only to tie the parts together and to help the concrete to develop the necessary tension in the embedded bars.

Example 5-2. Assuming that they are wanted, compute the required spacing for the shear connectors at the end of the composite beam of Example 5-1, using No. 8 bars because of the small space between the top of the slab and the I beam. Use the longitudinal shearing stress previously computed, and take all of it in the shear connectors. Let the length of the connectors = $5\frac{1}{2}$ in.

$$v_L = 102 \text{ psi} \quad \text{and} \quad AD = 11.5 \text{ in.}$$

The No. 8 bars may be somewhat equivalent to the bottom portion of a 3-in. channel. Using Eq. (5-1) for a 3-in. channel and $f'_c = 2{,}500$ psi,

$$Q_{uc} = 180(0.36 + 0.085)1 \sqrt{2{,}500} = 4{,}000 \text{ pli}$$

With a safety factor of 4, the allowable $Q_{uc} = 1{,}000$ pli. Then one connector can withstand a force of $5.5 \times 1{,}000 = 5{,}500$ lb. The total shear per inch of beam is $102 \times 11.5 = 1{,}170$ lb. The shear connectors should

---

[1] H. W. Birkeland, Differential Shrinkage in Composite Beams, *J. ACI*, May, 1960. Also, W. Zuk, Thermal and Shrinkage Stresses in Composite Beams, *J. ACI*, September, 1961.

therefore be spaced at $5,500/1,170 = 4.7$ in., but call it 4.5 in. c.c.   If
the fillet welds are $\frac{1}{4}$ in. and have a safe resistance of 2,000 pli, then
$5,500/2,000 = 2.75$ in. of weld will be needed, but use at least 4 in. so as
to hold the bar firmly.

The concrete of the slab is partly under the edges of the flanges of the
I beam.   This will hold the slab and steel together so that no other hold-
down device is necessary.

5-4. Design of Composite Beam for a Bridge.   The AASHO specifications are to be used for highway bridge design.   These specifications
are based on WSD.   Allowable stresses are shown in Table 1-9.   Some
special pertinent data applicable to composite beams made of structural
steel and concrete slabs are as follows:

1. $E_s = 29,000,000$ psi; $n =$ values in Table 1-9.

2. Shear connectors shall be attached by rivets or welds at least equal
to the required resistance Q of the connector, and they shall resist both
horizontal and vertical movement.   Top cover over connectors shall not
be less than 1 in.; edge distance from connectors to tips of flanges shall
not be less than 1 in.

3. If the dead load acts on the composite section, creep shall be considered, using n as in Table 1-9 or $3n$, whichever gives higher stresses and
shears.

4. The neutral axis should be below the top flange of the I beam; if
not, concrete below the neutral axis is to be omitted from calculations of I
and moments.

5. The effective flange width of the slab as a T beam shall not exceed
$L/4$, the distance c.c. of beams, or 12 times the least thickness of the slab.
If the flange is on one side only, the effective width shall not exceed $L/12$,
one-half the distance to the next beam, or six times the least thickness of
the slab.

6. If the beams and forms are not shored, the dead-load stresses in the
I beam acting alone shall be added to those caused by composite-beam
action.   If the beams are shored and the concrete has attained $0.75f_c'$
when the shores are removed, the stresses shall be based upon composite-
beam action only.

7. If the steel beams are continuous, the positive moment part may be
designed with composite sections as for simple spans.   Concrete is not to
be relied upon for negative moments except to develop any longitudinal
bars as negative reinforcement.   If the steel beam alone is to resist the
negative moment, no shear connectors are really needed.

8. Horizontal shear for design of shear connectors is to be

$$S = \frac{Vm}{I} \tag{5-6}$$

where S is the horizontal shear per inch at the junction of slab and steel beam, V is the external transverse shear at the point considered due to loads provided the concrete strength $= 0.75f'_c$, m is the statical moment of the transformed compressive concrete area *in terms of steel* about the neutral axis of the composite section or statical moment of the area of the bars embedded in the concrete for negative moment, and I is the moment of inertia of the transformed composite section found by dividing the effective concrete flange *width* by n.

9. The specified formulas for computing the values of shear connectors are the same as Eqs. (5-1) to (5-4), inclusive, with a maximum spacing of 24 in. and $A = 2.7$ only. However, if not calculated by Eq. (5-5), a safety factor of 4 may be used.

10. The ratio of span to *over-all* depth preferably should not exceed 25; that of span to depth of steel beam alone should not exceed 30. For continuous beams, L may be considered as the distance between dead-load points of contraflexure. If made shallower, the deflection should not exceed that which would be computed using the specified depths—meaning lower unit stresses.

11. Deflection from LL + I should not exceed $L/800$ for simple or continuous spans, $L/300$ for cantilever arms. If diaphragms of adequate strength to spread loads laterally are used, deflection may be computed on the basis that all beams or girders act together and have equal deflection, using standard loads. Deflection of steel beams carrying all of the dead load before the concrete strength is $0.75f'_c$ shall be computed on the basis of no composite action.

Example 5-3. Design a simply supported interior composite beam for a highway bridge having a span $L = 76$ ft, using WSD. The c.c. spacing of beams is 7 ft 6 in.; the slab is 8 in. thick; $f'_c = 3,000$ psi; $f_c = 0.4f'_c$; $f_s = 18,000$ psi; and $n = 10$. The dead load is to be supported by the steel beam alone. Shoring cannot be used because the bridge is to be over an important highway on which traffic must be maintained. Assume that the LL + I moment at mid-span is 750,000 ft-lb and the maximum LL + I end shear is 47,000 lb. Provide for a possible bituminous topping weighing 25 psf which may be put over the slab in the future.

Since the span is 76 ft and the steel will have to support a lot of dead load, assume a depth of about $L/20$. Try 3 ft 9 in. over all. Use welded construction. Next, assume that about 50 per cent of the stress in the steel beam will be due to the dead load when the steel is acting alone. Then a trial area of one flange may be found by assuming an effective depth of 3.6 ft and a unit stress of 9,000 psi. Using the LL + I moment given,

$$A_f = \frac{750,000}{3.6 \times 9,000} = 23.2 \text{ in.}^2$$

Try plates 14 × 1½ in. for the flanges and a web plate 42 × ⅜ in.   The web
will have some value in helping the flanges.   This might make a composite
section like that shown in Fig. 5-4(*a*).   Since the flange is so thick and the
shearing strength of the slab is important for highway loading, the bottom
of the slab will be set at the top of the steel as shown.   The shear connectors will provide lateral support in the future through the slab, but
because the bare steel has to carry such a large dead load, strong cross
frames and a light temporary lateral bracing system will be used to brace
the girders.

In computing the dead load to be supported by the steel alone, assume
5 psf for steel details and 10 psf for forms and construction loads.

The steel girder and its properties are shown in Fig. 5-4(*b*).   The dead
load per linear foot of girder is

Concrete:	$7.5 \times 100 =$	750
Girder:	$58 \times 3.4 =$	200
Miscellaneous:	$7.5 \times 15 =$	110
		1,060 plf

The dead-load moment at the center is

$$M_{DL} = \frac{1,060 \times 76^2}{8} = 760,000 \text{ ft-lb}$$

Then $f_s = \dfrac{760,000 \times 12 \times 22.5}{22,130} = 9,300 \text{ psi top and bottom}$

The transformed section of the composite member in terms of steel
is shown in Fig. 5-4(*c*).   Since the width of the slab flange c.c. of beams is
less than 12*t*, the full 90 in. is counted upon.   Then $b/n = \frac{90}{10} = 9$ in., as
shown in the sketch.   The properties of this transformed section are also
shown in (*c*).

The LL + I stresses with the composite section acting are

$$f_b = \frac{750,000 \times 12 \times 37.2}{45,000} = 7,430 \text{ psi}$$

$$f_t = \frac{750,000 \times 12 \times 15.8}{45,000} = 3,160 \text{ psi as steel}$$

or $f_{tc} = \dfrac{3,160}{n} = \dfrac{3,160}{10} = 316 \text{ psi as concrete}$

$$f_{t'} = \frac{750,000 \times 12 \times 7.8}{45,000} = 1,560 \text{ psi in top flange of girder}$$

The future moment due to the extra topping will be

$$M = \frac{7.5 \times 25 \times 76^2}{8} = 136,000 \text{ ft-lb}$$

This will cause a stress in the bottom of the composite section.

$$f_{b'} = \frac{136{,}000 \times 12 \times 37.2}{45{,}000} = 1{,}350 \text{ psi}$$

However, the removal of the 10 psf allowed for forms and equipment used in construction will relieve the composite section of the following moment:

$$M_r = \frac{7.5 \times 10 \times 76^2}{8} = 54{,}000 \text{ ft-lb}$$

Then there will be what amounts to an elastic springback in the composite member and a reduction of stress in the bottom equal to

$$f_r = \frac{54{,}000 \times 12 \times 37.2}{45{,}000} = 540 \text{ psi}$$

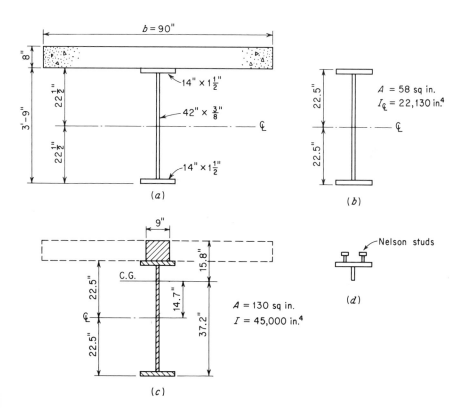

Fig. **5-4** Composite beam for a simply supported highway bridge.

The critical stress will be in the bottom flange of the girder.   Combining all of the different stresses given above, the tension in the bottom will be

Initial as I beam alone	$= \quad$ 9,300 psi
Reduction springback as composite section	$= \quad$ −540 psi
	8,760 psi
LL $+$ I as composite section	$= \quad$ 7,430 psi
	16,190 psi
Added topping as composite section	$= \quad$ 1,350 psi
	17,540 psi

This is close to the allowable 18,000 psi and will be accepted.

What type of connector should be used?   The spirals tend to interfere with the placing of the slab reinforcement.   It seems that $\frac{7}{8}$-in. Nelson studs will be suitable.   From Table 5-1, one of these connectors can be assumed to have a safe resistance of 15,600/4 = 3,900 lb.

The end reaction V for LL $+$ I was given as 47,000 lb.   The extra topping will add 25 $\times$ 7.5 $\times$ 38 = 7,100 lb, giving V = 54,100 lb.   The statical moment of the slab about the C.G. axis of the composite member is to be found next, using the transformed area in terms of steel, as shown in Fig. 5-4(c).

$$m = 9 \times 8(15.8 - 4) = 850 \text{ in.}^3$$

for the material above the girder flange.   Then, from Eq. (5-6),

$$S = \frac{Vm}{I} = \frac{54,100 \times 850}{45,000} = 1,020 \text{ pli}$$

at the end.   Therefore, 3,900/1,020 = 3.82 in. c.c. if only one connector is used.   However, try two studs in a row, as shown in Fig. 5-4(d), with a spacing of 2 $\times$ 3.82 = 7.64 in. Call it 7 in. c.c.

At the quarter point of the span, if the shear V = 28,000 lb, the spacing might be 7.64 $\times$ 54.1/28 = 14.7 in.   Call it 14 in. c.c.   The approximate number of sets of shear connectors per girder may be estimated as follows:

10 $\times$ $\frac{12}{7}$ = 17 spaces in the first 10 ft from the end
10 $\times$ $\frac{12}{10}$ = 12 spaces in the second 10 ft from the end
10 $\times$ $\frac{12}{14}$ = 9 spaces in the third 10 ft from the end
8 $\times$ $\frac{12}{16}$ = 6 spaces in the 8 ft near the center

Total spaces per one-half girder = 44.   Then the total number of studs required will be 44 $\times$ 2 $\times$ 2 + 2 = 178 per girder.   Of course, the spacing might be made uniform at 0.5L/44 = 38 $\times$ $\frac{12}{44}$ = 10.4 in. or perhaps 10 in. c.c. throughout.

The approximate LL + I deflection can be found by use of the usual formulas. With $w = 600$ plf and a concentrated load $P = 16,900$ lb at the center of the span,

$$\Delta = \frac{5WL^3}{384EI} + \frac{PL^3}{48EI} = \frac{5(600 \times 76)76^3 \times 12^3}{384 \times 29,000,000 \times 45,000}$$
$$+ \frac{16,900 \times 76^3 \times 12^3}{48 \times 29,000,000 \times 45,000} = 0.54 \text{ in.}$$

This is so much less than the $L/800 = 76 \times \frac{12}{800} = 1.14$ in. allowed that no greater refinement of calculation is needed.

In this case, someone might question the springback reduction of 540 psi in the bottom flange. This feature is ordinarily neglected. If so, the total tension in the bottom flange will be 18,080 psi, which is still satisfactory.

One should not get so busy designing the composite member that he forgets to design the steel girder itself properly for its service. In this case, it seems that, although the web is theoretically thick enough so that intermediate stiffeners are not needed, the heavy flange plates should be steadied somehow. One way to do this is to place the cross frames about 7 or 8 ft apart and to connect them to full-depth stiffeners. Of course, end stiffeners will be needed at the bearings.

5-5. Design of a Shored Composite Beam. In this case, the *Manual of Steel Construction*,[1] referred to as "AISC," will be used as the specifications to be followed. The following are some of the pertinent requirements which may differ from those already explained:

1. The projection of the slab on one side beyond the *edge* of the steel beam should not exceed $8t$ if it is to be counted fully in the composite section. The total flange width is then $16t$ + the flange width. The flange width is also limited to $L/4$ or the distance c.c. of beams. When the slab is on one side only, the projection is limited to $6t$ and the total width to $L/12$ or one-half the distance to the next beam.

2. Steel beams fully and integrally encased, as shown in Fig. 5-1(a), with 2-in. minimum cover on sides and bottom, with at least $1\frac{1}{2}$ in. of cover on top, completely wrapped on both sides and bottom with wire mesh or other reinforcing, can be considered to be composite without shear connectors. The stress must not exceed $0.66f_y$.

3. Unshored beams are to support all dead load prior to hardening of concrete as bare steel, also all DL + LL applied to the composite member after the concrete has set, without the stress exceeding $0.66f_y$. Any concrete below the neutral axis shall be neglected in computing I of the composite section. The encased steel beam may be designed to resist

[1] Published by American Institute of Steel Construction, Inc., New York.

all DL + LL as bare steel without shoring, but it is not to be stressed beyond $0.76f_y$.

4. Stress in laterally supported compact rolled shapes and built-up symmetrical members may be stressed to $0.66f_y$; other rolled shapes, built-up members, and plate girders, to $0.60f_y$.

5. I of composite section is to be computed as a transformed section in terms of steel $-A_s = A_c/n$.

6. When unshored, the section modulus of the composite beam for computation of stress in the bottom flange shall not exceed

$$S_{tr} = (1.35 + 0.35M_L/M_D)S_s \qquad (5\text{-}7)$$

where M_L = LL moment
  M_D = DL moment
  S_s = section modulus of the steel beam alone referred to the bottom flange
provided that the stress in the beam acting alone to hold the wet concrete is not stressed over that in item 4.

7. End shear is to be carried by the web of the steel beam only.

8. Except for item 2, shear connectors shall resist all horizontal shear along the top flange of the steel beam.

9. The *total* horizontal shear to be resisted between the point of maximum positive bending moment and the end of the beam (or the inflection point of continuous beams) is to be taken as the smaller of the following:

$$V_h = \frac{0.85f_c'A_c}{2} \qquad (5\text{-}8)$$

or

$$V_h = \frac{A_sf_y}{2} \qquad (5\text{-}9)$$

where A_c is the area of concrete as determined from item 1 and A_s is the area of the steel beam.

10. The number of shear connectors each side of a point of maximum bending moment shall not be less than

$$N = \frac{V_h}{q} \qquad (5\text{-}10)$$

where q is the allowable ultimate shear load for one connector or pitch of spiral as given in Table 5-1, which is the AISC specification, divided by the safety factor. These connectors may be uniformly spaced. They are to have a minimum cover of 1 in. in all directions.

11. The DL deflection of a steel beam acting alone should not exceed $1\frac{1}{2}$ in. Cambering may be used when needed.

12. The limiting over-all depth of the steel beam or girder plus the concrete slab under static loads = $L/24$ and of the steel beam or girder

plus the slab under heavy impact or vibratory loads = $L/20$. For the steel beam or girder itself, the depth should not exceed $L/30$.

13. Cover plates (not thin ones) may be used except when the concrete slab is on one side of the steel beam only.

Example 5-4. Design a simply supported steel-and-slab composite beam by AISC for the following WSD conditions: $L = 60$ ft; c.c. of beams = 10 ft; slab = 6 in.; LL = 200 psf; $f_s = 18,000$ psi; $f_y = 36,000$ psi; $f'_c = 3,000$ psi; and $n = 9$. The steel beam is to be shored. Stiffness is important.

First, an estimate has to be made to get a trial section. A limiting over-all depth might be $L/24 = 2.76$ ft, but try a 36-in. WF beam. It is probable that the composite member will be 20 to 30 per cent stronger than the steel alone, so try 20 per cent. A trial uniform load per foot is $w = (200 + 75)10 + 200 = 2,950$ plf. Call it 3,000 plf. Then

$$M = 3 \times 60^2/8 = 1,350 \text{ ft-k}$$

A trial section modulus is

$$SM = \frac{1,350 \times 12}{1.2 \times 20} = 675 \text{ in.}^3$$

From a steel handbook, this could be a 36WF230 ($SM = 835$ in.³). However, it is efficient to use a lighter beam with a bottom cover plate to offset the effect of the concrete slab somewhat and to get a more balanced section. Therefore, try the member shown in Fig. 5-5(a).

The width of the flange or slab is 120 in. This is greater than $16t$ + flange = $16 \times 6 + 12 = 108$ in. Therefore, the effective width must be 108 in., and the transformed width = $\frac{108}{9} = 12$ in. The transformed section and its properties are shown in Fig. 5-5(b).

The total load per foot is

Beam, cover plate, and details =		200 plf
Slab	= 10 × 75 =	750 plf
Live load	= 10 × 200 =	2,000 plf
	Total =	2,950 plf

The bending moment at the center is

$$M = 2,950 \times 60^2/8 = 1,330,000 \text{ ft-lb}$$

$$f_b = \frac{1,330,000 \times 12 \times 29.1}{28,500} = 16,300 \text{ psi}$$

This is somewhat conservative but will be accepted.

Now design the shear connectors. From Eq. (5-8),

$$V_h = \frac{0.85 f'_c A_c}{2} = \frac{0.85 \times 3,000(108 \times 6)}{2} = 826,000 \text{ lb}$$

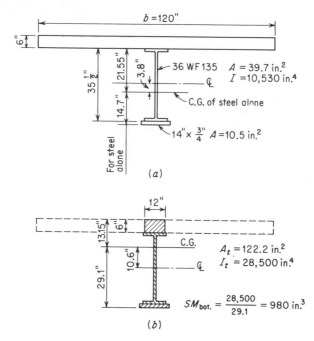

Fig. 5-5   Unsymmetrical steel section in a simply supported composite beam.

From Eq. (5-9),

$$V_h = \frac{A_s f_y}{2} = \frac{50.2 \times 36,000}{2} = 903,000 \text{ lb}$$

The smaller of these controls and gives the force to be taken care of by the connectors in one-half of the girder, but these figures are based upon ultimate strengths.   Using $\frac{3}{4}$-in. Nelson studs at 11,500 lb each, as shown in Table 5-1, for q in Eq. (5-10) gives

$$N = V_h/q = 826,000/11,500 = 72 \text{ studs for one-half of the span}$$

Use them in sets of two as shown in Fig. 5-4(d), making 36 pairs.   Therefore, space the pairs uniformly at $30 \times \frac{12}{36} = 10$ in. c.c.

Using the standard formula for deflection as an approximation without considering creep,

$$\Delta = \frac{5(2,950 \times 60)60^3 \times 12^3}{384 \times 29,000,000 \times 28,500} = 1.75 \text{ in. total}$$

The DL $\Delta = 0.56$ in. Therefore, it might be desirable to camber the girder about $\frac{3}{4}$ in.

5-6. Continuity. Consider what happens in a composite beam if it is made continuous. Tension will be in the top where the member passes over an interior support. The concrete cannot be relied upon for tension, hence it cannot help the beam's resistance to bending at the points where the largest moments generally occur. The situation is indicated in Fig. 5-6(a) to exaggerated scale. If the bars a are the usual transverse or temperature reinforcement, they are so small that they will be overstressed; then cracks like c are almost inevitable. The top flange of the steel beam will probably be badly overloaded.

Several heavy longitudinal bars might be used in the concrete throughout the region where any tension may occur in the top, as shown in Fig. 5-6(b). These cannot be stressed properly except as shear connectors between the structural-steel beam and the concrete are able to hold the concrete in place, so that bond on the bars can develop the necessary tensile stress in them. Obviously, with such large tensile stresses in the bars, the concrete will have hair cracks clear through its section.

For example, take the composite beam shown in Fig. 5-5(b). Assume that nine No. 9 bars are placed as shown in Fig. 5-6(b). The neutral axis of the steel beam and the bars as a combined section is so close to the center of the beam web that it will be assumed to be located there. Then

$$I = 7{,}796 + 10.5 \times 18.13^2 + 9 \times 20.75^2 = 15{,}120 \text{ in.}^4$$
$$SM \text{ for the top} \quad = 15{,}120/20.75 = 730 \text{ in.}^3$$
$$SM \text{ for the bottom} \quad = 15{,}120/18.5 = 820 \text{ in.}^3$$

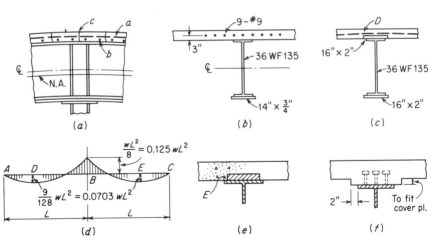

Fig. 5-6 Some problems in connection with continuous composite beams.

Thus, the bottom has less section modulus than the composite section based on the bottom flange of the latter—the beam in Fig. 5-5(b)—which is 980 in.³. The top is even weaker. The latter might be more heavily reinforced, but it is not practicable to place too many bars in the concrete.

Another possible procedure is to put a joint in the slab at c in Fig. 5-6(a) and to omit shear connectors in the tensile region, so that the slab can slip a little on top of the I beam. The latter would then have to be strengthened somewhat as shown in (c) to resist all of the moment.

Consider the bending-moment diagram of Fig. 5-6(d) for a two-span heavy continuous beam with uniform load all over and constant I. The negative bending moment at B is $0.125/0.0703 = 1.78$ times the positive moment at D and E. Now suppose that a beam similar to Fig. 5-5(b) is designed and able to resist the bending moment at D and E as a composite member and that its section modulus is 820 in.³ The section modulus needed at B will be $1.78 \times 820 = 1,460$ in.³ The steel beam shown in Fig. 5-6(c) has a section modulus of 1,510 in.³ so that it could be used, but it has been heavily reinforced. This shows that it is not desirable to use composite members too extensively as continuous beams, unless the steel beam is made strong enough to resist the negative moment. However, the case sighted here is for very heavy loads and long spans. Ordinary construction with much shorter spans and moderate loads will not cause such trouble, and bars in the concrete may serve to resist the tension at the top.

In Fig. 5-6(c), notice that the top cover plate causes a weak section in the slab at D, where the latter meets the edge of the plate. The reinforcement in the slab might be draped over the cover plate as shown, but if the slab is thin, there is little room for negative reinforcement in the slab, and the section is weak in resisting shearing forces. In general, the top cover plate ought to be narrower than the flange of the I beam, as shown in (e), so as to provide a small seat for the slab along the edges. This can be done usually in building work of typical dimensions if the use of bars as shown in (b) is not feasible or desirable. Another possible arrangement is shown in sketch (f), where the I beam is placed sufficiently below the bottom of the slab to allow for the thickness of the cover plate.

With the use of steel beams or girders and concrete slabs, and with the combinations of prestressed-concrete beams or girders and concrete slabs as explained in Chap. 13, there is a wide range of usefulness for composite construction.

PRACTICE PROBLEMS

For all problems, assume $f'_c = 3,000$ psi, $f_c = 1,350$ psi, $f_s = 18,000$ psi unless otherwise noted, $n = 9$, and WSD is to be used.

Building Construction

5-1 Compute the safe bending moment for a fully encased steel beam similar to that in Fig. 5-1 as a shored composite section if the following data apply: Code specifications, $t = 6$ in., $b = 84$ in., steel beam $= 24WF76$ ($A = 22.37$ in.^{2}, $I = 2,096$ in.^{4}, $S = 175$ in.^{3}), width of stem $= 14$ in., and top of steel beam is 3 in. below the top of the slab. Design the shear connectors if the span is 40 ft, the beam is simply supported, and the composite beam is fully loaded. Compare the answer with the safe bending moment for the steel beam alone.

5-2 A steel beam 18WF50 ($A = 14.71$ in.^{2}, $I = 801$ in.^{4}, and $S = 89$ in.^{3}) has a slab 5 in. thick resting on its top. The beams are 8 ft c.c. Compute the safe bending moment of the member as a shored composite section if the following conditions are applicable: AISC specifications, $f_s = 20,000$ psi, and the slab is resting on the top of the steel flange. What uniform load can it support if simply supported and the span is 32 ft? Design the shear connectors.

5-3 A shored, simply supported composite beam is similar to Fig. 5-3(a). The same steel beam and bottom cover plate are used, but the slab is 6 in. thick, and the beams are 8 ft c.c. Use AISC. What is the maximum live load which the beam can support if the span is 36 ft? Compare the moment resistance of the composite section with that of the steel beam alone. Design the shear connectors.

5-4 Design a simply supported composite beam to support a live load of 100 psf if the span is 42 ft, the beams are 8 ft c.c., the slab is 6 in., and the beams are unshored.

Bridge Construction. For all bridge problems assume a live load of 640 plf and a concentrated load of 18,000 lb for each 10-ft lane for maximum moment. For maximum end shear, increase the 18,000 lb to 26,000 lb concentrated. Impact $= 50/(L + 125)$. Modify lane loads to suit spacing of beams. Use AASHO. Design the shear connectors.

5-5 Design a simply supported composite highway bridge for a span of 50 ft if the beams are 7 ft c.c., not shored, and the slab is 7 in. thick.

5-6 Assume a simply supported composite girder like that in Fig. 5-4 except that the web is $40 \times \frac{3}{8}$ in. and the cover plates are 12×2 in. What is the maximum resisting moment if the girder is to be shored?

5-7 A simply supported highway bridge has the same steel member as that shown in Fig. 5-5. The beams are 7 ft c.c., the slab is 7 in., and $L = 48$ ft. What is the maximum safe LL if the beam is to be shored?

5-8 A simply supported highway bridge has a web $48 \times \frac{1}{2}$ in., a top cover plate $14 \times 1\frac{1}{2}$ in., and a bottom cover plate 16×2 in. The beams are 8 ft c.c., and the slab is $7\frac{1}{2}$ in. thick. The span is 84 ft. Is the girder a proper design if it is to be unshored? If not, what is a proper girder to use?

COLUMNS

6

6-1. Introduction. Members carrying direct axial loads which cause compressive stresses of such magnitude that these stresses largely control the design of the members may be included in the general classification called *columns*. Columns may also be subjected to bending moments as well as to axial loads. Therefore, the foregoing definition is given as a general one to differentiate between a column that resists bending and a beam that resists some longitudinal force as well as primarily bending moments. This chapter will consider columns with axial loads only.

Because of the nature of the material, concrete columns are generally relatively short. Longitudinal bars are added to assist in supporting the axial loads; also, hoops and spirals serve the same general purpose or at least enable the concrete and main bars to do so more effectively. Sometimes structural sections are considered as a sort of glorified reinforcement.

Columns need not be vertical, but, to avoid confusion, it will be advisable to consider them to be so, using the term *strut* to describe inclined or horizontal compression members.

Some additional symbols to be used in connection with columns are as follows:

A_c = area of core of a spirally reinforced column measured to the outside of the spiral, in.^{2}

A_g = gross area of a spirally reinforced or tied column, in.^{2}

h = actual unsupported length of a column, ft or in.

h' = effective length of a column, ft or in.

p_s = ratio of volume of spiral reinforcement to total volume of core of a spirally reinforced-concrete or composite column

r = radius of gyration of gross concrete area of a column, in.

6-2. General Discussion of Reinforced-concrete Columns. A simple square concrete column with eight vertical reinforcing bars is shown in Fig. 6-1(*a*). Under the action of the direct compression, the concrete bulges out laterally, as shown in exaggerated manner in Fig. 6-1(*b*). From a consideration of Poisson's ratio, this is to be expected when a material is placed under compression. It is also obvious that the bars themselves are somewhat like very slender columns. Naturally, they tend to buckle, but they cannot bend inward against the concrete. Therefore, they will buckle in the direction of least resistance, viz., away from

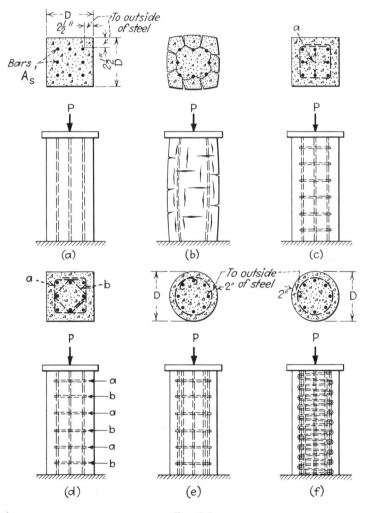

Fig. 6-1

the column's axis. This action causes tension in the outside shell of the concrete which, if the pressure becomes sufficient, will crack open somewhat as shown, the failure usually being sudden.

The way to overcome this trouble seems to be obvious. If the column is a square one, as pictured in Fig. 6-1(c), the apparent remedy is the placing of small bars around the longitudinal reinforcement, forming a series of bands or ties which are wired to the main bars and which are supposed to keep the latter from buckling, as well as to restrain the bulging action of the concrete. However, if the upper view of this figure is examined carefully, it will be seen that these bands are in the form of hollow squares. Therefore, when bars *a* try to bend sideways, they exert a lateral force which is normal to the straight sides of the bands. The bulging of the concrete does likewise. However, the bands are not effective in withstanding such beam action without bending outward so much that the concrete may crack. It is therefore advisable to use two kinds of ties, placing them alternately or adjacent so that one holds the corner bars while the other supports the intermediate ones, as shown in Fig. 6-1(d). This arrangement results in troublesome details, and it handicaps the "rodding," or compacting, of the concrete.

The next logical improvement seems to be the placing of the longitudinal bars in a circle, with circular hoops placed outside of, and wired to, them as pictured in Fig. 6-1(e). The buckling tendency of the main bars and the bulging of the inner portion of the concrete merely cause tensile stresses in these hoops, which means that they are really effective in restraining this action. However, the single hoops have to be spliced by lapping or by bending their ends around some of the main bars. This again is troublesome when a large number of hoops must be used and when they are of substantial size and therefore stiff.

The best way to support the longitudinal bars and the concrete is by means of spiral reinforcement as illustrated in Fig. 6-1(f). These spirals are merely long bars of small diameter which are bent around the main bars, forming a helix. In this way, small pieces are eliminated, the field work is decreased, and the waste of material in splices is avoided.

The arrangement of the main bars in a circular pattern with spiral reinforcement (or hoops) is advisable even when the cross section of the finished column must be square rather than round. Of course, the protective coating of concrete over the bars is needed to guard the steel against fire and corrosion, but the exterior surface can be shaped to suit the architectural requirements as long as a minimum cover of $1\frac{1}{2}$ in. or $1\frac{1}{2}$ times the maximum size of the aggregate is maintained over the outer surfaces of the bars, unless the specifications call for even more protection.

It was the general practice in the past to neglect the strength of the concrete covering that is outside of the hoops, ties, or spirals—the 2-in.

layer shown in Fig. 6-1(e) and (f). This was done because the covering was assumed to be for protection against possible fire and because, in that event, it might spall off, especially when heated and subsequently hit by cold water. Also, this covering is outside of the main portion that can be restrained by the hoops or spirals. The present tendency, however, seems to be to assume that the entire section of the concrete will participate in resisting the loads, or at least a substantial part of it can be relied upon, particularly under conditions of working loads.

This assumption seems to be entirely logical because all the material must be shortened when the member is compressed. Then all of it must also resist this deformation, especially within the range of ordinary working loads. There are differences of opinion regarding how much concrete outside of the hoops or spirals may be relied upon in design. Obviously, some practical limits must be set up. This is generally about $1\frac{1}{2}$ to 2 in.

The arrangement of reinforcement divides columns into two general classes, viz., "tied" columns which have longitudinal bars with intermittent hoops or ties; and "spirally reinforced" columns, which have longitudinal bars that are enclosed within steel spirals. The advisability of using bars of large diameter which will be stiff and strong as "little" columns is self-evident.

Some of the practical considerations and Code requirements that are applicable to columns with axial loads only are as follows:

1. Columns serving as the chief supports for a floor or roof shall have a diameter of at least 10 in. or, if rectangular, a gross area of not less than 96 in.² and a least lateral dimension of 8 in. Auxiliary columns at intermediate locations in a building and not more than one story high may be smaller but not less than 6 in. thick.

2. The unsupported length of a column shall be considered to be the distance between the top of the floor at the bottom and the lower extremity of the following at the top:

a. Capital, drop panel, or slab, whichever is least, for flat-slab construction.

b. Underside of the deeper beam framing into the column in each direction at the next higher floor.

c. The clear distance between consecutive struts framing into the column in each vertical plane, provided that two such struts shall meet at the column at approximately the same level and provided that vertical planes through the struts do not vary more than 15° from a right angle. Of course, these struts are to have adequate dimensions and are to be well anchored to prevent lateral deflection of the column.

d. When beams or struts have brackets equal to their width and at least one-half as wide as the column, the clear distance is to be measured from the floor to the bottom of the bracket.

3. The length of rectangular columns shall be considered as that which produces the greatest ratio of length to radius of gyration of the column.

4. The effective length h' of columns in structures where lateral strength or stability is provided by shear walls or rigid bracing, by attaching to an adjacent structure of sufficient lateral strength, or by other means which provide adequate lateral resistance, shall be assumed to be the unbraced length h.

5. In structures which depend upon the columns for lateral stability, larger effective lengths h' shall be used as shown by the following:

a. The end of a column shall be considered hinged in a plane if in that plane r' exceeds 25, where r' is the ratio of ΣK of the columns to ΣK of the floor members in a plane at one end of a column and K is the stiffness factor, $4EI/L$ or $3EI/L$, of the member concerned, depending on restraint.

b. For columns restrained against rotation at one end and hinged at the other end, the effective length shall be taken as

$$h' = 2h(0.78 + 0.22r')$$

which is to be equal to or greater than $2h$, and where r' is the value at the restrained end.

c. For columns restrained at both ends, the effective length h' shall be taken as $h' = h(0.78 + 0.22r')$ which is to be equal to or greater than h, and where r' is the average of the values computed at both ends of the column.

d. For columns that are fixed at one end and free at the other, the effective length h' shall be taken as twice the over-all length.

6. Lapped splices of column reinforcement shall be at least equal to the following:

a. Deformed bars:

f_c', psi	f_y, psi	Lap in bar diameters (A305)
3,000	50,000 or under	20
3,000	60,000	24
3,000	75,000	30
Any value greater than 3,000		Not less than 12 in.
Less than 3,000	Any value	$\frac{1}{3}$ greater than above

b. The minimum lap for plain bars shall be twice that specified for deformed A305 bars.

c. When successive columns differ in size so that the longitudinal bars are offset at a splice, the outer bars shall be sloped at an inclination not exceeding 1:6 with respect to the axis of the column, and the portions of the bars above and below the offset shall be parallel to the axis of the column. Metal ties, spirals, or the floor construction itself shall provide adequate horizontal support at the bend points. The ties and spirals should not be more than eight bar diameters from these bend points. For design, assume that the horizontal thrust at a bend point is $1\frac{1}{2}$ times the horizontal component of the nominal stress in the inclined portion of the bar. Such offset bars shall be bent before placing, not when in the forms or when partially embedded in the concrete. Lap to suit the larger bar.

7. Welded splices or other positive connections may be used instead of lapped splices, and such are preferred for bars larger than No. 11. In the case of bars used for compression only, it is possible to transmit the compressive stress by bearing of square-cut ends held in concentric contact by welded sleeves or other suitable devices, but this obviously depends upon the accuracy of the field and shop work. To be approved, a welded splice should be butted and welded so that it will develop a tensile resistance at least equal to 125 per cent of the specified f_y of the bar. Any other positive connection for critical tensile or compressive reinforcement should develop a resistance equal to an approved welded splice.

8. When one or more faces of columns are offset 3 in. or more, the splices of vertical bars adjacent to an offset face shall be made by the use of separate dowels properly overlapped.

9. When a spirally reinforced column is built monolithically with a concrete wall or pier, the outer boundary of the column shall be assumed to be a circle with its perimeter at least $1\frac{1}{2}$ in. or $1\frac{1}{2}$ times the maximum size of the coarse aggregate outside of the limit of the spiral, or to be a square or rectangle having its faces $1\frac{1}{2}$ in. outside of the limits of the spiral.

10. It is permissible to design a circular column in the usual manner and then to build it as a square, octagonal, or other shape having the same least lateral dimension, doing so for architectural reasons. In such a case, the A_g, p_g, etc., may be assumed to be those of the equivalent circular column.

11. If a tied column has a larger cross section than required by the loading, a reduced effective area, A_g, not less than one-half of the total area may be used when computing the steel area needed and the load-bearing capacity.

12. When the specified f'_c for the concrete of the columns exceeds that for the floor construction by more than 40 per cent, the transmission of the load through the weaker concrete shall be provided by one of the following:

   a. Concrete of the strength specified for the columns shall be placed in the floor at the bottom of the column for an area four times A_g of the column, and this concrete shall be well integrated into the floor concrete and shall be properly placed and compacted.   This is obviously a nuisance.   A workman may not be able to tell the difference in the strengths by visual inspection, and he may get the better concrete in the wrong place.   A uniform quality is preferred.

b. The capacity of the column may be computed on the basis of the weaker concrete, then dowels or spirals or both may be added to attain the necessary capacity of the column.

c. The capacity of columns laterally supported by slabs or beams on all four sides and of approximately equal depth may be computed by using an assumed concrete strength in the column formulas equal to 75 per cent of the strength of the column concrete plus 35 per cent of the strength of the floor concrete.

13. There is the possibility that, in case of a fire, cold water striking a heated column will cause the concrete to spall off.   On this account, some engineers assume that the effective gross area of a column is limited by that bounded by lines $1\frac{1}{2}$ in. outside of ties or spirals, as indicated in Fig. 6-2.   This is something which depends upon the engineer's judgment regarding the risks to life and property for any particular case.

14. In tied columns, the amount of reinforcing spliced by lapping should not exceed a steel ratio of 0.04 in any 3-ft length of column.

15. There is some question regarding the difference between a column and a pier.   Both may carry heavy longitudinal loads.   For convenience, one may assume that the member is a pier when its length is three or less times its least lateral dimension, or when the stress in the concrete acting alone is well below that permissible in a column, so that no reinforcement is actually needed to carry stress even though it is used to resist overturning moments or to knit the structure together.   In bridge piers having portals and those that are cantilevered from the footings, the bars should be designed to resist these actions instead of to act as column reinforcement.

16. The clear space between longitudinal bars shall not be less than $1\frac{1}{2}$ times the diameter of the longitudinal bars or $1\frac{1}{2}$ times the size of the coarse aggregate used, or $1\frac{1}{2}$ in.   These spacings should also apply to adjacent pairs of bars at a lapped splice, although two bars that are spliced by lapping may be in contact.   However, preferred spacings are shown in Table 11 in the Appendix.

17. At the bottoms of reinforced-concrete columns the depth of the footing, or of the footing plus a pedestal on it, should be sufficient to enable

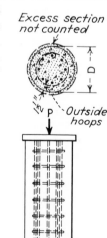

Fig. 6-2

dowels to develop the compressive stress in the column reinforcement through bond if the concrete at the bottom of the column is not to be overloaded. Hooks at the bottoms of such dowels are not effective, but 90° bends are often used merely as a means of resting the dowels on the footing reinforcement during the placing of the concrete. A pedestal should be larger than the column so as to reduce the stress in the concrete but also to form a shelf to support the column forms. The dowels should be set first and wired in place, not shoved into the wet concrete after pouring.

18. The Code requires that, for ultimate-strength design, any column should be designed for the longitudinal load plus an eccentricity of that load equal to $0.05t$ for spirally reinforced columns and $0.10t$ for tied columns, where t is the diameter or the dimension in the direction of bending about a principal axis. This combination will be explained in the next chapter. Therefore, this chapter will be confined to the illustration of WSD methods.

6-3. Design of Tied Columns. When a reinforced-concrete column like that in Fig. 6-1(d) first has a small load applied to it, the concrete and the steel will both be squeezed down slightly but equally. It would then seem that the stresses in the materials could be computed by the theory of the elasticity of the materials, and by the use of the transformed section. If the stresses are well below the "elastic limit" of the materials, one might assume that $f_s = nf_c$, where f_s and f_c are the unit stresses in the steel and concrete, respectively, and n is the modular ratio E_s/E_c. Then the applied load should be resisted by the area of the concrete times f_c plus the area of the longitudinal reinforcement times f_s.

As an example, refer to Fig. 6-1(d). Let A_g equal the gross effective area of the column in square inches and p_g equal the ratio of the total area of the longitudinal bars to this gross effective area. Then the area of the steel is

$$A_s = p_g A_g \tag{6-1}$$

The net area of the concrete is $A_g - A_s = A_g - p_g A_g = A_g(1 - p_g)$. Therefore,

$$
\begin{aligned}
P &= A_s f_s + (A_g - A_s)f_c = p_g A_g f_s + A_g(1 - p_g)f_c \\
&= p_g A_g n f_c + A_g(1 - p_g)f_c \\
&= A_g f_c[1 + (n - 1)p_g]
\end{aligned}
\tag{6-2}
$$

which gives the load in terms of the transformed section and of the stress in the concrete. In reality, however, this is not in accord with the facts or with modern practice.

Experience has shown that the action of a reinforced-concrete column under load seems to be rather peculiar.   At first, the conditions may be practically those given by Eq. (6-2); but if the load is large, and if it is continued for a long time, plastic flow or creep of the concrete takes place, resulting in a decrease of the unit stress in the concrete but an increase of the stress in the steel because of the latter's inability to "get out from under" the load.   Shrinkage of the concrete also tends to relieve the stress in the concrete and to shift load into the bars.   This action may continue (especially under heavy load) until the stress in the steel reaches the yield point f_y, whereupon the bars may deform appreciably without taking much further increase in stress, even though they cannot get rid of the stress already in them.   Then, at this stage, the shifting of the load from the concrete to the steel must practically cease because the latter is so much more compressible than the former.   If the load on the column is increased still more, the steel will continue to have a stress which is at or a little above the yield point, but the concrete, by itself, must carry the increase of the load until it fails.

According to these principles, a formula for the ultimate strength of a short reinforced-concrete column might be

$$P_u = (A_g - A_s)f_c' + A_s f_y \tag{6-3}$$

However, the Code does not permit the use of the ultimate load as given by Eq. (6-3) but requires that the reduction factor ϕ be included, that only $0.85f_c'$ be relied upon, and that a minimum moment as stated in item 18 of the preceding article be included.   The symbol P_o is used to denote the computed axial load without considering this minimum moment.[1] Then

$$P_o = \phi[(A_g - A_s)0.85f_c' + A_s f_y] \tag{6-4}$$

or, with $\phi = 0.7$ for tied compression members,

$$P_o = 0.7[(A_g - A_s)0.85f_c' + A_s f_y] \tag{6-5}$$

For WSD, the ultimate strength must be divided by a suitable safety factor.

From Eq. (6-5), neglecting the subtraction of A_s from A_g, a safety factor of a little over 2.5 applied to the first term will give approximately the first term of Eq. (6-6).   Similarly, calling $f_s = 0.4f_y$, a safety factor of a little over 2 applied to the second term of Eq. (6-5) will give approximately the second term of Eq. (6-6).   This is the Code formula for WSD.[2]

$$
\begin{aligned}
P &= 0.21f_c' A_g + 0.85 A_s f_s \\
  &= A_g(0.21f_c' + 0.85 f_s p_g)
\end{aligned}
\tag{6-6}
$$

---

[1] See Eq. (7-35).

[2] The corresponding AASHO formula for short tied columns for WSD only is $P_t = 0.8(0.225f_c' A_g + A_s f_s)$, where $f_s = 13,200$ psi for structural grade bars and 16,000 psi for intermediate grade.

where f_s is the allowable stress in the vertical column bars, which may be taken as 40 per cent of the minimum specified yield strength f_y, but not over 30,000 psi. Notice that the area of the steel A_s is not deducted from A_g because this would have a minor effect. It is evidently considered in determining the coefficients of f_c' and f_s in Eq. (6-6), which are empirical but based upon tests, experience, and judgment.

Equation (6-6) merely sets a maximum safe load for a column that has a given size and one specific makeup. It does not give the unit stresses in the steel and concrete when the column is not fully loaded. On the other hand, as long as the column is admittedly safe, the lesser unit stresses in it for partial loading are of little interest or importance.

The first form of Eq. (6-6) is convenient when a given column is to be analyzed—when A_g and A_s are known. The second form is more convenient when one is trying to determine the size of a tied column to hold a specified load.

The ratio of the longitudinal reinforcement p_g for a tied column should not be less than 0.01 or more than 0.04 for practical design. At least four bars should be used, and the minimum practicable and permissible size is No. 5 because the bars should be stiff enough for proper action and handling.

The ties which are wrapped around the main bars in Fig. 6-1(d) add considerable toughness to the member.[1] Of course, they should be adequate to brace the main bars properly. A practical guide is to use ties that are at least $\frac{1}{4}$ in. in diameter and spaced not over 16 times the diameter of the longitudinal bars, 48 times the diameter of the ties, or the least lateral dimension of the column. When there are more than four main bars, additional ties might be provided so that every one of the longitudinal bars is held firmly in its proper position and has a proper support. In very large columns which are heavily reinforced, it is not always practicable to meet this requirement completely. In such a case, the ties will generally be fairly large so that they will have considerable strength to resist lateral buckling of a main bar or two if properly held near by. Ties should hold every corner bar. At least alternate longitudinal bars along the sides should be held by ties having not over 135° angles at the bends, but no bar should be more than 6 in. from a supported bar.

The size of bars to be used as ties is a matter to be settled by good judgment. If they are too small, they will bend out of shape very easily. If they are too large, they will be difficult to bend to proper shape, they may be wasteful of steel, and the bends will not be so sharp as desired. Suggested rules for their size are as follows:

---

[1] See J. F. Pfister, Influence of Ties on the Behavior of Reinforced Concrete Columns, *J. ACI*, May, 1964. Also see D. E. Anderson and E. S. Hoffman, Column Details under the 1963 ACI Building Code, *J. ACI*, February, 1965.

1. One-fourth inch for small members where the ties are to be bent by hand in the field

2. No. 3 for columns having a least width of 10 to 18 in.

3. No. 4 for columns having a least width of 18 to 24 in.

4. No. 5 or larger for columns of larger size, and for piers where workmen may climb on the ties

If, for architectural or other reasons, a tied column is made larger than required for loading, a reduced effective area A_g as needed, but not less than one-half of the total area, may be used when computing the required area of longitudinal reinforcement.

The steps in the design of a tied column are as follows: (1) the determination of the quality of the concrete to be used and the allowable stresses in the materials for working-stress design or the load factors for ultimate-strength design; (2) the assumption of a size and shape; (3) the assumption of p_g or the number and size of bars to be used; and finally, (4) the test or analysis of the member to see that it can support the desired load.

Example 6-1. By means of Eq. (6-6), compute the safe load for the short tied column shown in cross section in Fig. 6-3(a). Assume $f'_c = 3,000$ psi and $f_s = 20,000$ psi. Let A_g equal the entire cross section.

The corners of the column of Fig. 6-3(a) are shown with chamfers in

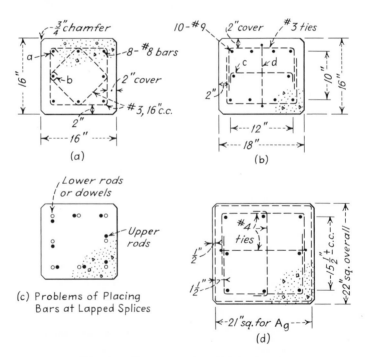

Fig. 6-3 Some examples of tied columns.

order to avoid sharp edges that may not be filled properly and that may chip off easily. Any reduction of area of the concrete that such chamfers may produce in A_g is customarily neglected. However, it will be considered here for purposes of illustration. Therefore,

$$A_g = 16^2 - 4(0.7^2/2) = 256 - 1 = 255 \text{ in.}^2$$
$$A_s = 8 \times 0.79 = 6.32 \text{ in.}^2$$

Using the first form of Eq. (6-6),

$$P = 0.21f'_c A_g + 0.85A_s f_s$$
$$= 0.21 \times 3,000 \times 255 + 0.85 \times 6.32 \times 20,000 = 160,500$$
$$+ 107,500 = 268,000 \text{ lb}$$

In this case, it can be seen that, even though p_g is only $6.32/255 = 0.0248$, the bars carry about 40 per cent of the total load.

Using the No. 3 ties shown and the limitations previously specified for ties, we find the following:

$16 \times 1 = 16$ in. based upon the longitudinal bars
$48 \times \frac{3}{8} = 18$ in. based upon the size of the ties used
16 in. = the minimum least lateral dimension of the column

Therefore, the 16-in. spacing is satisfactory.

Notice how the ties in Fig. 6-3(a) are arranged. One is square, with the ends hooked around the corner bar at a. A second tie is also square and bent around the middle bar at b. These No. 3 bars are too strong to be bent conveniently in the field by hand. However, when bent in advance at the shop, they are difficult to place. They generally have to be slid down the longitudinal bars or stacked at the bottom and pulled up after inserting the vertical reinforcement in the loops and corners. Both ties are placed in contact and wired to the main bars, and the two together constitute what is called one tie. They can also be alternated at one-half the spacing, as shown in Fig. 6-1(d).

A different arrangement of ties is pictured in Fig. 6-3(b). Two U-shaped bars are placed across the main reinforcement from the sides, lapped horizontally, and wired in place. Two other U-shaped bars with bent ends about 3 or 4 in. long are laid across the center with the bent ends horizontal and wired on top of the other ties, as shown by c, or with the bent ends vertical and crossing over the other ties, as pictured by d. As many cross bars are used as seem to be necessary. However, it is easy to see that an arrangement like that shown in Fig. 6-9 may be better than that shown in Fig. 6-3(b) because it gives more clear space than rectangular openings 6 or 8 in. in size so that an "elephant's trunk" (a pipe or hose on the bottom of a movable hopper) may be inserted for depositing the concrete.

Another problem is shown in Fig. 6-3(c). The open circles represent the bars in a lower column or the dowels projecting from the footing. How are the upper longitudinal bars to be set? One method is to place them as shown by the blackened circles. When these bars are large, it is obvious that they will interfere with the hooked ties of Fig. 6-3(a). The corner bars will not fit in the corners unless they are pulled over as soon as possible, which is what happens in the field. It will be advisable to place dowels so that they come inside of the main steel of the column itself. When columns are spliced, it often happens that the upper column is smaller than the lower one. In that case, the bars in the lower one must be pulled inward or dowels used to splice the steel. In any event, the reader can see that the bars in the field may not look just like the lines on a drawing.

As stated previously, the Code specifies that the reinforcement in a column should have a covering of concrete cast monolithically with the core (the portion inside the ties or spirals), this covering being at least $1\frac{1}{2}$ in. or $1\frac{1}{2}$ times the maximum size of the coarse aggregate used. This is for ordinary indoor construction. If the column is exposed to the ground or to the weather, the cover over No. 6 bars or larger should be at least 2 in., and $1\frac{1}{2}$ in. for No. 5 bars or smaller ones. In any case, the cover for resistance to fire or weathering must meet the requirements of any local and more restrictive code, and it should be what the engineer considers to be suitable for the intended service. The bars which determine this cover are naturally the ties or spirals because these should be outside of the longitudinal reinforcement.

In the case of tied columns, the bars near the corners are especially vulnerable since they can be affected by heat or moisture from two surfaces. It is therefore preferable to use 2 in. of cover over the ties in such cases.

The minimum number of bars in a tied column obviously should be four. The minimum size of the longitudinal bars should be No. 5, but preferably larger. The Code states that at least six such bars should be used in spirally reinforced columns. The Code also states that A_s should not be less than $0.01A_g$ nor more than $0.08A_g$. When p_g exceeds about 0.04, the amount of steel becomes impracticably great for proper detail with ordinary bars; this is also a sensible upper limit for tied columns in any case. However, the special large A 408 bars may permit a greater percentage p_g in some cases because each bar has such a large area. On the other hand, splicing of these big bars requires special attention.

It is possible to use high strength steels for column reinforcement. Some tests[1] of such columns seem to indicate that these special steels with

[1] C. E. Todeschini, A. C. Bianchini, and C. E. Kesler, Behavior of Concrete Columns Reinforced with High Strength Steels, *J. ACI*, June, 1964.

no pronounced change of the rate of deformation at the yield point may permit larger strains in the columns before failure than would steels with a yield plateau around 30,000 psi. However, in the author's opinion, one should be careful how far he trusts this greater deformation. In other words, it does not seem to be wise for the engineer to see how much he can reduce the margin between his design loads and the failure load.

Example 6-2. Compute the safe load for a short tied column having the cross section pictured in Fig. 6-3(b), using WSD methods. Determine the spacing of the ties if they are of the size and type shown on the drawing. Assume that $f'_c = 3,500$ psi, $f_s = 20,000$ psi, and A_g is to be computed on the basis of $1\frac{1}{2}$ in. over the ties.

Assuming the No. 9 bars to be $1\frac{1}{8}$ in. in diameter, the approximate dimension for computing the effective A_g is $1\frac{1}{8} + 2 \times \frac{3}{8} + 2 \times 1\frac{1}{2} = 4\frac{7}{8}$ in. larger than the c.c. spacing of the corner bars. Therefore,

$$A_g = 14.88 \times 16.88 = 251 \text{ in.}^2$$

$A_s = 10$ in.^{2}. From Eq. (6-6),

$$P = 0.21f'_c A_g + 0.85A_s f_s$$
$$= 0.21 \times 3,500 \times 251 + 0.85 \times 10 \times 20,000 = 354,500 \text{ lb}$$

Using the rules previously suggested for the spacing of the No. 3 ties, find $16 \times 1\frac{1}{8} = 18$ in., $48 \times \frac{3}{8} = 18$ in., or 16 in. Therefore, the least lateral dimension of the column controls this spacing.

Example 6-3. By working-stress methods, design a square, short, tied column to support a load of 425 kips, using 3,000-lb concrete and $f_s = 16,000$ psi.

One way to solve such a problem is to guess a size and then test it to see if it is satisfactory. Another way is to assume a percentage or ratio of reinforcement p_g somewhere between the acceptable ratios of 0.01 and 0.04, then to use this in the second form of Eq. (6-6) and to solve for A_g. Unless the column is to be as small as it is practicable to have it, a value of $p_g = 0.02$ to 0.03 is generally satisfactory.

Using $p_g = 0.025$ in this case,

$$P = A_g(0.21f'_c + 0.85f_s p_g)$$
$$425,000 = A_g(0.21 \times 3,000 + 0.85 \times 16,000 \times 0.025)$$
$$A_g = 425,000/970 = 438 \text{ in.}^2$$

If the column is 21 in. square, then $A_g = 441$ in.^{2}, which is satisfactory, but add $\frac{1}{2}$ in. for extra cover. Then

$$A_s = 0.025A_g = 0.025 \times 438 = 10.95 \text{ in.}^2$$

Preferably, the number of bars used in this column should be in multiples of four in order to have a symmetrical arrangement. Therefore, from Table 3 of the Appendix, the following can be assumed to be close enough for practical use: twenty No. 7 bars $= 12.0$ in.^{2}; sixteen No. 8 bars $= 12.6$ in.^{2}; twelve No. 9 bars $= 12.0$ in.^{2}; or eight No. 11 bars $= 12.5$ in.^{2} Before choosing one of these, look at Fig. 6-3(d). With a cover of $1\frac{1}{2}$ in. outside of the ties when computing A_g, the size of the core will be 18 in. Next, look at Table 11 in the Appendix. For square tied columns with a core dimension of 18 in., it is found that only sixteen No. 7 bars should be used, but any of the other three arrangements will be satisfactory as far as spacing of the bars is concerned. The choice will depend upon which size is available, which size is to be used in all columns if so ordered, and which scheme will make the most practicable construction. The eight No. 11 bars will provide the most simple arrangement for the ties, as shown in Fig. 6-3(d). These and the No. 4 ties will be adopted. The cover will be increased to 2 in. for fire protection, thus making the over-all size of the column 22 in. square.

The No. 4 ties can be used at the following spacing: $16 \times 1\frac{7}{16} = 23$ in.; $48 \times \frac{1}{2} = 24$ in.; or minimum width $= 22$ in. They will be spaced at 20 in. in order to be a bit conservative.

Figure 13A in the Appendix has been prepared to enable one to check or estimate column sizes quickly. To check the safe load for this column, use the central graph for 3,000-lb concrete. From the top, for a side of a square equal to 21 in., project down to a point midway between the diagonal lines for $p_g = 0.02$ and 0.03, then project to the left to find P. The procedure may be reversed by starting with the desired load, then finding the needed length of the side. If the value of f_s is to be larger than 16,000 psi, the actual column will be stronger or the side dimension found can be reduced slightly.

6-4. Design of Spirally Reinforced Columns.   Columns with spiral reinforcement as illustrated in Fig. 6-4(a) have already been referred to. This article is for the purpose of discussing methods of analysis and design of such members by WSD. Since the concrete and the longitudinal bars are supported more adequately by spirals than they would be by ties, spirally reinforced concrete columns are allowed to carry theoretically larger loads than can be safely applied to tied columns having the same amount of concrete and the same longitudinal bars.

The formula given by the Code for the safe load on a short spirally reinforced column for WSD is[1]

$$P = A_g(0.25f_c' + f_s p_g)$$
$$= 0.25A_g f_c' + A_s f_s \tag{6-7}$$

[1] The corresponding AASHO formula for WSD only for short spirally reinforced columns is $P_s = 0.225f_c' A_g + A_s f_s$, where $f_s = 13,200$ psi for structural grade bars and 16,000 psi for intermediate grade.

It will be noticed that Eq. (6-6) for tied columns is 0.85 times what is given by Eq. (6-7). As for other columns, f_s is not to exceed 40 per cent of the minimum specified value of the yield strength f_y, and never more than 30,000 psi. Again A_g is the gross effective area of the column, disregarding the area taken up by the longitudinal bars. The Code does not limit A_g to $1\frac{1}{2}$ in. outside of the spirals but some engineers prefer to do so as a matter of conservatism and safety.

The Code states that p_g should not be less than 0.01 nor more than 0.08, with a minimum number of bars equal to six and a minimum size of bar equal to No. 5. As stated for tied columns, the specified upper limit of p_g seems to be unreasonably large for A305 bars because of the difficulty of arranging the steel, splicing the bars, and placing the concrete.

An expression for the design of spirals,[1] as given by the Code for working-stress design, is

$$p_s = \frac{0.45(A_g/A_c - 1)f_c'}{f_y} \qquad (6\text{-}8)$$

where p_s is the ratio of the volume of the spiral reinforcement to the total volume of the core, and A_c is the cross-sectional area of the core. The yield strength f_y should not exceed 60,000 psi. As a general guide, one may use $f_y = 36,000$ to 40,000 psi for ordinary structural and intermediate grade bars, 50,000 psi for hard grade, and 60,000 psi for high-strength bars and cold-drawn wire. However, the designer should use whatever value

[1] The formula is based on tests showing the strengthening effect that spirals have on the enclosed concrete. The design criteria were set so that the spiral would give the core the same strength (or more) that the gross section has without the spiral. Supposedly, the spiral will not be highly stressed until the concrete spalls off.

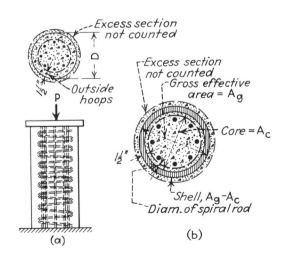

Fig. **6-4** Spirally reinforced column.

of f_y is specified by the manufacturer as long as the 60,000 psi maximum is retained.   Notice that in Eq. (6-8), p_s is the *volume* of the spiral.

The spirals should be made continuous and with even spacing except for at least $1\frac{1}{2}$ extra turns at the ends.   They should be held in place firmly and kept in line by vertical spacers, as pictured in Fig. 6-5.   At least two spacers should be used for spirals 20 in. or less in diameter, three spacers for sizes from 20 to 30 in., and four spacers for larger sizes and for spiral bars $\frac{5}{8}$ in. or larger in size.   The spirals are generally prefabricated. They should therefore be stiff and of a size and length that can be handled satisfactorily in the field.

The minimum size of spiral bar should be $\frac{1}{4}$ in. for rolled bars or No. 4 A. S. & W. gauge for drawn wire.   Splicing of these bars should be done by welding or by lapping $1\frac{1}{2}$ turns.

The spacing of spiral bars should not exceed one-sixth of the core diameter or about 3 in. clear spacing.   The minimum clear spacing formerly specified by the Code is $1\frac{3}{8}$ in. or $1\frac{1}{2}$ times the maximum size of coarse aggregate used.

The spiral should extend from the floor level in any story or the top of the footing or pedestal at the bottom to the lowest horizontal reinforcement in slab, drop panel, or beam at the top.   If a conical capital is used, the top of the spiral should extend at least to a level where the diameter of the capital is twice that of the column.

When partitions, block walls, or beams frame into spirally reinforced columns, especially those having conical capitals, it is often advisable to use a square shaft for the column and a frustum of a pyramid for the capital, with the spiral and column designed as though the section were circular.

Figure 14 in the Appendix has been prepared to give a graphical solution for the spiral bars of particular sizes in accordance with Eq. (6-8).

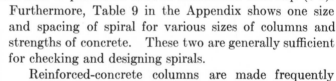

Furthermore, Table 9 in the Appendix shows one size and spacing of spiral for various sizes of columns and strengths of concrete.   These two are generally sufficient for checking and designing spirals.

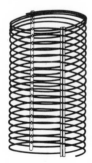

Reinforced-concrete columns are made frequently with light longitudinal bars but with strong spiral reinforcement.   Within reasonable limits, this is good construction, provided enough longitudinal bars are used to hold the spiral sufficiently to prevent it from collapsing vertically downward or bending sidewise during the deposition of the concrete, and provided that the column is not subjected to severe bending moments.

Fig. 6-5
A fabricated
spiral.

When one considers that the corrugations on a vertical deformed bar in a column will tend to produce a wedging action that is similar to that described in connection with Fig. 3-1, it would seem that the portion of a column where the bars are spliced by bond on overlapped lengths will be a critical section. In such a region the spirals are especially helpful compared with ties because they tend to keep the concrete from bursting; whereas the ties provide restraint at just a few localized points. Even at that, closer spacing of the ties at the laps of successive column bars or of bars and dowels might be helpful.

Example 6-4. Assume that one is designing short, circular spirally reinforced concrete columns for a large basement parking garage under a big commercial building. It is desired that all isolated columns be of the same size but with the longitudinal bars varying as required by the loads. Assume that the heaviest loads are 360 kips dead load and 270 kips live load; the lightest loads are 310 kips dead load and 240 kips live load. Assume $f'_c = 4,000$ psi, $f_s = 16,000$ psi, and $f_y = 40,000$ psi. Design these columns by WSD methods and Eq. (6-7). Let A_g be the area of the cross section with a cover of $1\frac{1}{2}$ in. outside of the spirals. Call the maximum $p_g = 0.04$.

The maximum $P = 360 + 270 = 630$ k. From Fig. 13*B* of the Appendix, it seems that a diameter of 23 in. may be suitable. Then, from Eq. (6-7),

$$P = \pi \times 11.5^2(0.25 \times 4,000 + 16,000 \times 0.04)$$
$$= 415(1,000 + 640) = 415,000 + 265,000 = 680,000 \text{ lb}$$

This is too large. Therefore, try a diameter of 22 in.

$$P = \pi \times 11^2(1,000 + 640) = 380(1,000 + 640) = 380,000$$
$$+ 243,000 = 623,000 \text{ lb.}$$

This is close enough.

$$A_s = 380 \times 0.04 = 15.2 \text{ in.}^2$$

Using $1\frac{1}{2}$ in. for cover as a trial, the core diameter will be $22 - 2 \times 1\frac{1}{2} = 19$ in. From Table 11 in the Appendix, twelve No. 10 bars can be used. This gives $A_s = 15.2$ in.^{2}, which is just right.

From Eq. (6-8),

$$p_s = \frac{0.45(A_g/A_c - 1)f'_c}{f_y}$$

The core area $A_c = \pi \times 9.5^2 = 283$ in.^{2} Then

$$p_s = \frac{0.45(380/283 - 1)4,000}{40,000} = 0.0153$$

Assuming a c.c. spacing of spirals equal to 3 in., the volume of concrete per spiral loop in the core is $A_c \times 3 = 283 \times 3 = 849$ in.^{3} The volume of the spiral bar for one loop is 849 × 0.0153 = 13 in.[3]   The cross-sectional area of the spiral bar is therefore approximately 13/(π × 18.5) = 0.223 in.[2]   This area is too much for No. 4 bars, and No. 5 at this spacing will be wasteful.   Therefore, try $2\frac{1}{2}$ in. for the spacing of the spiral.   Then A_c × 2.5 = 283 × 2.5 = 708 in.[3]   The volume of one loop will be 708 × 0.0153 = 10.8 in.[3]   The required cross-sectional area of the spiral bar is 10.8/(π × 18.5) = 0.186 in.[2]   Therefore, No. 4 spirals can be used. Table 9 in the Appendix also shows that No. 4 bars at $2\frac{1}{2}$ in. c.c. will be satisfactory.

The minimum P = 310 + 240 = 550 kips.   Keeping the 22-in. diameter, the second form of Eq. (6-7) would indicate that

$$P = 550,000 = 0.25 \times 380 \times 4,000 + A_s \times 16,000$$
$$A_s = (550,000 - 380,000)/16,000 = 10.6 \text{ in.}^2$$

Therefore, using the same No. 10 bars as before, nine will be required, giving A_s = 11.4 in.[2]   This will be accepted, as will the same No. 4 spirals at $2\frac{1}{2}$-in. spacing.

6-5.   Composite Columns.   The term *composite column* is used to denote a structural-steel column—or a cast-iron one—which is thoroughly encased in concrete that is reinforced with longitudinal bars and spirals. Such members may be encountered in the construction of large buildings. The cross section of the concrete is generally large, and it can be relied upon to assist the structural steel in resisting the applied load if properly detailed and built.

This classification should not include the ordinary H column which is encased in the minimum amount of concrete that can be used for the purpose of fire protection, as pictured in Fig. 6-7.   Therefore, composite columns will be considered to be those in which the strength of the reinforced-concrete portion is really substantial.   One such column is shown in Fig. 6-6, with longitudinal bars and spirals.

The Code limits the area of the metal core of composite columns to 20 per cent of the gross area of the column.   Those like Fig. 6-7 will be

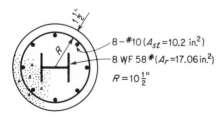

$8-\#10\,(A_{st}=10.2\text{ in.}^2)$

$8\text{ WF }58\,\#\,(A_r=17.06\text{ in.}^2)$

$R=10\frac{1}{2}{}''$

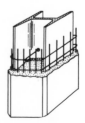

Fig. 6-6   **Fig. 6-7**

called combination columns, and they will be discussed in the next article. One possible use of the metal core is for the purpose of reducing the size of a normal concrete column to carry the same load. In the author's experience, cast-iron cores have seldom been used recently. In any composite column, a hollow metal core should be filled with concrete.

The transfer of load to the metal core should be positively provided for by the use of brackets, framed connections, bearing billets, or other reliable connections. These should be provided at the top of the metal core, at intermediate points where loads are applied, and at the bottom of the core. The metal core must also be designed to support safely whatever construction or other loads are to be applied to it prior to its encasement in concrete.

The amount of longitudinal reinforcement, spacing of bars, details of splices, thickness of cover, and design of spirals shall conform to the rules specified for a spirally reinforced column of the same over-all dimensions. The Code requires a clearance of at least 3 in. between the metal core and the spiral except when the core consists of a structural-steel H column, in which case the minimum clearance may be reduced to 2 in.

It is easy to see that the load-carrying capacity of a composite column is made up of three parts, viz., the strength of the structural steel or other core, that of the longitudinal bars, and that of the concrete. Therefore, the ultimate strength of the column, if there were no complications from buckling or deformation, might be assumed to be

$$P' = f'_c A_c + f_y A_s + f_y A_r \qquad (6\text{-}9)$$

where A_c is the net cross-sectional area of the concrete in square inches, and A_s and A_r equal the cross-sectional areas in square inches of the longitudinal bars and the core, respectively. However, USD is not permissible for such columns.

The Code has specified a WSD formula for the safe load-carrying capacity of a composite column. It is

$$P = 0.225 A_g f'_c + f_s A_{st} + f_r A_r \qquad (6\text{-}10)$$

where A_g = actual net cross-sectional area of the concrete

A_{st} = total cross-sectional area of the longitudinal reinforcement

A_r = area of the cross section of the metal core

f_r = allowable unit stress which may be applied to the metal core

The value of f_r should not exceed 18,000 psi for ASTM A36 steel, 16,000 psi for ASTM A7 steel, and 10,000 psi for cast iron. Equation (6-10) is to apply to any point of the composite column. If any loads are applied to the concrete portion between the structural brackets or other devices attached to the core, such loads shall not cause in the concrete area A_g a unit stress exceeding $0.35f'_c$.

Equation (6-10) is empirical and is used as a limiting formula.   It allows the same unit stress in the concrete as it does in a regular spirally reinforced column.   It does not enable one to compute the unit stresses in a column that is not loaded to its capacity, but the important question is whether or not the column will be safe for the required maximum load.

Occasionally, a reinforced-concrete structure may have a few structural steel beams or girders incorporated in it.   The cores of composite columns may then be useful in permitting the erection of the steelwork in advance by framing the steel girders into them, then following up with the concrete work later.

Example 6-5.   Compute the safe load on the short composite column shown in Fig. 6-6 if $f_c' = 4{,}000$ psi, $f_s = f_r = 18{,}000$ psi, and Eq. (6-10) is used.

$$A_{st} = 10.2 \text{ in.}^2 \qquad A_r = 17.06 \text{ in.}^2 \qquad A_g = \pi \times 12^2 - 10.2$$
$$- 17.06 = 424.74 \text{ in.}^2$$
$$P = 0.225 \times 424.74 \times 4{,}000 + 18{,}000 \times 10.2 + 18{,}000 \times 17.06$$
$$= 382{,}000 + 184{,}000 + 307{,}000 = 873{,}000 \text{ lb}$$

From the above figures, one can see the relative amounts of the load that are carried by the different parts of the composite column.   The concrete supports the largest portion.

6-6.   Combination Columns.   The name *combination column* refers to one which consists of a structural steel column which is encased in concrete at least $2\frac{1}{2}$ in. over all the metal except rivet heads, bolt heads, and nuts.   The difference between these and composite columns may be confusing, but the metal core in the latter is not to exceed 20 per cent of the total cross-sectional area.   However, the structural-steel member in a combination column is generally heavy; the strength of the concrete is small in comparison with that of the steel; little longitudinal reinforcement is used; the encasement is not restrained as it would be with spirals; the steel column may be subjected to a large compressive stress; considerable dead load may be on the H column before it is encased, and there may be considerable shrinkage of the concrete after placement.

A typical combination column is shown in Fig. 6-7.   Notice that the longitudinal reinforcement is negligible, whereas in a composite column it is substantial.

The Code states that the allowable load on a combination column shall be computed by the following WSD formula:

$$P = A_r f_r' \left(1 + \frac{A_g}{100 A_r} \right) \tag{6-11}$$

where A_r = cross-sectional area of the steel column

f'_r = allowable unit stress on the unencased steel column

A_g = total cross-sectional area of the concrete encasement

The concrete is to have f'_c equal to at least 2,500 psi. The concrete is to be reinforced by the equivalent of No. 10 A. S. and W. gauge welded wire mesh wrapped completely around the column and having the wires around the column not over 4 in. c.c. and those parallel to the column's axis not over 8 in. c.c. The wires should have a cover of 1 in. of concrete, and they should be lapped at least 40 diameters, and wired at the splices. Again, USD is not to be used for these columns.

A steel column with a cover of metal lath and plaster or Gunite is not to be classed as a combination column, but merely as one that has a fire-resistant cover.

Naturally, the steel column must be designed to support safely any construction or other loads applied to it prior to placing of the concrete encasement. Special brackets, framed connections, or other suitable structural materials are to be used to receive all of the loads applied at each floor level or elsewhere.

Example 6-6. Compute the safe load of a column like that of Fig. 6-7 if the steel section is a 14WF202-lb H section and the concrete is 22 in. square with 1-in. chamfers. Assume that the allowable unit stress in the bare structural-steel section is 18,000 psi.

$$A_r = 59.39 \text{ in.}^2 \qquad A_c = 22^2 - 4 \times \tfrac{1}{2} - 59.39 = 423 \text{ in.}^2$$

Equation (6-11) gives

$$P = 59.39 \times 18,000 \left(1 + \frac{423}{100 \times 59.39} \right) = 1,145,000 \text{ lb}$$

In this case, the area of the H is actually less than 20 per cent of the entire cross-sectional area, but the member is still classed as a combination column. The H alone would support $59.39 \times 18,000 = 1,070,000$ lb. Therefore, the concrete adds only $(1,145 - 1,070)100/1,070 = 7$ per cent to the strength of the unencased member.

6-7. Pipe Columns Filled with Concrete. Pipes may sometimes be used as columns; if filled with concrete, they may be even better for that purpose. However, except as end-bearing piles, their use is rather restricted. In effect, the empirical formula given by the Code for the calculation of the allowable load on such columns is

$$P = 0.25 f'_c \left(\frac{1 - 0.000025 h^2}{K_c^2} \right) A_c + f'_r A_r \qquad (6\text{-}12)$$

where h = unsupported length of column

K_c = radius of gyration of concrete area inside pipe

A_c = cross-sectional area of this concrete

A_r = cross-sectional area of pipe

and

$$f'_r = 17,000 - \frac{0.485h^2}{K_s^2} \tag{6-13}$$

when the pipe has a yield strength of at least 33,000 psi and an h/K_s ratio equal to or less than 120. The symbol K_s stands for the radius of gyration of the metal pipe itself. Again it is desirable to use a concrete with f'_c = 2,500 psi or more.

Example 6-7. Assume a 12-in. ID steel pipe $\frac{3}{8}$ in. thick with a height of 18 ft. Let f'_c = 3,000 psi and f_y = 40,000 psi.

$$A_r = 14.6 \text{ in.}^2$$

$$K_s = \frac{\sqrt{d^2 + d_1^2}}{4} = \frac{\sqrt{12.75^2 + 12^2}}{4} = 4.38 \text{ in.}$$

$$A_c = 113.1 \text{ in.}^2$$

and $\quad K_c = \sqrt{I_c/A_c} = \frac{R}{2} = 3 \text{ in.}$

Then $\quad \dfrac{h}{K_s} = \dfrac{18 \times 12}{4.38} = 49.2 \quad$ and $\quad \dfrac{h}{K_c} = \dfrac{18 \times 12}{3} = 72$

Therefore, from Eq. (6-13),

$$f'_r = 17,000 - \frac{0.485(18 \times 12)^2}{4.38^2} = 15,800 \text{ psi}$$

Using this in Eq. (6-12) gives

$$P = 0.25 \times 3,000(1 - 0.000025 \times 72^2)113.1 + 15,800 \times 14.6$$
$$= 750(0.87)113.1 + 231,000 = 74,000 + 231,000 = 305,000 \text{ lb}$$

6-8. Long Columns. Reinforced-concrete columns are generally wide (or thick) compared with their unbraced length h. The effective length h' has already been discussed in Art. 6-2.

The Code contains some requirements for computing the safe load on long columns where compression governs the design, but much depends upon the kind of bending to which the column is subjected along with the axial load. Referring to Fig. 6-8, the Code applies as follows:

1. When there is no lateral displacement of the ends A and B in sketch (a), when those ends are fixed or restrained as they usually are, when there is no appreciable eccentricity of the loads, and when the action of the connected floor systems causes a point of inflection at or near the middle of the column, no correction or reduction of capacity is required unless

h'/r exceeds 60.   The symbol r is the minimum radius of gyration of the cross section of the column.   For rectangular columns it is sufficient to assume that $r = 0.3$ times the minimum over-all depth of the column or, when there is definite bending in some direction, the depth is to be that in the plane of the bending moment.   For circular columns, r can be assumed to be 0.25 times the diameter of the column.   For other shapes, r should be computed.   When h'/r is between 60 and 100, an analysis of the effect of additional deflections on the column should be made, or the following factor should be used:

$$R = 1.32 - \frac{0.006h}{r} \tag{6-14}$$

where R is a factor by which the calculated load and bending moment to be resisted (if any moment exists) should be divided, but R should not exceed unity.   One can also say that the safe load is that of the same column, if short, multiplied by R.   If h/r exceeds 100, a special analysis is required.

It will be seen, after a bit of trial, that most columns in ordinary construction will have h'/r less than 60.   Therefore, no reduction of the capacity computed by the preceding formulas for short columns is required.

2. If no lateral displacement of the ends occurs but the column is bent in single curvature as shown in sketch (*b*), the factor R is to be

$$R = 1.07 - \frac{0.008h}{r} \tag{6-15}$$

where R again must not exceed unity.   Such a condition might be approximated when a column rests upon a narrow footing which does not offer sufficient restraint.

3. If lateral displacement of the ends is not prevented but the ends are fixed as illustrated in sketch (*c*),

$$R = 1.07 - \frac{0.008h'}{r} \tag{6-16}$$

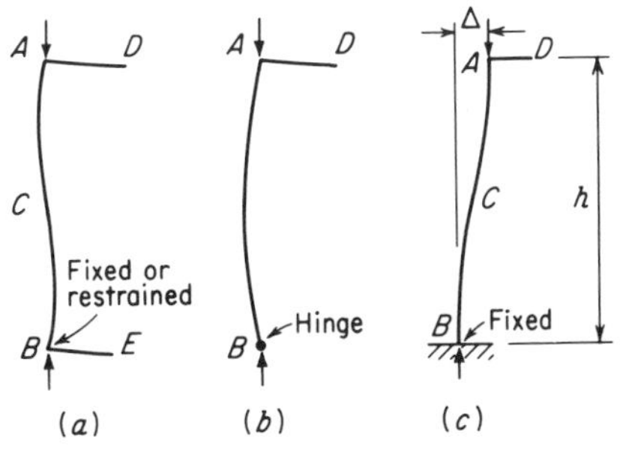

Fig. **6-8**   Deformation of columns.

where R again must not exceed unity.   However, if the lateral loads are caused by wind or earthquakes of short duration, Eq. (6-16) may be increased 10 per cent, thus making it become

$$R = 1.18 - \frac{0.009h'}{r} \qquad (6\text{-}17)$$

The formula for long columns given in AASHO is

$$P_{sl} = P_s \left(1.3 - \frac{0.03L}{d}\right) \qquad (6\text{-}18)$$

where P_s = safe load on a short column by WSD
  P_{sl} = safe load on the long column
  L = unsupported length of the column
  d = its diameter or least lateral dimension

This equation is to apply for columns having L/d greater than 10 but not exceeding 20.   Equation (6-18) is the same as that formerly specified in some of the earlier ACI Codes.

The reason for reducing the safe load for a long column compared with a short one is obviously because of the greater tendency to buckle and because of the serious effects of bending due to frame action, eccentricities, or other influences.   The first is illustrated in Fig. 6-8.   Of course, these examples show that there can be bending as well as axial load in column AB.   This combination will be considered in the next chapter.

Example 6-8.   Assume that a building column has the cross section shown in Fig. 6-9, and that the conditions conform to item 1 of Art. 6-8. The effective height h' is 25 ft.   Use WSD with $f_c' = 3,000$ psi, $f_s = 16,000$ psi, No. 8 bars for longitudinal reinforcement, and No. 4 ties 14 in. c.c. Compute the safe load on the column if the condition pictured in Fig. 6-8(a) applies.   Assume $h = h'$.

If h'/r is less than 60, Eq. (6-6) can be used; if not, compute R from Eq. (6-14).   The latter, with $r = 0.3 \times 15 = 4.5$ in. since this is a tied

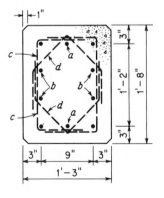

Fig. 6-9

column and $h' = 25 \times 12 = 300$ in., gives $h'/r = 67$.   Therefore, determine R first.

$$R = 1.32 - \frac{0.006h}{r} = 1.32 - 0.006 \times 67 = 0.918$$

The adjusted load for design then becomes greater than the actual load, or the safe load becomes less than it would be if the column were short. Now, applying the first form of Eq. (6-6),

$$P = 0.21f'_cA_g + 0.85A_sf_s$$
$$A_g = 15 \times 20 - 4 \times \tfrac{1}{2} = 298 \text{ in.}^2$$
$$P = 0.21 \times 3,000 \times 298 + 0.85(10 \times 0.79)16,000 = 295,000 \text{ lb}$$

This must be reduced by the factor R, and the safe load P' then becomes

$$P' = PR = 295,000 \times 0.918 = 271,000 \text{ lb}$$

If this column were the support of a portion of a highway viaduct, and if the same materials were to be used, compute the safe load in accordance with AASHO.   Since this is a tied column, the capacity as a short column would be

$$P_t = 0.8(0.225f'_cA_g + A_sf_s)$$
$$= 0.8(0.225 \times 3,000 \times 298 + 7.9 \times 16,000) = 262,000 \text{ lb}$$

From Eq. (6-18), with $L/d = 25 \times \tfrac{12}{15} = 20$,

$$P_{sl} = P_t\left(1.3 - \frac{0.03L}{d}\right) = 262,000(1.3 - 0.03 \times 20) = 183,000 \text{ lb}$$

PRACTICE PROBLEMS

Solve the following problems by means of working-stress design:

6-1   Compute the safe load for a short tied column 16 in. square with eight No. 7 bars.   The ties are No. 3 bars at 12 in. c.c. and have a cover of 2 in.   Assume $f'_c = 3,500$ psi and $f_s = 20,000$ psi.   Also assume (1) that A_g is to be based on $1\tfrac{1}{2}$ in. outside of the ties, then (2) that the entire section is included in A_g except for 1 in.² for corner chamfers.

Ans. (1) $P = 247,000$ lb.
(2) $P = 269,000$ lb.

6-2   Design a short square tied column to support a load $P = 480,000$ lb if $f'_c = 3,000$ psi, $f_s = 18,000$ psi, and $p_g = 0.02$.   Design the ties.

Ans. $D = 23$ in., with eight No. 10 bars and No. 4 ties at 16 in. c.c.

6-3   Design a short, square tied column to support a load of 600,000 lb. Let $f'_c = 4,000$ psi, $f_s = 20,000$ psi, and $p_g = 0.025$.   Design the ties and draw a sketch of the cross section.

6-4 A rectangular tied column has a length of 18 ft and a cross section 18 by 20 in., with ten No. 9 bars and No. 3 ties at 16 in. c.c. Using the entire cross section minus 2 in.² for chamfers, $f'_c = 3,000$ psi, and $f_s = 16,000$ psi, compute the allowable safe load.

6-5 The architect wants to use a column that is 16 ft long, 12 in. deep, and 24 in. wide. If $f'_c = 3,000$ psi, $f_s = 16,000$ psi, and the longitudinal reinforcement is twelve No. 8 bars, is the column safe for a total load of 250,000 lb? No chamfers are to be used because the column is to be fit into a 12-in. block wall. Use the entire section. *Ans.* Yes. $P = 310,000$ lb.

6-6 Compute the safe load on a short, circular spirally reinforced column having an outside diameter of 20 in., No. 3 spiral bars with $1\frac{3}{4}$-in. spacing, 2-in. cover over the spirals, and ten No. 8 bars. Assume $f'_c = 3,000$ psi, $f_s = 18,000$ psi, and A_g is based on 2 in. outside of the spirals.

Ans. $P = 378,000$ lb.

6-7 Find the safe load on the column of Prob. 6-6 if $f'_c = 3,500$ psi, $f_s = 20,000$ psi, the reinforcing is ten No. 9 bars, and the height of the column is 20 ft.

6-8 Compute the safe load on a short circular spirally reinforced column having an outside diameter of 18 in., a height of 25 ft, and eight No. 9 bars. Assume $f'_c = 4,000$ psi and $f_s = 20,000$ psi. Design the spiral, using $f_y = 40,000$ psi. *Ans.* $P' = 380,000$ lb.

6-9 Design a short, circular spirally reinforced column to support a centrally applied load of 400,000 lb, using $f'_c = 3,000$ psi, $f_s = 16,000$ psi, and $f_y = 40,000$ psi for the spiral. Assume $p_g = 0.025$.

6-10 A short spirally reinforced column is made 18 in. square for architectural reasons. It has eight No. 10 bars and No. 3 spirals at $2\frac{1}{2}$-in. spacing. Compute the safe load if $f'_c = 3,000$ psi, $f_s = 18,000$ psi, and f_y for the spiral = 36,000 psi.

6-11 Assume the column shown in Fig. 6-3(a) with eight No. 10 bars and No. 3 ties 15 in. c.c. Let $f'_c = 3,500$ psi and $f_s = 16,000$ psi. The effective length is 20 ft. Compute the safe load by the Code for the case of Art. 6-8, No. 1.

6-12 Compute the safe load for the column in Prob. 6-11 if the AASHO specifications are used.

COMBINED BENDING AND COMPRESSION

7

7-1. Introduction. In ordinary construction, there are many cases in which members are subjected to a combination of bending moments and direct axial loads. Generally, in reinforced-concrete work, the direct load in such combinations is a compressive force. Lateral earth pressures which act upon subway and foundation walls, columns to which beams are connected eccentrically, frame action between beams and columns whereby the deflections of the former compel the latter to bend also because of the rigidity of the connections, wind loads which force the columns of a building to bend sideways—all these are ordinary causes of combined compressive and flexural stresses in the members that are affected by them.

An experienced designer does not disregard these combined stresses. He tries to visualize what forces will exist and what deformations will occur in any proposed structure. He tries to find a way of designing the members to withstand the combined forces, and to make his computations with reasonable accuracy and without undue labor.

Problems involving compression and bending generally come into one of two classes: the first includes those members which have compression upon their entire cross section; the second covers those which have compression upon part of the section and tension upon the remainder. The former case will be considered first. The analysis of the latter situation generally involves considerable calculation. Because of limited space, this volume will deal only with a few methods of analysis.[1] Its purpose is to state fundamental principles and to illustrate the action of members that are subjected to combined compression and bending.

[1] For further treatment of this subject, see C. W. Dunham, "Advanced Reinforced Concrete," Chap. 1, McGraw-Hill Book Company, New York, 1964.

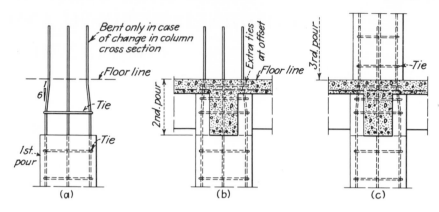

Fig. 7-1   Sequence of construction of columns and floor system.

7-2. Construction Procedure Affecting Columns. In attacking the problem of combined compression and bending, it is advisable first to analyze the probable construction conditions in so far as they may affect the design. The maximum bending moment in a column caused by frame action will occur at the top of the floor or at the bottoms of the beams that frame into the column—or at both. In the case of lateral pressure against a column without sidesway, there may be bending at the ends as well as near the middle of the member. Lateral displacement of the top with respect to the bottom will cause bending at both the top and bottom unless the former is hinged, which is seldom the case. Thus the largest bending moments are usually at the ends of columns.

However, as pictured in Fig. 7-1, in ordinary building construction it is customary to place the concrete in the column forms after the lower floor or footing has been constructed. Then the concrete of the column is deposited up to, or nearly up to, the bottoms of the beams of the next higher floor. The concrete of the beams and the floor is poured next; then that of the column in the next story is poured. In any case, each stopping point causes a plane of comparative weakness because the concrete of one "pour" is permitted to set before that of the following one is deposited. The bond between the two pours generally is much weaker than the bond between portions of a monolithic mass of concrete. Such a plane of division between successive pours is called a construction joint. Figure 7-1 shows that these joints generally occur at the points of maximum bending in the columns.

A construction joint may be roughened or keyed to lock the two sections together to resist shear when necessary. The full compressive strength can be relied upon, but the concrete at a construction joint is much weaker than usual in its resistance to tension. In fact, it seems reasonable to assume that the concrete of columns will not resist tension at the construction joints or at any cross section.

However, if one examines the conditions in the region where the dowels or lower bars lap the upper bars—as in the third pour indicated in Fig. 7-1(c)—he will realize that this a a point of relative weakness even though it may be a point of maximum moment. Some of the weakness may be caused by displacement of concrete by the laps; some, by the transfer of bond stresses; and some, by bond slippage.

7-3. Combined Compression and Bending without Resulting Tension upon the Section. Assume at first that the materials are really elastic and that working-stress methods are to be used in order to get a better understanding of the combined action. Let $ABCD$ of Fig. 7-2 represent a short piece of a square tied column with a symmetrical cross section. Let P equal the vertical load and M equal the bending moment, whatever its cause may be. Then $CEFD$ represents the pressure or stress diagram for the vertical load if P is applied at the center, the pressure is uniformly distributed, and the unit stress equals P/A_t, where A_t is the transformed area in terms of concrete. For the present, assume

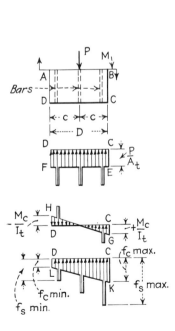

Fig. 7-2

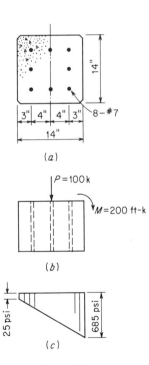

Fig. 7-3

that $A_t = A_g + (n - 1)A_s$ although the Code allows the use of the factor $2n$ instead of n for bars in compression because of the effect of plastic flow of the concrete when under steady pressure. Assuming linear distribution of the stresses, the figure $CDHG$ pictures the diagram for the internal stresses which resist the bending moment M, where $I_t = I_c + 2nI_s$, neglecting the area of concrete displaced by the bars. Then $CKLD$ is the diagram for the combined stresses. Of course there is no true neutral axis or point of zero stress within the section of the column.

It is apparent that a tensile stress like HD in Fig. 7-2 annuls an equal compressive stress, but the effect is unchanged as far as bending is concerned. Furthermore, the reinforcement is assumed to be conveniently symmetrical in this case. Therefore, from Fig. 7-2,

$$CK = \max f_c = \frac{P}{A_t} + \frac{Mc}{I_t} \tag{7-1}$$

where $c = D/2$—the distance from the center of gravity axis. This equation is in a general form and is primarily useful for making a rough check of a column section, and it gives results that are more conservative than is usually necessary. However, to illustrate its use, assume the column shown in Fig. 7-3(a), with $f_c' = 3,000$ psi, max $f_c = 900$ psi, and $n = 9$. Use P and M as given in (b).

$$A_t = 14^2 + 18 \times 8 \times 0.6 = 282 \text{ in.}^2$$

$$I_t = \frac{14 \times 14^3}{12} + 2 \times 3 \times 18 \times 0.6 \times 4^2 = 4,240 \text{ in.}^4$$

$$\max f_c = 100,000/282 + \frac{200,000 \times 7}{4,240} = 685 \text{ psi}$$

$$\min f_c = 100,000/282 - \frac{200,000 \times 7}{4,240} = 25 \text{ psi}$$

These results are shown in Fig. 7-3(c).

In situations involving direct compression combined with bending, it is generally desirable to obtain a fairly accurate preliminary design for a member to meet a specific set of conditions and then to analyze it. This may help to avoid fruitless analyses. Of course, one could "guess and test," but this is tedious. Figures 7-4 and 7-5 have been prepared to show approximately what direct loads P and simultaneous bending moments M can be withstood safely by certain square tied columns having equal reinforcement near all four sides and by circular spirally reinforced columns, using 3,000-lb concrete. The diagrams are based upon modifications of Eq. (7-1) and upon having no tension in the columns. They may be helpful in choosing trial sizes.

In making estimates of sizes when concrete of other than 3,000-lb strength is to be used, increase the loads and moments about 15 per cent for 2,500-lb concrete and decrease the design loads by approximately 15 or 20 per cent for 3,750-lb concrete. With the adjusted value of P as an ordinate, and with the adjusted M as an abscissa, find their intersection point on the diagram that most nearly meets the contemplated type of member and percentage of reinforcement. The estimated size can be interpolated from the diagram, or a somewhat larger size can be assumed. If the intersection of the P and M values falls beyond the ends of the inclined lines in Figs. 7-4 and 7-5, it indicates that the column will probably have tension on part of its cross section, and the curves are not applicable.

7-4. Combined Compression and Bending with Resulting Tension upon the Section. In order to illustrate general principles applicable to this type of problem, and to help the reader to visualize the general character of what happens, a numerical example will be solved by working-stress methods. Some general assumptions will be the following:

1. Regardless of the existence of the direct compressive load, the intensities of the stresses acting upon a cross section of the member are assumed to vary as the ordinates to a straight line.

2. The summation of all axial loads and the internal stresses upon the section must be zero in order to have equilibrium. Hence $\Sigma V = 0$. Also, $\Sigma H = 0$.

3. The summation of the bending moments caused by the axial loads and by the internal stresses must be zero about any axis in order to have equilibrium. Thus $\Sigma M = 0$.

4. The concrete is assumed to be unable to resist tension, but the bars in the tension zone will resist with an imaginary transformed area of nA_s. This is conservative because, if the concrete actually does resist tension, the column will theoretically be stronger in resisting a bending moment.

5. For illustrative purposes, the transformed area of bars in compression will be assumed to equal $(2n)A_s'$.

It is not always easy to determine by inspection whether there will be tension upon any section of a column that is subjected to a bending moment as well as to a longitudinal load. It is possible to assume that any bending moment M combined with a longitudinal load P is the same as the load P acting with an eccentricity e. Thus $e = M/P$. It is probable that there will be a tendency to have tension near one edge of a column of width D if e exceeds $D/6$ for a rectangular tied column, and if e exceeds $D/8$ for a circular spirally reinforced column.

Now assume the column shown in cross section in Fig. 7-6(*a*) having the load and bending moment shown in (*b*). What will be the maximum f_c at B if 3,000-lb concrete is used?

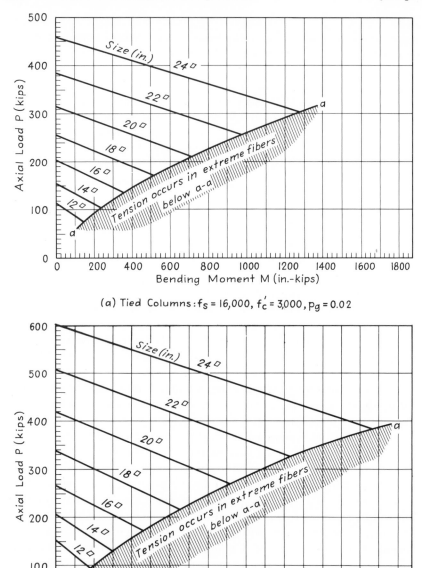

(a) Tied Columns: $f_s = 16,000$, $f_c' = 3,000$, $p_g = 0.02$

(b) Tied Columns: $f_s' = 16,000$, $f_c' = 3,000$, $p_g = 0.04$

Fig. **7-4**   Diagrams for estimating trial sizes of square tied columns with bending and compression.

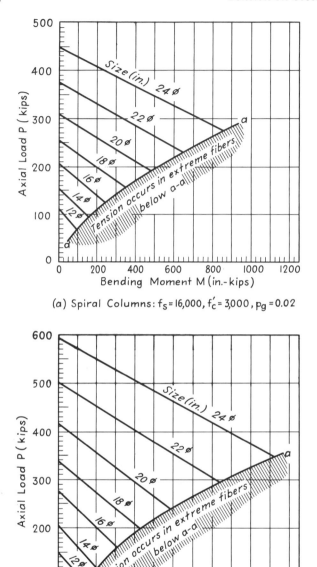

(a) Spiral Columns: f_s = 16,000, f'_c = 3,000, p_g = 0.02

(b) Spiral Columns: f_s = 16,000, f'_c = 3,000, p_g = 0.04

Fig. **7-5** Diagrams for estimating trial sizes of round spirally reinforced columns with bending and compression.

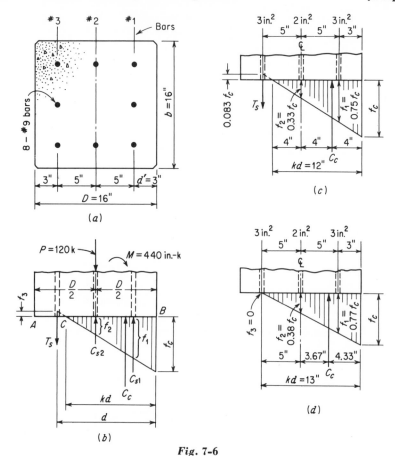

Fig. 7-6

Let Fig. 7-6(*b*) also illustrate the assumed resisting forces acting on the cross section at *AB*. The total force in the concrete will be

$$C_c = \frac{b(kd)f_c}{2} \qquad (7\text{-}2)$$

The force in the bars near *B* will be

$$C_{s1} = 2nA_{s1}f_1 \qquad (7\text{-}3)$$

The force in the bars at the center line will be

$$C_{s2} = 2nA_{s2}f_2 \qquad (7\text{-}4)$$

The force in the tensile steel (if there is any) will be

$$T_s = nA_{s3}f_3 \qquad (7\text{-}5)$$

In these equations, the various A_s's represent whatever the area of the steel at that location may be, and the various symbols f represent the assumed stress in the concrete at that particular location—that at the bars in this case. All of these forces added together should counteract the force P, or

$$P = C_c + C_{s1} + C_{s2} - T_s$$
$$= \frac{b(kd)f_c}{2} + 2nA_{s1}f_1 + 2nA_{s2}f_2 - nA_{s3}f_3 \qquad (7\text{-}6)$$

Then the summation of all of moments of these forces about the center line of the column must counteract the external bending moment Pe. Of course, the various forces and lever arms are dependent upon the values of kd and f_c. Then, for this case,

$$Pe = C_c\left(\frac{D}{2} - \frac{kd}{3}\right) + C_{s1}\left(\frac{D}{2} - d'\right) + C_{s2}\left(\frac{D}{2} - \frac{D}{2}\right) + T_s\left(d - \frac{D}{2}\right)$$
$$(7\text{-}7)$$

The next step is to estimate the value of kd and then to divide Eq. (7-7) by Eq. (7-6) to find e. If the estimate is correct, then the computed value of e should equal that found from M/P.

Now try this with the numbers from Fig. 7-6.

$$e = \frac{M}{P} = \frac{440}{120} = 3.66 \text{ in.} \qquad \frac{D}{6} = \frac{16}{6} = 2.67 \text{ in.}$$

Therefore, it looks as though there will be tension near point A. Try $kd = 12$ in. Then the lever arms from the center line of the column and the f stresses in terms of f_c are as shown in sketch (c). The substitution of these values, the transformed steel areas, $n = 9$, and $2n = 18$ in Eqs. (7-6) and (7-7), and dividing the latter by the former, gives

$$e = \frac{16 \times 12 \times f_c \times \frac{4}{2} + 18 \times 3 \times 0.75f_c \times 5 + 18 \times 2 \times 0.33f_c \times 0 + 9 \times 3 \times 0.083f_c \times 5}{16 \times 12 \times f_c/2 + 18 \times 3 \times 0.75f_c + 18 \times 2 \times 0.33f_c - 9 \times 3 \times 0.083f_c}$$

It will be noticed that f_c can be canceled out of all terms in this equation. Then the evaluation and summation of the terms give

$$e = 4.1 \text{ in.}$$

This exceeds the real value of $e = 3.67$ in., which means that the estimated value of kd was too small.

As a second try, assume $kd = 13$ in. Then $T_s = 0$, and $f_3 = 0$. Using the values in Fig. 7-6(d), and omitting f_c throughout, the new value

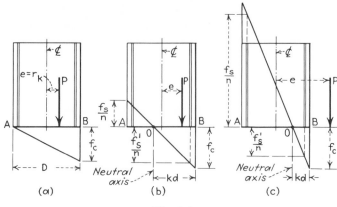

Fig. 7-7

of e from Eq. (7-7) divided by Eq. (7-6) is

$$e = \frac{16 \times 13 \times 3.67/2 + 18 \times 3 \times 0.77 \times 5 + 18 \times 2 \times 0.38 \times 0 + 0}{16 \times \frac{13}{2} + 18 \times 3 \times 0.77 + 18 \times 2 \times 0.38 - 0}$$
$$= 3.7 \text{ in.}$$

This is a lucky agreement.

Now substitute $kd = 13$ in. in Eq. (7-6) along with the values of A_s and f. Then

$$P = 210{,}000 = 16 \times 13 \times \frac{f_c}{2} + 18 \times 3 \times 0.77f_c + 18 \times 2 \times 0.38f_c$$
$$- 9 \times 3 \times 0$$

and $f_c = 753 \text{ psi}$

7-5. Approximate Method for the Design of Rectangular Columns with Tension on Part of the Section. The previously explained method of analysis of members having combined axial loads and bending moments requires considerable labor in its application. It is therefore desirable to have a reasonably accurate, approximate method of analysis which is easy to apply. Such a method will now be explained. It was originally developed and used by Frederick C. Lowy[1] and Erick M. Black.[2]

First consider the distribution of the stresses in a column supporting an eccentrically applied load. Figure 7-7(a) shows a rectangular column with a load P which has an eccentricity of sufficient magnitude to make the stress upon the cross section vary from zero at A to a maximum at B.

[1] Formerly designer, The Port of New York Authority, New York.
[2] Formerly assistant designing engineer, City Railway, Newark, N.J.

If there were no steel, e would equal $D/6$. In other words,

$$e = r_k = \frac{D}{6} = \frac{bD^2/6}{bD} = \frac{\text{section modulus}}{\text{area}}$$

where r_k will be called the *kern radius*. Similarly, for the reinforced-concrete member, let

$$r_k = \frac{\text{section modulus of the entire transformed section}}{\text{area of the entire transformed section}}$$

Let I'_c equal the moment of inertia of the complete section including $2n - 1$ times the area of the steel. The whole section must be used in this calculation. Therefore,

$$r_k = \frac{I'_c}{0.5D[A_g + (2n - 1)A_{sg}]} \qquad (7\text{-}8)$$

where A_{sg} is the total area of steel in the entire cross section. Thus, the value of r_k may be considered as the limit of the central portion, or kern (core), of the member. If the load acts within this kern, it will be assumed that there will be no tension upon the section; if it is applied at the limit of r_k, the stress at A will be zero, but at B, f_c = twice the average stress; if e exceeds r_k, the neutral axis O will move from A toward B, as shown in Fig. 7-7(b) and (c), causing tension near A and making f_c greater than twice the average stress in direct compression. However, when the magnitude of e becomes large compared with the depth of the section D, the member becomes primarily a beam, the direct stress being relatively insignificant. Usually, if $e = 2D$, the longitudinal load may be neglected; and for such a case, kd will approximate $D/3$.

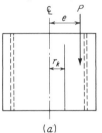

(a)

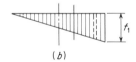

(b)

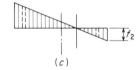

(c)

Considering the fact that r_k is about equal to $D/6$, it is seen that the kern radius comes approximately at the neutral axis of the section when the latter is acting as an ordinary beam. Therefore, it will be assumed that the stress diagram for any position of P, where e exceeds r_k, is made up of one part which is the triangle of Fig. 7-8(b), where P is assumed to be at r_k, plus a second diagram as shown in Fig. 7-8(c), the latter being caused by the load P acting, with a lever

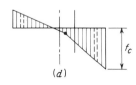

(d)

Fig. 7-8

arm of $e - r_k$, upon the section as a pure beam.   The resultant stress is shown in Fig. 7-8(d), being indicated in the diagram as bounded by a broken line merely for the reasons which will be explained later.

In the practical design of columns, it will be found that the stress in the concrete f_c is the critical one unless excessive bending causes yielding of the steel followed by crushing of the concrete.   Therefore, it is generally satisfactory to test the compression at the high-pressure side only, and it will be assumed that $f_c = f_1 + f_2$, as pictured in Fig. 7-8.

In any such problem, the procedure is simply the following:

1. Assume a section to be analyzed.

2. Find $r_k = \dfrac{I'_c}{0.5D[A_g + (2n - 1)A_{sg}]}$ \hfill (7-8)

3. Find $f_1 = \dfrac{2P}{A_g + (2n - 1)A_{sg}}$ \hfill (7-9)

4. Find $f_2 = \dfrac{P(e - r_k)kd}{I_c}$ \hfill (7-10)

5. $f_c = f_1 + f_2$ \hfill (7-11)

It should be noticed that I_c in step 4 is the moment of inertia of the transformed section of the member as an ordinary beam, allowing no tension in the concrete, using n for tensile steel and $2n$ for compressive steel.

This method still seems to require considerable work.   However, in any structure that has a large number of columns, it will be found advisable to simplify the work by using a moderate variety of sections. Therefore, the tentative members can be chosen; their areas, section moduli, and kern radii can be computed; and then the members can be used for any combination of P and M or P and Pe for which they are safe. Having the properties of the sections, the labor of finding f_1 and f_2 is relatively small.

If one wishes to compute the stresses in the steel, it is not advisable to use the results obtained from this approximate method.   That is why

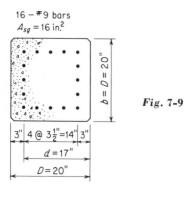

16 – #9 bars
$A_{sg} = 16$ in.²

Fig. 7-9

Fig. 7-8(d) shows a broken line so as to warn the reader not to attempt such a calculation.

For important structures it is desirable to make preliminary designs of the columns by this or another approximate method and then to check them by more exact means.

Example 7-1. Assume the column whose cross section is shown in Fig. 7-9. By the approximate method, compute the maximum f_c for the following cases if $n = 8$. Assume that 1,200 psi is a desired maximum for f_c. Carry the calculations to two significant figures.

   a. $P = 300$ kips, $M = 1,500$ in.-k, $e = 5$ in.
   b. $P = 200$ kips, $M = 1,200$ in.-k, $e = 6$ in.
   c. $P = 150$ kips, $M = 1,200$ in.-k, $e = 8$ in.
   d. $P = 200$ kips, $M = 1,400$ in.-k, $e = 7$ in.
   e. $P = 240$ kips, $M = 960$ in.-k, $e = 4$ in.

The first step is the calculation of A_t and I'_c.

$A_t = 20^2 + (16 - 1)16 = 640$ in.^{2}, neglecting the fillets

$$I'_c = \frac{bD^3}{12} + (16 - 1)I_s$$

$$= \frac{20 \times 20^3}{12} + (16 - 1)(10 \times 1 \times 7^2 + 4 \times 1 \times 3.5^2) = 21,400 \text{ in.}^4$$

Then, from Eq. (7-8),

$$r_k = \frac{I'_c}{0.5DA_t} = \frac{21,400}{10 \times 640} = 3.34 \text{ in.}$$

The next step is the calculation of I_c and S_c for the assumed cracked section as a beam, using n as the modular ratio for tensile steel and $2n - 1$ for the compression bars. The results are

$$I_c = 8,110 \text{ in.}^4 \qquad \text{and} \qquad S_c = 1,320 \text{ in.}^3$$

Case a. $c = 5$ in.

Then $$f_1 = \frac{2P}{A_t} = \frac{3 \times 300,000}{640} = 940 \text{ psi}$$

$$f_2 = \frac{P(e - r_k)}{S_c} = \frac{300,000(5 - 3.34)}{1,320} = 380 \text{ psi}$$

$$f_c = f_1 + f_2 = 940 + 380 = 1,320 \text{ psi (too high)}$$

Case b. $e = 6$ in.

$$f_1 = \frac{2 \times 200,000}{640} = 620 \text{ psi}$$

$$f_2 = \frac{200,000(6 - 3.34)}{1,320} = 400 \text{ psi}$$

$$f_c = 620 + 400 = 1,020 \text{ psi}$$

Case c. $e = 8$ in.

$$f_1 = \frac{2 \times 150,000}{640} = 470 \text{ psi}$$

$$f_2 = \frac{150,000(8 - 3.34)}{1,320} = 530 \text{ psi}$$

$$f_c = 470 + 530 = 1,000 \text{ psi}$$

Case d. $e = 7$ in.

$$f_1 = \frac{2 \times 200,000}{640} = 620 \text{ psi}$$

$$f_2 = \frac{200,000(7 - 3.34)}{1,320} = 560 \text{ psi}$$

$$f_c = 620 + 560 = 1,180 \text{ psi}$$

Case e. $e = 4$ in.

$$f_1 = \frac{2 \times 240,000}{640} = 750 \text{ psi}$$

$$f_2 = \frac{240,000(4 - 3.34)}{1,320} = 120 \text{ psi}$$

$$f_c = 750 + 120 = 870 \text{ psi}$$

These calculations indicate the ease with which any given column section can be tested for the approximate maximum stress when it is subjected to various combinations of axial load and bending.   With the maximum f_c limited to 1,200 psi, one can see at a glance that case *a* is slightly overstressed, case *e* is very lowly stressed, and the others seem to be reasonable.

7-6.   Code Formulas for Working-stress Design.   The strength of a member may be controlled by compression or by tension.   If the former, it means that, if the member is loaded to failure, the concrete in part or all of the section will fail before the stress in the steel becomes critical, although buckling of the bars in a tied column may be an initial cause of the failure.   If tension controls, it means that buckling or other bending action of the column causes the bars on one side to stretch so much that the concentration of compression on the other side causes the concrete to give way.   With ordinary columns, it is generally compression which limits the capacity.

Using the Code symbols, let N = any axial load normal to the cross section of the column but acting with some eccentricity that is real or determined by M/N.   Let N_b = the magnitude of N below which the permissible eccentricity about either principal axis is controlled by the tensile strength of the member and above which the permissible eccentricity about either principal axis is controlled by the compressive strength of the member.

In order to determine whether compression or tension controls, the Code states that the maximum permissible eccentricity e_b for N_b at the dividing point may be determined as follows:

1. For symmetrical spirally reinforced columns:

$$e_b = 0.43 p_g m D_s + 0.14t \qquad (7\text{-}12)$$

where p_g = ratio of the area of all the longitudinal steel to the gross area of the concrete = A_{st}/A_g

$m = f_y/(0.85f_c')$

D_s = diameter of a circle through the centers of the longitudinal bars

$t = D$ = over-all diameter of the column

2. For symmetrical tied columns:

$$e_b = (0.67 p_g m + 0.17)d \qquad (7\text{-}13)$$

where d is the distance from the extreme compression fiber to the centroid of the tensile steel. With bars on four sides of a tied column, as shown in Fig. 7-10(a), the tensile steel is supposedly considered to be concentrated at the location of the center of gravity of the bars on the tensile side with respect to the center line. The distance d would then be measured from the extreme compression fiber to this point.

3. For unsymmetrical tied columns:

$$e_b = \frac{p'm(d - d') + 0.1d}{(p' - p)m + 0.6} \qquad (7\text{-}14)$$

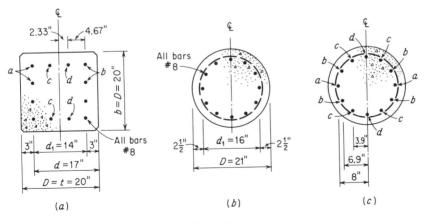

Fig. 7-10

where p = ratio of the tensile steel area to the effective area of concrete

p' = ratio of the compressive steel area to the effective area of concrete

d' = distance from the extreme compression fiber to the centroid of the compression reinforcement

Using USD philosophy, Eq. (7-12) might be derived as follows, referring to Fig. 7-12:

Assume $a = 0.72t$ for balanced design when both the tensile and compressive steel are stressed to f_y and the concrete is stressed to $0.85f_c'$. Then, with $A_s = A_s' = A_{st}/2$, the lever arm between the centroids of A_s and $A_s' = \frac{5}{8}D_s$, and $a = 0.72t$ or $0.72D$,

$$M_u = P_u e = 0.62D_s(0.5A_{st}f_y) + (0.5t - 0.36t)0.72(A_g \times 0.85f_c')$$

but $P_u = A_s'f_y - A_sf_y + 0.72(A_g \times 0.85f_c') = 0.72(A_g \times 0.85f_c')$ since $A_s'f_y$ and A_sf_y cancel out. Therefore,

$$e = \frac{M_u}{P_u} = \frac{0.31D_sA_{st}f_y + (0.14t)0.72(A_g \times 0.85f_c')}{0.72(A_g \times 0.85f_c')}$$

$$= 0.43D_sp_gm + 0.14t \qquad (7\text{-}12)$$

Similarly, Eq. (7-13) may be approximated as follows, with the lever arm between A_s and $A_s' = \frac{8}{9}d$ and a of Fig. 7-12 = $0.67d$:

$$M_u = P_u e = \frac{8}{9}d\left(\frac{A_{st}}{2}f_y\right) + (0.5d - 0.33d)0.67 \times 0.85(A_gf_c')$$

and $P_u = 0.67(A_g \times 0.85f_c')$

Then $\qquad e = \dfrac{M_u}{P_u} = \dfrac{(\frac{4}{9})dA_{st}f_y + (0.17d)0.67(A_g \times 0.85f_c')}{0.67(A_g \times 0.85f_c')}$

$$= (0.67p_gm + 0.17)d \qquad (7\text{-}13)$$

Again, Eq. (7-14) can be approximated as follows, using a of Fig. 7-12 as $0.6d$ and calling the moment resistance of the steel = $A_s'f_y(d - d')$:

$$M_u = P_u e = A_s'f_y(d - d') + (0.5d - 0.3d)0.6(A_g \times 0.85f_c')$$

and $\qquad\qquad P_u = A_s'f_y - A_sf_y + 0.6(A_g \times 0.85f_c')$

Dividing M_u by P_u and then dividing all terms by $0.85A_gf_c'$,

$$e = \frac{p'm(d - d') + 1.2d}{p'm - pm + 0.6} \qquad (7\text{-}14 \text{ approx})$$

In connection with Eq. (7-12), notice that, if there is no steel, then $e_b = t/7$, which is approximately the eccentricity which would give the stress condition shown in Fig. 7-7(a). Similarly, if there is no steel, Eq. (7-13) yields e_b = approximately $t/6$, so that there is zero stress at the tensile side.   Again, with no reinforcing, Eq. (7-14) gives $e_b = d/6$ when the stress at the tensile edge is zero.

If the resistance of the column is controlled by compression, the member should be proportioned so that

$$\frac{f_a}{F_a} + \frac{f_{bx}}{F_b} + \frac{f_{by}}{F_b} \leq 1 \qquad (7\text{-}15)$$

where f_a is the axial load divided by A_g, f_{bx} and f_{by} are components of the bending moments about the X and Y principal axes divided by the section modulus of the respective transformed uncracked sections, using $2n$ as the modular ratio for all longitudinal reinforcement, and

$$F_a = 0.34(1 + p_g m)f_c' = 0.34f_c'\left(1 + \frac{\text{force in steel at } f_y}{\text{force in concrete at } 0.85f_c'}\right) \qquad (7\text{-}16)$$

However, the computed allowable load N should not exceed the longitudinal load P permitted when the column supports only an axial load without any bending.   If there is no steel, Eq. (7-16) gives $F_a = 0.34f_c'$ as the maximum compression allowed in the concrete alone when under compression in a column.

Now call M_o equal to the bending moment which can be resisted by the member when subjected to pure flexure alone.   Call M_b equal to the bending moment when the axial load is N_b and there is an eccentricity of loading equal to e_b.   Then the allowable bending moment M in columns controlled by tension is to be assumed to vary linearly with the axial load from M_o to M_b.   The values of N_b and M_b are to be computed from the appropriate e_b and from Eq. (7-15).   The value of M_o should be calculated from one of the following:

1. For symmetrical spirally reinforced columns:

$$M_o = 0.12A_{st}f_yD_s \qquad (7\text{-}17)$$

2. For symmetrical tied columns:

$$M_o = 0.40A_sf_y(d - d') \qquad (7\text{-}18)$$

3. For unsymmetrical tied columns:

$$M_o = 0.40A_sf_yjd \qquad (7\text{-}19)$$

Of course, A_{st} = total longitudinal steel and A_s = area of tensile steel.

Where there is bending about both principal axes,

$$\frac{M_x}{M_{ox}} + \frac{M_y}{M_{oy}} \leq 1 \qquad (7\text{-}20)$$

where M_x and M_y are the bending moments about the X and Y principal axes, and M_{ox} and M_{oy} are the magnitudes of M_o for bending about these axes.

Example 7-2. Recompute the maximum safe load for the column shown in Fig. 7-10(a) for $P = 210$ kips, $M = 50$ ft-k about one principal axis, and $e = 50 \times 12/210 = 2.85$ in. Let $f_c' = 3,000$ psi, $f_y = 40,000$ psi, and $n = 9$.

From Eq. (7-13),

$$e_b = (0.67p_g m + 0.17)d$$
$$p_g = \frac{12 \times 0.79}{20^2} = 0.0237$$
$$m = \frac{40,000}{0.85 \times 3,000} = 15.7$$
$$e_b = (0.67 \times 0.0237 \times 15.7 + 0.17)17 = 7.1 \text{ in.}$$

Since this value of e_b far exceeds the actual $e = 2.85$ in., the strength of the column seems to be controlled by compression.

Then use Eq. (7-15), but first find the section modulus of the uncracked section, using $2n - 1$ for compression steel, and the axis at the center line.

$$I_c = \frac{20 \times 20^3}{12} + (2 \times 9 - 1)4 \times 0.79 \times 2 \times 7^2 + 17 \times 2 \times 0.79 \,.$$
$$\times 2 \times 2.33^2 = 18,900 \text{ in.}^4$$
$$S_c = \frac{18,900}{10} = 1,890 \text{ in.}^3$$
$$f_a = \frac{P}{A_g} = \frac{210,000}{20^2} = 525 \text{ psi}$$

From Eq. (7-16),

$$F_a = 0.34(1 + 0.0237 \times 15.7)3,000 = 1,400 \text{ psi}$$
$$f_{bx} = \frac{M}{S_c} = \frac{50,000 \times 12}{1,890} = 317 \text{ psi}$$
$$F_b = 1,350 \text{ psi for bending alone for 3,000-lb concrete}$$

Therefore, Eq. (7-15) gives, for bending about only one axis,

$$\frac{f_a}{F_a} + \frac{f_{bx}}{F_b} = \frac{525}{1,400} + \frac{317}{1,350} = 0.611$$

This is far less than unity, and the column is therefore safe.

Example 7-3. Investigate the safety of the column shown in Fig. 7-10(b) for $P = 225$ kips, $e = 6$ in., $M = 225 \times 6 = 1,350$ in.-k, $f_c' = 4,000$ psi, $f_y = 50,000$ psi, and $n = 8$.

For this case, construct an interaction diagram for the member.

$$A_g = \pi \times 10.5^2 = 347 \text{ in.}^2$$

$$p_g = \frac{12 \times 0.79}{347} = 0.0273$$

$$m = \frac{50,000}{0.85 \times 4,000} = 14.7$$

$$D_s = 16 \text{ in.} \quad \text{and} \quad t = D = 21 \text{ in.}$$

Then Eq. (7-12) gives

$$
\begin{aligned}
e_b &= 0.43 p_g m D_s + 0.14t \\
&= 0.43 \times 0.0273 \times 14.7 \times 16 + 0.14 \times 21 = 5.7 \text{ in.}
\end{aligned}
$$

This is a little less than the actual $e = 6$ in. Therefore, tension may control.

From Eq. (7-17),

$$M_o = 0.12 A_{st} f_y D_s = 0.12 \times 12 \times 0.79 \times 50,000 \times 16 = 910,000 \text{ in.-lb}$$

For a spirally reinforced column with an axial load only, the Code states that

$$
\begin{aligned}
P &= A_g(0.25 f_c' + 0.4 f_y p_g) \\
&= 347(0.25 \times 4,000 + 0.4 \times 50,000 \times 0.0273) = 536,000 \text{ lb}
\end{aligned}
$$

Plot this as point B in Fig. 7-11 and draw a horizontal line through it. This represents the largest permissible value of P or N, even when $e = 0$.

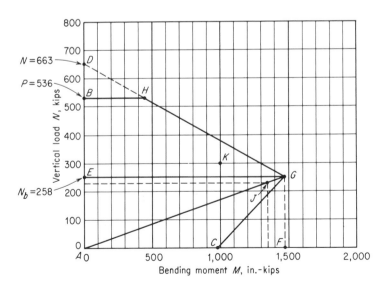

Fig. **7-11**   Interaction diagram.

Next, Eq. (7-15) becomes

$$\frac{f_a}{F_a} + \frac{f_b}{F_b} = 1 \text{ as a maximum} \tag{7-15}$$

Assuming this limit, what N goes with it?

Determine the various factors as follows, calling N the axial load:

$$f_a = \frac{N}{A_g} = \frac{N}{347}$$

$$F_a = 0.34(1 + p_g m)f'_c = 0.34(1 + 0.0273 \times 14.7)4{,}000 = 1{,}910 \text{ psi}$$

If F_a is multiplied by A_g, it gives a fictitious value of

$$N = 1{,}910 \times 347 = 663{,}000 \text{ lb}$$

Plot this as point D on the vertical axis in Fig. 7-11 for $M = 0$. This will be used later.

$$f_b = \frac{M}{S_c} = \frac{Ne}{S_c}$$

To find S_c, proceed as follows, using $2n$ for the modular ratio and using the dimensions shown in Fig. 7-10(c):

I_c: Concrete: $\pi \times 10.5^4/4 = $ 9,550 about center line
Bars a: $(2 \times 8 - 1)2 \times 0.79 \times 8^2 = $ 1,515
Bars b: $15 \times 4 \times 0.79 \times 6.9^2 = $ 2,255
Bars c: $15 \times 4 \times 0.79 \times 3.9^2 = $ 720
Bars d: $= $ 0
                                        $\Sigma = $ 14,040 in.^{4}

$$S_c = \frac{14{,}040}{10.5} = 1{,}340 \text{ in.}^3$$

Therefore, $f_b = \dfrac{Ne}{1{,}340}$

$$F_b = 0.45 \times 4{,}000 = 1{,}800 \text{ psi}$$

Equation (7-15) may be rewritten as

$$\frac{N/A_g}{F_a} + \frac{Ne/S_c}{F_b} = 1 \tag{7-21}$$

Substituting the computed values in Eq. (7-21), find

$$\frac{N/347}{1{,}910} + \frac{Ne/1{,}340}{1{,}800} = 1$$

$$\frac{N}{347 \times 1{,}910} + \frac{Ne}{1{,}340 \times 1{,}800} = 1$$

or $N = \dfrac{1}{1/(347 \times 1{,}910) + e/(1{,}340 \times 1{,}800)} = 1$

By definition, e_b from Eq. (7-21) is supposed to be the eccentricity at the dividing point between the value of N controlled by compression and that controlled by tension. Then substitute $e = e_b = 5.7$ in. in this equation for N and find N_b as follows:

$$N_b = \frac{1}{1/(347 \times 1,910) + 5.7/(1,340 \times 1,800)} = 258,000 \text{ lb}$$

Plot this as point E in Fig. 7-11, and draw a horizontal line through it. At $e_b = 5.7$ in. therefore, $M = N_b e_b = 258,000 \times 5.7 = 1,470,000$ in.-lb. Plot this as point F on the horizontal axis of Fig. 7-11. Then project vertically from F to the horizontal line through E, finding point G. Now draw line DG to locate point H where DG intersects the horizontal line through B. Connect C and G; also, A and G. Values of N and M which intersect inside the area $ABHGA$ of the interaction diagram will indicate control of resistance by compression; those in area $AGCA$, control by tension. An intersection falling outside these areas will indicate overloading.

For this case, with $N = 225,000$ lb and $M = 1,350,000$ in.-lb, plot the intersection J in Fig. 7-11. This intersection is just about at the border line between control by compression and control by tension. At any rate, the column is safe, but by a small margin.

Note particularly that this interaction diagram applies for this member alone. Assume that P were changed to 300 kips and M to 1,000 in.-k. The intersection K shows that it is safe and controlled by compression.

7-7. Analysis of Columns by Whitney's[1] Ultimate-strength Method.

Figure 7-12 shows a rectangular member that is subjected to a longitudinal load P with an eccentricity e which causes compression and bending in it. Assume a symmetrical concrete section with bars at two opposite sides. Assume A'_s and A_s are not necessarily equal.

Assume that the resistance of the member under flexural action is controlled by compression. Let Eq. (2-23) for beams, with the safety provision ϕ included, represent the moment resistance of the concrete alone about the tensile steel. Then, from Fig. 7-12, the moment resistance of the compressive reinforcement about the tensile steel will be $\phi A'_s f_y d_1$. The total resisting moment is therefore

$$M_u = \phi(A'_s f_y d_1 + \tfrac{1}{3} b d^2 f'_c) \tag{7-22}$$

Taking moments of P_u about the tensile reinforcement, Fig. 7-12 gives

$$M_u = P_u e' = P_u \left[e + \left(d - \frac{D}{2} \right) \right] \tag{7-23}$$

$$P_u \left(e + d - \frac{D}{2} \right) = \phi(A'_s f_y d_1 + \tfrac{1}{3} b d^2 f'_c) \tag{7-24}$$

[1] Charles S. Whitney, Plastic Theory of Reinforced Concrete Design, *Trans. ASCE*, Vol. 68, 1942.

If the cover over the bars d' in Fig. 7-12 is the same for both sides, as it usually will be in a column,

$$d = \frac{d_1 + D}{2} \tag{7-25}$$

Substituting Eq. (7-25) in the left side of Eq. (7-24) and solving for P_u gives

$$P_u = \phi \left[\frac{2A_s'f_yd_1}{2e + d_1} + \frac{2bd^2f_c'}{3(2e + d_1)} \right]$$

Then, dividing the last term by d^2 and multiplying by D,

$$P_u = \phi \left[\frac{A_s'f_yd_1}{e + d_1/2} + \frac{bDf_c'}{3eD/d^2 + (6Dd - 3D^2)/2d^2} \right] \tag{7-26}$$

Equation (7-26) applies when the eccentricities are large. For example, a qualitative picture of the situation may be obtained by referring to Fig. 7-13. Sketch (a) shows a case where e exceeds $\bar{x}$, the distance from the center of the member to the resultant of the assumed compressive resistances ΣC. Here T must be a tensile force in order to stop clockwise rotation. In sketch (b), e is assumed to equal $\bar{x}$. Then T will be zero for equilibrium. In (c), e is less than $\bar{x}$; hence a compressive force, pictured as ΣC_b, is needed to stop counterclockwise rotation. Since Eq. (7-26) is based upon compression in the right-hand side of the section in Fig. 7-12 and tension in the steel in the other side, it cannot apply to the condition in Fig. 7-13(b) and (c).

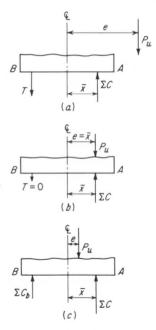

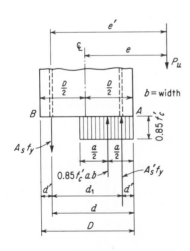

Fig. 7-12

Fig. 7-13

For spirally reinforced columns, $\phi = 0.75$; for tied columns, $\phi = 0.70$. The difference in these two values reflects the greater strength and reliability of spiral reinforcing over that of ties in its effect on the resistance of the concrete and bars.

For small eccentricities, Whitney modified Eq. (7-26) to make it reasonably applicable, and the results are reported to agree fairly well with tests. His procedure is the following:

1. Modify Eq. (7-26) so that, as e approaches zero, P_u approaches the proper value for the member as an axially loaded column.

2. If $A_s = A_s'$, as for a truly symmetrical member, the resistance of the total steel becomes $2A_s'f_y$.

3. The total strength of the concrete, using $0.85f_c'$ as a maximum, and not deducting the area occupied by the bars, becomes $0.85bDf_c'$ when $e = 0$.

4. With $e = 0$, $3eD/d^2$ in Eq. (7-26) $= 0$ and, to obtain the coefficient 0.85 for the last term of Eq. (7-26), the remainder of the denominator must be 1.178 in order to give $1/1.178 = 0.85$. Therefore, let

$$P_u = \phi\left(\frac{2A_s'f_y}{2e/d_1 + 1} + \frac{bDf_c'}{3De/d^2 + 1.178}\right) \qquad (7\text{-}27)$$

express the equation for P_u for eccentricities less than $\bar{x}$ of Fig. 7-13. This, of course, assumes that the ultimate load is controlled by the compressive strength.

For a round spirally reinforced column as shown in Fig. 7-10(b), substitute $0.8D$ for D. For a spirally reinforced round column and a square column with spiral reinforcement and bars arranged in a circle, assuming one-half of the steel to be effective on each side of the section, assume that $0.67d_1$ is to be substituted for d_1 in the preceding equations. Use these substitutions in Eq. (7-26), with $d = (D + d_1)/2$ and $bD = \pi R^2$. Then

$$P_u = \phi\left[\frac{A_s'f_y(0.67d_1)}{e + 0.67d_1/2} + \frac{\pi R^2 f_c'}{\dfrac{3eD}{(D + d_1)^2/4} + \dfrac{6D(D + d_1)/2 - 3D^2}{2(D + d_1)^2/4}}\right] \qquad (7\text{-}28)$$

The use of $0.67d_1$ is on the assumption that $A_s = A_s' = A_{st}/2$, and that each is concentrated at an extreme limit of $0.67d_1$.

Equations (7-26) and (7-28) are supposed to apply for cases where tension controls. However, the Code formulas to be given later are somewhat different.

Use the same substitutions in Eq. (7-27) to make it apply to round spirally reinforced columns. Let $d = (D + d_1)/2$, $2A_s' = A_{st}$,

$2e/d_1 = 2e/0.67d_1 = 3e/d_1,\ bD = \pi R^2,$

$$\frac{3De}{d^2} = \frac{3 \times 0.8De}{(0.8D + 0.67d_1)^2/4} = \frac{9.6De}{(0.8D + 0.67d_1)^2}$$

Then, Eq. (7-27) becomes

$$P_u = \phi\left[\frac{A_{st}f_y}{3e/d_1 + 1} + \frac{A_g f_c'}{9.6De/(0.8D + 0.67d_1)^2 + 1.178}\right] \quad (7\text{-}29)$$

This formula is for short circular columns with spirals and bars arranged in a circular pattern, and where the strength is controlled by compression.

How can one tell in advance whether compression or tension will control? For working-stress design, Eqs. (7-12), (7-13), and (7-14) have been specified in the Code for finding the dividing point e_b. These may still be a useful guide. If e_b as computed from the applicable formula is less than the actual eccentricity, it is probable that tension will control.

Example 7-4. The tied column shown in Fig. 7-10(a) is to support a vertical load of $P_{DL} = 120$ kips and $P_{LL} = 90$ kips. The bending moment near the ends of the column is 30 ft-k DL and 20 ft-k LL. Assume $f_c' = 3,000$ psi, $f_y = 40,000$ psi, $\phi = 0.70$, and the ultimate load must be $U = 1.5$ DL $+ 1.8$ LL. Is the column safe?

For design,

$$P_u = 1.5 \times 120 + 1.8 \times 90 = 342 \text{ kips}$$
$$M_u = 1.5 \times 30 + 1.8 \times 20 = 81 \text{ ft-k}$$
$$e = \frac{M_u}{P_u} = \frac{81 \times 12}{342} = 2.85 \text{ in.}$$

Using Eq. (7-13),

$$e_b = (0.67p_g m + 0.17)d$$
$$p_g = \frac{12 \times 0.79}{20^2} = 0.0237 \qquad m = \frac{40,000}{0.85 \times 3,000} = 15.7$$
$$e_b = (0.67 \times 0.0237 \times 15.7 + 0.17)17 = 7.15 \text{ in.}$$

Since the actual eccentricity is far less than this, compression seems to control. Therefore, use Eq. (7-27).

This is a case where the bars are spread around the periphery of the core of the column, whereas Eq. (7-27) is based on the assumption that the steel areas in compression and in tension are each concentrated in a single row as shown in Fig. 7-12. One approximation that can be made for the steel areas is the assumption that all of the bars c of Fig. 7-10(a) are partially effective at the location of bars a on the basis of the ratio of their relative distances from the center line of the member. Similarly, a portion of bars d may be considered to act at b. Thus

$$A_s = A'_s = \left(4 \text{ bars} + \frac{2 \times 0.5}{1.5}\right) 0.79 = 3.7 \text{ in.}^2$$

Equation (7-27) gives

$$P_u = \phi\left(\frac{2A'_s f_y}{2e/d_1 + 1} + \frac{bDf'_c}{3De/d^2 + 1.178}\right) \tag{7-27}$$

$$P_u = 0.7\left(\frac{2 \times 3.7 \times 40,000}{2 \times 2.85/14 + 1}\right.$$
$$\left. + \frac{20 \times 20 \times 3,000}{3 \times 20 \times 2.85/17^2 + 1.178}\right) = 620,000 \text{ lb}$$

This is considerably greater than the 342 kips required. Therefore, the column is very safe. If this is the worst loading condition, the column is much larger than it needs to be.

Example 7-5. The circular spirally reinforced column shown in Fig. 7-10(b) is to support a vertical load of $P_{DL} = 150$ kips and $P_{LL} = 75$ kips acting with an eccentricity of 5 in. The ultimate load is to be $U = 1.5 \text{ DL} + 1.8 \text{ LL}$. Assume $f'_c = 3,000$ psi, $f_y = 40,000$ psi, and $\phi = 0.75$. Is the column safe?

$$P_u = 1.5 \times 150 + 1.8 \times 75 = 360 \text{ kips}$$

Try Eq. (7-12) to see if compression or tension is likely to control.

$$e_b = 0.43 p_g m D_s + 0.14t \tag{7-12}$$

where D_s is the same as d_1.

$$p_g = \frac{9.48}{346} = 0.0274 \qquad m = \frac{40,000}{0.85 \times 3,000} = 15.7 \qquad D_s = 16 \text{ in.}$$
$$e_b = 0.43 \times 0.0274 \times 15.7 \times 16 + 0.14 \times 21 = 5.91 \text{ in.}$$

Since e_b is a little larger than $e = 5$ in., it looks as though compression may control. Therefore, try Eq. (7-29).

$$P_u = \phi\left[\frac{A_{st} f_y}{3e/d_1 + 1} + \frac{A_g f'_c}{9.6De/(0.8D + 0.67d_1)^2 + 1.178}\right] \tag{7-29}$$

$$P_u = 0.75\left[\frac{9.48 \times 40,000}{3 \times 5/16 + 1} + \frac{346 \times 3,000}{\dfrac{9.6 \times 21 \times 5}{(0.8 \times 21 + 0.67 \times 16)^2} + 1.178}\right]$$
$$= 457,000 \text{ lb}$$

This is more than the 360 kips required.

Example 7-6. Compute P_u for the short tied column shown in Fig. 7-10(a), which is the same one as that of Example 7-4. Provide for the minimum eccentricity of $0.1D$ previously stated in Art. 6-2, item 18. As in Example 7-4, $f'_c = 3,000$ psi, $f_y = 40,000$ psi, and $\phi = 0.70$.

Use Eq. (7-27), with $e = 0.10 \times 20 = 2$ in. Since there are bars along all four sides, assume $A_s = A'_s = 3.7$ in.^{2} as in Example 7-4. Then

$$P_u = \phi \left(\frac{2A'_s f_y}{2e/d_1 + 1} + \frac{bDf'_c}{3De/d^2 + 1.178} \right) \tag{7-27}$$

$$P_u = 0.70 \left(\frac{2 \times 3.7 \times 40,000}{2 \times 2/14 + 1} \right.$$

$$\left. + \frac{20 \times 20 \times 3,000}{3 \times 20 \times 2/17^2 + 1.178} \right) = 690,000 \text{ lb}$$

In Example 7-4, the eccentricity was 2.85 in. and P_u was 620,000 lb. This shows how sensitive P_u is to reduction because of bending.

Example 7-7. Compute P_u for the column of Fig. 7-10(b) for the minimum eccentricity of $0.05D$ specified in Art. 6-2, item 18. This is the same column as for Example 7-5. Let $f'_c = 3,000$ psi, $f_y = 40,000$ psi, and $\phi = 0.75$ as before. Use Eq. (7-29), with $e = 0.05 \times 21 = 1.05$ in.

$$P_u = \phi \left[\frac{A_{st} f_y}{3e/d_1 + 1} + \frac{A_g f'_c}{9.6De/(0.8D + 0.67d_1)^2 + 1.178} \right] \tag{7-29}$$

$$P_u = 0.75 \left[\frac{9.48 \times 40,000}{3 \times 1.05/16 + 1} \right.$$

$$\left. + \frac{346 \times 3,000}{9.6 \times 21 \times 1.05/(0.8 \times 21 + 0.67 \times 16)^2 + 1.178} \right]$$

$$= 770,000 \text{ lb}$$

Compare this value of P_u with the 457,000 lb found in Example 7-5 when $e = 5$ in.

7-8. Code Formulas for Ultimate-strength Design for Rectangular Sections. Inasmuch as the Code uses special symbols in the design of members having combined axial loads and bending moments when considered from the ultimate-strength standpoint, some of those to be used later are grouped here for convenience.

$a =$ depth of equivalent rectangular stress block assumed to have a uniformly distributed compressive stress of $0.85f'_c$ at ultimate strength. Also, $a = k_1c$ from the fiber of maximum strain.

$a_b =$ depth of equivalent rectangular stress block for balanced conditions, and equal to k_1c_b.

$c =$ distance from extreme compression fiber to the neutral axis (same as kd or kD often used elsewhere).

$c_b =$ distance from extreme compression fiber to the neutral axis for balanced conditions, and equal to $d(87,000)/(87,000 + f_y)$.

d'' = distance from plastic centroid to centroid of A'_s.

e' = eccentricity of axial load at end of member measured from the centroid of tensile reinforcement.

e_b = eccentricity of axial load P_b from the plastic centroid of the section.

k_1 = a factor to be assumed = 0.85 for concrete strengths f'_c up to 4,000 psi and reduced 0.05 for each 1,000 psi by which f'_c exceeds 4,000 psi.

m' = $m - 1$.

M_b = bending-moment capacity at simultaneous crushing of concrete and yielding of tensile steel at balanced conditions and equal to $P_b e_b$.

M_u = bending-moment capacity under combined axial load and bending.

p_t = A_{st}/A_g.

P_b = axial load capacity at simultaneous crushing of concrete and yielding of tensile steel at balanced conditions.

P_o = axial load capacity of actual member when loaded concentrically.

P_u = axial load capacity under combined axial load and bending.

The plastic centroid of a section is the centroid of resistance to load computed on the assumption that the concrete is uniformly stressed to $0.85f'_c$ and that the steel is uniformly stressed to f_y. If the member is symmetrical, the plastic centroid is the same as the centroid of the cross section.

It is assumed that balanced conditions exist at the ultimate strength of a member when the tensile reinforcement reaches f_y at the same time as the concrete in compression reaches its assumed ultimate strain of 0.003 in./in.

The Code states that, even though a column is supposed to be loaded axially, some eccentricity is to be provided for in the design, the amount being $e = 0.05t$ for spirally reinforced columns and $0.10t$ for tied columns about either principal axis (but not both). This will provide for slight curvatures and for small variations between the theoretical and actual point of application of the load. If the member has only a small axial compressive load, it may be designed for the maximum bending moment $P_u e$ only, disregarding the axial load. However, the member should have a capacity equal to P_b and greater than the applied compressive load.

The formulas given herein apply to short columns only, and to rectangular sections. The steel is assumed to be near and parallel to two opposite faces and parallel to the axis of bending.

When e is not greater than e_b, assume the conditions shown in Fig. 7-14(a), where compression is assumed to control the design. Then, using the reduction factor ϕ,

$$P_u = \phi(C_c + C_s - T_s)$$

or
$$P_u = \phi(0.85f'_c ba + A'_s f_y - A_s f_s) \qquad (7\text{-}30)$$

Taking moments about the tensile steel as shown in Fig. 7-14(*b*), and again including ϕ,

$$P_u e' = \phi \left[C_c \left(d - \frac{a}{2} \right) + C_s (d - d') + T_s \times 0 \right]$$

or
$$P_u e' = \phi \left[0.85 f_c' b a \left(d - \frac{a}{2} \right) + A_s' f_y (d - d') \right] \qquad (7\text{-}30a)$$

Notice that, in Eq. (7-30), the stress in A_s is not necessarily f_y.

If the design happens to be balanced, the stress in A_s will be f_y. Then Eq. (7-30) will become

$$P_b = \phi [0.85 f_c' b (k_1 c_b) + A_s' f_y - A_s f_y] \qquad (7\text{-}31)$$

where $k_1 c_b$ is substituted for a and the forces are as shown in Fig. 7-14(*c*). Also, for the balanced conditions as defined previously for a symmetrical member, the plastic centroid will be at the center line—the centroid of the section. Using this axis, and referring to Fig. 7-14(*d*), which shows the

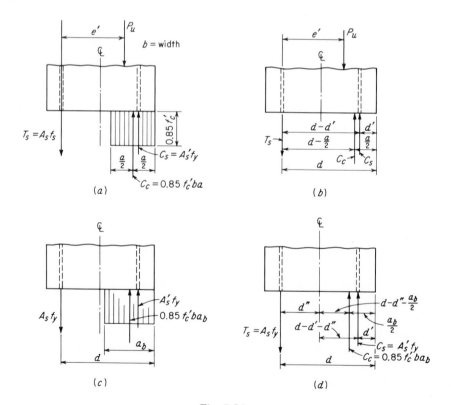

Fig. 7-14

lever arms, the balanced moment M_b becomes, from Eq. (7-30a),

$$M_b = P_b e_b = \phi \left[0.85 f'_c b a_b \left(d - d'' - \frac{a_b}{2} \right) \right.$$
$$\left. + A'_s f_y (d - d' - d'') + A_s f_y d'' \right] \quad (7\text{-}32)$$

It is obvious that, in practice, a column will seldom happen to be a balanced design. If P_u is less than P_b, or if e is greater than e_b, the ultimate capacity of the column will be controlled by tension in A_s. If P_u exceeds P_b, or if e is less than e_b, the ultimate capacity of the column will be controlled by compression in the concrete and A'_s.

When controlled by tension, and with A_s not equal to A'_s, the ultimate strength should not exceed the following:

$$P_u = \phi \left(0.85 f'_c b d \left\{ p'm' - pm + \left(1 - \frac{e'}{d} \right) \right. \right.$$
$$\left. \left. + \sqrt{(1 - e'/d)^2 + 2[(e'/d)(pm - p'm') + p'm'(1 - d'/d)]} \right\} \right) \quad (7\text{-}33)$$

If $A_s = A'_s$, Eq. (7-33) may be simplified to

$$P_u = \phi \left(0.85 f'_c b d \left\{ -p + 1 - \frac{e'}{d} \right. \right.$$
$$\left. \left. + \sqrt{(1 - e'/d)^2 + 2p[m'(1 - d'/d) + e'/d]} \right\} \right) \quad (7\text{-}34)$$

When controlled by compression, and with A_s not equal to A'_s, the ultimate load may be assumed to decrease linearly from P_o to P_b as the bending moment is increased from zero to M_b, where

$$P_o = \phi[0.85 f'_c (A_g - A_{st}) + A_{st} f_y] \quad (7\text{-}35)$$

Notice that, in this formula, the net area of the concrete is used. Based on this assumption, the ultimate strength may be given by either of the following:

$$P_u = \frac{P_o}{1 + [(P_o/P_b) - 1]e/e_b} \quad (7\text{-}36)$$

$$P_u = P_o - \frac{(P_o - P_b)M_u}{M_b} \quad (7\text{-}37)$$

With $A_s = A'_s$, and both in single layers parallel to the axis of bending, P_u may be approximated as follows:

$$P_u = \phi \left[\frac{A'_s f_y}{e/(d - d') + 0.5} + \frac{b t f'_c}{(3te/d^2) + 1.18} \right] \quad (7\text{-}38)$$

Notice that Eq. (7-38) is the same as Eq. (7-27) for all practical purposes when $d - d'$ is used for d_1, when the first term of Eq. (7-27) is

divided by 2, and when 1.18 is used instead of 1.178.   Thus the explanation given here is merely another way of arriving at the same equation as
that developed by Whitney and explained in the preceding article.

   Example 7-8.   Compute the ultimate load P_u for the column shown
in cross section in Fig. 7-15(a) if $f'_c = 3,000$ psi, $f_y = 40,000$ psi, and $e = 4$
in.   Also, compute the balanced load P_b and the moment M_b.

   From Eq. (7-38) for approximate design, with $\phi = 0.7$,

$$P_u = 0.7 \left[\frac{4 \times 40,000}{4/(17 - 3) + 0.5} + \frac{18 \times 20 \times 3,000}{(3 \times 20 \times 4)/17^2 + 1.18} \right] = 520,000 \text{ lb}$$

   From Eq. (7-31) for balanced conditions, with $\phi = 0.7$, $k_1 = 0.85$,
$A_s = A'_s$, and

$$c_b = \frac{d \times 87,000}{87,000 + f_y} \tag{7-39}$$

$$c_b = \frac{17 \times 87,000}{(87,000 + 40,000)} = 11.65 \text{ in.}$$

$$P_b = \phi[0.85f'_c b(k_1 c_b) + A'_s f_y - A_s f_y] \tag{7-31}$$

$$P_b = 0.7[0.85 \times 3,000 \times 18(0.85 \times 11.65)] = 318,000 \text{ lb}$$

Since P_u exceeds P_b, the section is controlled by compression.
   From Eq. (7-32),

$$M_b = P_b e_b = \phi \left[0.85 f'_c b a_b \left(d - d'' - \frac{a_b}{2} \right) \right.$$
$$\left. + A'_s f_y (d - d' - d'') + A_s f_y d'' \right] \tag{7-32}$$

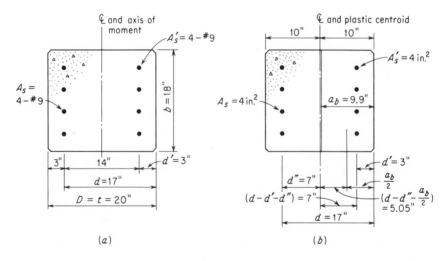

Fig. 7-15

for which the lever arms are shown in Fig. 7-15(b) and $a_b = 9.9$ in.

$$M_b = 0.7[0.85 \times 3,000 \times 18 \times 9.9 \times 5.05 + 4 \times 40,000 \times 7$$
$$+ 4 \times 40,000 \times 7] = 3,180,000 \text{ in.-lb}$$

Then
$$e_b = \frac{M_b}{P_b} = \frac{3,180,000}{318,000} = 10 \text{ in.}$$

Example 7-9. Assume the column shown in cross section in Fig. 7-15(a). Compute the ultimate load P_u if $f'_c = 3,000$ psi, $f_y = 40,000$ psi, $e = 12$ in., and $\phi = 0.7$.

In Example 7-8, e_b for balanced conditions for this column was found to be 10 in. Since e exceeds 10 in., the strength of the column is controlled by tension. Therefore, use Eq. (7-34) for $A_s = A'_s$.

$$m = \frac{f_y}{0.85f'_c} = \frac{40,000}{0.85 \times 3,000} = 15.7$$
$$m' = m - 1 = 15.7 - 1 = 14.7$$
$$e' = e + 7 = 12 + 7 = 19 \text{ in.}$$
$$p = \frac{A_s}{bd} = \frac{4}{18 \times 17} = 0.0131$$

Then $P_u = \phi(0.85f'_c bd\{-p + 1 - e'/d$
$$+ \sqrt{(1 - e'/d)^2 + 2p[m'(1 - d'/d) + e'/d]}\}) \quad (7\text{-}34)$$
$$P_u = 0.7(0.85 \times 3,000 \times 18 \times 17\{-0.0131 + 1 - 19/17$$
$$+ \sqrt{(1 - 19/17)^2 + 2 \times 0.0131[14.7(1 - 3/17) + 19/17]}\})$$
$$= 256,000 \text{ lb}$$

7-9. Code Formulas for Ultimate-strength Design for Round Columns with Spiral Reinforcement. The Code formulas for these columns with combined vertical loads and transverse bending moments are as follows:

1. Controlled by tension:

$$P_u = \phi\{0.85f'_c D^2[\sqrt{(0.85e/D - 0.38)^2 + p_t m D_s/(2.5D)}$$
$$- (0.85e/D - 0.38)]\} \quad (7\text{-}40)$$

2. Controlled by compression:

$$P_u = \phi\left[\frac{A_{st}f_y}{3e/D_s + 1} + \frac{A_g f'_c}{9.6De/(0.8D + 0.67D_s)^2 + 1.18}\right] \quad (7\text{-}41)$$

A comparison of the latter equation with Eq. (7-29) will show that they are basically the same when the symbol D_s is used for d_1.

Example 7-10. Assume the circular column shown in Fig. 7-16(a). Compute P_u if $f'_c = 3,000$ psi, $f_y = 40,000$ psi, $\phi = 0.75$, and $e = 10$ in. In Example 7-5, where the same column was used, e_b was found to be

5.91 in.   With $e = 10$ in., tension seems to control.   Therefore, use Eq. (7-40).

$$P_u = \phi \left\{ 0.85 f'_c D^2 \left[\sqrt{\left(\frac{0.85e}{D} - 0.38\right)^2 + \frac{p_t m D_s}{2.5D}} \right.\right.$$
$$\left.\left. - \left(\frac{0.85e}{D} - 0.38\right) \right] \right\} \quad (7\text{-}40)$$

$$P_u = 0.75 \left\{ 0.85 \times 3{,}000 \times 21^2 \right.$$
$$\left[\sqrt{\left(\frac{0.85 \times 10}{21} - 0.38\right)^2 + \frac{0.0274 \times 15.7 \times 16}{2.5 \times 21}} \right.$$
$$\left.\left. - \left(\frac{0.85 \times 10}{21} - 0.38\right) \right] \right\}$$

$$= 285{,}000 \text{ lb}$$

7-10. Code Formulas for Square Sections with Spirals and Bars Arranged in a Circular Pattern.   The Code formulas for these columns with combined axial load and bending again are for short columns. The specified formulas are as follows:

1. Controlled by tension:

$$P_u = \phi \left\{ 0.85 btf'_c \left[\sqrt{\left(\frac{e}{t} - 0.5\right)^2 + \frac{0.67 D_s p_t m}{t}} - \left(\frac{e}{t} - 0.5\right) \right] \right\} \quad (7\text{-}42)$$

2. Controlled by compression:

$$P_u = \phi \left[\frac{A_{st} f_y}{3e/D_s + 1} + \frac{A_g f'_c}{12te/(t + 0.67 D_s)^2 + 1.18} \right] \quad (7\text{-}43)$$

Notice that Eqs. (7-43) and (7-41) have the same first terms, but the denominators of the second terms differ because of the change in the shape of the concrete.

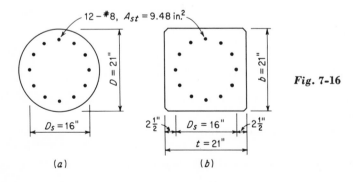

12 – #8, $A_{st} = 9.48$ in.^{2}

$D = 21''$ $b = 21''$

Fig. 7-16

$D_s = 16''$ $2\tfrac{1}{2}''$ $D_s = 16''$ $2\tfrac{1}{2}''$

$t = 21''$

(a)   (b)

Example 7-11. Assume that the column shown in Fig. 7-16(*b*) has the same reinforcement as that in (*a*) but is made square for architectural reasons. Compute P_u if $f_c' = 3{,}000$ psi, $f_y = 40{,}000$ psi, $\phi = 0.75$, and $e = 10$ in., as for Example 7-10.

Use Eq. (7-42) because this case seems to be controlled by tension.

$$p_t = 9.48/21^2 = 0.0215 \qquad m = 15.7 \text{ as before}$$

$$P_u = \phi \left\{ 0.85btf_c' \left[\sqrt{\left(\frac{e}{t} - 0.5\right)^2 + \frac{0.67D_s p_t m}{t}} - \left(\frac{e}{t} - 0.5\right) \right] \right\} \quad (7\text{-}42)$$

$$P_u = 0.75 \left\{ 0.85 \times 21 \times 21 \right.$$

$$\times 3{,}000 \left[\sqrt{\left(\frac{10}{21} - 0.5\right)^2 + \frac{0.67 \times 16 \times 0.0215 \times 15.7}{21}} \right.$$

$$\left. \left. - \left(\frac{10}{21} - 0.5\right) \right] \right\}$$

$$= 370{,}000 \text{ lb}$$

This shows that the extra concrete in the square section has increased the strength found in Example 7-10 from 285,000 to 370,000 lb, or 30 per cent.

Example 7-12. Again assume the column shown in Fig. 7-16(*b*), with $f_c' = 3{,}000$ psi, $f_y = 40{,}000$ psi, $\phi = 0.75$, and $e = 4$ in. Compute P_u.

It is obvious that compression will control. Therefore, use Eq. (7-43).

$$P_u = \phi \left[\frac{A_{st}f_y}{3e/D_s + 1} + \frac{A_g f_c'}{12te/(t + 0.67D_s)^2 + 1.18} \right] \quad (7\text{-}43)$$

$$P_u = 0.75 \left[\frac{9.48 \times 40{,}000}{\dfrac{3 \times 4}{16} + 1} + \frac{441 \times 3{,}000}{\dfrac{12 \times 21 \times 4}{(21 + 0.67 \times 16)^2} + 1.18} \right]$$

$$= 622{,}000 \text{ lb}$$

When this result is compared with that of Example 7-11, one can see the rapid reduction of the ultimate strength caused by an increase in the eccentricity. This should emphasize the fact that one should not neglect the effect of bending in columns when it may be important.

PRACTICE PROBLEMS

7-1 By the approximate method of Art. 7-5, compute the maximum stress in the concrete of the column shown in Fig. 7-17(*a*) if $f_c' = 3{,}000$ psi and $n = 9$ for the following load conditions:

     a. $P = 350$ kips, $e = 3$ in.
     b. $P = 250$ kips, $e = 12$ in.
     c. $P = 300$ kips, $e = 8$ in.

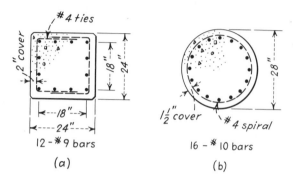

Fig. 7-17

7-2   By use of the Code formulas in Art. 7-6, construct an interaction diagram for the column shown in Fig. 7-17(a) if $f'_c = 3,000$ psi, $n = 9$, and $f_y = 40,000$ psi. Assume that one-third of the area of the inner bars is acting at the location of the outer ones. Then compare the results of the same load cases as given in Prob. 7-1 with this diagram.

7-3   By the use of the Code formulas in Art. 7-6, compute P_u, e_b, and P_b for the column shown in Fig. 7-18(a) if $f'_c = 4,000$ psi, $n = 8$, and $f_y = 50,000$ psi. Neglect the effect of the two bars at the center line.

7-4   Compute P_u for the cylindrical column shown in Fig. 7-17(b) if $f'_c = 3,000$ psi, $f_y = 40,000$ psi, $\phi = 0.75$, and $e = 4$ in.

7-5   Compute P_u for the cylindrical column shown in Fig. 7-17(b) if $f'_c = 3,000$ psi, $f_y = 40,000$ psi, $\phi = 0.75$, and $e = 12$ in.

7-6   Compute P_u for the square column shown in Fig. 7-18(a) if $f'_c = 4,000$ psi, $f_y = 50,000$ psi, $\phi = 0.7$, and $e = 3$ in.

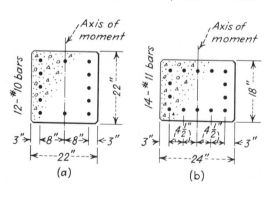

Fig. 7-18

7-7 Compute P_u for the square column shown in Fig. 7-18(a) if f'_c = 4,000 psi, f_y = 50,000 psi, ϕ = 0.7, and e = 10 in.

7-8 Compute P_u for the rectangular column shown in Fig. 7-18(b) if f'_c = 3,000 psi, f_y = 50,000 psi, ϕ = 0.7, and e equals the following:

    a. e = 3 in.

    b. e = 6 in.

    c. e = 12 in.

RETAINING WALLS

8

8-1. Introduction. Retaining walls are used to provide lateral support for a mass of earth or other material the top of which is at a higher elevation than the earth or rock in front of the wall, as shown in Fig. 8-1.

Gravity retaining walls such as that in sketch (*a*) depend mostly upon their own weight for stability. They are usually low in height. They are expensive because of their inefficient use of materials; sometimes they can be cheapened by using cyclopean concrete—concrete in which fairly large rocks are buried.

In contrast to them, Fig. 8-1(*b*) pictures an ordinary cantilever retaining wall. Part of its stability is obtained from the weight of the earth mass *ABCD*, but the wall's resistance to collapse depends upon the strength of its individual parts as cantilever beams. This action is pictured in Fig. 8-1(*c*).

Figure 8-2 shows some of this type of work as it looks in a large construction job. It is part of the Exit Plaza of the South Tube of the Lincoln Tunnel at New York City. In the background is part of a wall for which the concrete of the lower half has been placed and from which the forms have been stripped. Beside it is another portion for which the reinforcement has been placed and the forms are being built. The next part is a combined wall and pump room with the inside forms and part of the reinforcement clearly visible. The nearest portion shows the forms in place, braced, and ready for the pouring of the concrete.

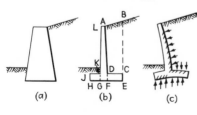

(*a*)   (*b*)   (*c*)

Fig. 8-1

***Fig.* 8-2** Construction of retaining walls, Exit Plaza of the Lincoln Tunnel, New York City.

The design of retaining walls requires a combination of theory and practical engineering sense. The designer must think of them in terms of the procedures that are involved in their construction. Therefore, in order to show these things clearly and to illustrate all the principles that may be involved in such construction, the problems that are used in this chapter are practical cases which are taken directly from such work as that in Fig. 8-2. The solutions of more simple problems will be relatively easy.

8-2. Definitions of Parts. The various portions of a typical rein-forced-concrete retaining wall are defined as follows, using Fig. 8-3(a) for reference:

Stem—the portion $ADKL$
Footing, or base—the part $JCEH$
Toe—the projecting part of the footing on the side toward which the wall tends to tip $JKGH$
Heel—the projecting portion of the footing on the side from which the wall tends to tip $CDFE$
Back—the surface AD
Front—the surface LK
Foundation—the material under the footing, below HE

8-3.   Types of Reinforced-concrete Retaining Walls.

T-shaped Wall.   A "T-shaped" retaining wall is shown in Fig. 8-3(*a*). This is the most simple and common type of cantilever wall.   The base should be from 0.4 to 0.6 times the total height *AH*, but it will vary somewhat with the position of the stem along the base and with the strength of the foundation.   The length of the toe *JK* should be about one-fourth to one-half of the base, the stem being located nearer the rear when it is desired to obtain foundation pressures that are as small and as nearly uniformly distributed as it is possible to have them—as for foundations upon clay.   Possible troubles from sliding because of the decrease in the dead load will be discussed later.

L-shaped Wall.   An "L-shaped" wall such as that in Fig. 8-3(*b*) is used when the wall is along a property line or in other situations where a toe cannot be provided.   Its disadvantages are excessive pressure at the front edge *B* and difficulty in resisting the bending moment at the junction between the stem and the heel.   The base should be about 0.5 to 0.55 times the height *AB*.

Reversed L-shaped Wall.   A "reversed L-shaped" wall is illustrated in Fig. 8-3(*c*).   It is usually difficult to make such a wall stable and to keep it from sliding if the height is great, because of the fact that the dead load is relatively small.   However, such walls are useful in cases where it is expensive or impossible to provide a heel.   The base *BC* should be about 0.5 to 0.6 times the height *AC*.

Counterforted Wall.   A "counterforted" wall such as that in Fig. 8-3(*d*) is a modification of the T- or L-shaped ones.   It has intermittent vertical ribs called counterforts.   This is advantageous for very high

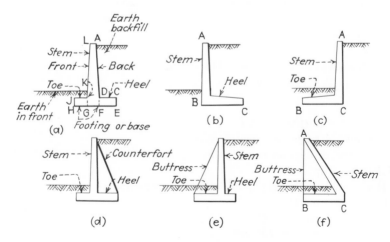

Fig. **8-3**   Types of retaining walls.

walls because the counterforts can be heavily reinforced so as to act as ties to connect the stem and the heel, really transforming the last two parts into continuous slabs which are supported by the counterforts. Although the stem and the heel can be relatively thin, the extra formwork and details may offset the economy in materials. The base should be about the same width as for a T- or an L-shaped wall.

Buttressed Wall. A "buttressed" wall such as that in Fig. 8-3(e) is like a reversed counterfort type with ribs or walls that serve the same general functions as the corresponding parts in a counterforted wall except that they are compression members instead of ties. The toe and the stem are continuous slabs. Such a wall may be built with an inclined stem as shown in Fig. 8-3(f). The base BC should ordinarily be about 0.5 to 0.6 times the height AB, depending upon the size of the heel.

8-4. Stability and Safety Factor. The stability of a retaining wall is its ability to hold its position and to perform its function safely. The safety factor is a measure of the magnitudes of the forces that are required to cause failure of the structure compared with the forces that are really acting upon it. Thus, if the safety factor is 1, the wall will be upon the point of failure. If, for any given design, it is 2, then the overturning moment or the horizontal forces may be doubled before the wall will fail. The magnitude of the safety factor to be used in a design will depend upon the engineer's judgment, the specifications, or the building code that is to be followed. In general, it may vary from 1.5 to 2.

A retaining wall may fail in one of four ways: by the collapse of its component parts, by overturning about the front of its toe, by excessive pressure upon its foundation, or by sliding upon its foundation. In a well-balanced design, the wall should be equally safe in all respects.

The bending of the individual parts and the pressures that act upon them are illustrated in Fig. 8-1(c). Each part must act as a cantilevered beam.

In order to illustrate overturning, let W in Fig. 8-4 represent the resultant of all the vertical forces, including the weight of the wall and the vertical component of the lateral earth pressure (if any) and of the earth on top of the base; let H = the resultant of all the horizontal forces. Then Hn represents a moment that tends to overturn the wall as an entity, rotating it about the point A, but Wm is the "righting," or "stabilizing," moment which resists this overturning. If Hn exceeds Wm,

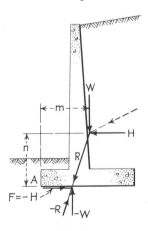

Fig. 8-4

the wall will tip over because the resultant of H and W will pass outside the base (beyond A).

Failure due to excessive pressure on the foundation will result in the tipping or overturning of the structure.   When the wall is founded upon rock, it may actually rotate about the corner A because the rock is very strong, but when it is on earth, the latter will settle, and the wall will tilt about a point to the right of A if the concentration of pressure becomes too great.   The effective value of m will be decreased.   If the overturning moment exceeds the reduced righting moment, the wall will overturn.

Sliding on the foundation may be demonstrated by considering Fig. 8-4.   The resultant of W and H is represented by R.   This, in turn, must be resisted by an equal and opposite reaction which may be called $-R$ and which will have components equal to $-W$ and $-H$.   The latter force is caused by the friction of the base upon the foundation so that $F = -H$.   Then, if f = the coefficient of sliding friction, $F = Wf$. When F can actually counteract H, the wall is said to be "stable against sliding."   Of course, when the wall is built upon an irregular rock surface, there is no difficulty about sliding.

A retaining wall needs weight in order to resist overturning and sliding. Therefore, it is not usually advisable to use high-strength concrete and excessively thin sections for ordinary walls because they will be too light in weight.   Web reinforcement in retaining walls is very troublesome.   It should be avoided by keeping the shearing stresses low.

It is often difficult to secure the desired safety factor against sliding in the case of earth-borne walls.   The ground in front of the structure may have considerable abutting power or passive resistance to being shoved away, but the wall should stand without depending upon this force. Absence of the earth in front of the wall when the backfill is placed behind the structure, thoughtless excavation of the earth along the toe by someone in the future, possible scouring or washing away of this material—all these are reasons for this statement.   Sometimes this passive resistance is relied upon, but this should be done with caution.

8-5.   Foundations.   Retaining walls that are founded upon earth present more of a problem than do those which are supported upon rock because the high pressures that can be applied to the rock are not permissible upon the earth.   Those pressures are the result of the combined action of the direct vertical load W and the overturning moment Hn.

Two different sets of pressure distributions are shown in Fig. 8-5. These differences are caused by the relative effects of the direct load and the overturning moment, but it is convenient to look upon them in terms of the position of the point at which the resultant of W and H intersects the bottom of the footing. If G is this point, its location with respect to W can be found by taking moments about G itself. Therefore,

$$a = \frac{Hn}{W} \tag{8-1}$$

This enables one to compute e, the eccentricity of the resultant with respect to C, the center of the base.

When G is at the right of L, it is inside the middle third of the footing. Therefore, the pressure diagram $AEKJ$ in sketch (a) is made up of a uniformly distributed pressure equal to EN which equals W/B for a 1-ft strip of wall, and a uniformly varying pressure which is caused by the moment

$$M = We$$

$$QJ \text{ or } KN = \frac{Mc}{I} = We \times \frac{6}{1 \times B^2} \tag{8-2}$$

Therefore,

$$p = \frac{W}{B} \pm 6\frac{We}{B^2} = \frac{W}{B}\left(1 \pm \frac{6e}{B}\right) \tag{8-3}$$

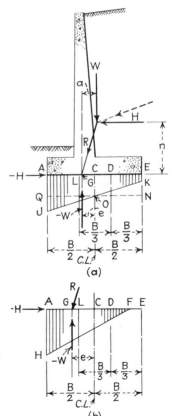

where p is the intensity of the greatest or the least pressures, and B is the width of the footing. The positive sign is to be used to find the maximum pressure AJ; the negative sign will give the minimum pressure EK. For earth-borne walls, the pressure diagram should be as nearly rectangular as it is practicable to make it.

When the resultant falls outside the middle third (left of L), the pressure diagram in Fig. 8-5(b) results. Since there can be no tension upon the base near E, Eq. (8-3) cannot be used. The pressure diagram is assumed to be the triangle AFH. Its total area must equal the direct load W, and its center of gravity must lie vertically below G. Therefore, for a strip 1 ft wide, $AH \times AF/2 = W$. Therefore, the maximum pressure AH is

$$AH = p = \frac{2W}{AF} = \frac{2W}{3AG} \tag{8-4}$$

The exact distribution of the resisting pres-

Fig. 8-5

Table 8-1   General Data Regarding Foundation Materials

Material	Safe bearing capacity, ksf	Angle of repose, deg	Maximum coefficient of friction of concrete on foundation
Sound rock	80		
Poor rock	30		
Gravel and coarse sand	10–12	37	0.6–0.7
Sand (dry)	6–8	33	0.4–0.6
Fine sand (wet but confined)	4	25	0.3–0.4
Clay and sand mixed	4–5	36	0.4–0.5
Hard clay	5–6	36	0.4–0.6
Soft clay	2	26	0.3

sure of the foundation may not vary as a straight line, but it is sufficient to assume that it does so.

The safe bearing value of any given soil and its probable deformation are so problematical and so dependent upon the qualities of the material itself that it is unwise to set any exact magnitudes for them, but the data of Table 8-1 may be used as a general guide. However, the conditions at the site must be examined and the bearing value of the soil should be tested before one designs an important retaining wall.

The frictional resistance of the foundation when a wall tends to slide upon it is also uncertain. For some soils, the resistance of the material against sliding upon itself is greater than that for the concrete sliding upon it. In order to take advantage of this greater frictional resistance, the footings of walls are often made with projections on the bottom, called *cutoff walls*, as shown in Fig. 8-6(a) and (b). Sometimes the bottom of the toe is sloped upward, as in Fig. 8-6(c). The first method is likely to cause disturbance of the earth as the result of digging the trench in it. Probably the real value of these measures comes from the increase of the shearing and abutting values of the confined earth when

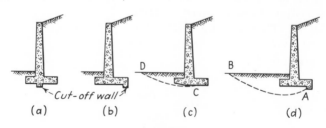

(a)   (b)   (c)   (d)

Fig. 8-6

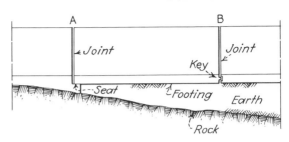

Fig. 8-7   Transition from
rock to earth.

it is subjected to the large pressures that exist under the forward portion
of the footing.   It seems to be best to put a cutoff wall at the rear of the
heel as shown in Fig. 8-6(*b*) because this is the region where the soil pressures under the footing are at a minimum.   Furthermore, it is probable
that the line of slippage may be somewhat as shown by the dotted line
AB in Fig. 8-6(*d*).   This requires the movement of a much heavier mass
of earth with more resistance from friction and cohesion than would occur
if the movement took place along *CD* in sketch (*c*).   However, when considering such cutoff walls, one must be careful to make certain that he is
not depending upon the simultaneous action of the abutting power and the
frictional resistance of the same mass of soil.   In other words, friction
enables the earth to develop its resistance to being shoved aside; hence the
same frictional forces should not be counted twice.

It is very important to found all walls upon undisturbed material.
The consolidation that has been produced by nature has probably given
the soil the best treatment that it can have for use as an ordinary foundation.   In no case should a wall be placed upon newly deposited fill when
avoidance of settlement is important.   Furthermore, the excavation for
the footing should be sufficient to remove organic matter and to get
below the frost line—about 4 to 5 ft in cold climates.

Another foundation problem which must not be overlooked in long
walls is that of founding them partly upon rock and partly upon earth.
If such conditions cannot be avoided, the structure should be designed
so that the last portion that rests upon rock has a seat to receive the footing of the adjacent earth-borne structure, the latter being a rather long
section which can act like a vertical beam as shown in Fig. 8-7.   The
compaction of the earth will then merely open the joint at *A* slightly,
and it will close the one at *B* somewhat, but the motions will not be very
apparent.   On the other hand, if the rock has only local gullies of shallow
depth, concrete piers without reinforcement may be built to carry the
loads to the surface of the rock as shown in Fig. 8-8(*a*).

If a retaining wall rests upon a shelf at one end as shown in Fig. 8-7,
or upon piers as in Fig. 8-8, one should remember that the presence of a
hard support may cause considerable local torsion in the footing because

of the tendency to concentrate the resistance at these points instead of at the adjacent but more yieldable soil.

If a retaining wall supports vertical loads upon its top, its heel, or its toe, these loads are to be included along with the weight of the wall, earth, etc., in finding the total load W and the location of its resultant. Similarly, if horizontal forces are applied to the top or anywhere else, they are to be included in the calculations for the magnitude and position of H. These resultants are then used as for any other wall.

Figure 8-8(b) illustrates the construction used at one wall of an industrial plant that is located on a hillside. The back of the wall laps under the siding. The heavy crane columns are supported partly on the top of the stem and partly on the local pilaster. Wind and cranes can cause lateral forces at the top of the wall. The toe should be wide enough to avoid excessive edge pressures. The local column load can be assumed to be spread over a length of wall that equals from one to two times the height of the stem. The horizontal projection or wing at the top of the wall is to spread the horizontal load from the column over a considerable

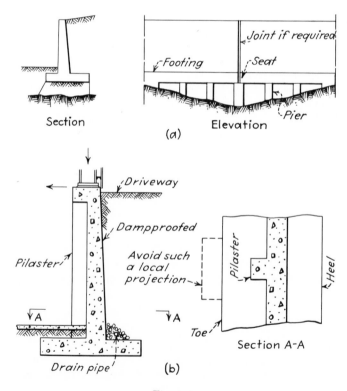

Fig. 8-8

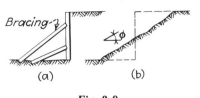

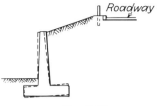

Fig. 8-9   Fig. 8-10

length of the wall. Incidentally, if local projections of the toe are made
at the columns, as shown by the dotted lines in Fig. 8-8(*b*), these must
be adequate to resist heavy localized pressures because, when the wall
tends to tip over, the pressures will be concentrated upon these projections which act somewhat as fulcrums.

8-6. Lateral Earth Pressure. When an excavation is made in earth
or when earth is piled up, the soil tends to slump and move sideways as
shown in Fig. 8-9. Force is required to prevent this motion. The force
exerted by the earth against any opposing structure is called the *active*
earth pressure in order to differentiate it from the *passive* pressure—the
resistance of the earth to being shoved aside by an outside force. The
greatest angle at which the earth slope will remain in equilibrium is
called the *angle of repose,* and it is usually denoted by ϕ. It ordinarily
varies from 30 to 40° from the horizontal.

The magnitude of the active earth pressure—called earth pressure
hereafter—is rather indeterminate. The soil behind any wall may vary
greatly in its characteristics from place to place; when the backfill is
deposited behind the wall in Fig. 8-10, it may exert a certain lateral pressure upon the structure; but when the wall deforms slightly, as shown
by the dotted lines, it may tend to relieve itself of some of the pressure
because of the cohesion and friction of the earth upon itself; if traffic or
some other force causes vibrations that break down this internal frictional resistance, it may cause an increase of pressure upon the wall;
when the earth dries out, it may shrink and settle as it dries; and when
the soil is saturated by a rain, it may expand again.

It is not possible to discuss in this text the various theories of earth
pressures that have been developed by others. However, it is necessary to adopt one theory of earth pressures for use as a basis of design.
Coulomb's theory[1] will be used in this text. This theory is based upon
the assumption that the wedge of earth that lies above the plane of rup-

[1] For brief explanation of both Coulomb's and Rankine's theories, see Milo S.
Ketchum, "Structural Engineers' Handbook," McGraw-Hill Book Company, New
York, 1918; or recent texts on soil mechanics.

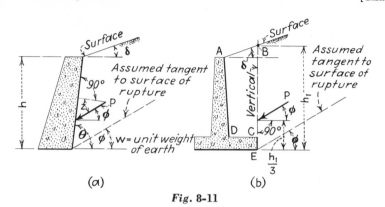

Fig. 8-11

ture—a plane at or above the one that is established by the angle of repose—will tend to slide downward and will shove the wall before it or will tip it over.   A general case is pictured in Fig. 8-11(*a*), where ϕ is the angle of repose of the soil and P is the total earth pressure per linear foot of wall, P being parallel to the plane of rupture.   The meaning of each of the other symbols is obvious.

Coulomb's general formula for the total thrust of the earth per foot of the length of the wall is

$$P = \frac{1}{2} wh^2 \frac{\sin^2 (\theta - \phi)}{\sin^2 \theta \sin (\theta + z) \left[1 + \sqrt{\dfrac{\sin (z + \phi) \sin (\phi - \delta)}{\sin (\theta + z) \sin (\theta - \delta)}} \right]^2} \quad (8\text{-}5)$$

where the meanings of all the terms are indicated in Fig. 8-11(*a*).   When applied to a reinforced-concrete cantilever wall, the conditions become as shown in Fig. 8-11(*b*).   The vertical line BE—labeled h_1—through the end of the heel can be taken as the effective back of the wall unit, of which $ABCD$ is earth.   Then $z = \phi$ and $\theta = 90°$.   Therefore, Eq. (8-5) becomes

$$P = \frac{1}{2} wh_1^2 \frac{\cos \phi}{(1 + \sqrt{2 \sin^2 \phi - 2 \sin \phi \cos \phi \tan \delta})^2} \quad (8\text{-}6)$$

The point of application of this force is $h_1/3$ above the bottom of the footing, assuming a triangular pressure diagram.   Roughly, this is about equivalent to the fluid pressure caused by a liquid weighing 30 to 35 pcf.

If the angle of repose $\phi = 30°$, then Eq. (8-6) becomes

$$P = \frac{1}{2} wh_1^2 \frac{0.87}{(1 + \sqrt{0.5 - 0.87 \tan \delta})^2} \quad (8\text{-}7)$$

It has been stated that the force P is assumed to be parallel to the plane of rupture.   This is a debatable question when the embankment or fill is subjected to vibrations due to trains, trucks, etc., which tend to break down the plane of rupture. In these cases, it is safer to assume that the earth pressure acts horizontally because its vertical component otherwise exerts a stabilizing effect theoretically.

8-7. Surcharge. "Surcharge" generally denotes a temporary or live load which is applied on top of the earth behind a retaining wall, tending to increase the earth pressure. These loads may be caused by trains, vehicles, or even piles of materials.

Surcharge diagrams are shown in Fig. 8-12. Sketch (a) pictures a highway upon an embankment; W' is the wheel load of a truck. Obviously, the earth spreads the load W' over increasingly large areas of soil as the depth is increased. The pressure diagram may be assumed to be a cone. The horizontal components of the pressure lines upon a vertical plane like AB in Fig. 8-12(a) must vary with the pressures and their directions. Experiments that have been made by M. G. Spangler[1] indicate that the diagram of horizontal pressures on AB which are due to the surcharge is somewhat as shown in Fig. 8-12(b)—a protuberance. They also indicate that the distance x has a great effect upon this diagram. When x is small, the magnitude of the resultant horizontal surcharge P_s is large, and its distance from the top of the wall is small. As x increases, the maximum intensity of pressure decreases as in Fig. 8-12(c); the forces are spread over a greater area, and y increases.

At the same time, the pressures from the surcharge have vertical com-

[1] Associate structural engineer, Iowa Engineering Experiment Station, Iowa State College, Ames, Iowa. The data have been published in *Paper* J-1, Vol. 1, p. 200, of *Proceedings* of the International Conference on Soil Mechanics and Foundation Engineering; also in *Iowa Eng. Exp. Sta. Bull.* 140.

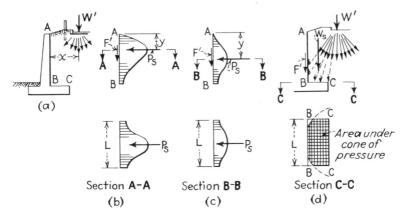

Section **A-A**   Section **B-B**   Section **C-C**
    (b)   (c)   (d)

Fig. **8-12**   Study of lateral pressures caused by live loads.

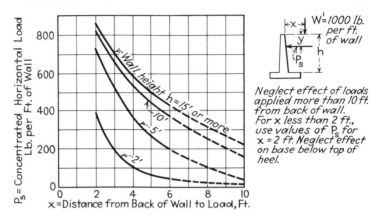

Fig. 8-13   Lateral pressure due to concentrated surcharge loads.

ponents as shown in Fig. 8-12(*d*).   Part of these cause downward stabilizing forces upon the heel *BC*; others may cause a downward frictional force *F'*.   However, the latter does not seem to be reliable because of the effect of vibrations.   Therefore, it will be allocated to those things which may increase the safety of the structure but are not included in the design.

With Spangler's report as a starting point, curves have been drawn as shown in Figs. 8-13 to 8-15.   The first two enable one to determine and to locate a single horizontal force which should be included in the equations for moment and for sliding.   On the other hand, Fig. 8-15 is prepared in order to give an approximate uniform load for the downward pressure on the heel as shown in Fig. 8-12(*d*).   In preparing these surcharge diagrams,[1]

[1] The curves have been prepared by Dr. A. H. Baker, formerly designer, The Port of New York Authority.

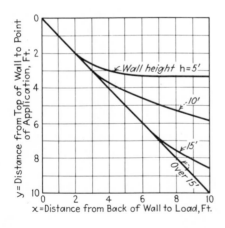

Fig. 8-14   Location of point of application of resultant lateral pressure due to concentrated surcharge loads.

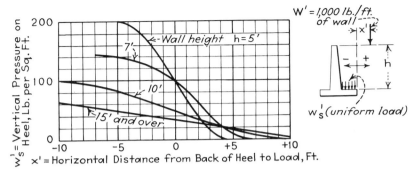

Fig. **8-15**   Downward pressure due to concentrated surcharge loads.

W' was assumed to be 18,000 lb.   The resultant of the horizontal components of the surcharge forces was used as a concentrated load.   A piece of wall 18 ft long was assumed to act as a unit, and the total force was then divided by 18 to get its average magnitude per foot of wall.   When x exceeds 10 ft, the effect of the surcharge is neglected; when h is greater than 15 ft, the effect of the surcharge is assumed to be constant in magnitude and position because the effect of the bottom of the diagram in Fig. 8-12(b) is negligible.   However, if the load W' differs from that assumed here—1,000 lb/ft over 18 ft—the values in Fig. 8-13 may be modified in proportion, but Fig. 8-14 should not be greatly affected.   It is unnecessary to include impact in these loads.   Of course, the real downward pressure at any point on the heel (Fig. 8-15) varies with the location of the load, the height of the wall, and the relative position of the point on the heel.   However, great refinement is not justified.   Therefore, a conical distribution at a maximum of 45° is assumed, and the curves have been made accordingly. This downward pressure becomes relatively so small when a wall is more than 15 ft high that one may neglect it, but it is included in problems here for purposes of illustration.

If the surcharge is a uniform load—say 300 psf—it may be converted into an equivalent depth of earth.   Using 100 pcf as the weight of earth, this surcharge is similar to an extra fill 3 ft deep.   The wall can be designed as though the top of the earth really came to a point 3 ft above its actual surface—as far as the earth pressure alone is concerned.

8-8.   Water Pressure.   Although a wall may have longitudinal drains behind it or small holes called weepers through the stem, as pictured in Fig. 8-16(a), the ground water behind the wall may become impounded if the outlets clog up or become filled with ice.   If the wall is built upon solid rock, the water may be almost entirely held back.   If the founda-

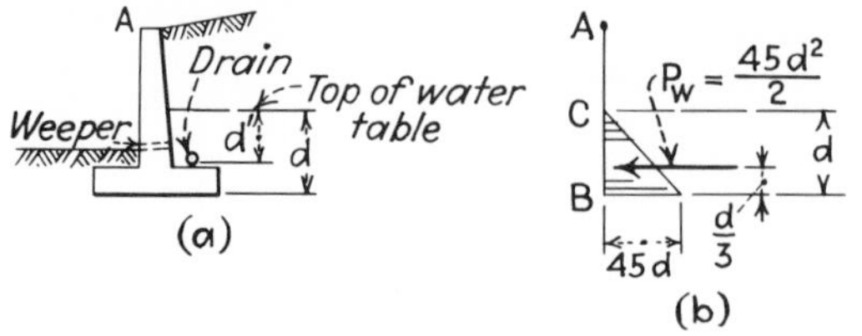

Fig. 8-16   Water pressure.

tion is composed of earth, water may be able to escape under the footing, but it is likely to require some head to force it through the soil.

The presence of water in the soil increases the lateral pressure on the wall. If the weight of saturated earth is assumed to be 120 pcf, and if the lateral pressure that it exerts is 60 per cent[1] of this weight, then this pressure equals 72 psf per ft of depth. However, using Eq. (8-7) for dry earth, assuming $\delta = 0$, and using $w = 100$ pcf, $P = 0.15wh_1{}^2$. If $h_1 = 1$ ft, $P = 0.15w = 0.15 \times 100 = 15$ lb. This shows that the pressure for the dry earth is 30 psf/ft of depth, since it varies uniformly from zero to a maximum. ($P = $ area of pressure triangle $= p \times 1/2$.) The saturated earth exerts a pressure that is 42 psf larger than that of the dry earth, but this excess will be called 45 psf/ft of depth. However, the top of the "water table" must be ascertained before one can determine the total pressure caused by the water.

Theoretically, the existence of water pressure on the back of the wall would be accompanied by hydrostatic uplift under the footing and downward pressure on top of it, all but the last of these pressures tending to tip over the wall. However, these forces are rather indeterminate, and the inclusion of them causes needless complication of the calculations. On this account, the horizontal hydrostatic pressure will be used alone; the maximum distance from the invert of the drain to the top of the water table—d' in Fig. 8-16(a)—will be assumed to have an arbitrary magnitude of 8 ft. (For small walls, the water will be assumed never to be less than 5 ft below the top of the earth even when it reduces d'.) The diagram of horizontal hydrostatic pressures will therefore be taken as shown in Fig. 8-16(b), the resultant being called P_w. If the back of the wall is sloped, the water pressure also has a vertical component, but it is usually negligible because of the steepness of the back.

8-9. Local Thrust at the Top of a Wall Due to Temperature.
Some retaining walls are likely to have a localized thrust (P_T in Fig. 8-17) which will act near their tops. One cause of such a horizontal thrust is the action of frost. If the soil were saturated and then it froze solid, no ordinary wall could withstand the resultant pressure. With properly

---

[1] Karl Terzaghi, Pressure of Saturated Sand, *Eng. News-Record*, Feb. 22, 1934.

drained soil, the pressure from freezing must be greatly reduced from that which would be caused by solid ice. However, it seems that the expansion of backfill upon freezing and then its contraction and settlement upon thawing may account for the failure of some retaining walls, particularly small ones, after years of service.

In order to make some arbitrary allowance for this pressure from ice, 700 plf of wall, applied at the top of the ground and parallel to the surface of the earth, will be assumed. This figure has no experimental verification. It is the product of experience and judg- ment, and is being used as indicated because of its relatively large effect upon small walls for which the thrust of the earth alone is small. If the ground is nearly level and the conditions facilitate the formation and collection of ice, this pressure will be included in the calculations; if the earth slopes and is thoroughly drained, it may be omitted. It should also be omitted from all calculations for sliding because the ground in front of the wall will be frozen also and it will resist the shear caused by ice pressure behind the structure.

Fig. 8-17

8-10. Preliminary Steps in the Design of a Cantilever Retaining Wall. In order to illustrate the theory of the design of a reinforced-concrete retaining wall, a practical example will be worked out in detail, using a problem that will illustrate most of the forces to be encountered in such work. For this, assume that the wall is to support a roadway with a 4-ft sidewalk; it is to have a concrete parapet to protect traffic; the top of the pavement is to be 20 ft above the adjacent ground. Assume also that the following data are specified:

w = weight of earth = 100 pcf
Safe bearing value of earth = 7,000 psf; its ultimate value = 14,000 psf
Angle of repose of earth = ϕ = 30°
Maximum coefficient of sliding friction of concrete on earth = 0.45
Maximum coefficient of shearing friction of earth on earth = 0.55

f_c' = 2,500 psi   and n = 10
f_c = 1,125 psi
f_s = 18,000 psi   elastic-limit stress = 36,000 psi
v_c = 55 psi
u = 175 psi (approx), assuming heavy top bars

Safety factor = 2 against sliding and overturning because the wall supports an important highway, and this value is very conservative. Its use will emphasize the relative seriousness of failure due to sliding, also the difficulty of securing the necessary safety factor against such action. Also, check the stresses when this safety factor is applied.

Equation (8-6) will be used to calculate the earth pressure.

The surcharge force and its position will be determined from Figs. 8-13 and 8-14; its downward pressure on the heel, from Fig. 8-15.

Water pressure will be 45 psf/ft of depth, using a height of 8 ft above the weepers.

Ice pressure will be 700 plf at the top of the roadway.

The design is to be based upon the working-stress theory.

In any problem like this, there generally are certain fundamentals and details that are desired by the engineer who is in charge. These automatically influence the design. In this case, some such points will be explained for the purpose of illustration, because these practical things are important and because they should be decided upon before the calculations are made. They are shown in Fig. 8-18.

1. The wall is to be the T type with a large heel and short toe—the latter being about 0.25B—so as to utilize the dead load of the earth.

2. The bottom of the toe is to be sloped up 2 in./ft to help resist sliding.

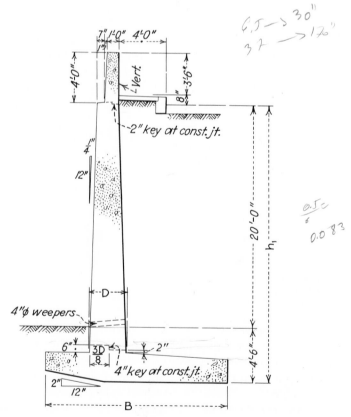

Fig. 8-18

Fig. **8-19** Footing of reversed L-shaped retaining wall, New York approach to Lincoln Tunnel.

3. The front face is to be battered $\frac{1}{4}$ in./ft for appearance.

4. The parapet is used as a guardrail. It is to be 12 in. thick at the top, and it is to be 3 ft 6 in. high.

5. The top of the main wall is to be at least 20 in. thick. This provides a shelf which may improve the appearance of the wall by reducing its apparent height. It also enables men to work inside the reinforcement and the forms during the placing of the concrete. (It is very difficult to avoid honeycombing when concrete is poured into forms for high thin walls, but it can be avoided if portions of the upper part of the forms are left off until the concrete is placed in the lower part of the structure.)

6. The footing is to be poured in sections about 30 to 40 ft long with

a construction joint for the stem as shown in Figs. 8-18 and 8-19, the latter being a picture of a footing for a heelless wall. The projections above the toe and heel are important as stops against which the forms for the stem can be held so as to keep them from being displaced or becoming wavy during the pouring of the concrete. The 4-in. key is made so as to resist the shear. The width of the forward raised portion of it should be from one-fourth to one-half of the thickness of the stem because the concrete that will be in the tensile side of the stem is not so effective in resisting shear as that which is in compression.

7. The stem is to be poured in one operation so as to avoid the marks and the change of appearance that may occur at horizontal construction joints. (Such deep lifts are difficult to make, and they are likely to have detrimental effects upon the quality of the concrete.)

8. The parapet is to be placed flush with the back of the wall with a 7-in. step at the outside as shown in Fig. 8-20(a). The step is placed below the sidewalk so that the top construction joint will not be visible. This arrangement with the offset on the outside face of the wall avoids the likelihood of settlement of the sidewalk away from the ledge, which might occur if it were constructed as shown in Fig. 8-20(b). On the other hand, if the sidewalk in the latter case is supported upon a shelf on the stem, settlement of the fill may tilt the slab or even crack it.

9. The bottom of the footing is to be 4 ft 6 in. below the finished grade in front of the wall so as to be below the frost line.

10. Weepers 4 in. in diameter and 10 ft c.c. are to be used at the level of the surface of the ground in front of the wall.

11. Since weight and stiffness are desirable in a retaining wall like this, use sections having more than the minimum thickness for balanced design. This means that the steel will govern the bending resistances. Therefore, use the working-stress method of analysis.

8-11. Design of the Stem. The effective height of the stem will be assumed to be 22 ft. The surcharge for trucks will be assumed to be about 1 ft from the curb. The loading diagrams are then as pictured in Fig. 8-21.[1]

[1] Note that the answers in such problems as these are not carried out to more than three significant figures in most cases.

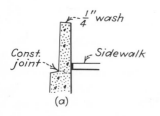

(a)

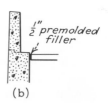

(b)

Fig. 8-20

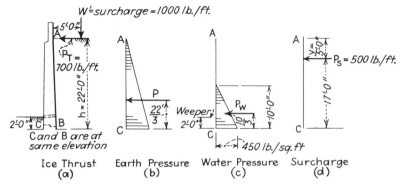

Fig. 8-21

From Eq. (8-7), with $\delta = 0$ and $w = 100$,

$$P = \frac{1}{2} wh_1{}^2 \frac{0.87}{(1 + \sqrt{0.5 - 0.87 \tan \delta})^2} = 15h_1{}^2 \qquad (8\text{-}8)$$
$$= 15 \times 22^2 = 7{,}260 \text{ lb}$$

where h is used instead of h_1. The force P is assumed to be horizontal because of the vibrations caused by traffic. When there are no such vibrations, the horizontal force may be assumed to equal $P \cos \phi$ (see Prob. 8-1).

Writing the equation for moments about C, the top of the toe in Fig. 8-21, since the stem is a vertical cantilever beam,

$$M = P_T h + P \frac{h}{3} + P_w \frac{(8 + 2)}{3} + P_s (h - y)$$
$$= 700 \times 22 + 7{,}260 \times \frac{22}{3} + 10 \times \frac{450}{2} \times \frac{10}{3} + 500 \times 17$$
$$= 84{,}600 \text{ ft-lb}$$

From Table 5 in the Appendix, with $f_s = 18{,}000$ psi, $f_c = 1{,}125$ psi, and $n = 10$, K for a balanced design is 189 and j is 0.872. Then, using Eq. (2-8), with $b = 12$ in.,

$$bd^2 = \frac{M}{K} \qquad \text{or} \qquad d = \sqrt{\frac{84{,}600 \times 12}{12 \times 189}} = 21.2 \text{ in.}$$

This is a large wall, and both weight and stiffness are desirable for such a structure. Therefore, use a thicker section than the minimum. Assume $d = 28$ in. and the cover over the bars $= 3$ in. (to provide good protection against the moisture in the adjacent earth). Then $D = 31$ in., giving a batter of $\frac{1}{4}$ in. in 12 in. for the back of the wall below the sidewalk. Assume $j = 0.89$.

$$A_s = \frac{M}{f_s j d} = \frac{84,600 \times 12}{18,000 \times 0.89 \times 28} = 2.26 \text{ in.}^2/\text{ft (approx)}$$

If No. 8 bars are used at 4 in. c.c., $A_s = 2.37$ in.^{2}; if $4\frac{1}{2}$ in. c.c., $A_s = 2.11$ in.^{2} Inasmuch as. the direct compressive load due to the weight of the stem will decrease the tension in the bars, the latter will be assumed.

The intensities of the shear and the bond stresses at the base of the stem are

$$v = \frac{H}{bd} = \frac{P_T + P + P_w + P_s}{bd} = \frac{700 + 7,260 + 2,250 + 500}{12 \times 28} = 32 \text{ psi}$$

$$u = \frac{H}{(\Sigma o)jd} = \frac{10,710}{3.14 \times 12/4.5 \times 0.89 \times 28} = 51 \text{ psi}$$

It is unnecessary to extend all these heavy bars for the full height of the stem. The required areas of steel at intermediate heights can be found with sufficient accuracy by assuming $j = 0.89$ and designing the wall for bending alone, neglecting any compressive steel. At mid-height,

$$M = P_T \frac{h}{2} + P \frac{h}{3} + P_s \left(\frac{h}{2} - y\right)$$

$$= 700 \times 11 + (15 \times 11^2) \times \tfrac{11}{3} + 500 \times 6 = 17,400 \text{ ft-lb}$$

$$A_s = \frac{M}{f_s j d} = \frac{17,400 \times 12}{18,000 \times 0.89 \times 22.5} = 0.58 \text{ in.}^2$$

From such calculations, the curve in Fig. 8-22 can be drawn, showing the permissible cutoff of the bars. It is customary to extend them beyond

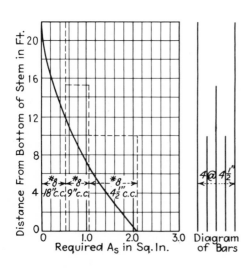

Fig. **8-22**   Diagram for determining the required lengths of reinforcement in stem.

the theoretical points a distance that need not exceed what is required to develop them through bond.   In this case, with $u = 175$ psi,

$$L_s = \frac{A_s f_s}{\Sigma o \times u} = \frac{0.79 \times 18,000}{3.14 \times 175} = 26 \text{ in.}$$

The bars should not be much farther apart at the top than the thickness of the wall at this point.

The stem has a safety factor of 2 because, if M is doubled, the stresses in the steel and the concrete will not exceed the elastic limit of the bars or the ultimate strength of the concrete.

The need for good judgment and common sense in engineering has been emphasized repeatedly.   One glaring case[1] of the lack of thinking and of a sense of proper proportion occurred in the building of an L-shaped retaining wall 20 ft high.   The main bars in the stem were to be $1\frac{1}{4}$-in. round at 12 in. c.c.   The figure 1 happened to be on a dimension line. The bars were ordered as $\frac{1}{4}$-in. round.   They were detailed, shipped, erected, and approved by an inspector.   As the backfill was placed, the wall progressively failed and had to be replaced.   It seems strange that no one would ask "What are $\frac{1}{4}$-in. bars doing in a 20-ft retaining wall?"

8-12.   Stability and Foundation Pressure.   Before the base of the wall is designed in detail, it is advisable to assume its size, to test the retaining wall for stability, and to see that the foundation pressure is satisfactory.   To do so for this case, assume that the thickness of the footing is about equal to that of the base of the stem, that $B = 0.6h_1$, and that the toe is about $0.25B$.

Thus, let $D = 30$ in. $B = 0.6(22 + 2.5) = 14.7$ ft.   Assume $B = 14$ ft 9 in. and the toe $= 3$ ft 6 in.

Figure 8-23(a) diagrammatically pictures the forces that are to be considered.[2]   The stabilizing effect of the surcharge W_s is found from Fig. 8-15 for

$$-8.67 - 0.46 + 5 = -4.13$$

It is 45 psf.   Also, from Eq. (8-8),

$$P = 15 \times 24.5^2 = 9,000 \text{ lb}$$

$$P_w = \frac{12.5^2}{2} \times 45 = 3,500 \text{ lb}$$

The weight of the earth on the toe will be neglected.

[1] *Eng. News-Record*, Feb. 8, 1951, p. 51.

[2] Notice that P is assumed to be horizontal so that $P_v = 0$.   This is conservative, but there is much difference of opinion among engineers as to whether or not P may be assumed to slope at the angle ϕ under such conditions.   If P is sloped, P_v is its vertical component.

The magnitudes and lines of action of the resultant of the vertical forces W and the resultant of the horizontal forces H are found as follows, taking moments about E because of convenience in finding the proper lever arms:

Vertical forces		*Weight*	*Lever arm*	*Moment*
Stem W_c	$=$	7,640	$\times\ 9.93\ =$	75,900
Footing $W_F = 2.5 \times 14.75 \times 150$	$=$	5,530	$\times\ 7.38\ =$	40,800
Earth $W_e = 8.67 \times 22 \times 100$	$=$	19,100	$\times\ 4.34\ =$	82,900
Earth $W_e' = 0.46 \times 22 \times 0.5 \times 100\ =$		500	$\times\ 8.82\ =$	4,400
Surcharge $W_s = 45 \times 8.67$	$=$	390	$\times\ 4.34\ =$	1,700
	$W\ =$	33,160 lb	$\Sigma M\ =$	205,700 ft-lb

$$x = \frac{\Sigma M}{W} = \frac{205,700}{33,160} = 6.2 \text{ ft}$$

Horizontal forces		*Force*	*Lever arm*	*Moment*
Temperature P_T	$=$	700	$\times\ 24.5\ =$	17,200
Surcharge P_s	$=$	500	$\times\ 19.5\ =$	9,800
Earth P	$=$	9,000	$\times\ 8.17\ =$	73,500
Water P_w	$=$	3,500	$\times\ 4.17\ =$	14,600
	$H\ =$	13,700 lb	$\Sigma M\ =$	115,100 ft-lb

$$n = \frac{\Sigma M}{H} = \frac{115,100}{13,700} = 8.4 \text{ ft}$$

These two forces are shown in Fig. 8-23(b).

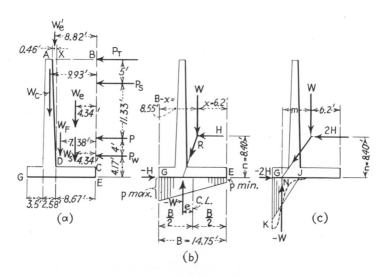

Fig. 8-23   Forces and foundation pressures.

The next step is the determination of the foundation pressures. The eccentricity of the resultant of W and H with respect to the center of the base is found by taking moments about G of Fig. 8-23(b). Thus,

$$e = \frac{B}{2} - \left[\frac{W(B - x) - Hn}{W} \right]$$

$$= 7.38 - \frac{283,500 - 115,100}{33,160} = 2.3 \text{ ft}$$

This is less than $B/6 = 2.46$ ft. Therefore, the pressures on the foundation are

$$\max p = \frac{W}{B}\left(1 + \frac{6e}{B}\right) = \frac{33,160}{14.75}\left(1 + \frac{6 \times 2.30}{14.75}\right) = 4,350 \text{ psf}$$

$$\min p = \frac{W}{B}\left(1 - \frac{6e}{B}\right) = 144 \text{ psf}$$

The maximum is quite conservative.

The safety factor against overturning must now be tested. Assume the overturning moment Hn to be doubled. In this case, let n remain unchanged, and replace H by $2H$ as in Fig. 8-23(c). Therefore

$$\Sigma M = 13,700 \times 2 \times 8.40 = 230,200 \text{ ft-lb}$$

Take moments about N [Fig. 8-23(c)].

$$m = \frac{2Hn}{W} = \frac{230,200}{33,160} = 6.94 \text{ ft}$$

$GN = 14.75 - 6.2 - 6.94 = 1.61$ ft. The point N is thus inside the toe G, but the pressure upon the foundation must be investigated. In Fig. 8-23(c), assume a triangular pressure diagram with the resultant W at its center of gravity. Therefore, $NJ = 2(GN) = 3.22$ ft, and the maximum pressure is

$$W = \frac{GK \times GJ}{2}$$

or

$$GK = \frac{2W}{GJ} = \frac{2 \times 33,160}{4.83} = 13,700 \text{ lb}$$

This value is less than twice the permissible pressure of 7,000 psf which was one of the conditions of the problem. If GK had exceeded the capacity of the earth, the latter would yield, and the pressure diagram might look like the dotted line in Fig. 8-23(c).

When large retaining walls are on earth foundations, it is probable that large edge pressures like GK in Fig. 8-23(c) will cause local compression of the soil and a resultant tilting of the wall. However, this is a test for the theoretical safety factor only, but high toe pressure is a feature which should not be overlooked.

Remembering that the thrust due to the effects of temperature P_T is to be omitted when one computes the forces that cause sliding, the required coefficient of friction is

$$f = \frac{H - P_T}{W} = \frac{13,000}{33,160} = 0.39$$

This seems to give a safety factor of the permissible coefficient divided by the required one, $0.45/0.39 = 1.15$, which is far below the desired value. However, assume that the passive resistance of the soil in front of the wall is 10 times its active pressure[1] so that $P = 10 \times 15h_1^2$. Then, for a depth of 4.5 ft, this gives 3,040 lb extra resistance per ft of wall, which increases the safety factor to

$$\frac{0.45}{(13,000 - 3,040)/33,160} = \text{about } 1.5$$

The bottom of the toe will be sloped upward as pictured in Fig. 8-18 in order to take advantage of the fact that the coefficient of shearing friction of earth on earth was assumed to equal 0.55 at the start of the problem. Since Fig. 8-23(c) shows that, before failure, most of the pressure diagram will be acting upon the toe, this higher coefficient will be used. Then the new safety factor is

$$\frac{0.55}{(13,000 - 3,040)/33,160} = 1.83$$

This is not quite up to the original requirements, but it will be accepted.[2] A cutoff wall like that shown under the back of the heel of the wall in Fig. 8-37 would be of some help if needed.

8-13. Design of the Heel. The footing, as it has been assumed in the preceding article ($D = 30$ in.), will now be designed for strength.

From Fig. 8-23(a), it is apparent that the weight of the earth and the heel and also the pressure from the surcharge are the forces that will affect the heel when the wall tends to tip over, causing the heel to bend as shown in Fig. 8-24(a). The pressure upon the foundation tends to relieve this bending. Figure 8-24(b) indicates the portion of the pressure diagram that acts in this way.

---

[1] An assumption that approximates Rankine's theory.

[2] Earth-borne retaining walls generally must be heavier than those which are on rock in order to prevent sliding. However, when one considers all probable forces, as has been done here, a safety factor of 2 is very conservative. If P_v is added to the vertical forces, the situation is improved.

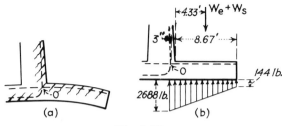

Fig. 8-24

The heel is a cantilever beam which projects from the line of resistance, the tensile steel in the stem. Therefore, taking moments about O,

$$M = (W_e + W_s) \times \text{lever arm} + \text{weight of heel} \times \text{lever arm}$$
$$- \text{area of pressure diagram} \times \text{lever arm}$$
$$= (19{,}100 + 390)(4.33 + 0.25) + \frac{2.5 \times 8.92^2}{2} \times 150$$
$$- \frac{144 \times 8.92^2}{2} - \frac{2{,}544 \times 8.92}{2} \times \frac{8.92}{3} = 64{,}700 \text{ ft-lb}$$

To find a trial value for the reinforcement, assume that $d = 27$ in. and $j = 0.87$. Then

$$A_s = \frac{M}{f_s j d} = \frac{64{,}700 \times 12}{18{,}000 \times 0.87 \times 27} = 1.84 \text{ in.}^2$$

The bars that are used should be spaced in multiples of $4\frac{1}{2}$ in. so as to match those coming down from the stem. Using No. 8 bars at $4\frac{1}{2}$ in. c.c. gives

$$A_s = \frac{0.79 \times 12}{4.5} = 2.11 \text{ in.}^2$$

Therefore,

$$p = \frac{A_s}{bd} = \frac{2.11}{12 \times 27} = 0.0065 \quad \text{and} \quad pn = 0.0065 \times 10 = 0.065$$
$$k = \sqrt{2pn + (pn)^2} - pn = 0.3 \quad \text{(see Fig. 10, Appendix)}$$
$$j = 1 - \frac{0.3}{3} = 0.9$$
$$f_s = \frac{M}{A_s j d} = \frac{64{,}700 \times 12}{2.11 \times 0.9 \times 27} = 15{,}100 \text{ psi}$$
$$f_c = \frac{2M}{kjbd^2} = \frac{2 \times 64{,}700 \times 12}{0.3 \times 0.9 \times 12 \times 27^2} = 660 \text{ psi}$$
$$V = W_e + W_s + \text{heel} - \text{pressure diagram}$$

Therefore,

$$V = 19{,}100 + 390 + 2.5 \times 8.92 \times 150 - 144 \times 8.92 - \frac{2{,}544 \times 8.92}{2}$$

$$= 10{,}200 \text{ lb}$$

$$v = \frac{V}{bd} = \frac{10{,}200}{12 \times 27} = 31.5 \text{ psi}$$

$$u = \frac{V}{(\Sigma o)jd} = \frac{10{,}200}{3.14 \times 12/4.5 \times 0.9 \times 27} = 50 \text{ psi}$$

Before calling these results satisfactory, the heel must be checked to see that, if the wall tips so as to relieve the upward foundation pressures under the heel, the resultant stresses will not exceed the elastic limit of the steel —$f_y = 36{,}000$ psi—or the ultimate strength of the concrete. For this case,

$$M = (W_e + W_s) \times \text{lever arm} + \text{weight of heel} \times \text{lever arm}$$

Therefore,

$$M = (19{,}100 + 390)(4.33 + 0.25) + \frac{2.5 \times 8.92^2 \times 150}{2} = 104{,}200 \text{ ft-lb}$$

Since this does not exceed twice the 64,700 ft-lb previously calculated, the bars in the heel will not be overstressed. The design will be accepted.

The heel will be sloped on the top from $D = 30$ in. at the stem to 18 in. at the back edge so as to save concrete, but the stresses will not be recomputed. Alternate bars may be cut off at about 7 ft from the stem, but the others will be extended full length and will be hooked as pictured in Fig. 8-29.

8-14. Design of the Toe. Part of the pressure diagram of Fig. 8-23(*b*) is reproduced in Fig. 8-25(*a*). Under its influence the toe will bend as shown in Fig. 8-25(*b*), but the weight of the concrete of the toe will be counted upon to counteract some of the bending moment. Technically, the frictional resistance along the base and under the toe will also tend to annul part of the bending moment, but it is too uncertain and too theoretical to rely upon. Then

$$M = (3{,}352 - 2.0 \times 150)\frac{3.5^2}{2} + 0.67 \times 998 \times \frac{3.5^2}{2} = 22{,}800 \text{ ft-lb}$$

A trial area of the bars is, assuming $d = 24$ in., because of the sloping bottom of the toe, and $j = 0.87$,

$$A_s = \frac{M}{f_s jd} = \frac{22{,}800 \times 12}{18{,}000 \times 0.87 \times 24} = 0.73 \text{ in.}^2$$

Since the stem reinforcement is No. 8 bars $4\frac{1}{2}$ in. c.c., assume that every other bar from the stem is bent around into the bottom of the toe. Then

$$A_s = 0.79 \times \frac{12}{9} = 1.05 \text{ in.}^2$$

Neglect the bars in the top of the toe.

$$p = \frac{A_s}{bd} = \frac{1.05}{12 \times 24} = 0.0036 \qquad pn = 0.0036 \times 10 = 0.036$$

$$k = \sqrt{2pn + (pn)^2} - pn = 0.235 \qquad j = 0.92$$

$$f_s = \frac{M}{A_s jd} = \frac{22{,}800 \times 12}{1.05 \times 0.92 \times 24} = 11{,}800 \text{ psi}$$

$$f_c = \frac{2M}{kjbd^2} = \frac{2 \times 22{,}800 \times 12}{0.235 \times 0.92 \times 12 \times 24^2} = 366 \text{ psi}$$

$$v = \frac{V}{bd} = \frac{0.5(4{,}350 + 3{,}352)3.5 - 300 \times 3.5}{12 \times 24} = 43 \text{ psi}$$

$$u = \frac{V}{(\Sigma o)jd} = \frac{12{,}400}{3.14 \times \dfrac{12}{9} \times 0.92 \times 24} = 134 \text{ psi}$$

These stresses will be accepted.

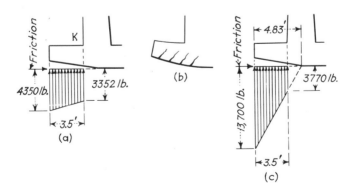

Fig. 8-25

The magnitudes of v and u are conservative, as they should be. Web reinforcement in such members is not desirable. The bars should be hooked to guarantee good anchorage. The toe will be shaped and the reinforcement will be arranged as shown in Fig. 8-29.

If the safety factor is tested by doubling H, and if the resultant pressure diagram in Fig. 8-23(c) is applied to the toe, the upward pressures will be as shown in Fig. 8-25(c). Then

$$M = (3{,}770 - 300) \times \frac{3.5^2}{2} + 0.67 \times 9{,}930 \times \frac{3.5^2}{2} = 62{,}000 \text{ ft-lb}$$

$$f_s = \frac{62{,}000 \times 12}{1.05 \times 0.92 \times 24} = 32{,}000 \text{ psi}$$

$$f_c = \frac{2 \times 62{,}000 \times 12}{0.235 \times 0.92 \times 12 \times 24^2} = 1{,}000 \text{ psi}$$

These two stresses are within the allowable values.

$$v = \left(\frac{13{,}700 + 3{,}770}{2} - 300\right) \frac{3.5}{12 \times 24} = 102 \text{ psi}$$

$$u = \frac{29{,}500}{3.14 \times \frac{12}{9} \times 0.92 \times 24} = 319 \text{ psi}$$

The shearing stress and bond are satisfactory since v can be

$$55 \times 2 = 110 \text{ psi}$$

and u can be $175 \times 2 = 350$ psi. Bars E in Fig. 8-29 are added for the reasons explained in Art. 8-16.

8-15. Design of the Parapet. The parapet at the top of the wall must be made sufficiently strong to withstand the effect of a vehicle colliding with it. The force of such a blow is problematical, but since it may be spread over a considerable length of the parapet, it will be assumed to be 500 lb/ft of wall, applied as shown in Fig. 8-26. It will not be considered in the design of the main wall because of the latter's safety factor and inertia.

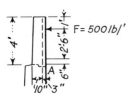

Fig. 8-26

Assuming $j = 0.9$,

$$A_s = \frac{M}{f_s j d} = \frac{500 \times 36}{18{,}000 \times 0.9 \times 10} = 0.11 \text{ in.}^2$$

Using No. 4 bars 12 in. c.c., $A_s = 0.2$ in.² These bars will be adopted and no further analysis is necessary.

Fig. 8-27   Partially completed retaining wall, Exit Plaza of the Lincoln Tunnel, New York City.

8-16.   Arrangement of Reinforcement and Other Practical Details.
Figure 8-27 is a photograph of the lower portion of the stem of a 35-ft retaining wall which is located on the south side of one of the plazas of the Lincoln Tunnel at New York City.   The work has been stopped at a horizontal construction joint.   It shows many of the features that have been or will be discussed, such as the division into sections by expansion joints, keyways, form ledges, dowels, flashing at expansion joints, and extensions of bars for the next pour.

Expansion (or contraction) joints should be located about 30 to 40 ft apart so as to eliminate shrinkage and temperature cracks.   Special horizontal reinforcement will be placed in the stem of each section to hold it together as a unit.   For this purpose, an area of steel equal to 0.2 per cent of the cross section of the stem will be used with about two-thirds of the steel near the front face because of its greater exposure.   The parapet will have reinforcement equal to 0.3 per cent.   Then

$$A_s = 0.002 \frac{(20 + 31)}{2} \times 12 = 0.61 \text{ in.}^2/\text{ft of height of stem}$$

Fig. 8-28 Placing reinforcement in the footing of a retaining wall, Lincoln Tunnel, New York City.

Use No. 6 bars 12 in. c.c. at front and 24 in. c.c. at back (A_s = 0.66 in.^{2}).

A_s = 0.003 × 12.5 × 12 = 0.45 in.^{2}/ft of height of parapet

Use No. 4 bars 10 in. c.c. front and back (A_s = 0.48 in.^{2}).

The longitudinal bars in the footing itself will be designed upon the basis of their use as ties; for a wall that is on earth, they should be 0.1 to 0.2 per cent of the cross section of the base in order to serve as adequate temperature reinforcement. Of course, when a footing is on rock, these bars are needed only as ties because the keying effect of the rough rock will not let the wall move.

Figure 8-28 is a picture of the base of the wall in Fig. 8-27—the small-heel and large-toe type. Note the way the toe reinforcement is bent up to serve as dowels for the stem, the cutting off of alternate bars in the toe, the tie bars, the heel reinforcement which the men are placing, the strips of concrete which will support the future chairs or spacers to be placed under the bars, and the timbers in the background for making the keys and form-supporting strips which are shown completed in Fig. 8-27.

For large walls, stagger the lengths of *A* and *B* to avoid splicing all bars at the same point.

A final sketch of the wall pictured in Fig. 8-18, as it would be made for contract purposes, is shown in Fig. 8-29. The explanations for certain practical details that have been shown in the sketch are as follows:

1. Bars A are extended as dowels for the stem, splicing bars C which are placed on the concrete of the construction joint after the base is set. (Bars should not be "hung in the air" if it is avoidable.)   If the spacing of the bars is small, bars A should be placed farther from the back so that bars C can be set at their rear instead of beside them, thus avoiding a "screening effect" which might seriously injure the development of bond on the backs of these bars.   The radius of the bend in bars A should be reasonable but not excessive.

2. Bars B are hooked into the footing and extended into the stem as

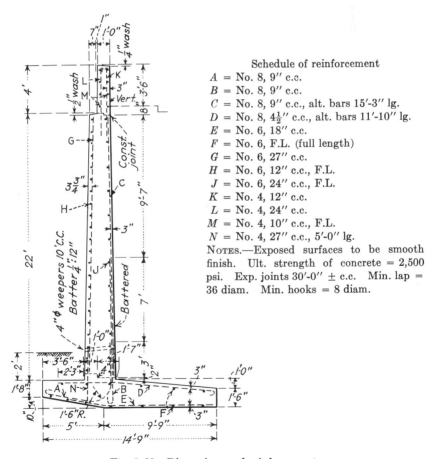

Schedule of reinforcement

A = No. 8, 9″ c.c.
B = No. 8, 9″ c.c.
C = No. 8, 9″ c.c., alt. bars 15′-3″ lg.
D = No. 8, $4\frac{1}{2}$″ c.c., alt. bars 11′-10″ lg.
E = No. 6, 18″ c.c.
F = No. 6, F.L. (full length)
G = No. 6, 27″ c.c.
H = No. 6, 12″ c.c., F.L.
J = No. 6, 24″ c.c., F.L.
K = No. 4, 12″ c.c.
L = No. 4, 24″ c.c.
M = No. 4, 10″ c.c., F.L.
N = No. 4, 27″ c.c., 5′-0″ lg.

NOTES.—Exposed surfaces to be smooth finish.   Ult. strength of concrete = 2,500 psi.   Exp. joints 30′-0″ ± c.c.   Min. lap = 36 diam.   Min. hooks = 8 diam.

Fig. 8-29   Dimensions and reinforcement.

far as they are required in order to avoid unnecessary laps.   However, these bars must not be too long, or they will become difficult to hold in place.   They are also useful in resisting the effect of the downward shear which comes from the loads on the heel.   If the heel is large and all the bars from the stem are bent like bars A, there is no reinforcement that extends directly into the region of the compressive forces caused by the cantilever action of the heel.

3. Bars D are extended into the toe to develop them beyond the neutral axis of the stem.   The designer should notice that there is considerable question regarding the bond stresses along bars D in front of the stem.   A glance at Fig. 8-1(c) shows the fundamental actions of the parts.   Both the heel and the toe tend to rotate in the same direction. A substantial fillet at the junction of the toe and the stem might help to develop the bond under the compressive side of the stem, but it would be somewhat troublesome to build.   In fact, bars A are also a possible source of weakness in bond, but one may look upon them as "cables" carried through the concrete to prevent closing of the angle between the top of the toe and the front of the stem.

4. Bars E are added arbitrarily in the footing to care for the bending stresses caused by the weight of the stem before the backfill is placed.

5. Bars F are below the main bars so as to simplify the supporting of the latter.

6. Bars G are set on the construction joint so as to support bars H which serve as temperature reinforcement.

7. Bars J are in front of A, B, and C because it is assumed that the back form for the stem will be erected first; then C will be placed.   Bars G will be set next; then bars H will be wired to them, after which the front forms will be erected.

8. Bars K are bent to knit the top of the parapet together and to develop them.

9. Bars L are stuck into the wet concrete and serve as ties for bars M, the latter being on the outside because they are placed later.   They might be placed inside of K and L.

10. Bars N are anchors to hold bars G and H so that they will not be displaced during the concreting.

8-17.   Stone-faced Retaining Walls.   A buttressed masonry retaining wall is pictured in Fig. 8-30.   Such a wall must be more or less of the gravity type.   However, walls of similar appearance may be built with stone facing which is backed by reinforced concrete as shown in Fig. 8-31.

If the masonry is applied on the face of the concrete and anchored thereto after the latter is set, the wall should be designed as an ordinary reinforced-concrete structure, neglecting the facing.   However, if the

Fig. 8-30   A buttressed masonry retaining wall, Weehawken, New Jersey.

Fig. 8-31   Construction of a stone-faced retaining wall, New Jersey Plaza of the Lincoln Tunnel.

stone facing is laid first, as it should be, a few feet at a time, then backed up with concrete which bonds to it thoroughly, the stones should not be disregarded in the design. Bearing in mind that stress and strain are coexistent, it is not correct to assume that the stones carry no load. Therefore, some fundamental principles will be explained, and a method for the design of such walls will be outlined.

What is the distribution of the stresses upon a cross section of the wall in Fig. 8-32(a)? If the section is taken through the mortar (E = about 1,500,000 psi, for cement and lime, f_c' = about 1,500 psi) the location of the neutral axis will depend partly upon n, which may be 12 to 15. Theoretically, the situation will be as shown by the solid lines in Fig. 8-32(b). If the section passes through the granite—E = say, 6,000,000 psi and for solid granite f_c' = about 12,000 psi—the theoretical position of the neutral axis will be very different because n may be 5. The dotted lines of Fig. 8-32(b) show this latter condition. Obviously, the neutral axis will not jump from O to O' and back again at each stone; the unit stress in the mortar cannot be one thing and that in the stone something else; the stress in the steel opposite the mortar cannot be thousands of pounds different from what it is an inch or two away, opposite the stone. The stresses must be consistent and reasonable.

If the wall is assumed to be a set of blocks as shown in Fig. 8-32(c), with a rubber band tied to their backs and with shear dowels near their fronts, and if a moment is applied to the set, they will distort as pictured. The compressive stresses will be concentrated at the front edges of the blocks. The bands will stretch. Then $k = 0$, and $j = 1$. This will approximate the action of the granite and the steel alone if the wall is entirely granite and steel with no concrete. On the other hand, if the wall is composed of concrete only, k may be somewhere around 0.25 to 0.4, and j may be 0.92 to 0.87. Furthermore, the mortar in the joints, being weaker than the granite, will control the allowable magnitudes of the compressive stresses.

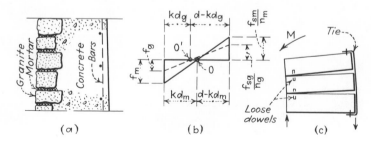

Fig. **8-32**

In general, the values of Table 8-2 may be used, assuming the best mortar and good workmanship:

Table 8-2   **Properties of Stone Masonry**

Kind of masonry	Weight, pcf	E, psi	n	Allowable compressive stress, psi, f_c	
				Cement-lime mortar	Cement mortar
Best granite ashlar	165	4,000,000	7.5	650	800
Medium granite ashlar	160	4,000,000	7.5	500	700
Rubble	150	2,000,000	15	250	350

The last column in this table may seem to indicate high values for f_c, but they will almost surely occur unless the wall is made unreasonably thick or the unit stress in the steel is kept very small. Unless the facing is relatively thin, the fact that some of the concrete is in compression may be neglected.

Therefore, in the design of masonry-faced walls, the following procedure is recommended:

1. Deduct at least 1 in. from the front face because of the weathering of joints. Then find d accordingly.

2. Use f_c and n as given in Table 8-2 for the materials and workmanship that are applicable.

3. Allow the usual maximum stress in the steel—18,000 to 20,000 psi— but the designer should notice that f_s will be low when a poor quality of masonry is used, because f_c must be low.

4. Design the wall as though it were concrete, but use the value of n from Table 8-2.

5. Be sure that the stones are well bonded into the concrete so as to resist the longitudinal shear.

6. Use expansion joints as for concrete walls.

7. Generally, unless the wall is very thick, use no front layer of steel because it will interfere with the work. Place all the temperature steel near the back.

A stone-faced wall supporting a side hill above a roadway cut is pictured in Fig. 8-33. The following features should be noticed:

1. The back of the stem is covered with membrane and asphalt waterproofing in order to avoid staining of the front by the leakage of water.

This waterproofing is protected from injury by a 4-in. layer of concrete. (Sometimes bricks or 2-in. precast-concrete blocks set in mortar are used.)

2. Since the wall rests upon rock, seepage water cannot readily pass under it. Weepers discharging onto the sidewalk will be objectionable; therefore a longitudinal drainage pipe with loose joints and surrounding gravel cover is placed on the heel. It has occasional outlets through the wall to a main drain under the roadway.

3. The toe is large and the heel is short so as to minimize excavation, but the latter must provide sufficient space for men to work in—about 21 in. clear.

4. The term *net line* means the theoretical line inside which no rock will be allowed to project. These lines are used as the theoretical dimensions of the base. The bottoms of the heel and toe are sloped upward in order to reduce excavation costs and because it is impossible and useless to require anyone to blast out rock so as to provide sharp reentrant angles or corners. A trench may look very nice on a drawing, but when the designer looks for it in the field he will probably see just a sort of gully.

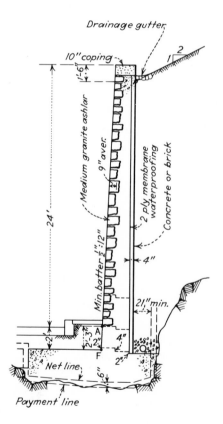

Fig. 8-33 A masonry-faced wall with membrane waterproofing.

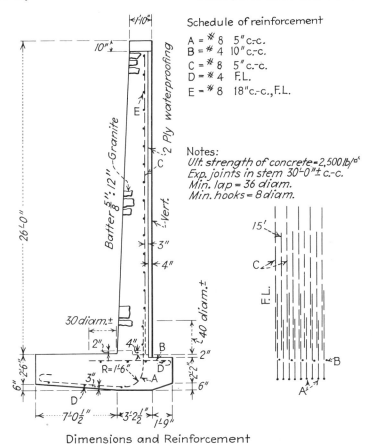

Schedule of reinforcement

A = #8 5" c.-c.
B = #4 10" c.-c.
C = #8 5" c.-c.
D = #4 F.L.
E = #8 18" c.-c., F.L.

Notes:
Ult. strength of concrete = 2,500 lb/^{a}
Exp. joints in stem 30'-0"± c.-c.
Min. lap = 36 diam.
Min. hooks = 8 diam.

Dimensions and Reinforcement

Fig. **8-34**

That is why it is usually customary to establish *payment lines* for excavation and concrete, these lines being beyond the net lines.   However, the foundation must be clean and free from loose material.   The concrete must be poured to sound bedrock as indicated in Fig. 8-33.

5. The batter on the front face of the wall is used to avoid the optical illusion of leaning forward—an effect that may result if a vertical face is viewed against the background of the sloping hillside behind it.

6. The toe is depressed below the curb and sidewalk in order to provide space for ducts and pipes.

7. The coping seals over the top of the masonry facing and the membrane waterproofing.   Both are thus protected against the penetration of water and loosening of the materials.

The reinforcement used in this wall is shown in Fig. 8-34.

8-18.   Miscellaneous Data.   The wall in Fig. 8-27 appears to be tapered—and it really is—becoming shallower in the background.   This varying of height is a common occurrence, and it introduces some problems in detail.   These problems vary with the particular situation, but some suggestions that may be helpful are the following:

1. Design the wall for a series of heights varying by 4 or 5 ft; then interpolate for intermediate heights if necessary.

2. Fix the alignment of the wall, with the front edge of the top or coping as the controlling line.

3. Use a constant width of coping and top thickness of the stem unless the range of variation of height is too great.   In such a case, make one or two definite breaks in the back of the coping and wall at contraction or expansion joints in order to decrease the thickness of the top.

4. Maintain a constant depth of coping or uniform top marking arrangement as shown in Fig. 14-5.

5. If the top of the wall is level whereas the bottom varies, keep intermediate markings and joints level also; if the top slopes and the bottom is level, keep them parallel to the bottom.   In neither case should markings or joints die out to feather edges.   It is better to make distinct breaks at vertical construction joints.   In the case of Fig. 14-5, when the top and lower V cuts approach within 1 or 2 ft of each other, omit the lower one beyond the vertical joint; if they are separating and exceed the adopted spacing by more than 1 or 2 ft, add another one.   In any such case, draw a perspective view as well as an elevation before accepting any pattern of markings.

6. Use a constant batter for the front and rear faces, starting from the top, and let the width of the stem vary at the bottom.

7. If the rate of slope of the bottom is small, the lower side of the footing may be set parallel to the grade whereas its top, if buried, may be in level steps as pictured in Fig. 8-35(a).   This uses more concrete but it simplifies the formwork and the reinforcing.

8. If the rate of slope of the bottom is large, the lower side of the footing may have to be stepped as in Fig. 8-35(b).   Note particularly the filler wall at the offset and the relation to the vertical joint.   The footings should not be made as shown in Fig. 8-35(c) because the differences in action at the junction of the higher and lower portions will almost certainly crack the wall.   Furthermore, be sure that the excavation can be stepped safely without weakening the bearing value of the soil too much.   It may be advisable to excavate as near the desired shapes as possible, then pour the footing directly on the soil, wherever it is.

9. Endeavor to use a uniform style of reinforcement and size of bars in the wall, varying the spacing in sections in order to simplify the work. Do not attempt to vary the spaces by less than 1 in.   However, make

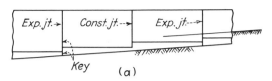

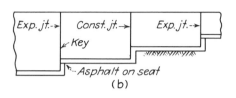

Fig. **8-35** Construction of footings for tapered retaining walls.

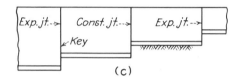

a complete change when necessary; e.g., do not try to use No. 11 bars in 6-ft walls.

PRACTICE PROBLEMS

8-1   Assume the concrete retaining wall shown in Fig. 8-36(*a*).   Let *w*, the weight of earth, = 100 pcf; ϕ, the angle of internal friction, = 34°. Using these data find the following:

a. If the earth is level and flush with point *A*, find the bending moment at the level of *K* for the earth pressure alone.

Discussion.   Use Eq. (8-6) with $\delta = 0$ to find *P*, which is inclined at the angle ϕ; then find the horizontal component of *P* in order to compute the bending moment.   *Ans. M* = 12,000 ft-lb.

b. Repeat part *a*, but include the effect of water pressure for an 8-ft head above *K* at 45 psf/ft of depth, plus ice pressure of 700 plf at *A*.

c. How much bending moment is caused at *K* by the surcharge effect of 1,000 lb/ft of wall at a point 5 ft behind *A*?   Combine this with *b* and *a*, but assume that the full magnitude of *P* is to act horizontally.

Discussion.   Use the data in Figs. 8-13 and 8-14 to find the magnitude and position of the lateral thrust.

8-2   For the retaining wall in Fig. 8-36(*a*) determine the safety factor against overturning and sliding, also draw the pressure diagram under the footing.   Assume *w* = 110 pcf, ϕ = 35°, allowable soil pressure = 7,000 psf, and the coefficient of friction = 0.6.   Include earth pressure, also

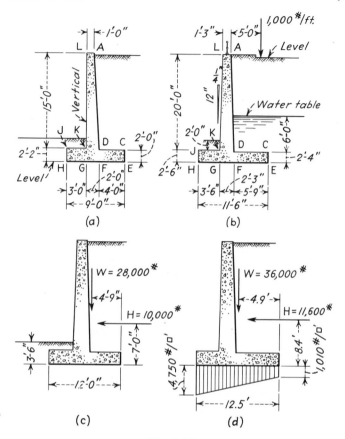

Fig. 8-36

water pressure for a head 10 ft above *HE*.   Neglect the effect of the earth
on top of the toe.   Also, neglect the abutting resistance of the earth in
front of the wall.   The earth pressure is inclined at 35°.

Ans. Safety factor against overturning about $H = 3.65$.   Safety
factor against sliding $= 1.73$.   Soil pressure at $H = 1,920$ psf and at *E*
$= 1,640$ psf.

Data for Probs. 8-3 to 8-10, inclusive (assume the earth pressure to be
horizontal):

$w = 100$ pcf
$\phi = 30°$
Allowable pressure on soil $= 6,000$ psf
Coefficient of sliding friction of soil $= 0.6$
$f'_c = 3,000$ psi $f_c = 1,000$ psi $f_s = 20,000$ psi

$v_c = 60$ psi $u = 185$ psi
Water pressure $= 45$ psf/ft of depth
Surcharge to be determined as in Art. 8-7

Consider the weight of earth on the toe.   Also include the abutting resistance of the soil in front of the wall, assuming this resisting pressure $= 300$ psf/ft of depth.

8-3 Design and detail the reinforcement for the stem of the retaining wall shown in Fig. 8-36(*b*). Consider pressures caused by earth, water, and surcharge as shown.

8-4 Determine the total vertical load W of the wall in Fig. 8-36(*b*), and compute the horizontal forces acting on it.   Then locate the position of their resultant at the base HE and find its eccentricity.

Ans. $W = 22,500$ lb 4.66 ft from E. $H = 9,700$ lb 7.25 ft above E.

$e = 2$ ft on the left side of the center line.

8-5 Recompute Prob. 8-4 if the height of the stem is reduced from 20 to 18 ft and the heel FE is 5 ft long.

8-6 With the resultant forces shown in Fig. 8-36(*c*), determine the pressure diagram under the footing.

8-7 Using the resultant forces shown in Fig. 8-36(*c*), determine the safety factor against overturning about a point 6 in. inside the edge of the toe.   Also determine the safety factor against sliding, including the effect of the soil in front of the wall.

8-8 Using the data shown in Fig. 8-36(*d*), design the reinforcement in the heel and toe.   Heel $= 6$ ft; toe $= 4$ ft; $d = 24$ in.; $h_1 = 25.2$ ft.

8-9 Using W and H as given in Fig. 8-36(*d*), see if this wall has a safety factor of 2 for overturning, sliding, and soil pressure.

8-10 Design a retaining wall for the conditions shown in Fig. 8-37.

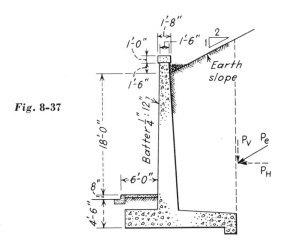

Fig. 8-37

Assume that the earth pressure is inclined at the angle ϕ as indicated. Assume that its components act as shown by the dotted lines.

8-11 Assume the heavy concrete retaining wall in Fig. 8-38. Analyze it completely if the following data are to be used: $w = 100$ pcf, $\phi = 34°$, $\delta = 0$; P is to act horizontally with its full magnitude (because of vibrations, its vertical component being assumed $= 0$); surcharge is to be as given in Figs. 8-13 to 8-15 for $W' = 1,000$ lb/ft of wall, the wheel load being applied 1 ft from the curb; ice pressure $= 500$ plf; water pressure $= 45$ psf/ft of depth with a head of 8 ft above the top of the toe; $f'_c = 3,000$ psi; $n = 9$; the maximum coefficient of friction $= 0.6$; the safety factor $= 1.75$.

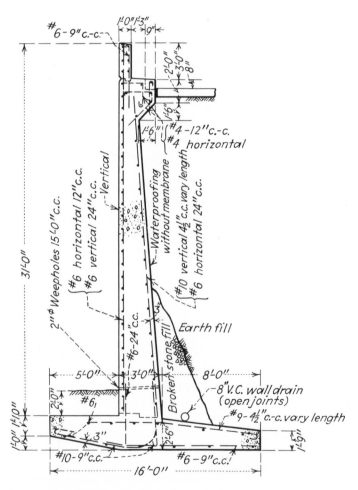

Fig. **8-38** Retaining wall designed for the connection to the Harlem River Speedway, Highbridge Park, New York City.

FOOTINGS

9

9-1. Introduction. A building or a bridge is generally considered to have two main portions—the superstructure and the substructure. The latter is often called the *foundation*. It supports the superstructure, but it may contain various parts or units of its own. There are many special kinds of foundations for which concrete is used, but this chapter will be confined mostly to reinforced-concrete footings.

The term *foundation* generally includes the entire supporting structure. Sometimes, as in the discussion of retaining walls, it is used to designate the material upon which the wall is supported. It must not be confused with the word *footing*, which generally applies only to that portion of the structure which delivers the load to the earth, as illustrated by AB in Fig. 9-1(*b*). These are called *spread footings* because they distribute the concentrated load over a large area which has a lower intensity of pressure. When the soil under a building is so weak that the footings are all joined together in one large slab or floor extending under the whole structure, the resultant footing is sometimes called a *reinforced-concrete mat*.

An *isolated footing*, which may also be called a spread footing, is an individual member that is used generally to distribute the load from the base of a column. If small, it may sometimes be made of plain concrete as in Fig. 9-5; if large, it may be reinforced like one of those shown in Fig. 9-6.

Footings are also used to support heavy walls, piers, abutments, machines, and many other structures whose large loads have to be spread over a considerable area of soil.

When concrete rests directly upon sound rock, the latter often has a strength equal to or greater than that of the concrete itself, so that there is no difficulty in obtaining sufficient support for the superstructure. Sometimes, concrete caissons or piles are used to carry the loads through inadequate material to rock or some other suitable stratum at a lower level. The illustrations here will be limited to footings that transmit the loads directly to earth or piles.

9-2. Some Fundamental Principles.   There are certain fundamental principles to be considered in the design of earth-borne footings.   When a load is placed upon earth, the latter is compressed somewhat.   The amount of this deformation depends upon the intensity of the load, the size of the loaded area, the properties of the soil, the depth at which the load is applied, and similar matters.   These settlements will not be investigated here, but they must be kept in mind because it is essential to plan any foundation so that the entire structure will settle equally.   If the unit pressures under the footings vary greatly, the settlements of different parts of the structure are likely to vary also, causing cracks to appear.   For instance, if a wall is loaded as shown in Fig. 9-1(*a*), the ends tend to settle and to cause a crack in the top,   It is best to have the pressure uniformly distributed; but if this cannot be done, it is desirable to have the greater deformations near the center so that there will be a tendency for the structure to compress near the top rather than to open up.   Of course, cracks at the bottom should also be prevented.

No footing should tilt harmfully.   The resultant of the resisting pressures under the footing will pass through the center of gravity of the applied loads, including those of the footing itself.   However, if the resultant of the loads is offside the center of gravity of the bearing area, the side having the higher pressures may settle more than the other side and cause the structure to lean in that direction.

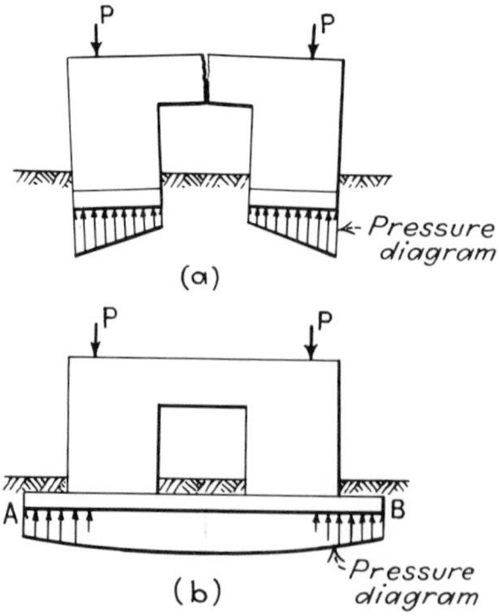

The effects of live loads and impacts are usually omitted in the study of the settlements of footings unless the structure is a warehouse or a building which may be subjected to large live loads for long periods of time.   The reason for this is the fact that the soil will not usually move or squeeze out quickly unless severely overloaded.   One-half or more of the live load may be considered in the case of warehouses, but the uniformity of the distribution of the foundation pressures for dead load alone must be investigated also.   However, the

Fig. 9-1

live load and impact must be included in the loads for which the individual footings are designed.

If the foundation for a building is partly on rock and partly upon earth, the situation is likely to be dangerous. This is obvious. In such a case, one should use caissons or piles down to rock under the portion of the structure that does not rest directly upon it, isolate the two portions of the structure so that the differential settlement will not cause trouble, or if the condition is unavoidable, use a very low intensity of pressure upon the soil.

It is advisable to place the bottoms of all footings below the frost line when on earth, usually $3\frac{1}{2}$ to 5 ft in cold climates.

The safe bearing value of any soil is a difficult thing to ascertain. Borings should be made at the site so that the properties of the soil can be determined sufficiently before any important structure is built; load tests of the soil in place may also be helpful. However, Table 8-1 may be used as a general guide. When a foundation goes to a depth of 10 ft or more below the surface or any adjacent excavation in natural undisturbed granular soil, these specified pressures may be increased somewhat.

The vertical component of hydrostatic pressure under a foundation is not considered as an additional load in the design of the footings. Since $\Sigma V = 0$, it makes no difference if the structure tends to float like a boat as a result of the hydrostatic uplift because the total pressure is dependent upon the weight of the entire structure. However, the side walls of basements in such special cases must be designed to withstand lateral pressures, and the basement floor will have to be of the mat type—or partially so.

As stated in connection with retaining walls, the pressure upon the soil should be computed by including the weight of the footing, or at least the excess weight of the footing over that of the earth originally above the bearing area, but the loads for which the footing itself must be designed need not include the weight of the footing itself because the wet concrete is already supported by the earth before the former sets. Backfill is also neglected when computing shears and bending moments in footings because its weight is generally fairly uniform over the footing and its direction is opposite to that of the bearing pressure. This reduced intensity of load for design purposes may be called the *net* pressure under the footing.

Table 1-8 gives some data regarding the allowable unit stresses in the concrete of footings.

Some general information about the action and design of isolated footings and wall footings on soil is as follows:

1. For concentric loads without accompanying overturning tendencies,

assume the load to be uniformly distributed over the bearing area of the column, pedestal, wall, or metallic column base and footing, and also over the soil or the piles underneath.

2. Reactions from piles are assumed to be concentrated at their centers.

3. All footings, whether having sloped or flat tops or pedestals, should be so proportioned that the allowable unit stresses in the materials are not exceeded at any point for working-stress design. For ultimate-strength design, there should theoretically be no point of comparative weakness with respect to the strength of other parts and of the footing as a whole if the design is to be efficient. However, footings should be thick and stiff. The reinforcement or the shearing strength may be the controlling factor, rather than the compression in the concrete. It often happens that the bond strength of the bars is critical, meaning that it is sometimes desirable to use a larger number of small bars instead of a lesser number of large ones.

4. The external bending moment on any section is to be determined on the basis of a vertical plane extending clear across the footing. Compute the moment of all the forces acting on one side of this plane. The location of this plane should be as follows:

a. At the face of a concrete column, pedestal, or wall, as indicated by *AB* in Fig. 9-2(*a*), *EF* in sketch (*c*), and *LM* in (*d*).

b. Halfway between the middle and edge of masonry walls, as shown by *AB* in Fig. 9-2(*b*).

c. Halfway between the face (or tips of flanges) of a steel column and the edge of the metallic base.

5. A footing under a long wall is primarily a beam that is cantilevered on both sides of the wall. This is called a *one-way* footing because the bending acts in one direction only. In this case, the bending moment can be computed on the basis of a 1 ft strip as a typical piece, with the maximum moment at *AB*, Fig. 9-2(*a*) and (*b*). A footing under a column as shown in Fig. 9-2(*c*) is a *two-way* footing because it bends in two directions and requires reinforcing accordingly.

6. The width resisting compression at any section is to be assumed as the entire width of the top of a flat-topped footing at that section, and the area resisting compression is the area above the neutral axis of the section. Sloped-top footings generally introduce some complications in this respect but these will be discussed later.

7. According to the Code, calculations for the shear to be used in the equation for computing the bond unit stress,

$$u = \frac{V}{(\Sigma o)jd} \tag{9-1}$$

are to be based upon the same section and loading as for bending moment.

Points of change of section or of reinforcement are also to be investigated similarly. However, refer to Fig. 9-3. If tested to failure, a section through a square footing under a square concrete column might punch through a "pyramid of rupture" as shown by $ABDC$ in sketch (a). This is indicated further by the dotted lines in (b). The corners G and H may not be well defined. The cracks in the bottom might look somewhat like those in sketch (c). In the case of a footing under a concrete wall, there is a tendency to punch out a "wedge of rupture" as indicated by the dotted lines JK and LM in sketch (d). It therefore seems to be desirable to be sure that the necessary bond strength can be developed outside of the imaginary surfaces AB and CD for those bars in the base of the pyramid of rupture. Hooks may or may not be needed, depending upon the available lengths, such as EB. Of course, the magnitude of V in Eq. (9-1) is that determined by the pressure under EB. It is desirable to check the bond at both the edge of the column and the edge of the pyramid of rupture. In general, the bond stresses are likely to be very important in footings because of the large intensity of pressure and the relatively short

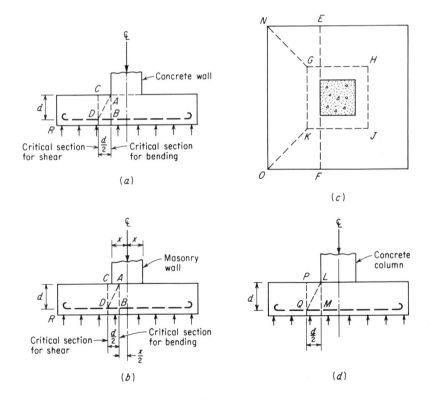

Fig. 9-2

lengths of the cantilevered portions of the footings. Round columns may be treated as though they were square columns having the same cross-sectional area.

8. The critical section for shear to be used in computing diagonal tension is assumed to be that determined by vertical planes through the base of an imaginary pyramid or wedge at the level of the reinforcement and which projects only $d/2$ beyond the section assumed for maximum bending, such as section CD in Fig. 9-2(a) and (b), and GK and PQ in (c) and (d). This is more conservative than the principles used in beam design. It is empirical but seems to agree with test results.[1] For a square footing under a square concrete column, the magnitude of V to be

---

[1] C. S. Whitney, Ultimate Shear Strength of Reinforced Concrete Flat Slabs, Footings, Beams, and Frame Members without Shear Reinforcement, *J. ACI*, October, 1957.

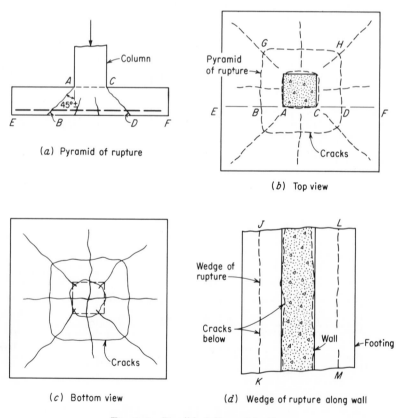

(a) Pyramid of rupture

(b) Top view

(c) Bottom view

(d) Wedge of rupture along wall

Fig. 9-3   Possible failure of footings.

used in

$$v = \frac{V}{b_o d} \tag{9-2}$$

for one side is the total net pressure under area $NGKO$ in Fig. 9-2(c). In Eq. (9-2), b_o is the width of the side, such as GK. However, v should not exceed $2\sqrt{f'_c}$ for WSD (or $4\phi\sqrt{f'_c}$ for USD) unless web reinforcement is used—a thing that is usually avoided. If v does exceed this value, web reinforcement should be provided but the steel stress allowed should be limited to 50 per cent of that used in ordinary beams. For a wall footing, the magnitude of V is the net pressure under a typical 1-ft strip having the length RD in sketches (a) and (b).

9. When a footing is supported on piles as illustrated in Fig. 9-11(a), with a critical section at HC, any pile whose center is 6 in. or more outside of HC is assumed to have its entire force effective in causing shear on HC, whereas any pile whose center is 6 in. or more inside of HC (toward the center of the footing) is assumed to have no effect on the shear on plane HC. For piles with their centers between the preceding two points, the portion of the pile reaction which is assumed to produce shear on HC may be interpolated on a straight-line basis between the entire load at 6 in. outside of the section to zero load at 6 in. inside the section. However, bearing in mind the fact that piles cannot be driven so exactly as one might wish, it is best to be somewhat conservative and to make the footing strong enough so that inaccuracies of pile locations will not endanger the structure. In the case of small footings on piles, the author prefers to assume that any pile may be 6 in. outside of the intended location; in large footings and mats, 12 in. outside.

10. The Code states that the minimum edge thickness above the reinforcement, even for sloped-top footings on soil, should be 6 in.; for footings on piles, 12 in. Plain concrete footings may be used under some walls and light loads, but they should not have an edge thickness of less than 8 in. when on soil or 14 in. above the tops of piles. However, plain concrete footings are seldom used when piles are necessary.

9-3. Concrete Pedestals. Even when a structure rests on rock, it is customary to use some sort of concrete footing under each column. Thick, chunky members like the footing in Fig. 9-5(a) are sometimes called *pedestals*. This term is also used to describe the upper part of the stepped footing in Fig. 9-6(a).

Some footings are primarily blocks rather than members which act as beams. For instance, suppose that a column is supported by a concrete footing which rests upon rock as pictured in Fig. 9-4(a). The reinforcement which is shown in the bottom is almost useless. If the rock is

properly prepared and roughened, the footing cannot possibly spread sideways so as to open up cracks and to stretch the reinforcement because of
the shearing resistance of the rock. It must be remembered that the
reinforcement in concrete cannot be effective in its ordinary function of
resisting tension unless the concrete can elongate or bend. Therefore,
since there is no strain in the bars in the case under discussion, there can
be no tension in the steel. However, rock cannot be carved out with a
knife. There is the likelihood of overbreakage, in which case the men in
the field may backfill the excess space and pour the concrete on the loose
material. Reinforcing may then be necessary unless the overbreakage is
filled with concrete.

The possible lines of action of the forces in a thick base are somewhat
as pictured in Fig. 9-4(*b*), these forces being a combination of direct compression and shear. The only reinforcement that will do much good in
such a case is that which will prevent the formation of cracks at the top
of the footing, such as the cracks near the upper corners in Fig. 9-4(*b*).
These cracks are not likely to occur under small loads if the footing or
pedestal has sufficient bearing area.

The Code states that the bearing pressure on the top of a footing or
pedestal under a concrete column or the base plate of a steel column for
WSD should not exceed $0.25f_c'$ if the entire area is loaded; $0.375f_c'$ if one-
third or less of the area is loaded; and proportionately between these
values when the loaded area is less than the full area but more than one-
third of that area, unless there is appreciable eccentricity of loading.
However, the increased bearing pressure above $0.25f_c'$ is allowed only when
the least distance between the edges of the loaded and unloaded areas is

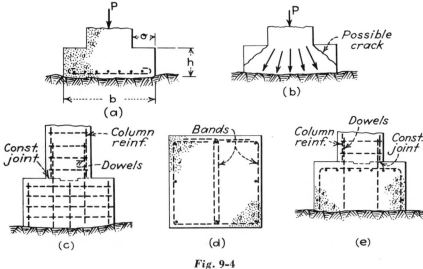

Fig. 9-4

at least one-fourth of the parallel side of the loaded area. For USD, these stresses might be doubled.

The depth and width of a pedestal or plain concrete footing should be such that the tension in the concrete in flexure does not exceed $1.6 \sqrt{f'_c}$ for working-stress design or twice that for ultimate-strength design.

Naturally, if the bearing pressure exceeds the limits stated previously, the base should be reinforced somewhat like a column. If the shearing stress is too high, the depth should be increased. When the bending strength is at all questionable, reinforcement should be added.

Figure 9-4(c) to (e) show some ways of arranging the reinforcement in pedestals and in thick bases resting on rock or on a concrete footing. The first two picture hoops or bands with a few supporting vertical ties, alternate bands being reversed in direction. However, these bands shown in (d) are not detailed properly; they should be made so as to get a better overlap at the center of the sides. Sketch (e) shows a two-way mat of reinforcement with the bars bent down. These arrangements are intended to prevent top bulging and cracking like that pictured in sketch (b).

Whenever a heavily reinforced column rests upon a concrete pedestal or footing, the stresses in the longitudinal bars of the column cannot suddenly vanish. If the steel is needed in the column shaft, it, or its equivalent, is also needed where the column joins the pedestal. It is therefore necessary to extend the column bars down into the pedestal or footing far enough to transfer their stress into the pedestal or footing by means of allowable bond stresses. However, this may be inconvenient in the field because of the difficulty of supporting the steel when the pedestal or footing is built. A more practical method is to have dowels installed in the pedestal or footing as indicated in Fig. 9-4(c). Of course, the total area of the dowels should not be less than that of the column reinforcement whose stress they are to transfer downward, and their length in both the column and the pedestal or footing should be enough to permit the transfer of the required forces by allowable bond stresses; and in fact, the bond should be able to develop the strength of the dowels. It is inadvisable to use less than four dowels, and the diameter of a dowel should not be more than $\frac{1}{8}$ in. larger than that of the column reinforcement. The compressive stress in the concrete at the bottom of the column itself is considered to be transferred by bearing into the pedestal or footing, but the unit stress in the latter should not exceed that allowed for the concrete below, based on the relative sizes of the loaded areas.

If a column or pedestal is round or octagonal, an imaginary square section having the same area as the column or pedestal may be taken as equivalent to the actual section when footings are designed.

The dimensions of a pedestal like that shown in Fig. 9-4(a) are largely a matter of judgment. If founded upon rock, the distance o should not be less than 3 to 6 in. so that the forms for the column can be supported upon the pedestal. The depth h must be sufficient to develop the dowels. In no case should an unreinforced-concrete pedestal have a depth that is less than the offset o, but it is desirable to have the depth h equal two times o or more.

9-4. Plain Concrete Footings. When a plain concrete footing is supported on earth, it must be large enough to spread the load without exceeding the allowable pressure on the soil. In such a case, too, a certain amount of bending will be produced in the footing itself, which then becomes a short thick cantilever beam acting to resist bending in two directions. It may be designed as a beam of homogeneous material as far as compression and tension are concerned, with a limiting tensile stress of $1.6 \sqrt{f_c'}$. The diagonal tension produced by shear is to be computed on the basis of a section $d/2$ from the column, with a limiting stress of $2.0 \sqrt{f_c'}$ for WSD or $4\phi \sqrt{f_c'}$ for USD. The principles will be illustrated by direct application to a problem.

Example 9-1. Find the stresses in the plain concrete footing of Fig. 9-5(a) if $f_c' = 2,500$ psi.

Although the intensities of the soil pressures may vary somewhat as shown by the dotted diagram in sketch (a), it is customary and satisfactory to assume that they are uniformly distributed. The rectangle $SVXW$ will therefore represent these pressures. Since the weight of the footing itself is excluded, the net upward pressure

$$p = \frac{P}{B^2} = \frac{30,000}{2.5^2} = 4,800 \text{ psf}$$

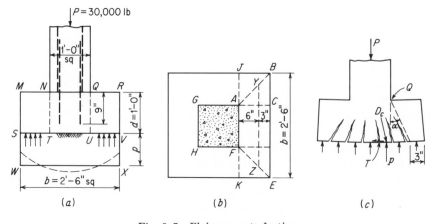

Fig. 9-5 Plain concrete footing.

The pressure under the column on the area $AFHG$ is

$$f_c = \frac{P}{A} = \frac{30,000}{144} = 208 \text{ psi}$$

This is safe because the area of the footing is more than three times that of the column and f_c is much less than $0.25f'_c$. Therefore, for a small load like this, no reinforcement is needed as far as the pressure on top of the footing is concerned.

The net upward pressures upon the bottom of the footing tend to cause compression in the top and tension in the bottom so as to try to split it apart, as shown in exaggerated fashion in Fig. 9-5(c). The concrete, since it is unreinforced, must act as a beam of homogeneous material in resisting this tension. Therefore, assume that a typical portion of the footing that is outside of the column area $AFHG$ in Fig. 9-5(b) is cut by the plane JK in sketch (b). The portion $JKEB$ acts as a cantilever about face AF in resisting the bending moment produced by the pressures from U to V of sketch (a). Therefore,

$$M = \frac{p \times JK \times AC \times AC}{2}$$

$$= \frac{4,800 \times 2.5 \times 0.75 \times 0.75}{2} = 3,370 \text{ ft-lb or } 40,500 \text{ in.-lb}$$

Taking the full section JK as effective in resisting this moment,

$$f_t = \frac{M}{Bd^2/6} = \frac{40,500 \times 6}{30 \times 12^2} = 56 \text{ psi}$$

This is less than $1.6\sqrt{2,500} = 80$ psi and will be accepted.

Now consider the diagonal tension in this footing. The upward pressures acting outside of a distance $d/2$ from the column are the ones to consider as effective in producing this diagonal tension, and they act on all four sides. The total depth is to be taken as d. Therefore, the trapezoidal area $YZEB$ in (b) shows a typical area to be considered for one side. $YZ = 2$ ft. Then

$$v = \frac{V}{b_od} = \frac{4,800 \times 0.5(2 + 2.5)0.25}{24 \times 12} = 9.35 \text{ psi}$$

which, as should be expected, is far less than $2.0\sqrt{2,500} = 100$ psi.

The depth QU in Fig. 9-5(a) should be able to develop the dowels extending into the column. If these are No. 5 bars, if any hooks or right-angle bends at their bottoms are neglected (as they should be when in compression), if there is a cover of 3 in. at the bottom, if the steel stress is 16,000 psi allowed and the maximum bond stress is 385 psi, then the length of bar required below NQ is

$$L_s = \frac{16,000 \times 0.31}{1.96 \times 385} = 6.6 \text{ in.}$$

Therefore, the footing is thick enough, but the dowels will be made so that they extend 9 in. into the footing.

In general, one should question the use of plain concrete footings when the loads are heavy and the structure is important.

9-5. Reinforced-concrete Footing for a Column. Column loads are usually large enough to require a reinforced-concrete footing to spread the pressure over a considerable area of soil. Such a footing may be made as one flat slab like that shown in Fig. 9-2(c) and (d). However, the thickness is controlled by the bending, bond, and shear conditions at and close to the column, with the result that the use of this thickness for the entire footing may be wasteful of concrete. The use of a stepped footing like that in Fig. 9-6(a) with a pedestal between the column and the footing proper is usually more economical in the use of concrete and steel because

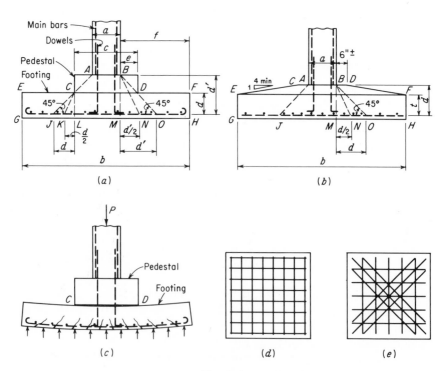

Fig. 9-6

it is deeper and stronger near the center and thinner near the outer cantilevered portions, where the bending moment and shear are decreased.
On the other hand, the footing might have the top sloped as in Fig. 9-6(*b*)
so as to have the greatest thickness near the center and to be progressively
thinner as the bending moment, bond requirements, and shear decrease
toward the edges.   However, the top should not slope at more than about
one vertical to four or five horizontal; otherwise the wet concrete may tend
to slump too much.

The design of a footing is not a very exact affair.   The members are
short and thick.   They do not curve in one direction like an ordinary
beam when under a column but tend to curve like a saucer; hence they are
structurally indeterminate.   Nevertheless, practical procedures have
given safe results during the past.

For the simple slab footing of Fig. 9-2(*c*) and (*d*), the maximum bending moment for a square footing may be computed as that due to the net
pressure under area *NEFO* acting about a vertical plane through *EF*.
The bond stress on the bars at *M* may be computed on the same basis and
then checked again to see that the ends beyond the pyramid of rupture are
sufficient to develop the bond required by the net pressures acting outside
of the 45° slopes.   The critical section for diagonal tension is at a vertical
plane through *PQ*, a distance $d/2$ out from the face of the column.   The
area to use in calculating *V* is shown as *NGKO*.

For the square stepped footing of Fig. 9-6(*a*), the width *c* of the
pedestal should probably be somewhere between two and three times the
width *a* of the column, and the thickness of the pedestal should be such
that the pyramid of rupture shown with the sides *AJ* and *BO* will be
contained within the concrete.   If these sides are outside of *C* and *D*, the
pyramid of rupture should be assumed to start at these points.   In other
words, the depth of the pedestal should be equal to or larger than the
projection *e* of the pedestal outside of the column.

The pedestal of a stepped footing should be poured monolithically with
the footing so as to make the two parts act as a unit and to avoid failure
through longitudinal shear at the junction between them.   This causes
trouble with the formwork and the deposition of the concrete because the
forms have to be held up and the concrete may slump from the higher to
the lower level.   However, the pedestal is useful not only in reducing the
stresses in the footing but in providing embedment for the dowels.   In
other words, referring to Fig. 9-6(*a*), the bending moment and bond stresses
may be computed on the basis of a vertical plane through *CL* instead of one
through *BM*, and the diagonal tension may be calculated as for a vertical
plane through *K* instead of one through *N*.   Again check the bond for
the ends *GJ* and *OH*.

The action of the pedestal of a stepped footing will be more clearly

understood by examining Fig. 9-6(c). If the footing bends under the
load, there will be compression in the concrete between C and D. The
curvature of the footing, if it could be so extreme, would produce a crack
between the footing and the pedestal, with the latter receiving a concentrated compressive load near C and D. These loads, in turn, might tend
to cause tension in the bottom of the pedestal, but it is impossible to have
the footing shorten from C to D while the pedestal elongates at the same
section unless the concrete fails along the junction. However, it is desirable to make the pedestal thick enough so that it will be sufficient for one
to assume that the pedestal is an unyielding support for the footing and
that the critical bending moments and shears will be in the cantilevered
portions of the footing beyond the pedestal.

For the square sloped-top footing shown in Fig. 9-6(b), the bending
moment and bond stresses may be computed on the basis of a vertical
plane through BM with the entire depth d; the diagonal tension, on the
basis of a section through N; and then the bond also checked for the
lengths of the bars beyond J and O, the limits of the assumed pyramid of
rupture. However, notice that if the flat top CD is made only wide enough
to support the forms for the column, then the effective depth will decrease
from d to whatever it is at the point in question. Furthermore, the
trapezoidal shape of the top when cut by a vertical plane requires that,
theoretically, one should compute the compressive stress f_c by means of
the transformed section. However, a very rough approximation for
computing the bending stress f_c at BM, sketch (b), may be made by
assuming a substitute rectangular section having the depth d and a width
equal to $(CD + JO)/2$. The thickness of footings is usually great enough
so that compression in the concrete is not critical until elongation of the
steel or diagonal tension progresses so far that the neutral axis rises close
to the top.

The reinforcement for a footing may be arranged in two layers as
shown in Fig. 9-6(d), or it can be placed in two normal and two diagonal
bands as pictured in Fig. 9-6(e). The latter seldom is worthwhile except
in very big footings like those under a chimney because of the complication
of the details of the reinforcement and the packing-up of the four layers at
the center.

For two-way footings of rectangular shape, the Code states that the
required bars in the long direction may be spaced uniformly across the full
width, the bending moment being computed on the basis of the full
section as for square footings. The bars in the short direction may also
be determined on the basis of the bending moment on the full section.
Then part of this reinforcement (across the short way) shown by Eq. (9-3)
may be uniformly distributed across a band width B centered with regard
to the column or pedestal and having a width equal to the length of the

short side of the footing, with the remainder of the bars uniformly spaced in the outer parts of the footing.

$$\frac{\text{Reinforcement in band width } B}{\text{Total reinforcement in short direction}} = \frac{2}{S + 1} \tag{9-3}$$

where S is the ratio of the long side of the footing to the short side. It may be that long, narrow footings should be designed and analyzed across the short direction in the same manner as for rectangular cantilevered beams, using the bending moment at the face of the column or pedestal and computing the shear for diagonal tension at a distance d from this same section, and employing the same allowable unit stresses as those specified in Table 1-8 for beams. In fact, the Code states that this method is one by which any footing should be tested.

Footings for walls like those shown in Fig. 9-2(a) and (b) should be designed for bending and bond as previously described, using typical strips 1 ft wide. For computing diagonal tension, they should be capable of withstanding the greater of the following requirements:

1. The shear computed about the plane CD of Fig. 9-2(a) and (b) located $d/2$ from the critical section for bending moment, with the safe unit stresses allowed in Table 1-8 for footings

2. The shear computed about a vertical plane through a section located a distance d from the critical section for bending moment, with the safe unit stresses allowed in Table 1-8 for beams without web reinforcement

If a footing is to be designed on the basis of working-stress methods, the footing should preferably have a safety factor of 2, or perhaps 2.5. This will usually be determined by the safety factor in the stresses used for the reinforcement. Hence, f_s should not exceed $f_y/2$. The locations of the critical sections for bending moment, bond, and diagonal tension will remain as previously described.

Whitney[1] has stated that tests have shown that the ultimate shear strength of footings tested to failure seems to agree fairly well with the formula

$$v_u = 100 + \frac{0.75}{d^2} M_u \sqrt{L_s} \tag{9-4}$$

for a section through the base of the pyramid of rupture. Here, M_u is the ultimate strength per inch of width, and L_s is the distance from the axis of bending moments to the point of application of the load on one side of this axis—one-half of the cantilevered portion of rectangular footings. He also recommends that the bars be more closely spaced under

[1] C. S. Whitney, Ultimate Shear Strength of Reinforced Concrete Flat Slabs, Footings, Beams, and Frame Members without Shear Reinforcement, *J. ACI*, October, 1957.

the pyramid of rupture in order to make the shearing and bending strengths more nearly equal.

Figure 9-6A may help the reader to a better understanding of the action of spread footings. Notice the following in the sketches:

(*a*) Let f represent the compressive stress at the top of a square footing along two edges of the column based upon the bending moment computed with JK and LM as the axes. The pressure under some differential area dA will cause moments about these axes with the lever arms y and x. It will also cause a resultant bending moment about some diagonal axis like that through the corner F at an angle θ normal to the lever arm $\sqrt{x^2 + y^2}$. If this axis NO is at 45° with the principal axes, the resultant of the two pressures f will be 1.4f as shown. The same applies for tensile stresses in the bottom.

(*b*) If ab and cd represent one of the bars in each 90° direction in the bottom of the footing, the force F_s in each will have a component equal to

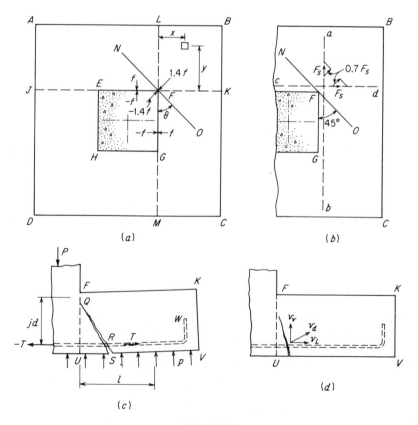

Fig. 9-6A

$0.7F_s$ normal to the 45° axis *NO*, the sum being the needed $1.4F_s$. Thus the two sets of bars at 90° really reinforce the footing so as to enable it to resist the effect of the forces anywhere under the footing.

(*c*) This shows how a crack *QS* might form when the footing is stressed highly. The tension *T* must be developed by bond at the right of *R*. The pressures *p* acting on part *SV* will cause bending about *FU* so that $Tjd = (\Sigma p)l$. It is easy to see the need for good anchorage beyond *R*. Whether some anchorage like *W* is needed depends upon the forces and the resistances involved.

(*d*) Here a tiny crack is assumed to be near section *FU*. The vertical shear on one side of the crack is pictured as v_v. At this point there must also be a horizontal or longitudinal shear v_L which is produced by the effect of the bond stresses in the pickup of tension by the steel. The resultant of these two shearing stresses is represented by v_d, the diagonal tensile stress at that point.

Example 9-2. Design a square footing like that in Fig. 9-2(*c*) to support a tied column 18 in. square which carries a load of 250 kips, using working-stress design. Assume that $f_c' = 3{,}000$ psi, $f_s = 20{,}000$ psi, $j = 0.9$, the safe net pressure on the soil = 4 ksf, and the depth of embedment = 4 ft.

The necessary footing area is $250/4 = 62.5$ ft². To allow a little for the extra weight of the concrete over that of the displaced earth, assume a footing 8 ft square as shown in Fig. 9-7. Then $p = 250/64 = 3.9$ ksf. Try a depth *d* a little over $BD/3 = 39/3 = 13$ in. Call it 14 in. and use a cover of 4 in. below the upper layer of bars in order to have the desired 3 in. under the lower layer. The upper layer has the smaller effective depth and will be used in making the design.

The bending moment about one face of the column is

$$M_{BG} = \frac{3.9 \times 8 \times 3.25^2}{2} = 165 \text{ ft-k}$$

$$A_s = \frac{M}{f_s jd} = \frac{165 \times 12}{20 \times 0.9 \times 14} = 7.85 \text{ in.}^2$$

Try ten No. 8 bars ($A_s = 7.9$ in.², $\Sigma o = 31.4$ in.²).

For *V* for computing bond stresses, use the rectangular area to the right of *BG*. Then

$$V_{BG} = 3.9 \times 8 \times 3.25 = 101.5 \text{ kips}$$

$$u = \frac{V}{(\Sigma o)jd} = \frac{101{,}500}{31.4 \times 0.9 \times 14} = 257 \text{ psi}$$

This is less than the 265 psi allowed for bottom bars in Table 13 of the Appendix. Incidentally, there is no doubt about these bars being bottom ones.

Now compute the diagonal tension about a vertical plane through HJ for a width KL.

$$v_{HJ} = \frac{V}{b_od} = \frac{3{,}900 \times 2.67(2.67 + 8)/2}{32 \times 14} = 124 \text{ psi}$$

This is greater than the 110 psi allowed in Table 1-8. Therefore, increase d to 15 in. Then $GH = 7.5$ in., $LK = 2$ ft 9 in., and $HF = 2$ ft $7\frac{1}{2}$ in.

$$v_{HJ} = \frac{3{,}900 \times 2.625(2.75 + 8)/2}{33 \times 15} = 111 \text{ psi}$$

This is close enough. Actually, the new A_s could be

$$A_s = \frac{165 \times 12}{20 \times 0.9 \times 15} = 7.33 \text{ in.}^2$$

Ten No. 8 bars are still needed, or will be used. The bond stress will be reduced somewhat.

Testing for f_c, an approximate value is

$$f_c = \frac{6M}{bd^2} = \frac{6 \times 165{,}000 \times 12}{96 \times 15^2} = 550 \text{ psi}$$

which, as expected, is very safe. Therefore, this design will be accepted with the bars spaced as shown in Fig. 9-7(*c*).

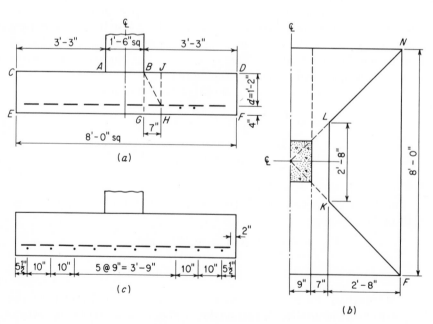

Fig. 9-7   Flat footing.

Table 9-1   Suggested Minimum Spacing c.c. in Inches
for Bars in Two-way Footings

Bar No.	f_c', psi					
	2,500	3,000	3,500	4,000	4,500	5,000
3	3	3	3	3	3	3
4	$3\frac{3}{4}$	3	3	3	3	3
5	$4\frac{1}{2}$	$3\frac{3}{4}$	$3\frac{1}{4}$	3	3	3
6	$5\frac{1}{2}$	$4\frac{1}{2}$	4	$3\frac{1}{2}$	$3\frac{1}{2}$	$3\frac{1}{2}$
7	$6\frac{1}{2}$	$5\frac{1}{4}$	$4\frac{1}{2}$	4	4	4
8	$7\frac{1}{4}$	6	$5\frac{1}{4}$	$4\frac{1}{2}$	$4\frac{1}{2}$	$4\frac{1}{2}$
9	$8\frac{1}{4}$	$6\frac{3}{4}$	6	$5\frac{1}{4}$	5	5
10	$9\frac{1}{4}$	$7\frac{3}{4}$	$6\frac{3}{4}$	6	6	6
11	$10\frac{1}{2}$	$8\frac{1}{2}$	$7\frac{1}{4}$	$6\frac{1}{2}$	6	6

In order to show that it is seldom necessary to check the shear in square footings at a distance d from the face of the column as for beams, using b equal to the entire width of the footing, test this footing in this manner, using $d = 15$ in.   Then

$$V = (3.25 - 1.25)8 \times 3,900 = 62,500 \text{ lb}$$
$$v = \frac{62,500}{96 \times 15} = 43.5 \text{ psi}$$

This is considerably less than the 60 psi allowed by Table 1-8 for beams without web reinforcement.

Example 9-3.   Assume that the stepped footing shown in Fig. 9-8(a) is to support a load $P = 300$ kips.   Check the design by USD if the load factor is 2.0, $f_c' = 3,000$ psi, $f_y = 40,000$ psi, $j = 0.9$, and the net ultimate strength of the soil is 10 ksf.   Use eleven No. 7 bars both ways.

$$P_u = 300 \times 2 = 600 \text{ kips}$$
$$p \text{ for the soil} = 600/8^2 = 9.4 \text{ ksf}$$

which is less than the allowed 10 ksf.

$$M_{FJ} = 9.4 \times 8 \times 2.75 \times 1.38 = 286 \text{ ft-k}$$

It is obvious that this footing is underreinforced as far as bending is concerned, since $p = 11 \times 0.6/(96 \times 15) = 0.0046$, which is far below p_b for balanced design.

Using Eqs. (2-16) and (2-18), with $\phi = 0.9$ for flexure, together with the data from Fig. 9-8(a), the ultimate resisting moment of the footing is found as follows:

From Eq. (2-16),

$$a = \frac{A_s f_y}{0.85 f'_c b} = \frac{6.6 \times 40,000}{0.85 \times 3,000 \times 96} = 1.08 \text{ in.}$$

From Eq. (2-18),

$$M_u = \phi \left[A_s f_y \left(d - \frac{a}{2} \right) \right] = 0.9 \left[6.6 \times 40,000 \left(15 - \frac{1.08}{2} \right) \right]$$
$$= 3,440,000 \text{ in.-lb or } 286 \text{ ft-k}$$

which is just right.

Now check the bond at the edge of the column.

$$u_{FJ} = \frac{V}{(\Sigma o) jd} = \frac{9,400 \times 2.75 \times 8}{30.2 \times 0.9 \times 15} = \frac{207,000}{407} = 510 \text{ psi}$$

Table 14 in the Appendix shows that 595 psi will be allowed for bottom bars.

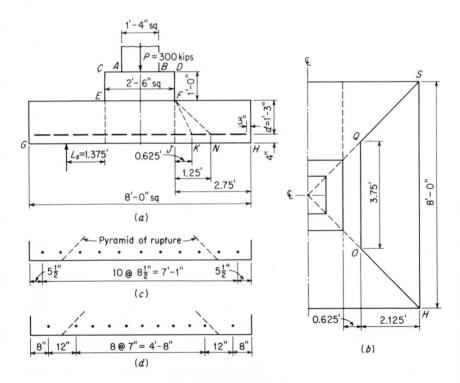

Fig. 9-8 Stepped footing.

Next, check the length of the bars beyond N, the limit of the pyramid of rupture.

$$M_N = 9.4 \times 8 \times 1.5 \times 2 = 226 \text{ ft-k}$$

$$f_s \text{ in the bars at } N = \frac{226}{6.6 \times 0.9 \times 1.25} = 30.4 \text{ ksi}$$

$$u_N = \frac{f_s A_s}{(\Sigma o) L_s} = \frac{30,400 \times 0.6}{2.75 \times 15} = 442 \text{ psi}$$

This is less than the 595 psi allowed. Therefore, no hooks are needed. Here L_s is the length of the bars beyond N. If the bars were to be fully developed beyond N, the bond stress would have to be

$$u = \frac{40,000 \times 0.6}{2.75 \times 15} = 580 \text{ psi}$$

This is also less than the 595 psi allowed.

Now check the diagonal tension. Using area $OQSH$,

$$V = 9,400 \times \frac{3.75 + 8}{2} \times 2.125 = 117,500 \text{ lb}$$

At $d/2$ from FJ,

$$v_K = \frac{V}{b_o d} = \frac{117,500}{3.75 \times 12 \times 15} = 174 \text{ psi}$$

According to the Code, the maximum v_c in the concrete without web reinforcement should not exceed $4\phi \sqrt{f'_c}$, where $\phi = 0.85$ for diagonal tension. Then $v_c = 4 \times 0.85 \sqrt{3,000} = 186$ psi, which is more than the computed 174 psi.

If one were to use Whitney's formula for ultimate shearing strength as given in Eq. (9-4), with M_u per inch of width equal to $286/96 = 2.98$ ft-k/in., and $L_s = 33/2 = 16.5$ in.,

$$v = 100 + 0.75 \times \frac{2,980}{15^2} \sqrt{16.5} = 100 + 40.3 = 140.3 \text{ psi}$$

In Fig. 9-8(c), there are seven bars through the base of the pyramid of rupture. If the bars are grouped more closely through this region so that there are nine bars included, as shown in Fig. 9-8(d), the computed M_u per inch of width in this region will be increased so that it is approximately

$$M_u = \frac{2.98 \times 9}{7} = 3.83 \text{ ft-k/in.}$$

Then
$$v = 100 + 0.75 \times \frac{3,830}{15^2} \sqrt{16.5} = 152 \text{ psi}$$

This indicates a little better strength but is not up to the 174 psi computed by the Code, [Eq. (9-2)]. If Whitney's recommendations are to be followed, it will be desirable to increase d to 16 in. At any rate, the bar arrangement shown in Fig. 9-8(d) will be adopted.

Example 9-4. Design a square sloped-top footing for a circular column as shown in Fig. 9-9, using the following data: $P = 275,000$ lb, diameter of column = 24 in., $f'_c = 2,500$ psi, $n = 10$, $f_c = 1,125$ psi, $f_s = 20,000$ psi, $v_c = 100$ psi, u = values given in Table 13 of the Appendix, and the allowable soil pressure = 6,000 psf. Use WSD.

In the case of a circular column, assume a square section having the same area.

$$a = \sqrt{\frac{\pi D^2}{4}} = 21.2 \text{ in., say 21 in.}$$

Assume the average depth of the footing to be 2 ft, and the weight to be 300 psf. Let the net pressure equal 6,000 psf less the excess weight of concrete over that of the earth = $6,000 - 2 \times 50 = 5,900$ psf. Then

$$b^2 = \frac{275,000}{5,900} = 46.6 \text{ ft}^2 \quad \text{and} \quad b = 7 \text{ ft (approx)}$$

Actually, $$p = \frac{275,000}{49} = 5,620 \text{ psf}$$

Figure 9-9(a) and (b) shows the footing as it will be assumed. $CA = 11$ in. $AE = 1$ ft $8\frac{1}{2}$ in. $QR = 3$ ft 7 in.

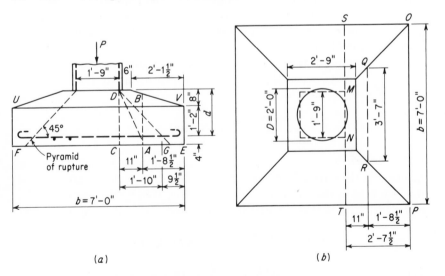

(a)   (b)

Fig. 9-9 Sloped-top footing.

Taking moments of the pressures about MN, the edge of the equivalent square column,

$$M = \left(\frac{p \times ST \times SO \times SO}{2}\right) 12 \text{ in.-lb}$$

$$= \left(\frac{5,620 \times 7 \times 2.625 \times 2.625}{2}\right) 12 = 1,625,000 \text{ in.-lb}$$

As seen from Fig. 9-9(a), the section through ST is not a rectangle. As a start in estimating the amount of steel needed, assume that the section is rectangular and that $j = 0.88$. Then

$$\text{Trial } A_s = \frac{M}{f_s jd} = \frac{1,625,000}{20,000 \times 0.88 \times 22} = 4.2 \text{ in.}^2$$

Try ten No. 6 bars, with $A_s = 4.4$ in.^{2} and $\Sigma o = 23.6$ in.^{2}/in.

Now analyze the footing by the transformed-section method to illustrate the procedure. Assume that the neutral axis is below the corners U and V. Then, taking the static moments of the area about this neutral axis,

$$\frac{84}{2}(kd)^2 - \frac{2 \times 25.5 \times 8}{2}\left(kd - \frac{8}{3}\right) = 12 \times 4.4(22 - kd)$$

$kd = 6.03$ in. This is above the level of the corners so that kd must be recomputed. Therefore,

$$\frac{33}{2}(kd)^2 + 2\left(\frac{kd}{2} \times \frac{25.5kd}{8}\right)\frac{kd}{3} = 10 \times 4.4(22 - kd)$$

$kd = 5.7$ in. (approx)   and $d - kd = 22 - 5.7 = 16.3$ in.

$$I_c = \frac{33 \times 5.7^3}{3} + \frac{2 \times 18.2 \times 5.7^3}{36} + \frac{2 \times 18.2 \times 5.7 \times 1.9^2}{2}$$
$$+ 10 \times 4.4(16.3)^2 = 14,300 \text{ in.}^4$$

$$S_c = \frac{I_c}{kd} = \frac{14,300}{5.7} = 2,510 \text{ in.}^3$$

$$S_s = \frac{I_c}{n(d - kd)} = \frac{14,300}{10 \times 16.3} = 88 \text{ in.}^3$$

$$f_c = \frac{M}{S_c} = \frac{1,625,000}{2,510} = 650 \text{ psi}$$

$$f_s = \frac{M}{S_s} = \frac{1,625,000}{88} = 18,500 \text{ psi}$$

As might be expected, the compressive stress is small. The tension in the steel is satisfactory.

Using the entire pressure under portion $SOPT$ in Fig. 9-9(b), the bond stress on the bars at section CD is

$$u = \frac{V}{(\Sigma o)jd} = \frac{5{,}620 \times 2.625 \times 7}{23.6 \times 0.88 \times 22} = 226 \text{ psi}$$

This is less than the 320 psi allowed in Table 13 of the Appendix. Of course, the same reinforcement will be used in both directions.

Now test the diagonal tension at section AB in Fig. 9-9(a). The assumed shear will be due to the pressure under area $QOPR$ in sketch (b). The depth d at AB is

$$d = 22 - \frac{(11 - 6)8}{25.5} = 20.43 \text{ in.}$$

$$v = \frac{V}{b_o d} = \frac{5{,}620 \times 1.71}{43 \times 20.43}\left(\frac{3.58 + 7}{2}\right) = 58 \text{ psi}$$

This design will be accepted. The bars will be spaced uniformly across the footing at 8 in. c.c. both ways because the pyramid of rupture comes so close to the edges of the footing. The ends of the bars will be hooked to provide good anchorage.

Example 9-5. Design a stepped footing for the conditions shown in Fig. 9-10(a). Assume $f'_c = 3{,}000$ psi, $f_s = 20{,}000$ psi, $j = 0.88$, and the allowable soil pressure $p = 8{,}000$ psf. Use the Code unit stresses and use working-stress design. An adjacent retaining wall limits the width of the footing to 6 ft.

As a start, assume that the footing adds 10 kips to the soil pressure at the bottom of the footing in excess of that caused by the original ground. Then a trial area of bearing is

$$A = \frac{550 + 10}{8} = 70 \text{ ft}^2$$

Call the dimensions 6 by 12 ft. A plan of this footing is shown in Fig. 9-10(b), and a section of one long end is pictured in (c). Notice that d is deliberately made equal to one-half of the cantilever length.

Check the excess weight of this trial footing at 50 pcf.

$$W' = (6 \times 12 \times 2.33 + 4 \times 4 \times 2.67)50 = 10{,}600 \text{ lb}$$

This is near enough. The net pressure for the design of the footing is

$$p = 550/72 = 7{,}640 \text{ psf}$$

Bending at FE:

$$M = 7{,}640(6 \times 4 \times 2)12 = 4{,}400{,}000 \text{ in.-lb}$$

$$A_s = \frac{4{,}400{,}000}{20{,}000 \times 0.88 \times 24} = 10.4 \text{ in.}^2 \text{ (approx)}$$

Try thirteen No. 8 bars at $5\frac{1}{2}$ in. c.c. uniformly spaced. $A_s = 10.3$ in.^{2};

$$\Sigma o = 40.8 \text{ in.}^2$$
$$u = \frac{7,640 \times 6 \times 4}{40.8 \times 0.88 \times 24} = 213 \text{ psi (approx) vs. 265 psi allowed}$$
$$\text{Approx } f_c = \frac{6M}{bd^2} = \frac{6 \times 4,400,000}{72 \times 24^2} = 640 \text{ psi (safe)}$$

Compute the diagonal tension at a vertical plane $d/2$ from the face HJ of the pedestal—a plane through GK.

$$V = 7,640 \times 6 \times 3 = 138,000 \text{ lb}$$
$$v = \frac{138,000}{72 \times 24} = 80 \text{ psi}$$

This is less than the 110 psi allowed in Table 1-8.

Now check the diagonal tension at a section d from HJ, treating the footing as a regular cantilevered beam.

$$V = 7,640 \times 6 \times 2 = 91,700 \text{ lb}$$
$$v = \frac{91,700}{72 \times 24} = 52 \text{ psi}$$

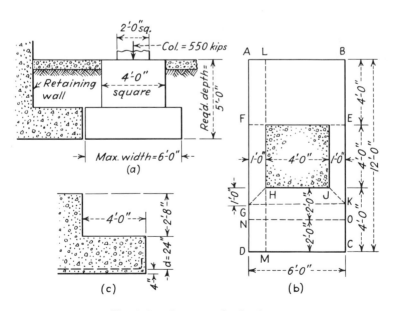

Fig. **9-10**   A rectangular footing.

This is only slightly less than the 60 psi allowed in Table 1-8 for beams without web reinforcement. However, these calculations show the desirability of making footings thick in order to reduce the diagonal tensile stresses and to avoid using web reinforcement.

Bending at LM:

$$M = 7,640(12 \times 1 \times 0.5)12 = 550,000 \text{ in.-lb}$$

$$A_s = \frac{550,000}{20,000 \times 0.88 \times 24} = 1.3 \text{ in.}^2$$

For a maximum value of $u = 265$ psi,

$$\Sigma o = \frac{V}{ujd} = \frac{7,640 \times 12 \times 1}{265 \times 0.88 \times 24} = 16.3 \text{ in.}^2 \text{ needed}$$

Use twelve No. 4 bars, the smallest size desired here, giving $A_s = 2.4$ in.^{2} and $\Sigma o = 18.8$ in.^{2}. These are satisfactory. This shows that the necessary bond resistance may control over the tensile resistance in heavily loaded footings. The ends of all bars will be bent up at 90° to provide good anchorage.

Inspection shows that it is not necessary to compute v and f_c for the long direction. From Eq. (9-3),

$$\frac{2}{S+1} = \frac{2}{\frac{12}{6}+1} = \frac{2}{3}$$

Therefore, place eight of these No. 4 bars in a 6-ft width centered on the pedestal, with two near each end at about 16-in. spacing.

Example 9-6. Design the reinforcement for and compute the bond and diagonal tensile stresses in the proposed footing shown in Fig. 9-11(a)

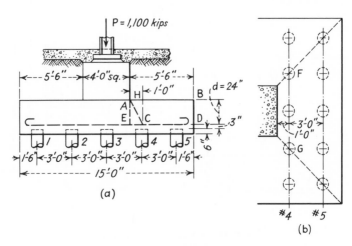

Fig. **9-11** Footing on piles.

if it is 15 ft square. Assume $f'_c = 3,000$ psi, $f_s = 22,000$ psi, and $j = 0.9$. The load P is supported by 25 wooden piles.

Piles can seldom be driven in exactly the positions shown on the drawings; hence conservatism in design is necessary. For example, there are five piles in rows 4 and 5, each having a net load of

$$p = \frac{1,100}{25} = 44 \text{ kips}$$

The theoretical bending moment about a vertical plane through AE, the face of the pedestal, is

$$M = 5 \times 44(1 + 4) = 1,100 \text{ ft-k}$$

But suppose that these piles are 6 in. too far to the right. The bending moment would then be $5 \times 44(1.5 + 4.5) = 1,320$ ft-k, an increase of 20 per cent. This is not improbable. Therefore, it seems that this should be provided for, even though the safety factor might cover it. The same comment applies to bond and shear. Therefore, make one design or analysis on the basis that the pile reactions are at the centers of the piles and that all are driven exactly as intended; then make a second computation on the basis that all piles are 6 in. to the right of where they are called for.

First Design: For $M = 1,100$ ft-k,

$$A_s = \frac{1,100}{22 \times 0.9 \times 2} = 27.8 \text{ in.}^2$$

Try twenty-two No. 10 bars ($A_s = 27.9$ in.^{2}, $\Sigma o = 87.8$ in.^{2}).

$$u = \frac{10 \times 44,000}{87.8 \times 0.9 \times 24} = 232 \text{ psi}$$

but only 205 psi is allowed for No. 10 bars by the Code. Therefore, use twenty-eight No. 9 bars ($A_s = 28$ in.^{2}, $\Sigma o = 99.2$ in.^{2}) and

$$u = \frac{10 \times 44,000}{99.2 \times 0.9 \times 24} = 205 \text{ psi}$$

which is satisfactory because the allowed u for No. 9 bars is 235 psi.

Now compute the diagonal tension at GF. Based upon item 9 of Art. 9-2,

$V =$ three piles No. 5 $+ \frac{1}{2}$(two piles No. 5 at corners)

               $+ \frac{1}{2}$(one pile No. 4) $+ \frac{1}{4}$(two piles No. 4 at F and G)

or $V = 44(3 + 1 + 0.5 + 0.5) = 220$ kips

Using $b_o = FG = 6$ ft,

$$v = \frac{220,000}{(6 \times 12)24} = 127 \text{ psi vs. } 110 \text{ psi allowed}$$

This indicates that the footing is not thick enough. Therefore, try $d = 27$ in. Then $EC = 13.5$ in. so that the piles in row 4 are 1.5 in. to the left of C. Then, according to the Code, the new

V = three piles No. 5 + $\frac{1}{2}$(two piles No. 5 at corners)
  + 4.5/12(one pile No. 4) + $\frac{1}{2}$ × 4.5/12(two piles No. 4 near F and G)
or $\qquad V = 44(3 + 1 + 0.375 + 0.375) = 209$ kips

$$v = \frac{209,000}{(6 + 72)27} = 99 \text{ psi}$$

which appears to be all right.

Changing A_s and u for $d = 27$ in.,

$$A_s = \frac{1,100}{22 \times 0.9 \times 2.25} = 24.7 \text{ in.}^2$$

Using No. 9 bars and $u = 235$ psi,

$$\Sigma o = \frac{V}{ujd} = \frac{10 \times 44,000}{235 \times 0.9 \times 27} = 77.3 \text{ in.}^2/\text{in.}$$

Therefore, use twenty-five No. 9 bars which will give $A_s = 25$ in.^{2} and $\Sigma o = 88.6$ in.^{2} The footing now seems to be satisfactory.

Second Design: Imagine that the piles in rows 4 and 5 are 6 in. further to the right than intended, but use the new $d = 27$ in. Then $M = 1,320$ ft-k and

$$A_s = \frac{1,320}{22 \times 0.9 \times 2.25} = 29.6 \text{ in.}^2$$

For No. 9 bars, the required

$$\Sigma o = \frac{10 \times 44,000}{235 \times 0.9 \times 27} = 77.3 \text{ in.}^2 \text{ as before}$$

Then thirty No. 9 or twenty-four No. 10 bars may be used.

Now check the diagonal tension at a vertical plane through the new location of C if this point is 15 in. from AE. The new GF will be 78 in.

V = three piles No. 5 + approx 0.8(two piles No. 5 near corners)
  + 0.75(one pile No. 4) + approx 0.67(two piles No. 4 near F and G)
or $\qquad V = 44(3 + 1.6 + 0.75 + 1.34) = 294$ kips

$$v = \frac{294,000}{78 \times 27} = 140 \text{ psi}$$

This is considerably above the 110 psi allowed. It shows that inaccuracies of pile locations may become serious. In this case, the author would have the pile locations checked after driving. If the inaccuracies are important and as assumed above, he would have the piles cut off at a lower elevation and the footing made deeper, until both the shear and the original reinforcement are satisfactory.

Example 9-7. By working-stress methods, design a spread footing for one of the steel columns shown in Fig. 9-12, using $f'_c = 3,000$ psi, $f_s = 18,000$ psi, and the unit stresses allowed by the Code. The soil is *caliche* and has a total safe bearing capacity of 5 tons/ft² at the level of the bottom of the footing. The bin can be full or empty; the wind and adjoining cranes cause overturning tendencies which, of course, are reversible. The two critical load combinations for design, besides the vertical loads alone, are

$$\text{max} = \text{max } P + \text{max } M + 5S = 725,000 \text{ lb} + 300,000 \text{ ft-lb}$$
$$+ 5 \times 15,600 \text{ ft-lb}$$
$$\text{min} = \text{min } P + M' + 5S' = 40,000 \text{ lb} + 200,000 \text{ ft-lb}$$
$$+ 5 \times 10,000 \text{ ft-lb}$$

where P = vertical load in the column

M = max overturning moment applied by the structure at the column base

M' = overturning moment at the column base due to wind alone

S, S' = corresponding horizontal forces at the column base

Fig. **9-12**   Suspended ore bin along north side of concentrator at the Morenci Reduction Works, Phelps Dodge Corporation, Morenci, Arizona.   View looking west.

For a combination of wind and live loads, assume that the allowable unit stresses and the bearing pressure may be increased 30 per cent.

When starting such a problem, the designer should see what limiting factors affect the case—interferences with width and length, use or omission of a pedestal, size of pedestal if used, depth to bottom of footing, etc. He should choose the type of footing that he thinks is best and then see if he can use it; if something interferes, then the design should be modified as necessary.

One way to obtain trial dimensions of such a rectangular footing is to determine the minimum area for the direct load only, adding an estimate of the weight of the footing and soil or flooring over it. Thus, in this case, let $P = 725,000$ lb $+ 50,000$ lb for the footing $= 775,000$ lb.

$$A = 775,000/10,000 = 77.5 \text{ ft}^2 \text{ (say } 8 \times 10 \text{ ft)}$$

Next, assume that, for direct loads and overturning, the pressures under the footing vary uniformly. Then, if the pressure diagram is trapezoidal, the maximum pressure under the footing is

$$p = \frac{P}{A} + \frac{Mc}{I} = \frac{P}{bh} + \frac{6M}{bh^2} \quad \text{or} \quad \frac{P}{bh}\left(1 + \frac{6e}{h}\right) \quad (9\text{-}5)$$

where b and h are, respectively, the width and length of the footing, the latter being in the direction of the overturning tendency, and e is the eccentricity, the total overturning moment divided by the total vertical load. Then assume h and solve for b.

Now assume $h = 10$ ft. The maximum moment at the bottom of the footing is $300,000 + 5 \times 15,600 = 378,000$ ft-lb. Using the maximum bearing pressure increased by 30 per cent, Eq. (9-5) gives

$$10,000 \times 1.3 = \frac{775,000}{10b} + \frac{6 \times 378,000}{100b} \qquad b = 7.7 \text{ ft}$$

This could be used, but it seems that a longer and narrower footing may be better. Try a footing 6.5×13 ft. Its area will be about the same as the broader one, but it will be more effective in resisting the overturning.

The trial footing is shown in Fig. 9-13(a) and (b). Since the footing is narrow, it is sloped two ways only, thus making a strong ridge clear across the top. An estimate of its weight, including the earth above it but no live load on the floor, is 52,000 lb. Therefore, the maximum vertical load becomes 777,000 lb. A 3×3.5-ft pedestal will be used under the steel base plate.

For the maximum load P without overturning, the bending moment about the edge of the pedestal at EF is

$$M_{EF} = 6.5\left(\frac{777,000}{13 \times 6.5} - 600\right)\frac{5^2}{2} = 6.5(9,200 - 600)\frac{5^2}{2} = 699,000 \text{ ft-lb}$$

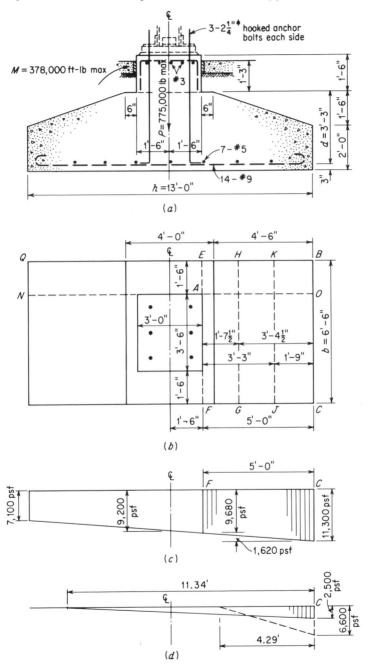

Fig. 9-13 Footing under column supporting ore bin, Morenci Reduction Works, Phelps Dodge Corporation, Morenci, Arizona.

where the 600-lb reduction is an approximation of the weight of the footing and soil, since the footing is to be designed for the net soil pressure.   Then, with $d = 39$ in. as shown in sketch (a), and $j = 0.88$,

$$A_s = \frac{699,000 \times 12}{18,000 \times 0.88 \times 39} = 13.6 \text{ in.}^2$$

Therefore, try fourteen No. 9 bars longitudinally, providing $A_s = 14$ in.^{2} and $\Sigma o = 49.6$ in.^{2} Then the bond stress is found to be

$$u = \frac{(9,200 - 600)5 \times 6.5}{49.6 \times 0.88 \times 39} = 164 \text{ psi (satisfactory, since 235 psi is}$$

$$\text{allowed on No. 9 bars)}$$

At GH, $d/2 = 19.5$ in. from EF, compute the diagonal tension as for footings, with the net pressure $p = 8,600$ psf.   At this point,

$$d = 39 - 18(19.5 - 6)/54 = 34.5 \text{ in.}$$
$$V_{GH} = 3.375 \times 6.5 \times 8,600 = 188,000 \text{ lb}$$
$$v_{GH} = \frac{188,000}{78 \times 34.5} = 70 \text{ psi vs. 110 psi allowed}$$

Also, compute the diagonal tension at JK, a distance d from EF, as a cantilevered beam with a maximum allowable value of 66 psi as for other beams.   Here

$$d = 39 - 18(39 - 6)/54 = 28 \text{ in.}$$
$$V_{JK} = 1.75 \times 6.5 \times 8,600 = 98,000 \text{ lb}$$
$$v_{JK} = \frac{98,000}{78 \times 28} = 45 \text{ psi (satisfactory)}$$

The approximate f_c at EF is

$$f_c = \frac{6M}{bd^2} = \frac{6 \times 699,000 \times 12}{78 \times 39^2} = 425 \text{ psi (very safe)}$$

Now investigate the conditions when there are maximum vertical and overturning forces.   First, compute the edge pressures under the footing for the load and bending moment shown in Fig. 9-13(a), using Eq. (9-5).

$$p_{max} = \frac{P}{bh}\left(1 + \frac{6e}{h}\right) \quad \text{and} \quad p_{min} = \frac{P}{bh}\left(1 - \frac{6e}{h}\right) \quad (9\text{-}5a)$$

For this condition, $e = 378,000/777,000 = 0.49$ ft.

$$p_{max} = \frac{777,000}{6.5 \times 13}\left(1 + \frac{6 \times 0.49}{13}\right) = 11,300 \text{ psf}$$
$$p_{min} = 9,200(1 - 0.226) = 7,100 \text{ psf}$$

These are shown in Fig. 9-13(c).

The bending moment will be computed at a section through *EF* as before, using the pressure diagram in (*c*), with 600 psf deducted to get the net pressures.

$$M_{EF} = 6.5(9{,}080 \times \tfrac{5^2}{2} + 1{,}620 \times \tfrac{5}{2} \times 0.67 \times 5) = 826{,}000 \text{ ft-lb}$$

With $j = 0.88$ and $d = 39$ in., and 30 per cent overstress,

$$A_s = \frac{826{,}000 \times 12}{(18{,}000 \times 1.3)0.88 \times 39} = 12.4 \text{ in.}^2$$

This is less than that needed for the vertical loads alone.

The preceding computations for A_s, and the fact that the maximum edge pressure shown in sketch (*c*) is less than $1.3 \times 9{,}200$ for vertical loads $= 12{,}000$ psf, indicates that the critical conditions for bond and diagonal tension will apparently be for the case of the maximum vertical load only.

The next step is to determine the reinforcement in the short direction, this being for vertical loads only. Take moments about *NO* through one edge of the pedestal, Fig. 9-13(*b*), using the average net pressure of 8,600 psf and $d = 36$ in. to allow something for the fact that the shape of the cross section is not rectangular.

$$M_{NO} = 8{,}600 \times 13 \times 1.5 \times 0.75 = 126{,}000 \text{ ft-lb}$$
$$A_s = \frac{126{,}000 \times 12}{18{,}000 \times 0.88 \times 36} = 2.65 \text{ in.}^2$$

Use nine No. 5 bars having $A_s = 2.79$ in.^{2} and $\Sigma o = 17.6$ in.^{2}

$$u = \frac{8{,}600 \times 13 \times 1.5}{17.6 \times 0.88 \times 36} = 300 \text{ psi}$$

This is satisfactory, since Table 13 of the Appendix allows 420 psi for these bars. However, hooks will be used on all bars in both direction to provide a reliable anchorage.

Obviously, the diagonal tension will not be important across the long direction because the $d/2$ intersection points beyond *NO* are mostly outside of *QB*.

The last thing to check is the overturning of the footing under the second load case previously listed, assuming a safety factor of 1.5. The abutting power of the soil and floor will not be relied upon. Including the weight of the footing and earth as 52,000 lb,

$$P = 40{,}000 + 52{,}000 = 92{,}000 \text{ lb}$$
$$M = 200{,}000 + 5 \times 10{,}000 = 250{,}000 \text{ ft-lb}$$
$$e = \frac{M}{P} = \frac{250{,}000}{92{,}000} = 2.72 \text{ ft}$$

This eccentricity means that the pressure diagram is triangular and that the length in contact with the bottom of the footing will be equal to $(h/2 - e)3 = (\frac{13}{2} - 2.72)3 = 11.34$ ft, as shown in Fig. 9-13(d). Then

$$\frac{p}{2} \times 11.34 \times 6.5 = 92,000 \qquad \text{and} \qquad p = 2,500 \text{ psf}$$

If the overturning moment is multiplied by 1.5 to test the safety factor,

$$M' = 250,000 \times 1.5 = 375,000 \text{ ft-lb}$$
$$e' = 375,000/92,000 = 4.07 \text{ ft}$$

Then $\quad \dfrac{p'}{2} \times 4.29 \times 6.5 = 92,000 \qquad$ and $\qquad p' = 6,600$ psf

This is indicated by the dotted lines in sketch (d). This is still very safe. The danger might come from an excessive eccentricity which causes the resultant pressure to be too close to one edge of the footing or even outside of it.

The footing will be accepted as designed. If the anchor bolts shown in Fig. 9-13(a) did not extend to the bottom to grip the footing, reinforcement would have to be used to keep the pedestal from being ripped off.

Suppose that the footing in Fig. 9-13(a) is subjected to overturning moments both lengthwise and crosswise of the footing. It will be sufficient to compute the bearing pressure for the vertical loads alone, to add to it that due to the moment in one direction $(6M/bh^2)$, and then to add also the pressure caused by the crosswise moment $(6M'/b^2h)$. The bending moments in the footing may be computed at one side of the pedestal for $P + M$ alone or at the adjacent side for $P + M'$ only.

Example 9-8. Compute the soil pressures under the footing shown in Fig. 9-14 for the loads and overturning moment given there, and design the reinforcement. The footing is notched out to clear a heavy machinery foundation which must be isolated from it. The maximum permissible soil pressure is 4 tons/ft². Assume $f'_c = 3,000$ psi, $f_s = 20,000$ psi, and allowable Code stresses. Let $j = 0.88$.

The weight of the footing is taken as the approximate total weight of the concrete and the earth over it, this weight being assumed to be at the center line of the column. The earth is often omitted from the calculations, especially when the footings are deep.

Since this footing is not rectangular, Eq. (9-5) must be used in its general form. An outline of a procedure yielding sufficiently satisfactory results is the following:[1]

---

[1] For an explanation of the use of principal axes for such a problem, see C. W. Dunham, "Foundations of Structures," 2d ed., McGraw-Hill Book Company, New York, 1962.

1. Locate the center of gravity of the assumed bearing area.

2. Compute the moment of inertia of this area about both rectangular axes.

3. Combine the moments due to eccentricity of loading with those due to other causes.

4. Compute the resultant pressure by adding that due to the direct load to those caused by the moments.

5. Find the critical portion of the footing, approximate the bending moment, and test the section.

Axis X-X:

Part	A	l	M	l^2	I_{CG}	Al^2	ΣI
$ABCG$	22	2.25	49.5	5.06	29.3	111.3	140.6
$GDEF$	33.8	2	-67.6	4	57	135.2	192.2
	55.8		-18.1				332.8

$$y = \frac{-18.1}{55.8} = -0.32 \text{ ft}$$

$$I_{PQ} = I_X - Ay^2 = 332.8 - 55.8 \times 0.32^2 = 327 \text{ ft}^4$$

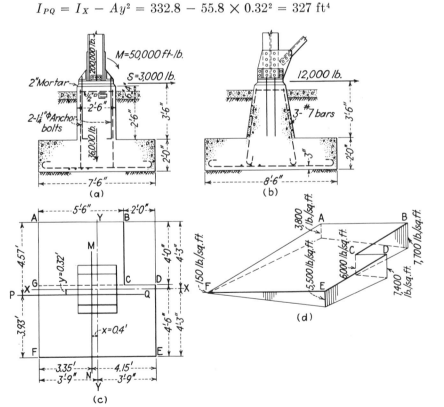

Fig. 9-14   Unsymmetrical footing for steel column in industrial plant.

Axis Y-Y:

Part	A	l	M	l^2	I_{CG}	Al^2	ΣI
$ABCG$	22	-1	-22	1	55.5	22	77.5
$GDEF$	33.8	0	0	0	158.2	0	158.2
	55.8		-22				235.7

$$x = \frac{-22}{55.8} = -0.4 \text{ ft}$$

$I_{MN} = I_Y - Ax^2 = 235.7 - 55.8 \times 0.4^2 = 227 \text{ ft}^4$

$P = 236{,}000$ lb (assume it at center of column)

M_X at bottom $= 12{,}000 \times 5.5 + 236{,}000 \times 0.32 = 142{,}000$ ft-lb

M_Y at bottom $= 50{,}000 + 3{,}000 \times 5.5 + 236{,}000 \times 0.4 = 161{,}000$ ft-lb

The greatest pressure is likely to be at B or D in Fig. 9-14(c). Letting the subscripts denote the point under consideration, then

$$p_B = \frac{236{,}000}{55.8} + \frac{161{,}000 \times 2.15}{227} + \frac{142{,}000 \times 4.57}{327} = 7{,}700 \text{ psf}$$

$$p_D = \frac{236{,}000}{55.8} + \frac{161{,}000 \times 4.15}{227} + \frac{142{,}000 \times 0.57}{327} = 7{,}400 \text{ psf}$$

The pressure diagram under the whole footing is shown in Fig. 9-14(d). The use of this in designing the footing generally involves broad approximations. A designer might resort to various detailed computations, but it is seldom worthwhile. Such footings are generally the exception, having little duplication. Therefore, the designer should think first whether or not his design is obviously safe. Each special problem requires the use of good judgment in deciding these matters.

The following computations show one way of designing this footing. It is based upon the philosophy that, if there is a system in the structure that is able to resist the loads, that system will act before the structure will fail.

1. Looking at Figs. 9-14(d) and 9-15(a), it seems that a strong band or beam might be made running lengthwise of the footing, such as $KBNP$ in Fig. 9-15(a). Make it strong enough to carry the shears and moments from the entire portion $JABL$ about TL, or $HMEF$ about SQ. Assume it to be 3 ft 6 in., a little wider than the pedestal.

2. Judge whether TL or SQ is the critical section. Test both if it seems to be necessary. Although the area $JABL$ is less than $HMEF$, it is subjected to larger pressures and appears to be the critical side.

3. From Fig. 9-14(d) and similar computations, get the pressures at each of the four corners, then average that at A and B, also at J and L. The resultant approximate pressure diagram is shown in Fig. 9-15(b). The average dead load of the footing and the material over it is

$36,000/55.8$ = about 600 psf. Then the approximate net pressure diagram is that shown in Fig. 9-15(c). The bending moment at TL is

$$M_{TL} = 5.5 \left(4,000 \times \frac{2.75^2}{2} + 1,200 \times 2.75^2 \times 0.67 \right) = 117,000 \text{ ft-lb}$$

$$V_{TL} = 5.5 \left(4,000 \times 2.75 + 1,200 \times \frac{2.75}{2} \right) = 70,000 \text{ lb}$$

4. With d = 21 in.,

$$A_s = \frac{117,000 \times 12}{20,000 \times 0.88 \times 21} = 3.8 \text{ in.}^2$$

Use twelve No. 5 bars at $3\frac{1}{2}$ in. c.c., hooked.

An approximate value of f_c is

$$f_c = \frac{6 \times 117,000 \times 12}{42 \times 21^2} = 450 \text{ psi (very safe)}$$

At a distance $d/2$ = 10.5 in. from TL,

$$v = \frac{5,800 \times 1.88 \times 5.5}{42 \times 21} = 68 \text{ psi (less than 110 psi)}$$

At a distance d from TL,

$$v = \frac{5,800 \times 1 \times 5.5}{42 \times 21} = 36 \text{ psi}$$

without considering any 45° diagonals, and using only a 42-in. width.

$$u = \frac{70,000}{23.5 \times 0.88 \times 21} = 161 \text{ psi (very safe)}$$

5. The strip $UDMR$ in Fig. 9-15(a) seems to have the greatest bending moment of any crosswise to $KBNP$. Arbitrarily assume a typical piece

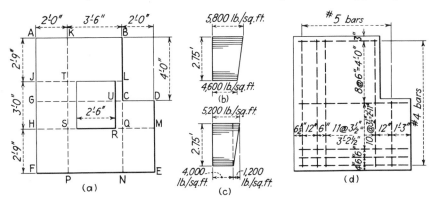

Fig. **9-15**   Reinforcement in an unsymmetrical spread footing.

12 in. wide with an average net upward pressure of 6,400 psf.   For this strip, M_{UR} = 20,000 ft-lb and A_s required = 0.65 in.^{2} Therefore, use No. 4 bars $3\frac{1}{2}$ in. c.c., for which u = 161 psi.

The final reinforcing plan is shown in Fig. 9-15(d).   The top hoop and the bars in the pedestal are shown in Fig. 9-14(a) and (b).

9-6. Combined Footings. In some cases, where structures are founded upon soils having low bearing capacities, it is advisable to com-bine the footings of two or more columns.   One common example of this occurs when the outside row of columns is close to the building line and it is impossible for the footing to spread over onto adjacent property to avoid troublesome eccentricity of loading.   The load upon the outer column is also likely to be larger than that on inner ones if heavy exterior walls are used.   The footing may be built somewhat as shown in Fig. 9-16(a), with an enlargement near the outer column so as to produce a uniform intensity of pressure.   Sometimes an elongated pedestal may be added in order to make the structure a sort of inverted T beam which is supported by the two columns.   On the other hand, the footing may be trapezoidal in plan as shown in (b).   The pedestals are generally desirable in order to avoid having the entire footing wastefully thick.   Of course, the fundamental objective is to secure uniform settlement and to avoid tipping of the footing.

Many industrial buildings are equipped with heavy cranes that are carried on columns along the exterior walls as well as on interior columns. The foundations of these may be made with spread footings for the columns and grade beams for the walls, or they may be continuous con-crete walls with pilasters and medium-sized footings at the columns.   In the latter case, the wall may be designed as a stiffening girder or a long narrow footing to spread the crane column's load.   This latter type of foundation may require more excavation and concrete, but it produces a stiff construction.

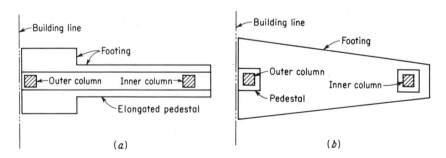

Fig. **9-16**   Combined footings.

Fig. 9-17   Reinforcement in invert of the New Jersey shaft of the Lincoln Tunnel at New York City.

In other cases, like the invert of the New Jersey shaft of the Lincoln Tunnel (Fig. 9-17), all the footings are combined into a thick, continuous reinforced-concrete slab or mat.   This was done here in order to make a boxlike structure which would withstand hydrostatic pressures.   In the case of the Merchants Refrigerating Company Building, Varick Street, New York City (Fig. 9-18), a solid 5-ft mat was used in order to spread the loads to the soil.   Figure 9-19 shows one of several continuous footings used to support the viaduct of the New Jersey approach to the Lincoln Tunnel at New York City.   The soil is deep, saturated clay, and there are 144 wooden piles under each footing.

Example 9-9.   Design a typical combined footing to support the two corridor columns of a multistory building in Fig. 9-20(*a*).   The soil is silty, and its safe net bearing capacity for the design of the footing is 2 tons/ft.[2]   Assume $f'_c = 3{,}000$ psi, $f_s = 20{,}000$ psi, $j = 0.9$, and the other stresses allowed by the Code.   Because of certain local piping, the ends *A* and *J* of the footing are not to project more than 3 ft beyond the column centers.

The trial area $= \frac{500}{4} = 125$ ft.[2]   Try a footing 16 ft long and 8 ft wide with pedestals 3 ft square and 15 in. deep.   Two separate, elongated footings 3 × 10.5 ft could be used, but a combined footing like that in Fig. 9-20(*a*) is preferred.

Fig. 9-18   Merchants Refrigerating Company Building, New York City.   (*Turner Construction Company*)

The pressure on the soil is $p = 500/(16 \times 8) = 3.91$ ksf and the load per foot w is 31.25 kips.   Assuming uniform pressure, the net longitudinal shear diagram is as shown in Fig. 9-20(b), assuming that the pedestals are so strong that they reduce the critical shears near and under the columns.

The bending-moment diagram may be found as follows:

$$M_B = 31.25 \times 1.5 \times 0.75 = 35 \text{ ft-k}$$
$$M_C = 31.25 \times 3 \times 1.5 - \tfrac{250}{3} \times 1.5 \times 0.75 = 47 \text{ ft-k}$$

treating the column load as spread uniformly over the 3-ft pedestal.

$$M_D = 31.25 \times 4.5 \times 2.25 - 250 \times 1.5 = -58 \text{ ft-k}$$
$$M_E = 31.25 \times 8 \times 4 - 250 \times 5 = -250 \text{ ft-k}$$
$$M_I = 31.25 \times 6 \times 3 - 250 \times 3 = -187 \text{ ft-k}$$

This diagram is symmetrical and is shown in Fig. 9-20(c).

The first step in designing this footing is an assumption of a total depth.   Since the diagonal tension is likely to be critical, test this first.   Try a total depth of 15 in. and $d = 12$ in.   On this basis, the maximum shear will be at $d/2 = 6$ in. from the edge of the pedestal KL [Fig. 9-20(d)]. Then NO would equal 4 ft and

$$V = \left[\frac{(4+8)2}{2} + 8 \times 1 \right] 3.91 = 78 \text{ kips}$$

$$v = \frac{78{,}000}{48 \times 12} = 136 \text{ psi vs. 111 psi allowed}$$

This is too much. Then try $d = 15$ in., with $NO = 4$ ft 3 in. For this case,

$$V = \left[\frac{(4.25 + 8)}{2} 1.88 + 8 \times 1 \right] 3.91 = 76 \text{ kips}$$

$$v = \frac{76{,}000}{51 \times 15} = 100 \text{ psi (safe)}$$

The steel required at E with $d = 15$ in. is

$$A_s = \frac{250 \times 12}{20 \times 0.9 \times 15} = 11.1 \text{ in.}^2$$

Use twelve No. 9 bars ($A_s = 12$ in.^{2}, $\Sigma o = 42.5$ in.^{2}). Place these bars 7 in. c.c. in the top, centered about the longitudinal axis of the footing.

***Fig.* 9-19** Foundation of six-lane viaduct, New Jersey approach to the Lincoln Tunnel, New York City.

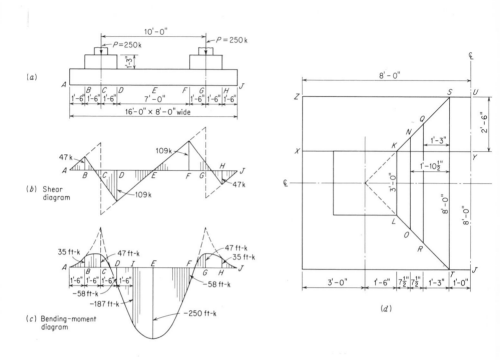

Fig. 9-20   Combined footing.

Use all bars about 13 ft long so that they will be effective in resisting bond.

In checking the bond stress near the ends of these bars, it is best to investigate the pick-up of bending moment as shown in sketch (c). The most rapid change of moment seems to be in the space from D to I. Here it is $(187 - 58)/18 = 7.2$ ft-k/in. of length. The change in stress in the bars per inch is therefore found from

$$M = A_s f_s jd \qquad \text{or} \qquad 7{,}200 \times 12 = 12 \times f_s \times 0.9 \times 15$$

Then $f_s = 532$ psi/in. The required bond stress to develop this tension is $f_s / \Sigma o = u$. Thus, $532/3.54 = 150$ psi needed vs. 165 psi bond allowed by Table 13 in the Appendix.

As a matter of interest, check the diagonal tension at a section through QR as a rectangular beam with $d = 15$ in. and $b =$ the full width of the footing. Use the entire pressure under the portion of the footing from the center JU to a cross section through NO − 1 ft $10\frac{1}{2}$ in. plus 1 ft from JU.

$$V = (8 \times 2.88)3.91 = 90 \text{ kips}$$
$$v = \frac{90,000}{96 \times 15} = 62.5 \text{ psi}$$

This is greater than the 60 psi allowed by Table 1-8. However, the assumption used is ultraconservative because the shearing forces outside $TLKS$ will be resisted along the adjacent sides of the pedestal. The diagonal tension at the other three sides of each pedestal will be satisfactory because the shears will be less than at KL.

The maximum compression in the concrete will be at the center E. The approximate f_c here is

$$f_c = \frac{6 \times 250,000 \times 12}{96 \times 15^2} = 835 \text{ psi (very safe)}$$

The bars in the bottom at C and G should be in proportion to the bending moment, or $\frac{47}{250}$ = about $\frac{1}{5}$ of the steel at E. Use eight No. 5 bars (A_s = 2.48 in.^{2}, Σo = 15.7 in.^{2}) spaced 12 in. c.c. with alternate bars 6 ft long and the others full length. For these,

$$u = \frac{1.5 \times 31,250}{15.7 \times 0.9 \times 15} = 221 \text{ psi vs. 235 psi allowed}$$

To determine the crosswise bending moment at a vertical plane through XY, assume that the section extends to the center line of the footing in this case, even though it might include only the width ZS = 7 ft.

$$M_{XY} = 3.91 \times 8 \times 2.5 \times 1.25 = 98 \text{ ft-k}$$
$$A_s = \frac{98 \times 12}{20 \times 0.9 \times 15} = 4.35 \text{ in.}^2$$

Use fourteen No. 5 bars (A_s = 4.34 in.^{2}, Σo = 27.4 in.^{2}).

$$u = \frac{3.91 \times 8 \times 2.5}{27.4 \times 0.9 \times 15} = 212 \text{ psi vs. 235 psi allowed}$$

This design will be accepted even though the footing may be a little thicker than the possible minimum. In such a design, if the ends were not restricted, it would have been desirable to make the footing longer and narrower so as to more nearly equalize the bending moments under the columns and near the center.

PRACTICE PROBLEMS

For all problems, assume that f'_c = 3,000 psi, f_s = 18,000 psi, and j = 0.9. Use the Code stresses specified in Table 1-8 and Table 13 in the Appendix for working-stress design.

9-1   Assume a plain concrete pedestal or footing like that shown in Fig. 9-5.   It is 3 ft square and 18 in. deep.   It rests upon earth and supports a column that is 14 in. square and carries 40,000 lb.   What is the tensile stress in the concrete?

9-2   A masonry wall 16 in. thick supports a load of 16 kips/ft.   It is to be supported on a footing similar to that in Fig. 9-2(*b*).   The bearing capacity of the soil is only 2 ksf.   If the width is 8 ft, $d = 12$ in., and the reinforcement is No. 7 bars 8 in. c.c., is the footing safe?

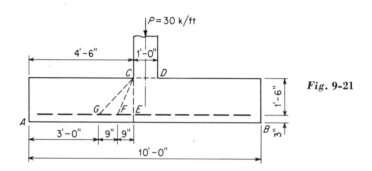

Fig. 9-21

Discussion.   Use a typical strip 12 in. wide supporting 16 kips.   Take moments about point *A* of the sketch—4 in. from the center line in this case.   Check the bond on the basis of a vertical section through *A*.   Check the diagonal tension at a point $d/2$ from *A* as a footing and at *d* from *A* as a cantilevered beam.   *Ans.* Everything is safe.

9-3   Design a square reinforced-concrete footing similar to that in Fig. 9-2(*c*) and (*d*) to support a 15-in. square column which carries a load of 150,000 lb.   The allowable net bearing capacity of the soil is 5,000 psf.

9-4   Design a square reinforced-concrete footing with a sloped top to support a 16-in. square column which carries a load of 175 kips.   The bearing capacity of the soil is 4 ksf.   Assume a flat top 24 in. square and slopes of one vertical to four horizontal.

9-5   Design a square stepped footing similar to that shown in Fig. 9-6(*a*) to support a column load of 300,000 lb.   The column is 20 in. in diameter, the pedestal is 3.5 ft square, and the allowable pressure on the soil is 5,000 psf.   Detail the reinforcement.

9-6   A stepped footing is to support a column load of 450 kips.   Its width is limited to 7 ft.   The pedestal is to be 4 ft square and 2 ft high, with its top at ground level.   The allowable soil pressure is 6 ksf.   Design the footing and detail the reinforcement.

9-7   Design the reinforcement for the wall footing shown in Fig. 9-21, and check the footing for diagonal tension at vertical planes through *F* and *G*.   Compute the moment about *CE*.

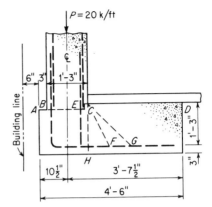

Fig. 9-22   Eccentric wall footing.

9-8   The footing shown in Fig. 9-22 is eccentrically loaded but is reinforced so as to transmit bending moment into the wall. If the bottom bars are No. 8 at 6 in. c.c., is the footing safe in all respects?

Discussion.   Assume that the center of moments is at the center of the wall, since the construction is unbalanced but continuous—somewhat like a stub-end frame. Test the shear at $d/2$ from C as a footing and at d from C as a cantilevered beam.

Ans. Yes. $f_s = 16,500$ psi, v at $d/2 = 59$ psi, v at $d = 43$ psi.

9-9   The stepped footing shown in Fig. 9-24(a) is to be designed for the following conditions: $P = 250$ kips, $M = 50$ ft-k, $S = 10$ kips, and adequately strong soil. What is the maximum soil pressure? Is the footing safe if the reinforcement is thirteen No. 6 bars both ways?

Ans. Yes. $f_s = 17,900$ psi, $u = 181$ psi, v at $d/2 = 51$ psi.

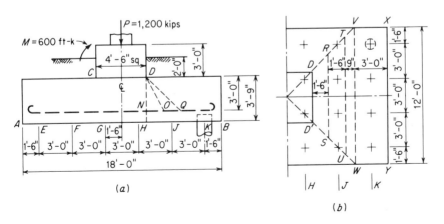

(a)

(b)

Fig. 9-23

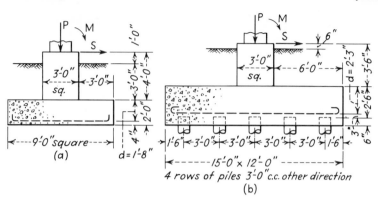

Fig. 9-24

9-10   Compute the total pile reactions under the footing shown in Fig. 9-23 if the column load, the overturning moment, and the dimensions are as given in the sketch.   Design the reinforcement and check for the bond and diagonal tensile stresses.

Discussion.   Include the weight of the footing and the earth over it when computing the gross pile load, but exclude them when computing the net pile reactions and analyzing the footing.   The moment acts in the long direction only.   Determine I for the pile group about the center line of the column and footing, neglecting the areas of the piles themselves but treating them as units.   Then $I = 4 \times 2(1.5^2 + 4.5^2 + 7.5^2) = 630$ piles-ft^2.   Any pile load $= \Sigma W/24 \pm Mc/I$ for this case.   Compute the moment about DN, also the bond stress.   Find the diagonal tension at O with a width RS, and finally at Q with the width TU.

Ans. p_K gross $= 64$ kips, bars $=$ twenty-three No. 11, $u_{DN} = 135$ psi, $v_O = 121$ psi.

9-11   The footing shown in Fig. 9-24(*b*) has four rows of five piles each. Design the reinforcement, and check the footing if $P = 800$ kips, $M = 120$ ft-k, and $S = 20$ kips.

9-12   Assume the footing shown in Fig. 9-14(*a*) to (*c*).   Compute the pressure diagram if $P = 260,000$ lb, $M = 60,000$ ft-lb, and $S = 4,000$ lb.

LARGE SLABS

10

10-1. Introduction. Large slabs are a very useful type of reinforced-concrete construction. The types to be studied herein are one-way slabs, two-way slabs supported by beams or walls along the column lines forming square or rectangular panels, and flat slabs supported at localized points by columns, except for outer walls or spandrel beams. It is the purpose of this chapter to serve as an introduction to the analysis and design of such slabs by empirical methods specified by the Code and by AASHO. In general, uniform loading will be used because this text is an elementary one designed to help the reader to get started in his study of reinforced concrete. In a second volume[1] an entire chapter is devoted to a further study of large slabs; therefore this chapter will not duplicate what is covered there.

Although large slabs are highly indeterminate, they are encountered so frequently in practice that it seems proper to include a method of analysis for some ordinary types in this text. Although the empirical methods specified by the Code may seem at first to be complicated, they really are not too difficult to follow, and they seem to yield safe results. The Code methods and recommendations are the result of much study, experience, and discussion.

10-2. One-way Slabs. Under the name of one-way slabs will be included both those of uniform depth and those composed of joists or miniature T beams similar to those shown in Fig. 10-1. As the name indicates, the bending moments are assumed to be resisted by beam action between two or more lines of support so that all of the main bars will be parallel to each other and perpendicular to those supports.

[1] C. W. Dunham, "Advanced Reinforced Concrete," chap. 6, McGraw-Hill Book Company, New York, 1964.

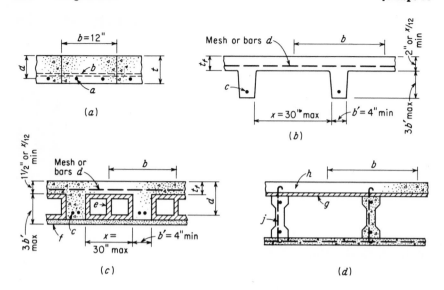

***Fig.* 10-1**   Some types of large slabs.

In Chap. 2, one-way slabs with uniform live load like that shown in Fig. 10-1(*a*) were treated as though they could be cut longitudinally into a number of slices 1 ft wide, then designed as though each slice were an independent rectangular beam which might be simply supported or continuous.   In the sketch, such a slice is pictured for a section having a positive bending moment resisted by bars *a* in tension.   However, transverse reinforcement represented by bars *b* is required in all cases to resist cracking from shrinkage and temperature changes, the minimum required area being 0.002 times the entire cross-sectional area of the concrete perpendicular to those bars.   It is obvious that for uniform loads alone, all imaginary 12-in. slices will behave alike and that there will be no shifting of load from one slice to its neighbor.   However, when a concentrated load is applied to such a slab, the concrete is not in slices, but it, together with bars *b* in Fig. 10-1(*a*), will try to spread load laterally as the slab tends to sag downward locally under the load.   The slab, under this type of loading, is then a highly indeterminate structure.

Some of the applicable AASHO requirements for solid, monolithic slabs for bridge floors are as follows:

1. The effective span S = the clear span if monolithic with concrete walls or beams; S = the clear distance between the edges of I-beam flanges + $\frac{1}{2}$ the flange width when on steel stringers; S = the clear span + $\frac{1}{2}$ the width of wooden stringers.

2. Slabs are to be designed for the loads of truck wheels. E = the width of slab in feet over which a wheel load is considered to be distributed;

P_{15} = 12,000 lb for one rear wheel load of H15-S12 loading; P_{20} = 16,000 lb for one rear wheel load of H20-S16 loading.

3. Main reinforcing perpendicular to traffic:

a. Wheel loads are distributed over a width

$$E = 4 + 0.06S, \text{ 7 ft max} \tag{10-1}$$

b. H20-S16 loading: live-load moment for simple spans (not including impact) = $P_{20}(S + 2)/32$ ft-lb/ft of width of slab for spans 2 ft to 24 ft, inclusive. (10-2)

c. H15-S12 loading: Live-load moment for simple spans (not including impact) = $P_{15}(S + 2)/32$ ft-lb/foot of slab width for spans 2 ft to 24 ft, inclusive. (10-3)

d. Live-load moments for spans continuous over three or more supports = 0.8 times the values found in *b* and *c* above for both positive and negative moments.

4. Main reinforcing parallel to traffic:

a. Wheel loads are distributed over a width = $2E$.

b. H20-S16 loading: Live-load moment for simple spans up to 50 ft = $900S$ ft-lb/ft of width of slab. (10-4)

Live-load moment for simple spans 50 ft to 100 ft = $1,000(1.3S - 20.0)$ ft-lb/ft of width of slab. (10-5)

c. H15-S12 loading: Live-load moment for simple spans = 0.75 times the values in *b.*

d. Live-load moments in continuous spans are to be found by suitable analysis.

5. Reinforcement perpendicular to main bars:

a. Main bars perpendicular to traffic:

$$\text{Percentage compared to main steel} = \frac{100}{\sqrt{S}}, \text{ max } 50\% \tag{10-6}$$

b. Main bars parallel to traffic:

$$\text{Percentage compared to main steel} = \frac{220}{\sqrt{S}}, \text{ max } 67\% \tag{10-7}$$

6. Bond and shear are considered adequate for slabs designed for bending moments specified in preceding paragraphs.

7. Edges and breaks in continuity are to be supported by diaphragms or equivalent.

The joist or T-beam construction shown in Fig. 10-1(*b*) and (*c*) can be designed for uniform loading as though it were an independent unit having one joist or rib each and having a flange of width b. However, such slabs are primarily for use with uniform loading because the thin flanges cannot sustain or transfer heavy concentrated loads.

Minimum thicknesses and maximum depths for monolithic joist construction are shown in Fig. 10-1(b) and (c). Bars d are to be designed for all loading needed (including any concentrations) but must be at least equal to shrinkage and temperature requirements although with bars spaced not over $5t_f$ or 18 in. c.c.   Allow $\frac{1}{2}$ in. extra concrete over t_f if the wearing surface of warehouses and industrial construction is made monolithic with the slab, but this extra is not to be relied upon for design. Of course, conduits and pipes should be so placed as not to endanger the structure and preferably with 1 in. of cover on both sides.

These joist slabs are light and relatively stiff. However, if they are continuous, the narrow ribs may be overstressed in compression at the supports unless they are thickened or heavily reinforced. The ribs may also have rather large diagonal tension, but the Code allows v_c to be increased by 10 per cent. The tiles or blocks e do not help resist positive moments, but if their compressive strength is as good as that of the concrete, the vertical shells in contact with the concrete of the joists may be relied upon for negative moment and shear, but no other part can be counted upon. The fillers provide a flat surface on which the ceiling f can be attached.

The floor shown in Fig. 10-1(d) is sometimes made with precast joists on which some type of reinforcement like Hy-rib g is placed. The thin concrete slab h is then poured on top. There should be some kind of wire or other ties j to hold the parts together, to steady the joists, and to make at least a partially composite beam.   A hung ceiling may be attached as shown or placed farther down to allow space for ducts and piping. This is not a type for resisting heavy loads, but it has considerable value for small structures.

Example 10-1. A one-way rectangular slab has the typical cross sections shown in Fig. 10-2. It is continuous over a series of spans for which the effective span is $L' = 12$ ft. Is the slab safe for an interior span for the following assumed conditions? Use USD; LL = 300 psf

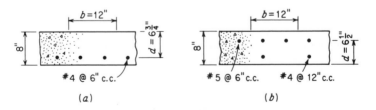

Fig. 10-2

uniform all over; $f'_c = 3,000$ psi; $f_y = 36,000$ psi; $U = 1.5\text{DL} + 1.8\text{LL}$; and $\phi = 0.9$.

$$U = 1.5 \times 100 + 1.8 \times 300 = 690 \text{ psf}$$

Use this for w in the formulas from Table 15-1 for a typical interior span.

$$M \text{ at midspan} = \frac{wL'^2}{16} = \frac{690 \times 12^2}{16} = 6,200 \text{ ft-lb}$$

$$M \text{ at support} = -\frac{wL'^2}{11} = \frac{690 \times 12^2}{11} = 9,000 \text{ ft-lb}$$

$$\text{Shear at support} = \frac{wL'}{2} = \frac{690 \times 12}{2} = 4,140 \text{ lb}$$

Use Eqs. (2-16) and (2-18) for tensile steel only.

$$a = \frac{A_s f_y}{0.85 f'_c b} \qquad (2\text{-}16)$$

$$M_u = \phi \left[A_s f_y \left(d - \frac{a}{2} \right) \right] \qquad (2\text{-}18)$$

Positive moment, sketch (a):

$$a = \frac{0.4 \times 36,000}{0.85 \times 3,000 \times 12} = 0.47$$
$$M_u = 0.9[0.4 \times 36,000(6.75 - 0.47/2)] = 84,500 \text{ in.-lb or } 7,050 \text{ ft-lb}$$

Negative moment, sketch (b), neglecting A'_s:

$$a = \frac{0.62 \times 36,000}{0.85 \times 3,000 \times 12} = 0.73$$
$$M_u = 0.9[0.62 \times 36,000(6.5 - 0.73/2)] = 123,000 \text{ in.-lb or } 10,250 \text{ ft-lb}$$

These are satisfactory. Notice that the bars in the top in (b) have more cover than the bottom ones.

The diagonal tension at d from the support is

$$v = \frac{V}{bd} = \frac{4,140 - 690 \times 0.54}{12 \times 6.5} = 48 \text{ psi (safe)}$$

Example 10-2. Assume that a bridge floor or deck is supported by steel stringers spaced so that the effective span S of the slab is 9 ft 8 in. Is the slab safe for H20-S16 loading with 30 per cent impact? Use WSD and AASHO with the following conditions: the slabs are continuous; $f_c = 1,200$ psi; $f_s = 18,000$ psi; $n = 10$; $v_c = 90$ psi; $t = 8$ in.; $d = 6$ in.; and $j = 0.88$. The reinforcement is perpendicular to traffic, and $A_s = $ No. 6 at 6 in. c.c.

From   Eq.   (10-1), $E = 4 + 0.06S = 4 + 0.06 \times 9.67 = 4.58$ ft.
From Eq. (10-2), for H20-S16, and 30 per cent impact,

$$M_{LL} = \frac{(9.67 + 2)}{32} \, 20{,}800 = 7{,}600 \text{ ft-lb/ft}$$

Because of continuity,

$$M_{LL} = 0.8 \times 7{,}600 = 6{,}080 \text{ ft-lb/ft}$$

Let the dead-load moment at a stringer be called

$$M_{DL} = \frac{wS^2}{11} = \frac{100 \times 9.67^2}{11} = 850 \text{ ft-lb/ft}$$

Then the maximum moment is $6{,}080 + 850 = 6{,}930$ ft-lb/ft.   Using this,

$$f_s = \frac{6{,}930 \times 12}{0.88 \times 0.88 \times 6} = 17{,}900 \text{ psi (safe)}$$

$$f_c \text{ approx} = \frac{6 \times 6{,}930 \times 12}{12 \times 6 \times 6} = 1{,}160 \text{ psi (safe)}$$

10-3. Two-way Slabs.   These are large rectangular slabs that are supported at all four edges and that have reinforcement in two perpendicular directions parallel to the supporting edges as shown in Fig. 10-3. They are frequently used, especially in large, continuous monolithic floor systems.   They may be solid slabs, two-way ribs or joists with some sort of tile or block filler between them and with a thin concrete slab poured monolithically on top of them, or "waffle" construction having two-way ribs with a thin monolithic top slab and air spaces between the ribs. Some of these are similar to those shown in Fig. 10-1(*b*) and (*c*) except that the ribs or joists extend in both directions.

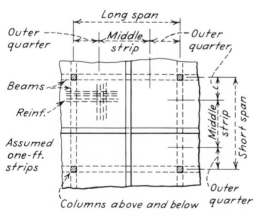

Fig. 10-3   Assumed distribution of uniform loads on two-way slabs.

First of all, it is well for one to try to visualize how a two-way slab behaves under load. Imagine that two planks are laid at right angles over the beams as shown by the two strips in Fig. 10-3. If a weight is placed at their intersection, the shorter, stiffer plank will support more of the load than will the longer one, but their deflections will be equal. When a uniform load is placed all over the slab, it will sag in the central portion also. However, the beams around the edges of the slab are relatively stiff and will not deflect much. The slab then tries to deliver the load to these beams as efficiently as possible. It seems obvious that if both directions are reinforced equally, the short spans will be the stronger.

Naturally, the central portions of the slab will deflect the most. Therefore, the middle strips shown in Fig. 10-4(a) will be the critical ones whether the edges of the slabs are simply supported or continuous over the beams. The portions near the beams are affected to a lesser degree than the middle strips, but they are also affected by the T-beam action of the beams themselves.

The slabs should not be so thin that they will sag excessively. The Code limits the over-all thickness t to not less than $3\frac{1}{2}$ in. nor less than the perimeter of the slab divided by 180. In such slabs as these, the bars should not be spaced farther apart than $3t$, and the minimum A_s must not be less than that needed for shrinkage and temperature reinforcement $- 0.002A_c$.

The Code gives three empirical methods for the design and analysis of two-way slabs. Method 3 will be used here in its entirety, with the necessary accompanying tables. This method has been selected because it enables one to readily take into account the effect of the restraint conditions at the supporting edges.

The special symbols to be used are as follows:

A = length of clear span in the short direction.

B = length of clear span in the long direction.

C = moment coefficients for two-way slabs given in Tables 10-1, 10-2, and 10-3. Coefficients will be used with identifying subscripts: for example, $C_{A\,\text{neg}}$, $C_{B\,\text{DL}}$, $C_{A\,\text{LL}}$.

m = ratio of short span to long span.

w = uniform load per square foot. For negative moments and shears, w is the total DL + LL for use in Table 10-1. For positive moments, separate w into DL and LL for use in Tables 10-2 and 10-3.

w_A, w_B = percentages of the load w in the A and B directions according to Table 10-4. These are to be used for computing shears in the slabs and also for the calculation of loadings on the supports.

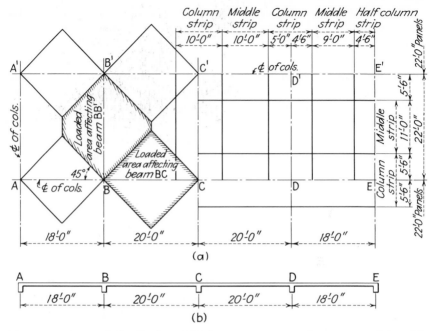

Fig. 10-4   Assumed distribution of loads to beams supporting two-way slabs with uniform loads.

For purposes of calculation, a two-way slab is to be considered to con-sist of strips in each direction as follows:

1. A middle strip one-half panel wide, symmetrical about the panel center line, and extending across the panel in the direction in which the moments are considered, as shown in Fig. 10-4(a).

2. A column strip one-half panel in width, occupying the two quarter-panel areas outside of the middle strip, as shown in Fig. 10-4(a).

The following limitations are specified:

1. Where the ratio of the short span to the long span is less than 0.5, the slab is to be designed as a one-way slab spanning across the short direction, except that some negative reinforcement, as required for a ratio $m = 0.5$, should be provided along the short edges.   This is to avoid local cracking as the slab tries to deflect alongside a support.

2. At discontinuous edges, assume a negative moment equal to one-third of the positive moment to be used in that panel (and in the direction perpendicular to that edge).

The critical sections for the calculation of bending moments are the following:

1. For negative moments: along the edges of the panel at the faces of the supports.

2. For positive moments: along the center lines of the panels.

The bending moments for the middle strips are to be computed by the use of Tables 10-1 to 10-3 from the following:

$$M_A = CwA^2 \quad \text{and} \quad M_B = CwB^2 \tag{10-8}$$

where the bending moments are for 1 ft of width of the strip.

The bending moments in the parallel column strips should be gradually reduced from the full value of M_A or M_B at the edge of the middle strip to one-third of these values at the edge of the parallel side of the panel. If the negative moment on one side of a common support is less than 80 per cent of that on the opposite side, the difference should be distributed in proportion to the relative stiffnesses of the slabs.[1]

The shear stresses in the slab may be calculated on the assumption that the uniform load is distributed to the supports in accordance with Table 10-4. This table is also useful in making a rough preliminary estimate of the size of slab needed.

The loads on the beams supporting a rectangular two-way slab should be computed by using Table 10-4 for the percentages of loads in the A and B directions. However, the load on the beam along the short edge should never be less than that of an area bounded by the intersection of 45° lines from the corners, as shown for BC in Fig. 10-4(a). These loads may be considered to be converted into equivalent, uniformly distributed loads per foot of the short beam by the use of the following formula:

$$\text{Equivalent uniform load per foot} = \frac{wA}{3} \tag{10-9}$$

Example 10-3. Assume a warehouse building composed of several bays or panels 22 ft wide, as shown in Fig. 10-4(a). In the other direction there are four panels having the dimensions shown in (b). By USD, design a typical portion of floor across the building for the following conditions: LL = 200 psf; f'_c = 3,000 psi; f_y = 36,000 psi; U = 1.5DL + 1.8LL; ϕ = 0.9 for bending and 0.85 for shear; and the slab is to be rectangular in section.

Since the building is symmetrical about the longitudinal center line along CC', design panels $CC'D'D$ and $DD'E'E$ only. The slabs are continuous across all supports except AA' and EE'.

It is first necessary to estimate the slab thickness. The 22 ft is a long distance for a slab, but the loads are carried in two directions, and there is considerable continuity. On the other hand, because the building is a warehouse, the slabs should be reasonably stiff. Therefore, assume $t = L/35 = 22 \times \frac{12}{35} = 7.5$ in. Add $\frac{1}{2}$ in. on top to allow for a wearing surface. Then the total depth is 8 in., giving a DL of 100 psf, but design the slab as though $t = 7.5$ in.

---

[1] $3I/L$ if one end (the other one) is simply supported; $4I/L$ if the far end is continuous.

Table 10-1   Coefficients for Negative Moments in Two-way Slabs*

$$M_{A\ neg} = C_{A\ neg} \times w \times A^2 \left.\right\}$$
$$M_{B\ neg} = C_{B\ neg} \times w \times B^2 \left.\right\}$$ where w = total uniform dead load plus live load

Ratio $m = \dfrac{A}{B}$		Case 1	Case 2	Case 3	Case 4	Case 5	Case 6	Case 7	Case 8	Case 9
1.00	$C_{A\ neg}$		0.045		0.050	0.075	0.071		0.033	0.061
	$C_{B\ neg}$		0.045	0.076	0.050			0.071	0.061	0.033
0.95	$C_{A\ neg}$		0.050		0.055	0.079	0.075		0.038	0.065
	$C_{B\ neg}$		0.041	0.072	0.045			0.067	0.056	0.029
0.90	$C_{A\ neg}$		0.055		0.060	0.080	0.079		0.043	0.068
	$C_{B\ neg}$		0.037	0.070	0.040			0.062	0.052	0.025
0.85	$C_{A\ neg}$		0.060		0.066	0.082	0.083		0.049	0.072
	$C_{B\ neg}$		0.031	0.065	0.034			0.057	0.046	0.021
0.80	$C_{A\ neg}$		0.065		0.071	0.083	0.086		0.055	0.075
	$C_{B\ neg}$		0.027	0.061	0.029			0.051	0.041	0.017
0.75	$C_{A\ neg}$		0.069		0.076	0.085	0.088		0.061	0.078
	$C_{B\ neg}$		0.022	0.056	0.024			0.044	0.036	0.014
0.70	$C_{A\ neg}$		0.074		0.081	0.086	0.091		0.068	0.081
	$C_{B\ neg}$		0.017	0.050	0.019			0.038	0.029	0.011
0.65	$C_{A\ neg}$		0.077		0.085	0.087	0.093		0.074	0.083
	$C_{B\ neg}$		0.014	0.043	0.015			0.031	0.024	0.008
0.60	$C_{A\ neg}$		0.081		0.089	0.088	0.095		0.080	0.085
	$C_{B\ neg}$		0.010	0.035	0.011			0.024	0.018	0.006
0.55	$C_{A\ neg}$		0.084		0.092	0.089	0.096		0.085	0.086
	$C_{B\ neg}$		0.007	0.028	0.008			0.019	0.014	0.005
0.50	$C_{A\ neg}$		0.086		0.094	0.090	0.097		0.089	0.088
	$C_{B\ neg}$		0.006	0.022	0.006			0.014	0.010	0.003

* A crosshatched edge denotes slab is continuous or fixed; unmarked edge denotes a simply supported edge with negligible torsional resistance.

Table **10-2   Coefficients for Dead Load Positive Moments in Two-way Slabs***

$$M_{A \text{ pos DL}} = C_{A \text{ DL}} \times w \times A^2$$
$$M_{B \text{ pos DL}} = C_{B \text{ DL}} \times w \times B^2$$
where w = total uniform dead load

Ratio $m = \dfrac{A}{B}$		Case 1	Case 2	Case 3	Case 4	Case 5	Case 6	Case 7	Case 8	Case 9
1.00	$C_{A \text{ DL}}$	0.036	0.018	0.018	0.027	0.027	0.033	0.027	0.020	0.023
	$C_{B \text{ DL}}$	0.036	0.018	0.027	0.027	0.018	0.027	0.033	0.023	0.020
0.95	$C_{A \text{ DL}}$	0.040	0.020	0.021	0.030	0.028	0.036	0.031	0.022	0.024
	$C_{B \text{ DL}}$	0.033	0.016	0.025	0.024	0.015	0.024	0.031	0.021	0.017
0.90	$C_{A \text{ DL}}$	0.045	0.022	0.025	0.033	0.029	0.039	0.035	0.025	0.026
	$C_{B \text{ DL}}$	0.029	0.014	0.024	0.022	0.013	0.021	0.028	0.019	0.015
0.85	$C_{A \text{ DL}}$	0.050	0.024	0.029	0.036	0.031	0.042	0.040	0.029	0.028
	$C_{B \text{ DL}}$	0.026	0.012	0.022	0.019	0.011	0.017	0.025	0.017	0.013
0.80	$C_{A \text{ DL}}$	0.056	0.026	0.034	0.039	0.032	0.045	0.045	0.032	0.029
	$C_{B \text{ DL}}$	0.023	0.011	0.020	0.016	0.009	0.015	0.022	0.015	0.010
0.75	$C_{A \text{ DL}}$	0.061	0.028	0.040	0.043	0.033	0.048	0.051	0.036	0.031
	$C_{B \text{ DL}}$	0.019	0.009	0.018	0.013	0.007	0.012	0.020	0.013	0.007
0.70	$C_{A \text{ DL}}$	0.068	0.030	0.046	0.046	0.035	0.051	0.058	0.040	0.033
	$C_{B \text{ DL}}$	0.016	0.007	0.016	0.011	0.005	0.009	0.017	0.011	0.006
0.65	$C_{A \text{ DL}}$	0.074	0.032	0.054	0.050	0.036	0.054	0.065	0.044	0.034
	$C_{B \text{ DL}}$	0.013	0.006	0.014	0.009	0.004	0.007	0.014	0.009	0.005
0.60	$C_{A \text{ DL}}$	0.081	0.034	0.062	0.053	0.037	0.056	0.073	0.048	0.036
	$C_{B \text{ DL}}$	0.010	0.004	0.011	0.007	0.003	0.006	0.012	0.007	0.004
0.55	$C_{A \text{ DL}}$	0.088	0.035	0.071	0.056	0.038	0.058	0.081	0.052	0.037
	$C_{B \text{ DL}}$	0.008	0.003	0.009	0.005	0.002	0.004	0.009	0.005	0.003
0.50	$C_{A \text{ DL}}$	0.095	0.037	0.080	0.059	0.039	0.061	0.089	0.056	0.038
	$C_{B \text{ DL}}$	0.006	0.002	0.007	0.004	0.001	0.003	0.007	0.004	0.002

* A crosshatched edge denotes slab is continuous or fixed; unmarked edge denotes a simply supported edge with negligible torsional resistance.

Table 10-3 Coefficients for Live Load Positive Moments in Two-way Slabs*

$$\left.\begin{array}{l} M_{A \text{ pos LL}} = C_{A \text{ LL}} \times w \times A^2 \\ M_{B \text{ pos LL}} = C_{B \text{ LL}} \times w \times B^2 \end{array}\right\} \text{ where } w = \text{total uniform live load}$$

Ratio	Case 1	Case 2	Case 3	Case 4	Case 5	Case 6	Case 7	Case 8	Case 9
$m = \dfrac{A}{B}$									
1.00 $C_{A \text{ LL}}$	0.036	0.027	0.027	0.032	0.032	0.035	0.032	0.028	0.030
$C_{B \text{ LL}}$	0.036	0.027	0.032	0.032	0.027	0.032	0.035	0.030	0.028
0.95 $C_{A \text{ LL}}$	0.040	0.030	0.031	0.035	0.034	0.038	0.036	0.031	0.032
$C_{B \text{ LL}}$	0.033	0.025	0.029	0.029	0.024	0.029	0.032	0.027	0.025
0.90 $C_{A \text{ LL}}$	0.045	0.034	0.035	0.039	0.037	0.042	0.040	0.035	0.036
$C_{B \text{ LL}}$	0.029	0.022	0.027	0.026	0.021	0.025	0.029	0.024	0.022
0.85 $C_{A \text{ LL}}$	0.050	0.037	0.040	0.043	0.041	0.046	0.045	0.040	0.039
$C_{B \text{ LL}}$	0.026	0.019	0.024	0.023	0.019	0.022	0.026	0.022	0.020
0.80 $C_{A \text{ LL}}$	0.056	0.041	0.045	0.048	0.044	0.051	0.051	0.044	0.042
$C_{B \text{ LL}}$	0.023	0.017	0.022	0.020	0.016	0.019	0.023	0.019	0.017
0.75 $C_{A \text{ LL}}$	0.061	0.045	0.051	0.052	0.047	0.055	0.056	0.049	0.046
$C_{B \text{ LL}}$	0.019	0.014	0.019	0.016	0.013	0.016	0.020	0.016	0.013
0.70 $C_{A \text{ LL}}$	0.068	0.049	0.057	0.057	0.051	0.060	0.063	0.054	0.050
$C_{B \text{ LL}}$	0.016	0.012	0.016	0.014	0.011	0.013	0.017	0.014	0.011
0.65 $C_{A \text{ LL}}$	0.074	0.053	0.064	0.062	0.055	0.064	0.070	0.059	0.054
$C_{B \text{ LL}}$	0.013	0.010	0.014	0.011	0.009	0.010	0.014	0.011	0.009
0.60 $C_{A \text{ LL}}$	0.081	0.058	0.071	0.067	0.059	0.068	0.077	0.065	0.059
$C_{B \text{ LL}}$	0.010	0.007	0.011	0.009	0.007	0.008	0.011	0.009	0.007
0.55 $C_{A \text{ LL}}$	0.088	0.062	0.080	0.072	0.063	0.073	0.085	0.070	0.063
$C_{B \text{ LL}}$	0.008	0.006	0.009	0.007	0.005	0.006	0.009	0.007	0.006
0.50 $C_{A \text{ LL}}$	0.095	0.066	0.088	0.077	0.067	0.078	0.092	0.076	0.067
$C_{B \text{ LL}}$	0.006	0.004	0.007	0.005	0.004	0.005	0.007	0.005	0.004

* A crosshatched edge denotes slab is continuous or fixed; unmarked edge denotes a simply supported edge with negligible torsional resistance.

Table 10-4   Ratio of Load w in A and B Directions for Shear in Two-way Slab and Load on Supports*

Ratio $m = \dfrac{A}{B}$		Case 1	Case 2	Case 3	Case 4	Case 5	Case 6	Case 7	Case 8	Case 9
1.00	w_A	0.50	0.50	0.17	0.50	0.83	0.71	0.29	0.33	0.67
	w_B	0.50	0.50	0.83	0.50	0.17	0.29	0.71	0.67	0.33
0.95	w_A	0.55	0.55	0.20	0.55	0.86	0.75	0.33	0.38	0.71
	w_B	0.45	0.45	0.80	0.45	0.14	0.25	0.67	0.62	0.29
0.90	w_A	0.60	0.60	0.23	0.60	0.88	0.79	0.38	0.43	0.75
	w_B	0.40	0.40	0.77	0.40	0.12	0.21	0.62	0.57	0.25
0.85	w_A	0.66	0.66	0.28	0.66	0.90	0.83	0.43	0.49	0.79
	w_B	0.34	0.34	0.72	0.34	0.10	0.17	0.57	0.51	0.21
0.80	w_A	0.71	0.71	0.33	0.71	0.92	0.86	0.49	0.55	0.83
	w_B	0.29	0.29	0.67	0.29	0.08	0.14	0.51	0.45	0.17
0.75	w_A	0.76	0.76	0.39	0.76	0.94	0.88	0.56	0.61	0.86
	w_B	0.24	0.24	0.61	0.24	0.06	0.12	0.44	0.39	0.14
0.70	w_A	0.81	0.81	0.45	0.81	0.95	0.91	0.62	0.68	0.89
	w_B	0.19	0.19	0.55	0.19	0.05	0.09	0.38	0.32	0.11
0.65	w_A	0.85	0.85	0.53	0.85	0.96	0.93	0.69	0.74	0.92
	w_B	0.15	0.15	0.47	0.15	0.04	0.07	0.31	0.26	0.08
0.60	w_A	0.89	0.89	0.61	0.89	0.97	0.95	0.76	0.80	0.94
	w_B	0.11	0.11	0.39	0.11	0.03	0.05	0.24	0.20	0.06
0.55	w_A	0.92	0.92	0.69	0.92	0.98	0.96	0.81	0.85	0.95
	w_B	0.08	0.08	0.31	0.08	0.02	0.04	0.19	0.15	0.05
0.50	w_A	0.94	0.94	0.76	0.94	0.99	0.97	0.86	0.89	0.97
	w_B	0.06	0.06	0.24	0.06	0.01	0.03	0.14	0.11	0.03

* A crosshatched edge denotes slab is continuous or fixed; unmarked edge denotes a simply supported edge with negligible torsional resistance.

1. *Panel CC'D'D.* Call $A = 20 - 1 = 19$ ft and $B = 22 - 1 = 21$ ft. This assumes that the stems of the supporting T beams are only 12 in. wide. This is probably too narrow, but it is conservative in computing the clear spans. Then

$$m = \frac{A}{B} = \frac{19}{21} = 0.9$$

$$U = (1.5 \times 100 = 150 \text{ psf for DL}) + (1.8 \times 200 = 360 \text{ psf for LL})$$

Then let DL = 150 psf, LL = 360 psf, and total $w = 510$ psf.

For this panel, all four edges are continuous. Therefore, use Case 2, with $m = 0.9$ for finding the applicable numbers in all four tables.

a. Negative moments: Refer to Table 10-1, with $w = 510$ psf. Then Eq. (10-8) gives

$$M_{A\,neg} = C_{A\,neg} \times w \times A^2 = 0.055 \times 510 \times 19^2 = 10,100 \text{ ft-lb/ft}$$

for CC' and DD'.

$$M_{B\,neg} = C_{B\,neg} \times w \times B^2 = 0.037 \times 510 \times 21^2 = 8,300 \text{ ft-lb/ft}$$

b. Positive dead-load moment: Use Table 10-2, with $w_{DL} = 150$ psf.

$$C_{A\,DL} = 0.022 \qquad C_{B\,DL} = 0.014$$
$$M_{A\,pos\,DL} = 0.022 \times 150 \times 19^2 = 1,200 \text{ ft-lb/ft}$$
$$M_{B\,pos\,DL} = 0.014 \times 150 \times 21^2 = 930 \text{ ft-lb/ft}$$

c. Positive live-load moment: Use Table 10-3, with $w_{LL} = 360$ psf.

$$C_{A\,LL} = 0.034 \qquad C_{B\,LL} = 0.022$$
$$M_{A\,pos\,LL} = 0.034 \times 360 \times 19^2 = 4,400 \text{ ft-lb/ft}$$
$$M_{B\,pos\,LL} = 0.022 \times 360 \times 21^2 = 3,500 \text{ ft-lb/ft}$$

d. Maximum positive moment:

$$M_A = 1,200 + 4,400 = 5,600 \text{ ft-lb/ft}$$
$$M_B = 930 + 3,500 = 4,430 \text{ ft-lb/ft}$$

e. Shear: Use Table 10-4, with $w = 510$ psf.

$$w_A = 0.60 \qquad w_B = 0.40.$$

Therefore, $0.6 \times 510 = 306$ psf can be considered going to CC' and DD' of Fig. 10-4—the short direction. Then, from Table 15-1,

Shear along CC' and $DD' = 306 \times \frac{19}{2} = 2,900$ lb/ft
Shear along CD and $C'D' = (510 - 306)\frac{21}{2} = 2,140$ lb/ft

f. Preliminary check of slab: The maximum negative moment is 10,100 ft-lb/ft crossing CC' and DD' in Fig. 10-4. Before the slab can be checked, a percentage of steel p or the number and size of bars must be assumed.

In this case, try No. 5 bars at 6 in. c.c., giving $A_s = 0.62$ in.^{2}/ft. Using Eqs. (2-16) and (2-18), gives

$$a = \frac{A_s f_y}{0.85 f'_c b} = \frac{0.62 \times 36,000}{0.85 \times 3,000 \times 12} = 0.73$$

$$M_u = \phi \left[A_s f_y \left(d - \frac{a}{2} \right) \right] = 0.9 \left[0.62 \times 36,000 \left(6 - \frac{0.73}{2} \right) \right] = 113,000,$$
$$\text{in.-lb or } 9,400 \text{ ft-lb/ft}$$

This shows that M_u is smaller than the 10,100 ft-lb needed but that the bars can easily be used closer together. The concrete will undoubtedly be safe in compression.

The shear along CC' and DD' was found to be 2,900 lb/ft. Then, even using this at full value instead of a distance d from the edge,

$$v_u = \frac{V_u}{bd} = \frac{2,900}{12 \times 6} = 40 \text{ psi}$$

This is far less than that permitted by $2\phi \sqrt{f'_c} = 2 \times 0.85 \sqrt{3,000} = 93$ psi allowed for 3,000-lb concrete. Therefore, continue to use the assumed thickness of the slab and $d = 6$ in.

2. *Panel DD'E'E.* Call $A = 18 - 1 = 17$ ft and $B = 22 - 1 = 21$ ft. Then

$$m = A/B = \tfrac{17}{21} = 0.81 \quad \text{Call it } 0.8.$$

Use the same loading as before.

a. Negative moments: This panel is simply supported along EE', the long side, but is continuous elsewhere. Therefore, use case 8 with $m = 0.8$ in all of the tables. Then, from Table 10-1, with $w = 510$ psf,

$$M_{A \text{ neg}} = 0.055 \times 510 \times 17^2 = 8,100 \text{ ft-lb/ft for } DD'$$
$$M_{B \text{ neg}} = 0.041 \times 510 \times 21^2 = 9,200 \text{ ft-lb/ft for } DE \text{ and } D'E'$$

b. Positive dead-load moment: From Table 10-2, with $w_{DL} = 150$ psf,

$$M_{A \text{ pos DL}} = 0.032 \times 150 \times 17^2 = 1,400 \text{ ft-lb/ft}$$
$$M_{B \text{ pos DL}} = 0.015 \times 150 \times 21^2 = 1,000 \text{ ft-lb/ft}$$

c. Positive live-load moment: From Table 10-3, with $w_{LL} = 360$ psf,

$$M_{A \text{ pos LL}} = 0.044 \times 360 \times 17^2 = 4,600 \text{ ft-lb/ft}$$
$$M_{B \text{ pos LL}} = 0.019 \times 360 \times 21^2 = 3,000 \text{ ft-lb/ft}$$

d. Maximum positive moment:

$$M_A = 1,400 + 4,600 = 6,000 \text{ ft-lb/ft}$$
$$M_B = 1,000 + 3,000 = 4,000 \text{ ft-lb/ft}$$

e. *Shear:* From Table 10-4, with $w_A = 0.55$ and $w_B = 0.45$, the proportional load carried in the short direction is $0.55 \times 510 = 280$ psf, and that in the long direction is $0.45 \times 510 = 230$ psf. Then from Table 15-1, for one end simply supported, the reaction across DD' in Fig. 10-4(a) is

$$V_u = \frac{1.15wL'}{2} = \frac{1.15 \times 280 \times 17}{2} = 2{,}700 \text{ lb/ft}$$

In the other direction,

$$V_u = \frac{wL'}{2} = \frac{230 \times 21}{2} = 2{,}400 \text{ lb/ft}$$

These shears are less than in the other panel and are therefore safe.

f. *Moments along DD':* For the interior panel, the negative moment along DD' was 10,100 ft-lb; for the exterior panel, 8,100 ft-lb. This is just about the 80 per cent difference for the limit where adjustment of the moment is necessary. For illustration, the adjustment will be made.

Stiffness factors: I equal on both sides:

Exterior panel: $\dfrac{3I}{L}$ or $\dfrac{3}{L} = \dfrac{3}{17} = 0.177$

Interior panel: $\dfrac{4I}{L}$ or $\dfrac{4}{L} = \dfrac{4}{19} = 0.21$

Ratios:

Exterior panel: $\dfrac{0.177}{0.177 + 0.21} = 0.46$

Interior panel: $\dfrac{0.21}{0.387} = 0.54$

The difference in the moments is $10{,}100 - 8{,}100 = 2{,}000$ ft-lb/ft. However, which way does it go? If torsional resistance of the beam is neglected, the moments on each side of beam DD' should balance. With 2,000 ft-lb to adjust, then $0.54 \times 2{,}000 = 1{,}080$ ft-lb is the correction for the interior side of DD' and $0.46 \times 2{,}000 = 920$ ft-lb is the correction for the exterior side. The result is to reduce the negative moment for the interior side to $10{,}100 - 1{,}080 = 9{,}020$ ft-lb and to add the 920 ft-lb to the exterior side, giving $8{,}100 + 920 = 9{,}020$ ft-lb so that both sides now balance.

g. *Negative moment for the exterior edge EE':* The negative moment specified for a simply supported edge like EE' is one-third times the positive moment in the middle strip across it—from DD' to EE' in this case, or $6{,}000/3 = 2{,}000$ ft-lb/ft, as found in paragraph **2d**.

3. *Final Design of Reinforcement.* Since a uniform depth of slab will be used throughout, it is perhaps desirable to make a diagram like that

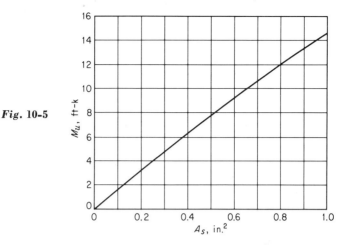

Fig. **10-5**

shown in Fig. 10-5 which is based upon Eqs. (2-16) and (2-18) for USD, with $b = 12$ in. and $d = 6$ in. in all cases.   It is then easy to adjust the bars to suit the moments at various sections.   These can be summarized as shown in Table 10-5, using the lettering from Fig. 10-4.

The spacing of the bars in the column strips can be made to suit the variations from the adjacent middle strips to a spacing not over three times that of the middle strips.   These adjustments are not shown here.

Table 10-5   **Data Regarding Slab Reinforcement, for Middle Strips**

Member	Moment, ft-lb	A_s, in.²	Size of bars	Spacing of bars, in.	Approximate length of bars
			Negative		
CC'	10,100	0.67	5	$5\frac{1}{2}$	0.3L each side of beam
DD'	9,020	0.58	5	6	0.3L interior; 0.33L exterior side
$CD, C'D'$	8,300	0.52	5	7	0.3L both sides
$DE, D'E'$	9,200	0.59	5	6	0.3L both sides
EE'	2,000	0.12	4	12	0.25L inside of beam
			Positive		
CC'-DD'	5,600	0.36	4	6	Alternate 0.75L and full length
CD-$C'D'$	4,430	0.27	4	9	Alternate 0.75L and full length
DE-$D'E'$	4,000	0.25	4	9	Alternate 0.75L and full length
DD'-EE'	6,000	0.37	4	6	Alternate 0.9L and full length

10-4. Slabs Made of Lightweight Concrete. Slabs are usually underreinforced, so that the compressive strength of normal stone concrete is not fully utilized, and the heavy material adds much to the weight of the structure, especially in the case of multistory buildings. The use of lightweight concrete can be particularly advantageous for floors and roofs. However, special care and skill are needed to obtain good, uniform results. There are various types of lightweight aggregates, each with its own special properties, so that the engineer should get proper guidance concerning the best available one that will serve his purpose, and he should obtain expert advice about how and when to use it in making lightweight concrete.

Lightweight aggregates are likely to be angular and to have a rough surface texture, tending to cause difficult placing and finishing. If not properly prewetted well in advance, their absorption of water may vary enough to cause difficulty in getting uniform quality and workability. Air entrainment (4 to 8 per cent) should always be used because it aids so much in producing plasticity and reduces bleeding and segregation. Substitution of sand for part of the lightweight fine aggregate may help to get workability, but it adds to the unit weight. A greater than usual percentage of lightweight fine aggregate (plus the necessary extra cement to coat the particles) may also help in this respect. Finishing should not be done too soon after placing of the wet concrete because of the danger of working fines to the top. The aggregates should be well graded, probably with more fines than used for normal stone concrete.[1]

The effect of the water-cement ratio on the quality of lightweight concrete is similar to that for normal stone concrete. Naturally, the type and properties of the lightweight aggregates themselves affect the unit weight and compressive strength which can be obtained. In general, good compressive strength can be obtained by the use of the proper materials if the concrete is cured the same as normal stone concrete should be. Bond strength and creep are about the same as for stone concrete; shrinkage may be somewhat greater (perhaps 25 per cent), but the resistance to diagonal tension probably will be considerably smaller. The slump should generally be not over 2 or 3 in. for best finishing because the material is so light.[2]

Lightweight concretes may have a thermal conductivity of about 3 Btu/(hr)/(in.) of thickness compared with about 10 Btu for normal stone concrete, giving a real advantage for insulation.

[1] See ASTM C 330, Tentative Specification for Lightweight Aggregates for Structural Concrete.

[2] G. H. Nelson, and O. C. Frei, Lightweight Structural Concrete Proportioning and Control, *J. ACI*, February, 1958.

Example 10-4. An architect wants to use a waffle type of floor in a multistory, fireproof school building, making it of lightweight concrete weighing 110 pcf with reinforcing. He proposes to use a module having the dimensions shown in Fig. 10-6(a). The floor will be simply supported on the exterior wall CD but continuous across the corridor wall AB and across the beams AD and BC. These beams are to be used so that the space below can be used for such purposes as a library, cafeteria, or small assembly room—eliminating cross partitions and columns. The beams are not to project more than 12 in. below the bottom of the waffles. Assume the following conditions: WSD; $f_c' = 3,000$ psi; $f_c = 1,200$ psi; $f_s = 18,000$ psi; $j = 0.88$; $n = 9$; LL = 75 psf. There must be some allowance for a possible acoustical ceiling and a floor finish.

The first step is to select dimensions for the beams and waffles. The beams will be assumed to be 24 in. wide, as shown in Fig. 10-6(a). The clear space EF between them is 22 ft. If standard steel-pan forms 24 in. square and 10 in. deep are used, with one edge flush with the edge of each beam, nine such pans would occupy 18 ft of the panel. The remaining distance is 4 ft. With eight ribs or joists between the pans, each joist will be 6 in. wide. In the other direction, with the edge of a pan at EF and GH, ten pans will occupy 20 ft, leaving 4.5 ft for joists. Then nine joists 6 in. wide can be used. These fit nicely. The slab will be made $3\frac{1}{2}$ in. thick to allow for conduits. A typical trial section is shown in Fig. 10-6(b). This will be a two-way slab.

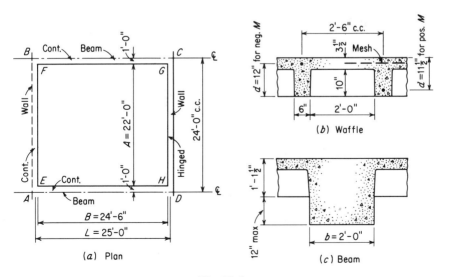

Fig. 10-6

1. *Design of Waffles.*   The dead load of the waffle floor can be found as follows:

Area of one waffle $= 2.5^2 = 6.25$ ft^2

Volume of one waffle $= \dfrac{6.25 \times 13.5}{12} - \dfrac{2^2 \times 10}{12} = 3.72$ ft^3

Weight of one waffle $= 3.72 \times 110 = 410$ lb

Dead load per square foot $= \dfrac{410}{6.25} = 66$ psf

Assume the total loads to be

$$
\begin{array}{lcr}
\text{Waffles} & = & 66 \text{ psf} \\
\text{Ceiling, etc.} & = & 9 \text{ psf} \\
\text{Total DL} & = & 75 \text{ psf} \\
\text{LL} & = & \underline{75 \text{ psf}} \\
w & = & 150 \text{ psf}
\end{array}
$$

Referring to Fig. 10-6(*a*), one short side is simply supported, but all others are continuous.   Also,

$$m = \frac{A}{B} = \frac{22}{24.5} = 0.9$$

Therefore, this design will be in case 9, with $m = 0.9$, in all of the tables.

a. *Negative moment:* From Table 10-1, with $w = 150$ psf,

$$M_{A\,\text{neg}} = 0.068 \times 150 \times 22^2 = 4{,}900 \text{ ft-lb/ft at } FG$$
$$M_{B\,\text{neg}} = 0.025 \times 150 \times 24.5^2 = 2{,}300 \text{ ft-lb/ft at } EF$$

b. *Positive DL moment:* From Table 10-2, with $w_{\text{DL}} = 75$ psf,

$$M_{A\,\text{pos DL}} = 0.026 \times 75 \times 22^2 = 940 \text{ ft-lb/ft}$$
$$M_{B\,\text{pos DL}} = 0.015 \times 75 \times 24.5^2 = 670 \text{ ft-lb/ft}$$

c. *Positive LL moment:* From Table 10-3, with $w_{\text{LL}} = 75$ psf,

$$M_{A\,\text{pos LL}} = 0.036 \times 75 \times 22^2 = 1{,}300 \text{ ft-lb/ft}$$
$$M_{B\,\text{pos LL}} = 0.022 \times 75 \times 24.5^2 = 990 \text{ ft-lb/ft}$$

d. *Total positive moment:*

$$M_{A\,\text{pos}} = 940 + 1{,}300 = 2{,}240 \text{ ft-lb/ft across } FG \text{ and } EH$$
$$M_{B\,\text{pos}} = 670 + 990 = 1{,}660 \text{ ft-lb/ft across } EF \text{ and } GH$$

e. *Shear:* From Table 10-4, with $w = 150$ psf,

$$V = \frac{0.75 \times 150 \times 22}{2} = 1{,}240 \text{ lb/ft along } FG$$

$$= \frac{1.15 \times 0.25 \times 150 \times 24.5}{2} = 530 \text{ lb/ft along } EF$$

$$= 0.25 \times 150 \times 24.5 - 530 = 390 \text{ lb/ft along } GH$$

f. Design of waffles: Assume that the effective depths are as shown in Fig. 10-6(*b*).

Middle strip across FG and EH:

Positive moment per waffle = 2,240 × 2.5 = 5,600 ft-lb.

$$A_s = \frac{5,600 \times 12}{18,000 \times 0.88 \times 11.5} = 0.27 \text{ in.}^2$$

Use one No. 4 bar full length and one No. 4 bar 18 ft long across the center. The compression in the top will be negligible.

Middle strip across EF and GH:

$$A_s = \frac{1,660 \times 2.5 \times 12}{18,000 \times 0.88 \times 11.5} = 0.27 \text{ in.}^2$$

Use one No. 5 bar full length.

Section along FG and EH:

$$A_s = \frac{4,900 \times 2.5 \times 12}{18,000 \times 0.88 \times 12} = 0.77 \text{ in.}^2$$

Use one No. 8 bar 15 ft long across the beams for negative reinforcement.

$$\text{Approx } f_c = \frac{6M}{bd^2} = \frac{6 \times 4,900 \times 2.5 \times 12}{6 \times 12^2}$$

$$= 1,040 \text{ psi } (1,200 \text{ psi allowed})$$

$$v = \frac{V}{bd} = \frac{1,240 \times 2.5}{6 \times 12} = 43 \text{ psi}$$

However, the Code allows only

$$v_c = 0.17 F_{sp} \sqrt{f_c'} \quad \text{for lightweight concrete having no web reinforcing}$$
$$= 0.17 \times 4 \sqrt{3,000} = 37 \text{ psi}$$

The value of F_{sp}—the ratio of the splitting tensile strength to the square root of the compressive strength—may be taken as 4 if not determined otherwise. Therefore, the webs should have some stirrups or should be made thicker. The latter will be adopted, and the thicker webs will be obtained by using pans 20 in. wide × 24 in. long in the first set next to the beams, making the webs 10 in. wide.

$$u = \frac{1,240 \times 2.5}{3.14 \times 0.88 \times 12} = 94 \text{ psi } (180 \text{ psi allowed by Table 13, Appendix})$$

Section along EF:

$$A_s = \frac{2,300 \times 2.5 \times 12}{18,000 \times 0.88 \times 12} = 0.36 \text{ in.}^2$$

Use one No. 6 bar across the corridor and 8 ft beyond the wall. By

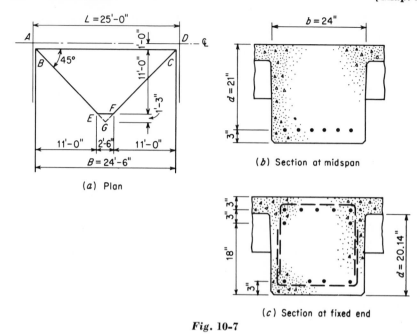

Fig. 10-7

comparison with the other middle strip, f_c, v, and u will all be safe without widening the joists.

Section along GH: Since the negative reinforcement at the outer wall should be at least one-third that of the positive reinforcement in the middle strip across *EF* and *GH*, this will be at least $0.27/3 = 0.09$ in.[2] Use one No. 4 bar per joist projecting into the panel about 6 ft and bent down into the concrete at the edge.

2. *Design of Beam BC.* Figure 10-7(a) shows in plan the area of floor which can be considered to contribute to the loading on one side of the beam *BC* which is continuous across *A* and simply supported at *D*. The span is called 25 ft; one-half of the width, 1 ft; the over-all depth, 24 in. The 45° lines from the corners of the clear space show the loaded area to be *BCFE*. This is not a triangle, hence Eq. (10-9) cannot be used directly. However, in this case, it will be sufficient to assume that the loaded area is the triangle *BGC*, then to deduct the approximate effect of the triangle *EGF*. Thus, Eq. (10-9) will give the equivalent uniform load[1] $= wBC/3 = 150 \times \frac{24.5}{3} = 1,225$ plf. The one-half beam itself,

---

[1] For a simple span *A* with a triangular loaded area, the total load $W = wA \times A/4 = wA^2/4$. The max moment $M = WA/6 = wA^3/24$. If the equivalent uniform load $wA/3$ is used, then $M = (wA/3)(A^2/8) = wA^3/24$. This is the basis for Eq. (10-9).

with an LL of 75 psf on it, will add $75 + 1 \times 2 \times 110 = 295$ plf. The total for both sides will then be $2(1,225 + 295) = 3,040$ plf. Call it 3,000 plf. However, the triangle EGF is excess and equal to a weight of $(2.5 \times 1.25/2)150 = 235$ lb at the center. Call this a concentrated load. This will produce a relieving negative moment at A equal to

$$3PL/16 = 3 \times 235 \times \tfrac{25}{16} = 1,100 \text{ ft-lb}$$

also a relieving positive moment at the center equal to

$$5PL/32 = 5 \times 235 \times \tfrac{25}{32} = 900 \text{ ft-lb}$$

These are very small but are shown for illustration.

The beam is continuous across the corridor at A, but call it fixed at this point. Then the maximum negative moment at A, using $L = 25$ ft instead of 24.5 ft, is

$$M_A = \frac{wL^2}{8} - 1,100 = \frac{3,000 \times 25^2}{8} - 1,100 = 233,000 \text{ ft-lb}$$

$$M \text{ near center} = \frac{9wL^2}{128} - 900 = \frac{9 \times 3,000 \times 25^2}{128} - 900$$
$$= 131,000 \text{ ft-lb}$$

For the positive moment at midspan, with $d = 21$ in.,

$$A_s = \frac{131,000 \times 12}{18,000 \times 0.88 \times 21} = 4.7 \text{ in.}^2$$

Use six No. 8 bars with four full length and two about 20 ft long.

$$\text{Approx } f_c = \frac{6 \times 131,000 \times 12}{24 \times 21^2} = 890 \text{ psi}$$

For the negative moment at A, with $d = 20.14$ in.,

$$A_s = \frac{233,000 \times 12}{18,000 \times 0.88 \times 20.14} = 8.8 \text{ in.}^2$$

Use seven No. 10 bars ($A_s = 8.89$ in.^{2}). By using the transformed section and Fig. 10-7(c), find $f_c = 1,180$ psi and $f_s = 16,900$ psi. These are satisfactory.

The critical shear near A will be approximately

$$V = \frac{5wL}{8} - 3,000 \times 1.75 = \frac{5 \times 3,000 \times 25}{8} - 5,200 = 41,600 \text{ lb}$$

$$v = \frac{41,600}{24 \times 20.14} = 86 \text{ psi } (v_c = 37 \text{ psi allowed})$$

Use No. 4 stirrups at 6 in. c.c.

Near D, the critical shear is

$$V = \frac{3wL}{8} - 5,200 = 23,000 \text{ lb}$$

$$v = \frac{23,000}{24 \times 20.14} = 48 \text{ psi}$$

Use No. 4 stirrups at 9 in. c.c.

10-5. Flat Slabs.   The term *flat slabs* will be used here to denote large rectangular slabs of approximately uniform thickness which are supported on columns but which have no beams or girders to carry these slabs— except possibly at the outside of the structure or at openings.   Decreased height of each story, excellent lighting and ventilation, better fire resistance because of the absence of projecting corners, easier formwork, and better economy for heavy uniform loading—these are some of the advantages of flat-slab construction.   Figure 10-8 shows its use for both an elevated railway and a warehouse.

The column capital, as shown in Fig. 10-9, is an enlargement of the end of a column which is designed and constructed to become an integral part of the column and flat slab.   The drop panel is a thickened portion of a flat slab surrounding the column, capital, or any brackets that are

Fig. 10-8   Construction of Lackawanna Terminal Warehouse, Jersey City, New Jersey.   (*Turner Construction Company.*)

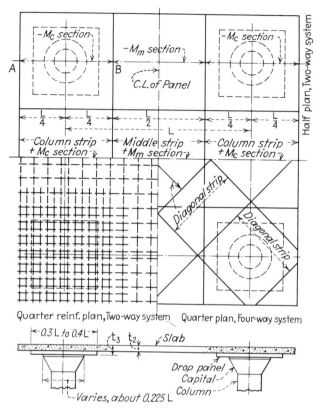

Fig. **10-9**   Flat slab.

used. For design purposes, the slab is divided each way into column strips, which serve the purpose of beams between the columns; and middle strips, which may be regarded as suspended spans that are carried by the column strips for two-way systems with reinforcement parallel to the column rows in both directions. The column strips are assumed to have a width equal to one-half of the width of the panel and centered on the column lines. A middle strip is the remaining one-half of the panel between the adjacent half-column strips.

In a four-way system, as indicated in Fig. 10-9, the standard imaginary column and middle strips are crossed by two sets of diagonal reinforcement in other imaginary strips which are supposed to carry the loads directly to the columns. However, the four-way system is more complicated, and the "packing up" of the four layers of steel is objectionable if the slab is relatively thin. Figure 10-10 should be studied carefully because it shows a two-way system that is under construction.

Fig. 10-10   Construction of Lackawanna Terminal Warehouse, Jersey City, New Jersey. (*Turner Construction Company.*)

When a flat slab is loaded, it deflects on all sides of the columns which tend to punch through the floor.   In fact, an exaggerated example can be seen when one props up a large canvas with a system of poles.   Negative bending moments exist around the edges and across the tops of the columns or capitals.   The latter are flared out in order to reduce these moments and the shears—the same being the function of the drop panels. Flat slabs are highly indeterminate, but an empirical method of analysis recommended by the Code will be explained and illustrated.

The construction to be illustrated is that used for floors and roofs. Drop panels, and even capitals, may be omitted in some cases.   If so, structural crossed members or rings of bent reinforcement (called *shear heads*) may be used to improve the strength of the slab in the vicinity of the tops of the columns.   The heavy flat slabs that are used for foundation mats[1] may be from 1 to 5 ft thick, and their analysis may be made by other means with sufficient accuracy.

The following is an attempt to present the Code's empirical method of analysis and the specifications in brief form with a few extra ideas added: *A*. Limitations for applicability of specifications.

1. Slabs must be rectangular and monolithic with columns.
2. There should be three or more rows of panels (bays) in each direction.

[1] See C. W. Dunham, "Foundations of Structures," 2d ed., chap. 7, McGraw-Hill Book Company, New York, 1962.

3. Maximum ratio of length to width of panel = 1.33.
4. Dimensions of adjacent panels should not vary by more than 20 per cent of the longer span.
5. Slabs may be solid or ribbed.
6. Columns should not be offset more than 10 per cent of the span in the direction of the offset from either axis between center lines of successive columns.
7. For structures of moderate height and floor-to-floor distances, bending moments caused by lateral forces from wind or earthquakes may be combined with critical moments and may be distributed between column and middle strips with allowable stresses increased by one-third, but resultant sizes must not be less than required by dead and live loads.

B. Principal design sections.
1. Negative moments—along edges of panel. Notice that these sections for column strips cross the columns. For convenience, $-M_c$ will be used to denote the negative bending moment in column strips, and $-M_m$ will denote the negative bending moment in middle strips.
2. Positive moments—along center lines of panel. For convenience, $+M_c$ will denote positive bending moment in a column strip, and $+M_m$ will denote positive bending moment in a middle strip.
3. For computing compression due to bending, use three-fourths of width of strip as effective. If section passes through drop panel, use three-fourths of width of drop. The latter may be too severe a restriction when the drop panels are relatively small and are considerably less than the width of the column strips. However, it may allow for the nonuniformity of resistance across the section. Make reductions for recesses that weaken the section.
4. In two-way systems, the reinforcement to resist the computed bending moment in a strip must lie within the strip itself, and all tensile bars in the strip may be included.
5. In four-way systems, reinforcement to resist computed bending in a strip equals the sum of normal areas of all bars in the strip times the cosine of the angle between each bar and the longitudinal axis of the span.

C. Slab thickness.
1. Slabs with specified drop panels at all supports should not be less than $L/40$, t_2 as given in Eq. (10-11), or 4 in. thick.
2. Slabs without drop panels as in item 1, or where a drop panel is omitted at any corner of the panel, should not be less than $L/36$, t_1 as given in Eq. (10-10), or 5 in. thick.
3. Total thickness t_1 in inches for slabs without drop panels, or

through drop panels if used, should not be less than

$$t_1 = 0.028L \left(1 - \frac{2c}{3L}\right) \sqrt{\frac{w'}{f_c'/2,000}} + 1\tfrac{1}{2} \qquad (10\text{-}10)$$

where w' is the total load in pounds per square foot, L is the span considered in feet, and c is in feet.

4. Total thickness t_2 in inches beyond drop panels for slabs with drop panels should be at least

$$t_2 = 0.024L \left(1 - \frac{2c}{3L}\right) \sqrt{\frac{w'}{f_c'/2,000}} + 1 \qquad (10\text{-}11)$$

5. Where exterior supports provide negligible restraint to the slab, increase the value of t_1 or t_2 for the exterior panel by 15 per cent.
6. Minimum thickness of slabs where drop panels at wall columns are omitted should be equal to $(t_1 + t_2)/2$ provided c complies with proper requirements.

D. Drop panels.
   1. The thickness of the downward projection of a drop panel below the slab should be appreciable—usually 3 to 4 in.
   2. The side or diameter of a drop panel should be at least 0.33 times the span in the parallel direction.
   3. The maximum total thickness at a drop panel used for computing negative reinforcement for column strips should be $1.5t_2$.

E. Capitals.
   1. The effective portion of a capital is limited to that inside a frustum of a right circular cone with a 90° vertex angle. If no capital is used, the faces of the column are considered to be the edges of the capital.
   2. Where column capitals are used, c should be the diameter of the circle formed by the intersection of the bottom of the slab or drop panel and the capital, not counting any ineffective area outside of the assumed 90° cone.
   3. With no capital, c should be the diameter or thickness of the column in the direction considered. Allow for effective structural steel or other detail embedded in the slab or drop panel if it serves somewhat as a capital.
   4. When a reinforced-concrete beam frames into a column with no capital or bracket on the same side as the beam, to compute bending in strips parallel to the beam, use c as the width of the column plus twice the projection of the beam below or above the slab or drop panel.
   5. Use the average of c at two supports at the ends of a given column strip to compute the slab thickness t_1 or t_2.

6. At exterior columns, brackets able to transmit the negative moments and shears from column strips to the columns may be substituted for capitals. Take the value of c as twice the distance from center of column to the point where the bracket is $1\frac{1}{2}$ in. thick but not more than the thickness of the column plus twice the depth of the bracket.

F. Bending moments.

1. For rectangular panels, the total of the positive and negative bending moments M_o in the direction of either side should be assumed to be not less than

$$M_o = 0.09WLF\left(1 - \frac{2c}{3L}\right)^2 \tag{10-12}$$

where W is the total load on the panel and $F = 1.15 - c/L$ but not less than unity.

2. Unless provided otherwise, the bending moments at the critical sections of column and middle strips should be at least those in Table 10-6, 10-7, or 10-8.

3. Use the average values of c at the two supports at the ends of a column strip to compute M_o for that strip.

4. Use the average values of M_o as computed from Eq. (10-12) for two parallel half-column strips in a panel (when they differ) in computing the bending moments in a middle strip.

5. The bending moments in a middle strip parallel to a discontinuous edge are to be assumed the same as in an interior panel.

6. The bending moment determined from Tables 10-6 to 10-8 may be varied for design purposes by not more than 10 per cent, but the sum of the positive and negative moments in a panel should not be less than the amount specified therein.

G. Shearing stresses.

1. Referring to Fig. 10-9, compute shearing stresses on the following vertical sections:

   a. A section which follows a periphery b_o located a distance $d/2$ beyond the edges of the column, capital, or drop panel and concentric with them.

   b. A reduced section when required by paragraph $K2$.

2. Assume $v = V/b_o d$ where d is the depth of slab or drop panel minus $1\frac{1}{2}$ in.

H. Reinforcement.

1. Arrange bars for intermediate as well as critical sections.

2. Determine A_s from M at critical sections.

3. Space bars uniformly across strips generally. At critical sections,

Table 10-6   **Bending Moments in Interior Flat-slab Panel**

With drop panel:		
Column strip	Negative moment	$0.50M_o$
	Positive moment	$0.20M_o$
Middle strip	Negative moment	$0.15M_o$
	Positive moment	$0.15M_c$
Without drop panel:		
Column strip	Negative moment	$0.46M_o$
	Positive moment	$0.22M_o$
Middle strip	Negative moment	$0.16M_o$
	Positive moment	$0.16M_o$

the spacing should not exceed twice the slab thickness except for cellular and ribbed slabs.

4. Minimum A_s at 90° to main reinforcement = $0.002A_c$, where A_c is the gross area of the concrete at that section.

5. At least 25 per cent of the required negative reinforcement in a column strip should be within a width determined by the distance d on each side of the column or capital.

6. At least 50 per cent of the required negative reinforcement in a column strip should cross over the drop panel, if any.

7. Spacing of the remainder of the bars in a column strip may vary uniformly from that required by items 5 and 6 to that needed at the middle strip.

8. It is desirable to extend a few of the top and bottom bars of all strips full-length to knit the structure together and to allow for the effect of localized heavy loads.

9. In slabs over cellular and ribbed construction, and in solid slabs, the minimum reinforcement should be that required by shrinkage and temperature reinforcement $(0.002A_c)$.

10. At exterior supports, extend bars perpendicular to edge for positive moment so as to secure at least 6-in. embedment in supports and columns, except at the drop panel itself. Anchor negative reinforcement into edge beams, walls, and columns.

11. If shear reinforcement is used, start the first stirrup or bend at or less than $d/2$ from the face of the support, column, etc.

12. Splice bars as needed by allowing proper bond but preferably not at points of maximum moment.

13. For computing reinforcement required, never assume thickness of drop panel below slab to be more than one-quarter of the distance from the edge of the drop panel to the edge of the column or capital.

I. Columns.
1. Minimum size should be 10 in.
2. Minimum I_c of gross concrete section above and below slab should be

$$I_c = \frac{t^3 H}{0.5 + W_D/W_L} \tag{10-13}$$

where t need not be greater than t_1 or t_2 as determined by Eqs. (10-10) and (10-11), H is the average story height of columns above and below the slab, W_D is the total dead load, and W_L is the greater value of the live loads on any two adjacent spans under consideration.
3. Columns smaller than required by Eq. (10-13) may be used if the bending-moment coefficients given in Tables 10-6 to 10-8 are increased in the following ratios:
    For negative moment:

$$R_n = 1 + \frac{(1 - K)^2}{2.2(1 + 1.4 W_D/W_L)} \tag{10-14}$$

    For positive moment:

$$R_p = 1 + \frac{(1 - K)^2}{1.2(1 + 0.10 W_D/W_L)} \tag{10-15}$$

    where K is the ratio of I of column used to I_c required by Eq. (10-13). The required slab thickness shall be modified as needed.
4. Columns are to be designed for bending moments produced by unequally loaded spans and unequal spacing of columns. Such moments should be the maximum computed from

$$\frac{W L_1 - W_D L_2}{f} \tag{10-16}$$

    where L_1 and L_2 are lengths of adjacent spans and $f = 30$ for exterior columns and 40 for interior ones. $L_2 = 0$ when considering exterior columns. Divide the moment between columns directly above and below the floor or roof being considered in direct proportion to their stiffness (I/L), and apply these moments directly as computed to the critical sections of the columns.

J. Beams.
1. Beams or the equivalent should be used at all discontinuous edges unless cantilevered.
2. Edge beams should support one-fourth of uniform load on adjacent panel plus all loads directly on top of them.
3. Interior beams are to support one-fourth of uniform load on both

panels—plus any loads directly on them—as T beams.   If the
panels are elongated, use the principles shown in Fig. 10-4(a).
  4. If interior beams or walls off column center lines interfere with flat-
    slab action, it may be desirable to frame entire panel as beam-and-
    slab construction.
K. Openings.
  1. Openings may be used in slab in area common to two intersecting
    middle strips if the required positive and negative steel areas are
    provided.
  2. In an area common to two column strips, not over one-eighth of
    width of a column strip in any span should be cut by openings.
    Provide extra steel on all sides to equal all bars interrupted.   The
    part of the periphery b_o for shear resistance which is cut by radial
    projections of the openings to the center of the column should be
    considered ineffective.
  3. In an area common to one column strip and one middle strip,
    openings may interrupt one-quarter of the bars in either strip, but
    an A_s equivalent to the bars cut should be provided by extra ones
    on all sides of the opening.
  4. Openings which eliminate more of any strip than stated herein
    should be completely framed with beams to transmit the loads to
    the columns.[1]

Example 10-5.   Assume that an industrial building is to be 6 panels
wide and 15 panels long.   It is to be two-way flat-slab construction with
drop panels.   All spans are 20 ft both ways; the live load = 250 psf; the
drop panels are 8 ft square; and c for the capitals is 4.5 ft.   Design the
slabs for the exterior and the first interior panels of a typical intermediate
bay as shown in the key plan in Fig. 10-11.   Assume $f'_c = 3,500$ psi and
$f_s = 20,000$ psi.   Use the Code's empirical procedure.   Assume that there
is an edge beam at the exterior wall and that the slab is restrained.[2]
  1. *Interior Panel.*   Using $t_2 = L/40$, the slab thickness is

$$t_2 = 20 \times \tfrac{12}{40} = 6 \text{ in.}$$

However, this seems to be rather thin.   Try 7 in.   Then, using w for the
total load, $w = 250 + 88 = 338$ psf for the slab, and perhaps 400 psf at

---

[1] The reader may wish to investigate the design of flat slabs by elastic analysis
which is given in the Code.   This method assumes that the structure is divided into
a number of bents consisting of a number of columns in a row and wide strips of floor
bounded on each side by the center lines of the floor panels.   The junctions of floor
and columns are considered to be fixed, and the bents are to be analyzed as a continuous structure.
[2] NOTE: The numbers and letters given along the right-hand margin of the calculations refer to the applicable item in the preceding outline.

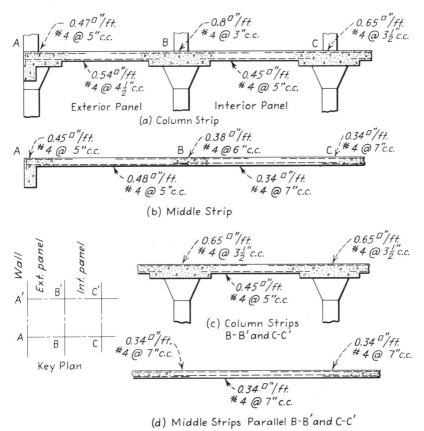

Fig. 10-11   Schematic diagram of flat-slab reinforcement.

the drop panel.   However, the latter's weight is not important and the average load will be called 350 psf for design purposes.   C1

Assume that the thickness of the drop panel = 4 in.   Then

$$t_3 = 7 + 4 = 11 \text{ in.}$$

and d at the edge of the panel = $11 - 1.5 = 9.5$ in.

$$M_o = 0.09(350 \times 20^2)20 \times 1\left(1 - \frac{2 \times 4.5}{3 \times 20}\right)^2 = 182,000 \text{ ft-lb} \quad F1$$

where F is unity.

Column strip:

$$-M_c = -0.50 \times 182,000 = -91,000 \text{ ft-lb} \qquad F2$$
$$+M_c = +0.20 \times 182,000 = +36,400 \text{ ft-lb} \qquad F2$$

Middle strip:

$$-M_m = -0.15 \times 182{,}000 = -27{,}300 \text{ ft-lb} \qquad F2$$
$$+M_m = +0.15 \times 182{,}000 = +27{,}300 \text{ ft-lb} \qquad F2$$

The critical section will be at the capital. The assumed effective width b through the drop is

$$b = \tfrac{3}{4} \times 8 \times 12 = 72 \text{ in.} \qquad B3$$

The effective width through the center of this strip, and through the middle ones, will be

$$b = \tfrac{3}{4} \times 10 \times 12 = 90 \text{ in.} \qquad B3$$

Then an approximate check of the compressive stress at the critical section for $-M_c$ gives

$$f_c = \frac{6M}{bd^2} = \frac{6 \times 91{,}000 \times 12}{72 \times 9.5^2} = 1{,}010 \text{ psi (safe)}$$

At the critical section for $+M_c$,

$$f_c = \frac{6 \times 36{,}400 \times 12}{90 \times 5.5^2} = 960 \text{ psi (safe)}$$

The middle strips will be even safer since M is smaller.

Now determine the reinforcement, using $d = 9.5$ in. for $-M_c$, $d = 5.5$ in. elsewhere, and $j = 0.88$. The imaginary strips are 10 ft wide.

Column strip:

$$\text{For } -M_c: \quad A_s = \frac{91{,}000 \times 12}{20{,}000 \times 0.88 \times 9.5} = 6.54 \text{ in.}^2$$

Use No. 4 bars at $3\tfrac{1}{2}$ in. c.c. ($A_s = 6.9$ in.^{2})

$$\text{For } +M_c: \quad A_s = \frac{36{,}400 \times 12}{20{,}000 \times 0.88 \times 5.5} = 4.51 \text{ in.}^2$$

Use No. 4 bars at 5 in. c.c. ($A_s = 4.7$ in.^{2})

Middle strip:

$$\text{For } -M_m \text{ and } +M_m: \quad A_s = \frac{27{,}300 \times 12}{20{,}000 \times 0.88 \times 5.5} = 3.38 \text{ in.}^2$$

Use No. 4 bars 7 in. c.c. ($A_s = 3.4$ in.^{2})

Next, check the slab for shear. At a section $d/2 = 4.75$ in. beyond the capital (c), the load on the two adjacent quarters of the panel beyond this section must be withstood by one-half of a circle of radius $2.25 \times 12 + 4.75 = 31.75$ in.

$$V = \left[10 \times 20 - \frac{\pi}{2} \frac{(31.75)^2}{12^2} \right] 350 = 66,000 \text{ lb} \qquad\qquad G$$

$$b_o = \pi r = \pi \times 31.75 = 99.5 \text{ in.}$$

$$v = \frac{V}{b_o d} = \frac{66,000}{99.5 \times 9.5} = 70 \text{ psi vs. } 1.2 \sqrt{f_c'} = 71 \text{ psi allowed} \qquad G2$$

This is satisfactory. More than 50 per cent of the bars will be over the 8-ft drop panel.

At a section $d/2 = 5.5/2 = 2.75$ in. beyond the drop panel, the load on two adjacent quarters of the panel beyond this section must be withstood by one-half of the total perimeter of the section. Hence,

$$b_o = (96 + 2.75 \times 2) + 2(48 + 2.75) = 203 \text{ in.}$$

$$V = \left(10 \times 20 - \frac{101.5 \times 50.75}{144} \right) 350 = 57,500 \text{ lb} \qquad\qquad G$$

$$v = \frac{57,500}{203 \times 5.5} = 52 \text{ psi (safe)} \qquad\qquad G2$$

Check the bond stresses as follows:

At edge of capital:

$$V = \left(10 \times 20 - \frac{\pi}{2} \times 2.25^2 \right) 350 = 67,200 \text{ lb}$$

$$\text{Perimeter} = \pi \times 2.25 = 7.06 \text{ ft}$$

$$\text{Average } V \text{ per ft} = \frac{67,200}{7.06} = 9,500 \text{ lb}$$

$$\text{At } 3\tfrac{1}{2} \text{ in. c.c., } \Sigma o = 5.39 \text{ in.}^2/\text{ft}$$

Then $\qquad u = \dfrac{V}{(\Sigma o)jd} = \dfrac{9,500}{5.39 \times 0.88 \times 9.5} = 211 \text{ psi (safe)}$

At edge of drop panel:

$$V = (10 \times 20 - 8 \times 4)350 = 58,800 \text{ lb}$$

$$\text{Perimeter} = 8 + 2 \times 4 = 16 \text{ ft}$$

$$\text{Average } V \text{ per ft} = \frac{58,800}{16} = 3,700 \text{ lb}$$

$$\text{At } 3\tfrac{1}{2} \text{ in. c.c., } \Sigma o = 5.39 \text{ in.}^2/\text{ft}$$

Then $\qquad u = \dfrac{3,700}{5.39 \times 0.88 \times 5.5} = 142 \text{ psi (safe)}$

If the point of inflection should occur before the edge of the drop panel is reached, the bar spacing will be 5 in. Then

$$u = \frac{142 \times 5}{3.5} = 203 \text{ psi}$$

which is still safe.

Table 10-7   **Bending Moment in Exterior Flat-slab Panel**

		A	B	C
With drop panel: Column strip*	Exterior negative Positive moment Interior negative	$0.44M_0$	$0.36M_0$ $0.24M_0$ $0.56M_0$	$0.06M_0$ $0.36M_0$ $0.72M_0$
Middle strip	Exterior negative Positive moment Interior negative	$0.10M_0$	$0.20M_0$ $0.20M_0$ $0.17M_0\dagger$	$0.06M_0$ $0.26M_0$ $0.22M_0\dagger$
Without drop panel: Column strip*	Exterior negative Positive moment Interior negative	$0.40M_0$	$0.32M_0$ $0.28M_0$ $0.50M_0$	$0.06M_0$ $0.40M_0$ $0.66M_0$
Middle strip	Exterior negative Positive moment Interior negative	$0.10M_0$	$0.20M_0$ $0.20M_0$ $0.18M_0\dagger$	$0.06M_0$ $0.28M_0$ $0.24M_0\dagger$

A = no edge beams or beam with depth $1\frac{1}{4}t$ or less.
B = beams $3t$ or more deep or integral concrete bearing walls.
C = masonry or other walls, negligible restraint.
* For half column strip adjacent to beam or wall, approx coefficients = $\frac{1}{2}$ values shown.
† Increase 30 per cent when continuous across supports type B or C.

2. *Exterior Panel.*   From Table 10-7, with type B and drop panels, the bending moments at critical sections, with M_o = 182,000 ft-lb, are
Column strip:

$$\text{Exterior} \ -M_c = 0.36 \times 182{,}000 = 65{,}500 \text{ ft-lb}$$
$$\text{Interior} \ -M_c = 0.56 \times 182{,}000 = 101{,}900 \text{ ft-lb}$$
$$+M_c = 0.24 \times 182{,}000 = 43{,}700 \text{ ft-lb}$$

Middle strip:

$$\text{Exterior} \ -M_m = 0.20 \times 182{,}000 = 36{,}400 \text{ ft-lb}$$
$$\text{Interior} \ -M_m = 0.17 \times 182{,}000 = 30{,}900 \text{ ft-lb}$$
$$+M_m = 0.20 \times 182{,}000 = 36{,}400 \text{ ft-lb}$$

The required reinforcement, using No. 4 bars as before, is the following:
Column strip:

$$\text{For exterior} \ -M_c: \quad A_s = \frac{65{,}500 \times 12}{20{,}000 \times 0.88 \times 9.5} = 4.7 \text{ in.}^2$$

Use No. 4 bars at 5 in. c.c. ($A_s = 4.8$ in.^{2})

$$\text{For interior } -M_c: \quad A_s = \frac{101{,}900 \times 12}{20{,}000 \times 0.88 \times 9.5} = 7.33 \text{ in.}^2$$

Use No. 4 bars at 3 in. c.c. ($A_s = 8.0$ in.^{2}). It might be better to use No. 5 bars at larger spacing, but it is desirable to keep the same size throughout where this is practicable.

$$\text{For } +M_c: \quad A_s = \frac{43{,}700 \times 12}{20{,}000 \times 0.88 \times 5.5} = 5.4 \text{ in.}^2$$

Use No. 4 bars at $4\frac{1}{2}$ in. c.c. ($A_s = 5.3$ in.^{2}—near enough).

Middle strip:

$$\text{For exterior } -M_m: \quad A_s = \frac{36{,}400 \times 12}{20{,}000 \times 0.88 \times 5.5} = 4.5 \text{ in.}^2$$

Use No. 4 at 5 in. c.c. ($A_s = 4.8$ in.^{2}).

$$\text{For interior } -M_m: \quad A_s = \frac{30{,}900 \times 12}{20{,}000 \times 0.88 \times 5.5} = 3.84 \text{ in.}^2$$

Use No. 4 bars at 6 in. c.c. ($A_s = 4.0$ in.^{2}).

For $+M_m$, this is the same as for exterior $-M_m$, or No. 4 bars at 5 in. c.c.

For the one-half column strip along the edge beam at the outside of the structure, use Table 10-8 for types B and 3 with drop panels. Since the bay considered is an interior one, the coefficient to use here for $-M_c$ is $C5$. Then

$$-M_c = 0.13 \times 182{,}000 = 23{,}700 \text{ ft-lb}$$
$$A_s = \frac{23{,}700 \times 12}{20{,}000 \times 0.88 \times 9.5} = 1.7 \text{ in.}^2$$

Use No. 4 bars at 12 in. c.c. for both ends. ($A_s = 2.0$ in.^{2}.)

For $+M_c$, use Table 10-8, types B and 3, with the coefficient $C4$. Then

$$+M_c = 0.05 \times 182{,}000 = 9{,}100 \text{ ft-lb}$$
$$A_s = \frac{9{,}100 \times 12}{20{,}000 \times 0.88 \times 5.5} = 1.13 \text{ in.}^2$$

It would be possible to use No. 4 bars at 21 in. c.c., but the bars should not be over twice the thickness of the slab apart—14 in. ($H3$). Therefore, use No. 4 at 12 in. c.c., the same as for $-M_c$.

Table 10-7 shows that, when the outside edges of the slab are not restrained, the bending moments in the exterior panel will be larger than those in an interior panel. In such a case, it is often desirable to make the end span a little shorter than the interior ones in order to more nearly equalize the moments in different parts of the slab.

Table 10-8   Bending Moments in Panels with Marginal Beams or Walls
(coefficient for use with M_o)

End support C1 C2 C3 C4 C5

Type of end support	Coef. C	Type of side support					
		1		2		3	
		Drop	No drop	Drop	No drop	Drop	No drop
A	C1	−0.22	−0.20	−0.17	−0.15	−0.11	−0.10
	C2	+0.12	+0.14	+0.09	+0.11	+0.06	+0.07
	C3	−0.28	−0.25	−0.21	−0.19	−0.14	−0.13
	C4	+0.10	+0.11	+0.08	+0.09	+0.05	+0.06
	C5	−0.25	−0.23	−0.19	−0.18	−0.13	−0.12
B	C1	−0.18	−0.16	−0.14	−0.12	−0.09	−0.08
	C2	+0.12	+0.14	+0.09	+0.11	+0.06	+0.07
	C3	−0.28	−0.25	−0.21	−0.19	−0.14	−0.13
	C4	+0.10	+0.11	+0.08	+0.09	+0.05	+0.06
	C5	−0.25	−0.23	−0.19	−0.18	−0.13	−0.12
C	C1	−0.03	−0.03	−0.03	−0.03	−0.03	−0.03
	C2	+0.18	+0.20	+0.14	+0.15	+0.09	+0.10
	C3	−0.36	−0.33	−0.27	−0.25	−0.18	−0.17
	C4	+0.10	+0.11	+0.08	+0.09	+0.05	+0.06
	C5	−0.25	−0.23	−0.19	−0.18	−0.13	−0.12

A = no edge beams or beam with depth $1\frac{1}{4}t$ or less.
B = beams $3t$ or more deep or integral concrete bearing walls.
C = masonry or other walls, negligible restraint.
1 = no beams.
2 = beams with depth $1\frac{1}{4}t$ or less.
3 = beams $3t$ or more deep or restraining walls.

10-6. Miscellaneous Slabs. There can be various modifications of the two-way and flat-slab construction. For two of these, refer to Fig. 10-12.

Sketch (a) shows in plan a portion of a paneled floor which is adaptable for office and apartment buildings. It consists of wide, shallow beams along the column lines, such as ABC, DEF, AD, etc. Either one-half of a similar wide beam or a deep spandrel beam can be used to support the exterior wall, the latter being preferable when heavy masonry walls are used because of its greater stiffness. A typical section is shown in (c). For design purposes, this is a system of continuous beams with two-way slabs spanning between them. Capitals, drop panels, or both could be used but are generally undesirable from the standpoint of appearance. This construction has the advantage of being shallow and not requiring a separate ceiling finish. The critical point for stresses is at the columns.

Sketch (*b*) shows in plan a modification of the preceding floor, using definite column strips as continuous beams both ways along the column lines but with waffle construction for the central two-way portions. This is useful when the loads are heavy, and the main grid of beams has to be fairly thick anyway. The waffles are stiff but much lighter than an

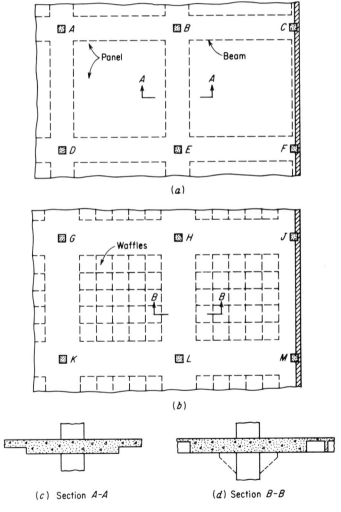

Fig. **10-12** Two-way beam system with thin slabs or waffle construction.

equivalent solid slab would be. A section is shown in (*d*). If used for industrial construction, capitals could be added as shown dotted in (*d*), and the waffles can be left without a hung ceiling cover. In such a system, the two-way grid of beams serves as the backbone of the construction.

For spans of about 20 ft or less, flat-plate construction may be used for light loads. It consists of a continuous slab of uniform thickness supported by columns without capitals or drop panels. It is a sort of modified flat-slab construction. Of course, the punching and shearing conditions in the vicinity of the columns are severe. Shallow shear heads made of welded crossed members of structural steel or of reinforcing bars can be embedded in the slabs at the columns to serve as a sort of hidden capital. When spans are much above 20 ft and the loads are heavy, look out for trouble. Naturally, a system like this is rather flexible, both vertically and horizontally. The transfer of lateral moments from the floor to the columns may cause very large local stresses.

Lift slabs are generally of the flat-plate type with embedded shear heads having an opening in their centers. They are built on the ground in a series of isolated layers, the columns being steel and previously erected as cantilevers from the footings. The slabs are later raised bodily and slid up the columns to points where the shear heads are attached to the columns at the proper floor level. Of course, the economy of this construction comes from the saving in formwork and the efficiency of the field operations.

Another modification of the flat-slab construction is the use of capitals and solid rectangular portions like drop panels with waffle construction throughout the remainder of the floor area—column strips as well as middle strips. This might be looked upon as a sort of strong umbrella construction around the columns, with waffles serving as the remainder of the column and middle strips.

Slabs spanning in one direction can be used in some cases for long spans, but they are likely to be very heavy. They are sometimes used for small bridges where waffle construction or joists would not be suitable because of the truck loads. Another type which may be used for long-span, one-way floors is a tubular slab. This consists of long fiber tubes embedded in the concrete in the direction of the main reinforcement. This produces a light, stiff construction with a smooth ceiling, as well as the floor itself.

A few special features relating to these large slabs are shown in Fig. 10-13. Notice the following:

(*a*) An exterior corner like *A* tends to lift up as the center of the slab sags downward. If not tied down, it may crack the walls or crack across the top as indicated by *BC*. Top bars may be put in to cross this crack as shown, or a grid parallel to *AC* and *AB* will serve the same purpose.

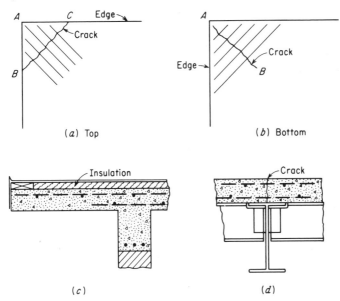

Fig. 10-13   Miscellaneous details.

This reinforcement should cover at least one-fifth of the longest span in each direction. The area of steel here should be equivalent to that used for the maximum positive moment. It is also desirable to extend perhaps one-quarter of the top steel across to a discontinuous edge and bend it down into the edge beam or wall.

(*b*) In the bottom of an exterior corner, a crack like *AB* may form. There should be bottom bars as indicated, or they may be used in two directions parallel to the sides of the slab, equivalent to the steel used for the maximum positive moment, and incorporated in the regular system of bottom bars.

(*c*) Cantilevered edges of large slabs like this are affected so much by a decrease in temperature compared with the heated interior of the building that they are likely to shrink seriously and to crack. Contraction joints should be provided to minimize this cracking.

(*d*) Adjacent large slabs on a system of continuous concrete beams will generally be satisfactory, but when on steel beams that are not continuous (even when encased in concrete), the deflections of the steel are likely to open up cracks as indicated. Making a contraction joint here, adding special top bars, or making the steel continuous will help to avoid unsightly cracking.

Large concrete-slab roofs on wall-bearing structures are likely to cause difficulty if not exceedingly well insulated. On a hot, sunny day the roof

will become hotter than the rest of the building.[1]   Therefore, the roof
will expand and tend to push the walls outward.   At the corners, any
masonry walls directly under the roof are likely to crack since the motion
is usually diagonally outward.   On a cold day, the action is reversed.
After a few seasons of alternate heat and cold, the corners of the walls are
likely to be in bad shape.

This expansion effect is especially bad in the case of masonry parapets
because they are subject to extreme temperature effects.   In one case, the
parapets of a building about 150 ft long had been pushed outward about
$\frac{1}{2}$ in. at all corners.

The best remedy for this condition is the use of a skeleton framing
with curtain walls.   Then the cracks will not be noticeable in most cases.
As for the parapets, they should have expansion joints approximately 40
ft c.c., even though these joints are troublesome to build.

When the edges of these large slabs are to be supported on masonry
walls, upward curling of the corners may be remedied by the use of concrete "collar" beams which look like spandrel beams and which extend
around the structure.   The slabs can be tied down to these beams.

PRACTICE PROBLEMS

For these problems assume $f'_c = 3,000$ psi, $f_y = 36,000$ psi, $f_s = 20,000$
psi, and stresses allowed by the Code.   Use WSD or USD as desired.

10-1   A two-way slab is 20 ft square, continuous at all four edges, and
8 in. thick.   It is to support a uniform live load of 150 psf.   Determine the
reinforcement for it.   Is it safe?

10-2   A two-way slab is to be 18 by 20 ft; one 20-ft edge is simply supported; the other three are continuous.   Design the slab for a live load
of 200 psf.

10-3   A floor is to be flat-slab construction with capitals and drop panels.
The column spacing is 20 ft both ways.   Design the floor for a live load
of 200 psf.   Assume strong restraining edge beams at the outside wall.

10-4   A building with a flat-slab floor is to have the outer panels 18 ft
wide and a series of interior ones 20 ft wide.   In the other direction, all
columns are 20 ft c.c.   Assume that the outer edges are supported on
masonry walls.   Design a typical exterior and interior panel of the floor
for a live load of 200 psf.

10-5   Design a floor of the type shown in Fig. 10-12(b) if the column
spacing is 20 ft both ways and the live load is 75 psf.

[1] Remember that insulation slows down the rate of transfer of heat but does not
prevent eventual heating of the roof below unless the space below the roof slab is
air-conditioned so as to remove excess heat.

FORMS

11

11-1. Introduction. The purpose of this chapter is to point out some of the important principles to bear in mind when planning the forms for cast-in-place concrete construction. It is especially important for the designer to consider them when making his plans so that he may attain the optimum economy and practicability. A thorough knowledge of how his structure is to be built, and provisions in the design for efficiency in the field work, are attributes of the expert in contrast to the novice. Furthermore, it is not conducive to good relations between the contractor and the design engineer when the former has to come to the latter with various explanations as to why certain features of the design should be changed in order to make the field work feasible.

Of course, the designer should not dictate to the contractor exactly how the latter is to conduct his operations. If he does so then, in the eyes of the law, the designer, or the owner whom he represents, becomes an employer and thereby assumes the responsibility for the operations. The design should show the results desired, but it should be such that these results can be secured efficiently and economically.

Details of forms will vary widely. The character of the structure, the availability of form materials, the amount of duplication, the equipment and supplies owned by or available to the contractor, and the methods formerly employed by the contractor—all these will affect what is to be done in constructing a particular job.

It is obvious that the forms are to support the plastic concrete until it has hardened and attained sufficient strength to support itself. They are also to give to each member the desired shape, dimensions, and surface finish.

11-2. Strength of Forms Holding Vertical Loads. Forms that support the weight of plastic or "green" concrete can be designed to hold up these fairly definitely determinable loads. However, a very important feature is that of stiffness. Lack of this quality may cause unexpected and unfortunate results.

The forms in some cases can be made with enough camber so that, after the loads are applied and the forms have deflected, the members will have the right shape and position. In other cases, deflection is to be minimized so that no damage will occur to the member itself or to other portions of the structure.

The forms should be designed for all the weights that are likely to affect them. Among these are the following:

1. The dead load of the forms themselves, including their supports.

2. All the plastic concrete that is above them, using a unit weight of 150 pcf to allow for the weight of some reinforcement.

3. The weight of any additional pours placed over the first one if the latter cannot safely support this weight itself.

4. The weight of a 200-lb man standing in any probable position. This may affect light forms for thin floors.

5. The weight of any equipment that may be placed upon the forms; e.g., buggies full of concrete, especially the motorized ones that have been used so efficiently on some modern jobs; a concreting bucket that is likely to be set down temporarily; and piles of materials placed on green concrete that merely transfers the load to the forms. Of course, abuses like some of these should be guarded against.

6. The effect of impact due to vibrators. This is difficult to determine but should be minimized by proper supervision.

The design of the forms to hold vertical loads is a matter of planning more for efficient fabrication, erection, and removal than for strength, at least in many cases. Experience in such work is extremely valuable. It is not practicable to attempt to show in this chapter the vast variety of details used for such work. Some are pictured merely to suggest ideas and arrangements.

Forms have failed. Probably this trouble has been due to attempts to economize a little too much because forms are often very costly. On the other hand, some inadequate detail that was overlooked can cause the failure of otherwise excellent work.

One illustration of such a failure[1] is that of a railroad overpass in Tennessee which failed just after the concrete was poured. The accident killed one man and injured three others.

As an illustration of the need for adequate stiffness, refer to Fig. 11-1(a). This arch is to be poured in seven sections in the sequence

---

[1] *Eng. News-Record*, May 22, 1952, p. 24.

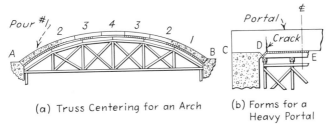

(a) Truss Centering for an Arch   (b) Forms for a
Heavy Portal

Fig. 11-1   Examples showing need for consideration of deflection of forms.

indicated in order to minimize the effect of shrinkage.   The main support for the forms is steel trussing.   It is used in order to maintain traffic below the forms.   Pours No. 1 will adjust themselves to any deflection of the trusses and their supports if they are completed fast enough. When the next sections are poured, each successive pour will cause more deflection of the trusses and the end posts.   These deformations tend to cause shear in the first pours near A and B and rotation about these ends. As the work progresses, this effect at the ends, and distortion of the weak concrete of other pours, increases.   The effect may be very harmful. One remedy is that of leaving a foot or two open between the abutments and pours No. 1, and between each successive section.   When the trussing has deflected under the weight of these portions, the closing pieces may be poured without serious harm.

Another illustration is pictured in Fig. 11-1(b).   These were low heavy piers with large portals.   The shafts were poured first up to the construction joints shown.   The portal forms were made of wood and were theoretically strong enough, but they were highly stressed.   However, no consideration was given to the deflection of the posts or to the crushing of wooden wedges on top of the posts.   Each portal required several hours for pouring.   Shortly after the forms were removed, cracks appeared somewhat as shown.   Although shrinkage may have aggravated the cracking, it seems that, when the portion CD was rigidly supported but DE was on yielding supports (with probable increase of deformation with time), there may have been a shearing failure of the weak concrete near D which made a plane of weakness that later opened up.

It is obvious, therefore, that the portal in Fig. 11-1(b) should be poured quickly so that all the concrete is placed, compacted, and adjusted to any settlement of the forms before the initial set takes place.   Another and better scheme is to use very big and stiff members as shores under the forms so that settlement will be negligible.   Remember that it is easy to plan rapid work in the office, but it may not be easy or even possible to conduct the field operations so expeditiously.   Therefore, a plan that is safe for adverse conditions will probably be advisable, and it certainly will not cause trouble if everything goes smoothly.

On the other hand, if one tried to pour the shafts and portals in one big rapid operation, wet shrinkage of the shafts might cause settlements there, whereas the portal would be "hung up" on the central forms. This might also be harmful, and it certainly would be more expensive to erect all the reinforcement, to build all the forms, and to place concrete way down in the bottom of such a "mess."

To reiterate, remember that, unless they are oak or other suitable hardwood, timber caps, sills, and wooden wedges may crush because of high pressures perpendicular to the grain. In fact, their crushing may proceed slowly, and it may continue for several days. In the meantime, the concrete has set, the settlement causes shearing deformations in the weak concrete, and these may constitute planes of weakness. Thus the forms may be very economical, but they may ruin the structure.

Not only must forms be strong vertically; they must be well braced, as indicated for the columns in Fig. 11-2. A long retaining wall was being built without the projecting portions shown at the base of the stem in Fig. 8-18 to hold the bottoms of the forms in line. The braces at one side of the bottom of one portion slipped about 1 in. The entire bottom bowed outward. This was discovered after the concrete had been placed, but it was impossible to push it back again. After the

Fig. 11-2 Some heavy construction for an industrial plant. (*International Smelting and Refining Co., Perth Amboy, New Jersey.*)

forms were removed, the bulge looked so bad that the contractor was compelled to replace that portion of the wall.

In another case, a long steel bin and some building columns were to be supported upon a very heavy concrete floor resting upon a series of walls and concrete columns. The formwork was not adequately braced laterally. Somehow one end swayed sideways about 2 in. at the top. This was not noticed until the forms were removed. The cost of replacement would be so great that the owner finally accepted the defective structure. However, the contractor had to pay for revising the steelwork to fit the displaced anchor bolts.

11-3. Lateral Pressure on Forms. The intensity of the lateral pressure produced by newly poured concrete is uncertain. It is generally assumed to be hydrostatic in character, but to a limited extent. The unit pressure seems to depend upon the rate of pouring, the temperature of the concrete, the nature of the coarse aggregate, the richness of the concrete in cement, the relative amount of fine aggregate, the slump, the use or absence of vibrators, and the use of retarding admixtures.

The setting of the cement soon causes plastic concrete to stiffen so that it becomes a weak solid. It no longer acts as a fluid when more concrete is deposited on it. Probably the temperature of the concrete is influential only when it is high enough to cause more rapid setting of the cement or so low that it retards this action. Furthermore, it seems that the coarse aggregate, especially if it is angular, tends to form a somewhat rigid mass near the bottom of a pour, so that it supports itself to a considerable extent. This probably applies in the case of sections under a foot or two in width more than it does in massive construction.

It also seems that a lot of cement with enough water to produce a slump of 2 to 4 in., and perhaps a large amount of sand, may tend to increase the fluidity of the newly poured concrete, thereby causing more pressure on the forms. On the other hand, too wet a mix will cause the grout or mortar to cease to lubricate the mixture and to permit the coarse aggregate to settle out, with a resultant decrease in the lateral pressure near the bottom of the pour.

The effect of vibrators in augmenting lateral pressures is likely to be important only in plastic concrete that has been poured for less than an hour or two. They probably will not have power enough to move concrete in which the formation of the cement gel has stiffened the mass.

An illustration of the utilization of this stiffening effect is the practice of pouring the bottom of a stepped footing, letting it stand for 30 to 40 min, then pouring the pedestal with side forms only, and doing this without having the second pour squeeze out the concrete under the edge of the forms.

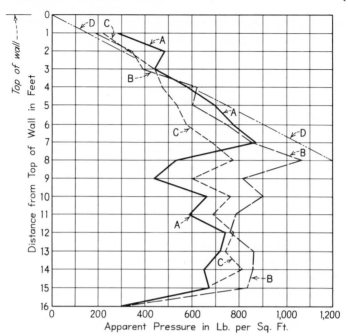

Average rate of pour = 4.25 ft.; max. rate = 8.5 ft. per hour
D = Theoretical pressure line at 150 pcf per ft. fluid pressure
A, B & C = Computed pressures; from deflection readings
Mix = approx. $1:2\frac{1}{2}:4$; slump = approx. 4 in.
No apparent effect from vibrators if depth exceeded 8 ft.

Fig. 11-3   Pressures on wooden forms with plywood lining.   (*Courtesy of Charles Macklin, Architect and Structural Engineer, Springfield, Illinois.*)

The pressure caused on forms for tremie concrete, or concrete otherwise deposited under water, will probably be similar to that of concrete deposited in air, but the intensity may be about two-thirds of that for the latter.[1]

Of course, any predictions of pressures indicated here are made upon the basis of proper deposition of the concrete.   If the concrete is dropped as a large gob falling through a height of several feet, the lateral pressure may be increased considerably.

The height of pours or lifts and the time required to elapse between successive pours are questions about which there is considerable debate. The following considerations may well be borne in mind in this connection:

1. Pressure on forms will not increase as the hydrostatic head if the lower portion has been deposited and undisturbed for 2 or 3 hr.   See Fig.

[1] P. J. Halloran and K. H. Talbot, The Properties and Behavior Underwater of Plastic Concrete, *J. ACI*, June, 1943.

11-3 as an illustration of this.   The lower portion of the forms need not be unduly strong if the rate of pouring is not too rapid.

2. Bleeding may be troublesome if deep pours are made too quickly and vibrated excessively.   Slower progress of the pour will permit the setting of the concrete so as to avoid serious bleeding.

3. If one lift is allowed to set for several days, considerable heating and expansion will occur initially; cooling and shrinking will then follow. When another pour is placed on top of the first one, the former is in its warmer and expanded condition as it sets.   Then, when it tries to shrink, the lower pour tends to prevent this shrinkage and to cause cracks in the upper lift.   It would seem, therefore, that a massive, tall structure should be poured full height by increments that are added at intervals of a few hours in order to have each lift placed before the one under it has set too long and shrunk too much.

4. The cleaning of horizontal joints—removal of laitance—is slow and costly.   Continuous but slow pouring of deep sections will often minimize or eliminate such cleaning.

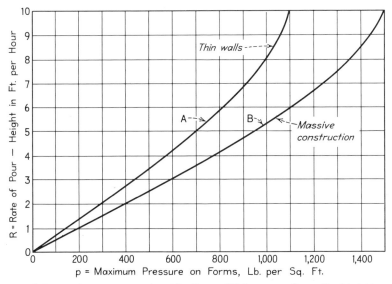

Ordinary columns are poured rapidly. Use p = 150 lb. per sq. ft. per ft. of height
A = Walls 12 in. or less in thickness
B = Walls 24 in. or more in thickness
Assumptions: ordinary mixes, slump approx. 2 to 4 in., moderate vibration, 60°
        to 70° F, friction decreases pressure in case of thin walls
        and columns, and vibration negligible below 6 or 7 ft.

Fig. 11-4   Curves for estimating pressure of concrete on forms (for preliminary use when lacking better data).

CHART BASED ON FORMULA DEVELOPED BY E.B. SMITH

$$P = H^{0.2} R^{0.3} + 0.12C - 0.3S$$

P = Resultant pressure on forms in lbs. per square inch

R = Rate of vertical fill per hr. expressed in ft.

H = Effective head of concrete of the depth from the surface of the concrete to the point where the cement has begun to stiffen, feet

C = Ratio by volume of cement to aggregate in per cent

S = Slump in inches

Concrete = 150 lbs. per cu. ft.

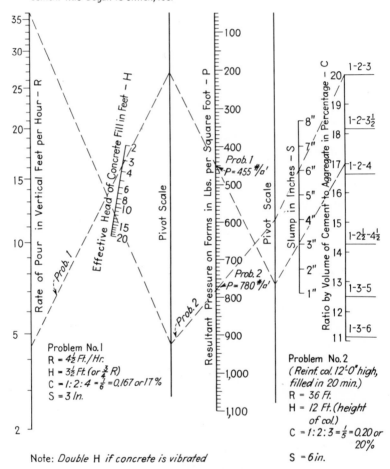

Note: *Double* H *if concrete is vibrated*

Fig. 11-5 Pressure on forms. (*A chart devised by C. E. Thunman, Springfield, Illinois.*)

5. Vibration or other methods of compaction should be conducted so as to avoid harm to set or partially set concrete. Good judgment used in this connection will generally permit slow but rather continuous pouring.

Figures 11-3 to 11-5 are given to show the results of some tests that have been made to ascertain the pressures on particular forms and to make some recommendations. These do not agree with the data given in Art. 1-16, but they are nevertheless very instructive. It seems that experimental data vary tremendously and that this whole question of lateral pressures on forms is not subject to exact theoretical determination. Conditions for any particular job will differ from those of another job. The design of forms is therefore primarily an art that is developed through experience, but it still should be developed as good engineering.

Form ties are customarily used to resist the lateral pressure of the concrete when there are forms on opposite sides of the section, as for a wall. If a form is to be used on one side only, as when concreting against a side of rock, earth, or previously set concrete, a strong set of lateral bracing will be needed to resist the thrust on the forms.

Some varieties of form ties are shown in Fig. 11-6. It is generally desirable to avoid having any steel at or very close to the surface of the concrete. That is why the ties are made so that the outer ends can be removed or broken off and the holes pointed up after removal of the forms. If the embedment of the weak section or the connection point is too great, it may be difficult to remove the outer portions and to fill the deep void completely with mortar when the end is taken out.

Another feature to avoid is "hydrostatic" uplift on the forms. This results when the forms are sloped considerably. When such shapes are necessary, it is often desirable to embed bolts in the previous pours to which the forms can be attached, to brace the forms against something

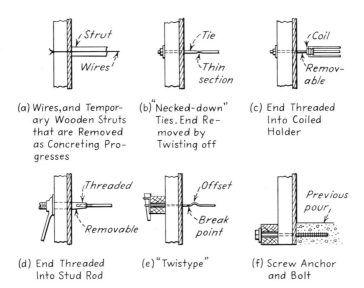

Fig. **11-6** Some principles of forms ties. (Wires are not to be used when work will be exposed. Point up holes with mortar after ends are removed.)

heavy and strong, or to make the pour slowly and in small lifts so that the initial set or some stiffening can occur in a lower portion before more concrete is placed, thus minimizing the uplift.

Where possible, it is advantageous to have ledges to support the forms, as in the case of footings under walls. It is difficult to build wall forms on irregular ground or to hold them high up in the air.

A special case that deserves some consideration is that of the pouring of concrete walls directly against the steel sheet piling of a cofferdam that is subjected to unbalanced water pressure from the outside. It is probable that there will be some leakage at the interlocks. Therefore, when concrete is poured, the water may force its way into the concrete and thereby cause weakness and bleeding.

Figure 11-7 illustrates this. Sketches (*a*) and (*b*) show a heavy concrete wall that is to be poured between piling *A* and an inner form *B*. Assume that there is a leak at *D* and that the water pressure outside the cofferdam exceeds the hydrostatic pressure of the plastic concrete. Water may then force its way into the neighboring concrete, causing a porous pocket; it may force a channel through the concrete to some outlet through form *B*, thus washing out the cement and fine aggregate through this

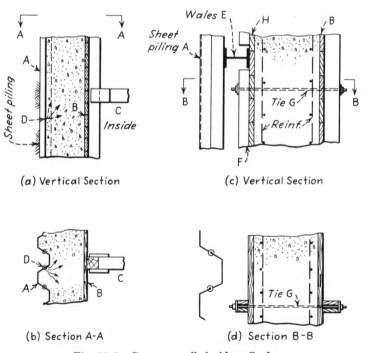

(a) Vertical Section   (c) Vertical Section

(b) Section A-A   (d) Section B-B

Fig. **11-7** Concrete walls inside cofferdams.

channel; or it may work its way up alongside the piling, thereby washing out the cement along some upward channel.

Furthermore, Fig. 11-7(*a*) and (*b*) indicates that the form *B* must be braced in some manner to resist the hydrostatic pressure of the plastic concrete. This bracing may be costly and troublesome. As an alternate, form ties might be welded to the sheet piling. Any wales and bracing to support the cofferdam itself are likely to interfere with the reinforcement and with the placing of the concrete.

Another arrangement is pictured in Fig. 11-7(*c*) and (*d*). Here the inner form *B* is set farther from the sheet piling *A* so that an outer form *H* can be built past wales *E*. It may also be desirable to extend form braces *F* past the wales. The form ties *G* then resist the pressure of the plastic concrete and permit the use of a minimum of bracing to steady the forms. Any leakage through the piling can then be collected at the bottom of the space between the piling and the back form; it can be pumped out, or provisions can be made to drain it away temporarily. The forms *H* are generally left in place.

If the cofferdam is in impervious material, the leakage may be small, so that the method of Fig. 11-7(*a*) will be satisfactory. The wall may then be made thick enough to extend the outer reinforcement past the wales if these are to be embedded permanently. However, large holes should be made in the webs of wales *E* to permit the concrete to fill the space below the webs and between the flanges because this space would otherwise form an air pocket. If the soil is porous and the external water pressures are high, the scheme of Fig. 11-7(*c*) may be preferable.

In some cases, if the sheet piling leaks, it may be possible to close the leaky places by calking, to lower the water table by means of well points, or even to box out locally and conduct the leakage to a temporary sump and pumps. If the sheet piling is to be removed later, bond of the cement to it should be prevented by painting the piling with oil or some suitable coating or by covering it with plywood, tar paper, old boards, or even cardboard. In one case about a 10-ft depth of concrete was poured against 40-ft sheet piling that was supposed to be pulled. However, the difficulty and cost of breaking it away from the concrete were so great that most of the piling was abandoned.

If the soil can be excavated so as to stand vertically, or nearly so, without the use of sheet piling or sheathing, one might think that he could pour the concrete of a wall directly against the ground without the use of backforming. For example, assume that a conveyer tunnel is to be built underground and that the soil seems to be able to hold on a vertical face temporarily. Unless the ground is a stiff clay, pouring of concrete against it is likely to disturb enough granular material so that it will fall and produce pockets of dirt in the concrete, causing weakness and leakage. Also,

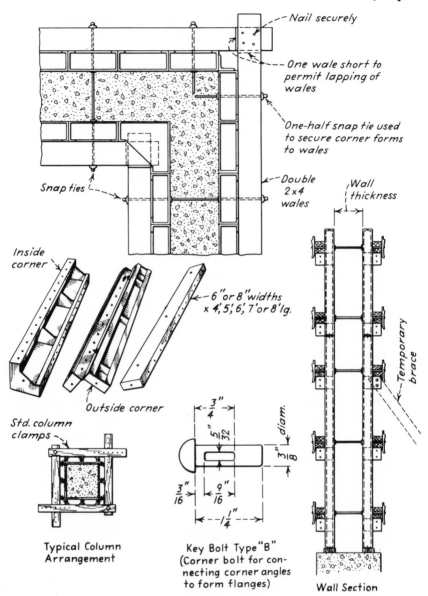

Fig. 11-8   Standard construction details, Atlas Speed Forms.   (*Irvington Form and Tank Corporation, New York, New York.*)

bracing one-sided forms may not be as easy as using form ties with double forms.

11-4. Details and Miscellaneous Data.   Forms are generally made of lumber and plywood or of steel.   The first two are very common, although, if there is to be enough duplication to permit extensive reuse, steel forms may be really economical.   Figure 11-8 shows steel forms that are made in standard units that can be assembled as needed to produce a vast variety of sizes and shapes.   It is sometimes advisable to plan a concrete structure with dimensions chosen to fit these forms.

When beams are planned, for example, it is advisable to consider how the forms are to be made, especially when they are to built of wood.[1]   It is important for economy that the carpentry work be minimized by the use of standard lumber without a lot of cutting.   As an illustration, the width of a beam might be made $9\frac{1}{2}$ instead of 10 in. so as to use a standard 1 by 10 to form the bottom, or the width might be 15 instead of 16 in. so as to use two 1 by 8 pieces.   Correspondingly, the depth of haunch or

[1] Formwork Simplified for Apartment House, *Eng. News-Record*, Aug. 16, 1951.

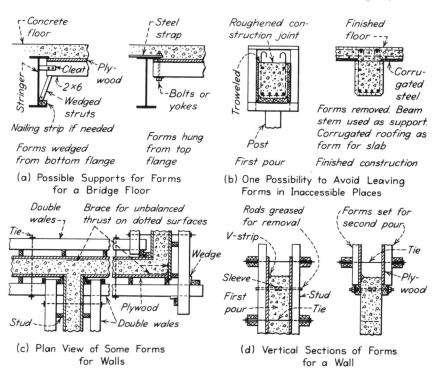

Fig. **11-9**   Some details of formwork.

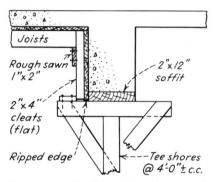

Fig. 11-10   Partial section of beam-and-slab form.

stem might well be made in multiples of standard sizes. Some illustrative sketches are pictured in Figs. 11-9 and 11-10.

Furthermore, it is economical to plan a job so as to use a series of various sizes that differ by 2 or 4 in., not by fractions or odd dimensions. This simplifies the formwork and permits a maximum of reuse. The reinforcement can be varied easily as the loads require.

It is sometimes worthwhile to make elaborate forms that are mounted on tracks or rollers so that they can be moved quickly from point to point. Forms for side walls and roofs of long tunnels are examples of this.

For tall structures, such as bridge piers, "slip forms" are used so that they can be raised slowly by jacks or screws as the concrete is poured. If the shafts are tapered, this causes complications in adjustments. Stepped piers may be used instead of those with battered sides.[1] Of course, various details will have to be invented to suit each particular job.

The special forms developed by the Vacuum Concrete Corp. have great possibilities for substantial economies,[2] especially if a structure is designed in advance for their use. The drying and stiffening effect produced by the vacuum enables side forms to be removed in from a few hours to a day instead of a few days after pouring of the concrete.

Removal of forms may be troublesome unless this operation is planned in their design. Oiling the forms or coating them with some other suitable material will help to avoid bond between the forms and the concrete. Adequate cleaning of a used form is essential prior to its reuse.

Heavy centering is usually supported upon wedges, shims, or jacks so that it can be "struck" or lowered away from the underside of the structure. Sand jacks are boxes filled with sand that will support the posts when desired but will lower them when the boxes are broken or the sand removed.

When one pour is made on top of a lower one from which the forms have been removed, special care is needed to prevent leakage of mortar where the upper forms rest upon, or lap over, the completed concrete.

---

[1] Ralph Holt, Slip Forms Reduce Cost of Tall Bridge Piers, *Civil Eng.*, December, 1950.

[2] K. P. Billner, Applications of Vacuum Concrete, *J. ACI*, March, 1952.

Pinching of the forms against the concrete, calking of joints, or mortaring the junction may remedy the difficulties.

After forms are set, or before the last side is closed in, all debris, dirt, and laitance should be removed. Reinforcement often has to be erected before the forms are completed unless the member is so large that a man can work inside the forms. This is shown in Fig. 11-2.

In pouring columns or other members with considerable height, panels of one side of the form may be left off until the top of the concrete approaches their location. Then they can be attached and the pouring continued.

In ordinary construction it is customary to build the concrete work of the main structure prior to completion of many of the minor parts, erection of equipment, installation of piping and wiring, finishing of floors, etc. Steelwork, anchor bolts, and various materials to be embedded often cause delays because they are not on hand or the details for their use cannot be determined soon enough. As an example, curb angles that are to be held by welded anchors buried in the concrete of floors might well be redesigned to provide attachment by means of anchor bolts that can be set beforehand. Expansion bolts are also useful. Similarly, holes can be left for piping as shown in Fig. 11-11. Plans that are made carefully to avoid delays from such troubles may result in substantial economies.

Obviously the position and details of forms should be carefully checked before any concrete is poured in them. In one case, some 20 ft of the bottom of a bridge pier about 100 ft high had been poured and set before

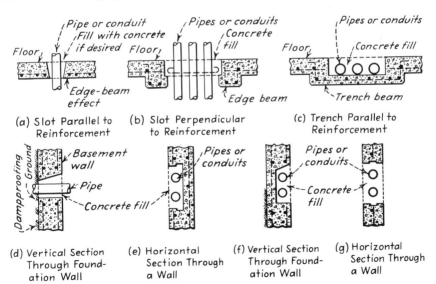

Fig. 11-11   Studies of provisions for installation of utilities.

it was noticed that the forms for the pier had been placed off center by 1 ft. The steel superstructure was fabricated and shipped to the job. What could be done about it? Should the bearings be set close to one edge, should the batters of the sides be changed so as to bring the top in proper position (causing an appearance of leaning), or should the completed work be demolished and the base rebuilt? Such questions cause many a headache.

Forms may be lined with various materials to produce a desired texture for the finished surface of the concrete. Cloth, masonite, steel sheets, plywood, and even clapboards may be used to get a surface varying from glazed for steel forms to make-believe clapboards. The forms for the Harlem River Speedway Connection of the New York Approach to the George Washington Bridge in New York were made with matched boards that had one half of the sides of the groove removed. This was done in order to make ridges on the surface that would collect dust and make the structure appear rough and aged. Even plastic-faced plywood is being offered as a material for forms.

There seems to be some question regarding the best material to use for forms in order to obtain the optimum durability of the surface of the finished concrete. Naturally, steel forms do not absorb the water in the concrete, whereas wooden forms do so to a certain extent, even when they are oiled somewhat for easy removal. A moderate amount of such absorption automatically produces a lower water-cement ratio near the surface, and this may be beneficial if it cannot proceed so far that it interferes with proper curing. One job reportedly constructed in 1905 had the forms lined with duck for the purpose of absorbing surface water. This structure is said to be in such good condition that the marks of the threads of the duck are still clearly visible.

In many cases it will be found to be desirable to coat form lumber with one of the plastic compounds especially prepared for that purpose. This prevents warping due to drying of the lumber, and it also prevents swelling of the wood and the possible creation of cracks in the concrete caused by such distortions. Such swelling may actually rupture green concrete if the situation is such that the expansive forces can be transmitted to the concrete.

When adequate fire protection will be provided, steel beams under concrete slab or ribbed floors may be housed in lath and plaster instead of being encased in concrete. This is especially advantageous when hung ceilings are used under ribbed floors that rest on top of the beams. The savings in weight, concrete, and forms may be considerable.

Steel pans placed on planks are used for one type of floor that is good and that minimizes the formwork.[1] It produces a one-way or two-way

[1] R. L. Reid, Steel-pan Forms Provide Economical Long-span Roof, *Civil Eng.*, September, 1951.

system of small T beams that provides a light but stiff floor—often called *waffle construction*.

The Dox floor construction illustrated in Chap. 12 has been developed to eliminate forms. The precast members automatically serve this purpose. They are useful for moderate spans and light loads when covered with an adequate mesh-reinforced topping.

Recesses in the faces of concrete construction are easy to make. Pieces of the right shape can be attached easily to the insides of the forms. They should be beveled or rounded at the edges so as to facilitate removal of the forms without spalling the corners of the offsets in the finished concrete. The details should be made so that air pockets and honeycombing of the concrete are guarded against. Remember that it is difficult to make concrete completely fill the spaces at undercut horizontal offsets. The edges are likely to be irregular and rounded off.

Projecting panels, brackets, corbels, and similar details are the cause of trouble and expense in building forms. If they are necessary, the details should be designed so as to simplify the formwork as much as possible. Sections tapered on four sides, conical shapes, and warped surfaces require costly forms, but slopes and circular or other simple curves in one direction are not too difficult to build.

It is desirable in many cases to have the forms prefabricated at some convenient shop where all the necessary equipment is available. They are made in panels that can be assembled on the job. A subcontract for all the forms for an industrial plant in the Southwest was let to a shop approximately 40 miles from the site. The forms for about 50,000 yd³ of concrete were shipped from the shop to the job.

In some cases, the forms, or a large portion of them, can be assembled as big units that are placed by cranes. This is especially helpful in the case of underwater construction.

When concrete floors are to be placed upon steel framing, the steelwork itself can often be used to support the forms, thus avoiding costly shoring. Precast members may be used similarly.

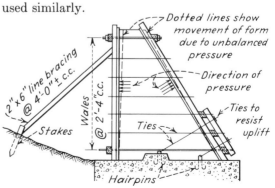

Fig. 11-12 Illustration of movement of forms for a retaining wall because of unbalanced pressure. (*Courtesy of C. E. Thunman, Springfield, Illinois.*)

The forms for thin concrete members may cost more than the materials for the concrete itself. In such cases, and in most concrete structures, excellent planning in the design can yield big savings in cost. To a large extent, designers might well plan their structures directly in terms of the economies in formwork as well as for architectural and engineering requirements.

In one large multistory building the contractor made braced steel-pipe scaffolding in 16-post units with prefabricated panels on top. The posts had adjustable bases. An entire unit could be placed on jack-equipped wheeled dollies and moved from one place to another. By using several of these units, the floors could be poured quickly and the forms reused many times, resulting in very large economies in the cost of the forms. However, the forms for one floor were to be supported by the floor previously poured below it, so that the latter had to be cured enough to attain the necessary strength—perhaps 7 days.

Corrugated steel forms (such as Cofar) can be used as forms for floors, particularly for buildings with structural-steel framework. Precast-concrete panels can also be very useful in some cases. Figure 11-14 shows the principles of how they were used in a waterfront dock to avoid costly formwork over water 30 ft or more deep.

11-5. Miscellaneous. The design of forms,[1] their supports, and their bracing apparently does not always receive the care and attention which it deserves, as seemingly evidenced by the numerous failures which have occurred. Each problem has its own special features, but one imaginary case will be illustrated in order to show some of the forces and details involved.

Let Fig. 11-13(a) represent the construction for a portion of a large basement that is to be in a rock excavation. The bottom is a mat that is designed to resist hydrostatic uplift. The concrete "sandwall" of the bottom and sides has been built, the membrane waterproofing applied, and the structural mat completed. The reinforcement is not shown in order to simplify the drawing. The forms and bracing for the wall from A to B are to be designed. Notice that the waterproofing prevents the use of form ties.

Assume that the wall forms are made with $\frac{3}{4}$-in. plywood lining, 3×6 studs 16 in. c.c., and 3×6 longitudinal timbers top and bottom. A 4×6 wale is to be placed against the studs as shown in (a), with 4×6 diagonal braces 4 ft c.c. from C to a kicker at D. Since a topping is to be used over the mat, it is assumed that the anchor bolts used to hold D and

---

[1] See Formwork for Concrete, SP-4, by M. K. Hurd under the direction of ACI Committee 347, 1963. Also, see Recommended Practice for Concrete Formwork, ACI 347-63.

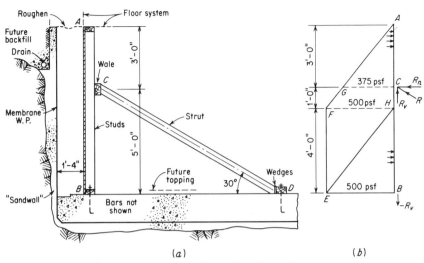

Fig. 11-13   Study of lateral bracing of forms.

B can be burned off later.   The wedges at D are for use in adjusting the position of the forms if necessary.   The struts are toenailed at C and D to hold them in position, and so is the wale.

Assume that the concrete is a $1:2:4$ mix; the slump is 3 in.; the rate of pouring is to be 4 ft/hr, and vibrators are to be used.   Use Thunman's diagram in Fig. 11-5.   Start with $R = 4$ on the left-hand scale.   Because of vibration, let the head $H = 8$ ft.   Draw a line from 4 on R through 8 on H to the left pivot line.   Next, draw a line from 1-2-4 on the C line through 3 in. on the S line to the right pivot line.   Connect the two points thus found on the pivot lines, and find the resultant pressure P equal to approximately 500 psf.   This means that the rate of pouring is so fast that, for the first hour, the lateral pressure is almost hydrostatic.   Considering the character of the pressure diagrams in Fig. 11-3, assume that the triangle BEH in Fig. 11-13(b) represents the pressure diagram applied to the forms at the end of the first hour and that $BEFA$ represents the pressure diagram at the end of the second hour—the end of the pour.   Notice that the pressure near B at the bottom of the forms is not increased, but it is not decreased either, because once the forms have been loaded, they cannot spring back even though the concrete is losing its plasticity.

The maximum horizontal pressure at C will come with the entire diagram $BEFA$ acting.   Taking moments about B,

$$R_c = \frac{500 \times 4 \times 2 + 500 \times 2 \times 5.33}{5} = 1{,}870 \text{ lb/ft}$$

$$R_b = 500 \times 4 + 500 \times 2 - 1{,}870 = 1{,}130 \text{ lb/ft}$$

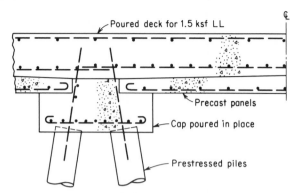

Fig. 11-14 Deck construction of relieving-platform dock at ocean harbor.

Of course, the plywood and the studs have to be designed to resist the pressures. It is probably satisfactory to use a bending stress of 1,500 psi in good lumber employed for such temporary purposes.

The pressure R_h applied at each strut will be

$$R_h = 1,870 \times 4 = 7,480 \text{ lb}$$

Then
$$R = \frac{7,480}{0.866} = 8,650 \text{ lb}$$

$$R_v = 8,650 \times 0.5 = 4,325 \text{ lb}$$

Similar but opposite forces will act at D, but these are horizontal and downward. The upward force R_v at C has to be counteracted by something. It is assumed that the anchor bolts at B will resist both R_h and R_v, and that the studs and the 3 × 6 bottom member are connected so as to be strong enough for this purpose.

There are other matters to be borne in mind which have not been previously mentioned. Some are as follows, part having been taken from the Code and AASHO:

1. Do not overlook the effect of unbalanced thrusts at the ends of pours and at vertical corners.

2. Projecting corners should be filleted.

3. Provide draft (bevels) for pulling off forms from projections such as girders and copings.

4. If the bottom sections of wall forms are left loose temporarily, this will expedite cleaning the bottom before depositing concrete, and it will help in the application of a mortar coating on top of existing concrete just before the upper pour is made.

5. Forms must be mortar tight.

6. Pipes or conduits embedded in columns should not displace more than 4 per cent of A_g.

7. Embedded piping should not have a temperature of over 150°F nor a pressure exceeding 200 psi. Except for snow removal, such pipes in solid slabs are to be placed between top and bottom reinforcement. In any case, the minimum cover is to be 1 in. Provide reinforcement normal to piping so as to have an area equal to at least 0.002 times the cross section of the concrete.

8. Forms should be oiled or saturated with water immediately before placing concrete.

9. Forms to be reused should be true to size and shape, thoroughly cleaned, and oiled.

10. The AASHO requirements regarding removal of forms may be a useful guide when the temperature is not below 40°F and normal portland cement is used:

Arch centers...........................	14 days
Centering under beams.................	14 days
Floor slabs...........................	7–14 days
Walls................................	12–24 hr
Columns..............................	1–7 days
Sides of beams and all other parts........	12–24 hr

11. Be sure that the structure can support itself and any applied loads before forms are removed, and continue proper curing in any case.

PRECAST CONCRETE

12

12-1. Introduction. Precast concrete is not a special kind of concrete but a method of fabrication. The term means any reinforced-concrete member or even one of plain concrete that is cast in forms somewhere other than its final position, then erected in place. Usually the casting is done on the ground near the site or in a shop that makes a specialty of such work.

The use of precast members is increasing considerably. As engineers become more familiar with the possibilities of such construction and with the great economies that can be made by proper use of it, they will undoubtedly resort to it more and more. However, it is seldom practicable to substitute precast members piece by piece for those which have been designed for pouring in place. Basically, a proper use of precasting involves an adequate knowledge of what can be done in this line, then the planning of a structure to use these products.

Prestressed concrete, which will be explained in the next chapter, is especially well adapted to combination with precasting. The two are so closely interrelated—or probably soon will be—that it is a question as to which one should be explained first. This chapter will deal with precasting in general, and it will leave most of the adaptation of precasting with prestressing to the next chapter.

12-2. Economics. Precasting may be a means of making considerable savings when compared with cast-in-place construction. This is primarily because of economies in the cost of forms. Working on the ground with all the necessary equipment handy, and with multiple use of forms, is far easier than building heavy forms, pouring concrete, finishing surfaces, sprinkling during the curing period, and removing forms when all these operations must be done above ground. This is obvious.

On the other hand, there are costs that may tend to offset these savings. Among these are the following:

1. Large area and accommodations required for production in volume.

2. Large storage space and facilities for curing.

3. Handling in the casting yard from casting position to the curing room or yard, and perhaps moving again to storage.

4. Large equipment for the transportation of long heavy members that must be handled carefully.

5. Large equipment for raising such heavy members into position.

6. Special details, particularly at junctions of members.

The designer should have intimate knowledge of what can and cannot be done practicably in precasting, or he should obtain expert and reliable advice regarding such matters. Then he should design his structure for the use of precast members in the first place. In fact, he should work out all important details and prepare the contract and specifications on that basis. He should let the contractor decide how and where he will make the members and what means he will use to erect them, but the plans should be very specific and complete in showing what the final result is to be. The contractor should not have to try to adapt the design to precasting.

The use of the vacuum can be very helpful in shortening the time that forms must be left in place. It, with the special equipment that has been or may be developed, can also cut down the cost and difficulty of handling precast members, especially when they are thin panels or long slender pieces. Steam curing can also be used to shorten the time between casting and erection or shipping of members.

Precasting in large volume is an assembly-line job. Space, equipment, and operations should be planned in great detail just as they would be when laying out the production line for a factory. This is properly the function of the contractor or of the manufacturer of concrete products.

12-3. Design Features. In general, precast members are designed as structurally determinate ones; i.e., they are not continuous beams or frames. However, in some cases, rigid frames, continuous beams of moderate length, columns, and miscellaneous pieces can be used even though they employ the principles of continuity. Again, precast parts may be incorporated in structures that appear to be indeterminate, and continuity may even be secured to a certain extent in some cases by means of poured-in-place junction pieces which tie the parts together. The variety of construction is so great that the best one can do in a book is illustrate different instances where precasting has been or can be used. From these, the reader can get an idea of the great possibilities of precast concrete.

The design of a precast member itself to withstand bending will ordinarily be similar to that of a corresponding part cast in place.   Provisions to support the end reaction may be considerably different, and special reinforcement may be required because of stresses that will exist during the handling and erection of the piece.   This is obvious in the case of a long precast pile in which the critical bending moments will probably occur when the pile is lifted from a horizontal to a vertical position.   Lifting may require special devices like eyebolts, looped bars, holes, or other special details.

As stated previously, the design of a precast-concrete structure requires thinking and planning which is different from that required for a conventional concrete structure.   It is almost as though one were using a different material.   In some ways, it approaches the character of the planning involved in heavy timber construction.   It is the planning of a structure to use standard premade articles, plus the planning of those prefabricated articles themselves, in many cases.

The end details,[1] provision for supporting the end reaction, shearing stresses in the ends, provisions for proper fit, provisions for seating or insertion of members, means for transferring end or sideward shears and moments when they are desirable, and provisions for tying the entire structure together—these are some of the special features that call for ingenuity and engineering skill.   They are special problems of precast-concrete construction.

Provisions for conduits, piping, fixtures of various sorts, windows, and doorways may be troublesome when walls, roofs, and floors are precast.   All must be planned in advance, and all details incorporated in them.

Bearings of simply supported members may be narrow.   Therefore, special care should be used to be sure that bond, end anchorage of reinforcement, and bearing pressures are adequately provided for.   One case of this is shown in Fig. 12-1(a), where precast rectangular purlins are supported upon shelves on the sides of a main precast girder.   The bearing of the girder on a column is pictured in sketch (c).   Both drawings show the utilization of embedded bars on top of the purlins and the girder to knit the structure together both ways.   This may not be necessary, but some kind of positive tie is desirable.   Bolts might be used, as shown in Fig. 12-9.

Precast members are to be designed in such a way as to facilitate the use and reuse of formwork.   Members may be poured in multiples or individually.   In any case, recesses, holes, ribs, and other details should be made with bevels in order to make it easy to remove the members from the forms, or to remove the forms from them.   The members should be

[1] Suggested Design of Joints and Connections in Precast Structural Concrete, ACI-ASCE Committee 512, *J. ACI*, August, 1964.

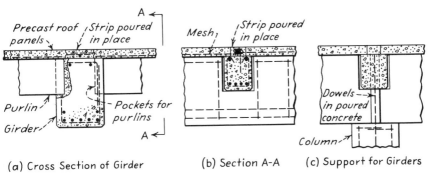

(a) Cross Section of Girder   (b) Section A-A   (c) Support for Girders

Fig. 12-1   One type of precast-concrete construction for a factory roof.

shaped so as to facilitate placing of the concrete and complete filling of the forms.   They may be cast in normal position (right side up), on their side, or even bottom side up.

For precast members it is often advantageous to make up the reinforcement as prefabricated cages.   The bars may be wired together or even tack-welded at their junctions.

In order to expedite the work it is often desirable to use quick-setting cement.   The concrete should be designed for at least $f_c' = 3,000$ to 4,000 psi.   It should have a small slump, and it should be compacted well by vibration or other means.   As indicated in Fig. 12-1, the compressive stress in the concrete is likely to be higher than in ordinary members because of the narrow width of the top.   Diagonal tension and shears due to handling generally make it desirable to use an effective system of web reinforcement when the members are long and heavy.

It is also desirable to use small-sized aggregate in order to have the bars closely spaced because a relatively high percentage of steel and close spacing of bars are customary.

Some of the Code requirements affecting precast members for buildings are as follows:

1. Maximum size of aggregate should not be over one-third of the least dimension of the member.

2. Minimum cover for parts not exposed to weather is the nominal diameter of the bars, but not less than $\frac{5}{8}$ in.

3. Lifting eyes or bolts should be designed for 100 per cent impact.

4. Rapid steam curing is allowed, but the compressive strength when loaded must be equal to the design strength for that load.

5. Members must not be overstressed in storage, transportation, or erection; and adequate temporary vertical and lateral support must be provided until permanent connections are made.

6. Welded splices may be used if they develop 125 per cent of the

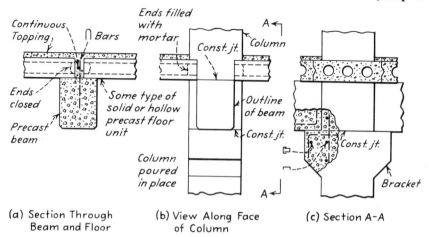

(a) Section Through   (b) View Along Face   (c) Section A–A
    Beam and Floor   of Column

Fig. 12-2   Building construction using precast members except for columns and topping.

yield strength, or 25 per cent excess steel should be added if the welds develop only the yield strength.

7. Wall panels may be thin but must be able to span horizontally and vertically as required.

8. Minimum column thickness is 6 in., and minimum A_g is 48 in.[2] if an analysis shows them to be safe.

12-4. Small Bridges.[1] Precast construction can be very useful for the superstructures of small highway bridges. The parts can be handled from the casting plant to the site by trucks, then lifted into place by one or two truck cranes, or erected by any other hoisting equipment that is suitable and available. Since the members are already cured, interference with traffic is often minimized. Where old deck spans are to be replaced, it is also practicable in many cases to maintain traffic on one longitudinal half of the old structure while the new half is being erected, then to transfer traffic to the latter during the completion of the job.

Figure 12-3 pictures one type of construction that might be used for spans of 15 to 25 ft, or possibly longer, if the depth, width of ribs, reinforcement, and other features are made adequate. The longitudinal channel-shaped members are made so that their bottoms are close together at the joints A and B, but their tops are about 1 to $1\frac{1}{2}$ in. apart. The keyway can thus be grouted from the top, and it will lock the mem-

---

[1] See Standard Specifications for Highway Bridges, The American Association of State Highway Officials, 1220 National Press Building, Washington, D.C., for design data.

bers together vertically.   The sheet asphalt is used to serve as a finished
pavement and to protect the structure from water to a considerable extent.

It is possible to use these sections in Fig. 12-3(*a*) without the poured-
in-place keyways if the spans are short and if leakage through the joints
is not objectionable.   However, even ordinary keys will help to distribute
wheel loads to adjacent members.

The bearings of precast bridges are likely to be troublesome details.
In Fig. 12-3(*b*) there is a keyway *DE* and projections *CD* and *EF* between
the ribs.   The ribs at *CF* extend straight across.   These two give a two-
way keying effect that locks the members in position longitudinally and
transversely.   However, the formwork and accuracy of finishing of the
abutment are difficult.   At the expansion bearing, keyway *DE* is omitted
and the top of the abutment is coated with asphalt, oil, or some other
compound to prevent bond.   Neoprene pads are sometimes used.

The diaphragms *G* are used to distribute loads between the two ribs of
each member.   The bulkhead strip *H* is used to terminate the sheet
asphalt.

Another detail of a bearing is pictured in Fig. 12-3(*c*).   The abutment
is finished level except for intermittent depressed keyways *KL* under the
end diaphragms.   The diaphragms are shallower than the ribs, and they
have a projecting key *OP*.   After erection the space *NOPGMLKJ* is dry-
packed with a stiff 1:2 mortar which serves to lock the deck and the
abutment together horizontally.

Figure 12-4 shows another arrangement that was used to replace a
series of old timber bridges in Florida.   Here the precast members are
separated and the transverse bars extended into the open spaces, where
they are connected by means of cable clamps as well as by the bond produced when the open strips are concreted.   Dowels projecting from the
abutments and piers into these open strips are concreted in.   They then

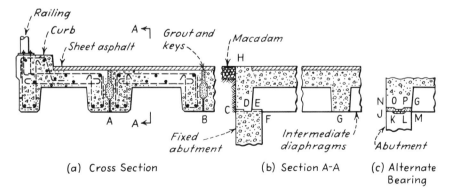

Fig. **12-3**   Some details for short precast-concrete highway bridge.

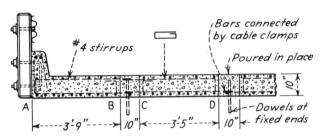

Fig. 12-4   Precast units 15 ft long used for some bridges built by Florida State Road Department.

serve to hold the superstructure in place.   The transfer of vertical shear between adjacent members depends upon the bond between them and the 10-in. strips, but keys might be used.   Generally, it is necessary to provide some means of attaching hoisting devices to these precast members, especially when they are to be placed close together.   One method is to provide two holes near each end so that eyebolts can be inserted.   The holes are later grouted.

Another arrangement for bridges of moderate spans—from 25 to 50 ft—is shown in Fig. 12-5.   The precast longitudinal stringers are designed as rectangular simply supported beams that are strong enough to support the weight of the forms, the concrete of the slab, and any concreting equipment that is to be used.   They may be handled by means of padded slings, embedded eyebolts, or transverse bolts placed through holes in the webs.   The last is good because of its cheapness and because it is not necessary to get under the members to lift them off the bottom forms.   After erection, the forms are supported upon the shelves at the bottoms of the beams.   The slab is then poured and finished.   The completed structure acts as a series of composite T beams.   The tops of the precast members are roughened in order to develop the necessary resistance to longitudinal shear.   They are also gripped laterally by the slab.

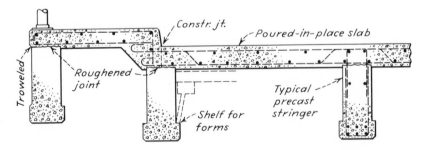

Fig. 12-5   Girder bridge with precast stringers and poured-in-place deck.

12-5. Building Construction. Buildings, if properly designed, constitute a field in which precast concrete can be very useful. They are often composed of slender members and thin panels that are high above ground, so that forms and shoring are relatively costly when the concrete is poured in place.

On the other hand, some building types are suitable primarily for cast-in-place construction. Among these are the following:

1. Flat-slab construction with drop panels and capitals of the type shown in Fig. 12-6(a). Here continuity is vital, and it is difficult to secure otherwise when large floor areas are involved. Shearing stresses around the drop panels and capitals may also be critical, so that a monolithic floor is preferable.

2. Flat "plate" floors that are a modification of the flat-slab type but without drop panels and capitals, as in sketch (b). Continuity and shearing resistance again may eliminate precasting when the floor area must be large.

3. Very large multistory buildings of beam-and-slab construction where continuity, frame action, shallow depths, and relatively small-sized columns are desired. A study of Fig. 12-1 will show how these could be accomplished better if the floor were monolithic. Furthermore, with precast construction, the column loads from above cause difficulties with the details of end bearings and column sizes. However, certain precast parts may be incorporated advantageously in such buildings. When the loads are light enough, the ends of the beams may be supported upon brackets on the columns, as in Fig. 12-2.

4. Any floor construction that must be self-supporting but must carry heavy machines or large live loads, and where a continuous well-integrated structure is necessary.

5. Large arches, domes, barrel-type construction, and other special cases. These generally require that the structures be poured in place, although it is sometimes possible to incorporate some precast parts into them.

Besides Figs. 12-1 and 12-2, which show a sort of imitation of heavy timber construction, there are many ways of using precast concrete in

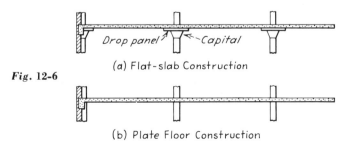

Fig. **12-6**

(a) Flat-slab Construction

(b) Plate Floor Construction

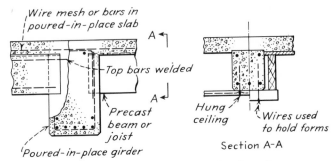

Fig. 12-7 Combination construction for a floor.

whole or in part for buildings. One is the combination of precast beams with poured-in-place floors as pictured for the bridge in Fig. 12-5. If a hung ceiling is to be used, it may be preferable to hang the forms from the beams rather than to use shelves for their support. Any wires or bolts that are visible when the forms are removed will be concealed by the ceiling. The girders that hold the beams, and the columns also, might have to be poured in place, somewhat as pictured in Fig. 12-7. This construction is not particularly economical for buildings.

One way is to use large precast panels for floors, roofs, and side walls, with the supporting beams poured in place after the slabs are erected on a few posts or bents. Figure 12-8 shows one application of this idea. The rigid frames are cast in place and serve as the "backbone" of the structure. The hollow panels were made on wooden forms on a concrete casting slab as follows:

1. Place reinforcement for one thin side.
2. Pour concrete of first thin side, screed it, and let it develop its initial set.

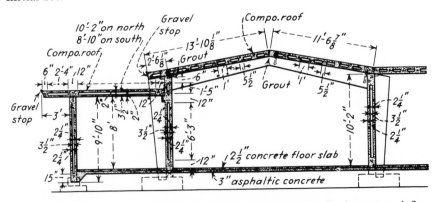

Fig. 12-8 Hollow precast-concrete slabs used for walls and roofs of one-story infirmary buildings at Los Angeles County Hospital. (*Eng. News-Record*, Mar. 8, 1951.)

3. Place strong waterproofed corrugated-paper cartons about $\frac{1}{8}$ in. thick to form hollow spaces.

4. Pour concrete around and over cartons.

5. Place reinforcement for other thin side.

6. Pour last concrete and finish the surface.

After erection, the joints between roof and wall panels are mortared. Some sort of tie between the two sides is desirable. Wires through the ribs between cartons may be sufficient if attached to the reinforcement of both faces.

Two large single-story warehouses[1] at the Naval Supply Depot, Mechanicsburg, Pennsylvania, were made with bolted channel-shaped sections as shown in Fig. 12-9. Even the columns were precast and hollow. The welding of reinforcement and the concreting of the joints shown in the figure transformed the structure so that it could act as a rigid frame. The roof panels are 5 ft wide and 20 ft long, of channel type, and with intermediate cross ribs.

The forms were hard-surfaced smooth-troweled concrete with hinged wooden sides. A vacuum process was used with high-early-strength cement. The sections were removed from the forms within 24 hr after pouring, using vacuum lifters.

Notice that the roof panels are held by welded straps. The struts that connect the various rigid frames are also attached by welding at one end.

It is possible to cast rigid frames in pieces as solid units of the same general shape as those parts shown in Fig. 12-9. However, the columns will have to be supported laterally until the structure is framed together. The columns can have square ends that rest in pockets in the foundations. Struts to provide longitudinal stiffness of the structure may have their main portion precast, but the junctions with the frames may have to be concreted in place so as to engage proper keys and dowels, or some other positive attachment may have to be invented.

Precast wall construction has been used considerably. Tilt-up walls are one adaptation. These walls are cast on the ground in one or more sections, then lifted or tilted upright and attached to each other and to whatever other parts may be necessary. Sometimes these panels are made with foam glass or other insulating material inside. They are built similarly to those explained in connection with Fig. 12-8. In the structure, the panels may be connected to the floor, to partitions, to columns, and to each other by welding steel details together, by bolting, by mortaring, or by any other suitable and positive means.

Precast wall panels can also be used in large multistory structures. When made with an insulating material in the middle, they are often

[1] W. Mack Angas, Precast Structural Members Facilitate Speedy Erection of Rigid-frame Buildings, *Eng. News-Record*, Apr. 18, 1946.

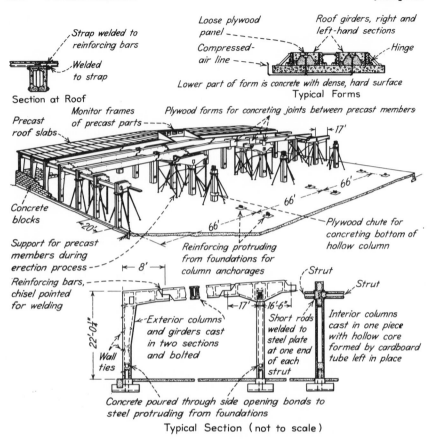

Major Dimensions

	Length	Width	Height	Thickness
Warehouse	600'	200'	24' av.	
Ext. girders	42½'	1'-8"	3' ⎫	2" or 2½" web
Int. girders	32'	1'-8"	2'-9" ⎭	3½" flge.
Ext. columns	22'	1'-8"	2'	⎰ 2", 2½" or 3" web ⎱ 3½" flge.
Int. columns	17'	1'-8"	1'-8"	3" min.
Roof panels	20'	5'		1¼"
End panel beams (2)	5'		6"	2" to 3"
Int. panel beams (3)	5'		6"	1½" to 2½"
Panel girder (2)	20'		8"	2½" to 3"
Struts	20'	1'	1'-4"	⎰ 1¾" min. web ⎱ 3¾" min. flge.

Fig. **12-9** Schematic drawing of a precast warehouse. (*Eng. News-Record*, Apr. 18, 1946.)

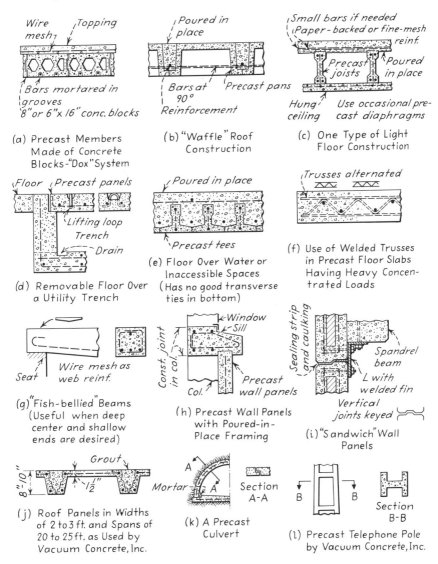

Fig. **12-10**   Some miscellaneous details of precast-concrete construction.

called *sandwich* walls.[1]   Much experimental and developmental work in this line is under way.   Condensation, durability, watertightness, manufacturing procedures, and structural details are among the problems that

[1] S. B. Roberts, Sandwich Walls Precast for Pulp Mill, *Eng. News-Record*, Feb. 22, 1951; also, P. M. Grennan, Precast Sandwich Wall Makes Costs Tumble, *Eng. News-Record*, Jan. 24, 1952.

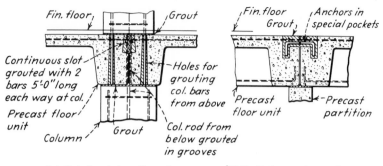

(a) Details at Junction of   (b) Tie Between Large Floor Units
    Floor and Columns

Fig. 12-11   Some details of precast construction for multistory buildings.   (*Patterned after work of Vacuum Concrete, Inc., Philadelphia, Pennsylvania.*)

are being tackled.   As stated previously, if a structure is to utilize such precast parts, it must be designed accordingly, and in great detail.

Devices have been invented for casting roofs or floors on the ground or on another floor in very large panels, then raising them into place by "jacking" them up the columns or by using other special hoisting devices.

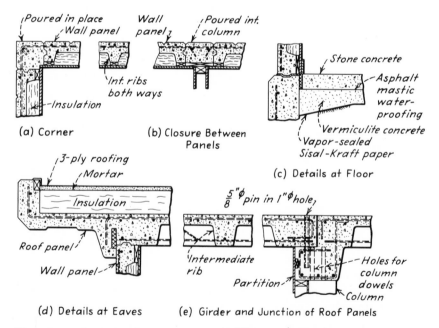

(a) Corner   (b) Closure Between
                        Panels

(c) Details at Floor

(d) Details at Eaves   (e) Girder and Junction of Roof Panels

Fig. 12-12   Some details for one-story buildings.   (*Patterned after work done by Vacuum Concrete, Inc., Philadelphia, Pennsylvania.*)

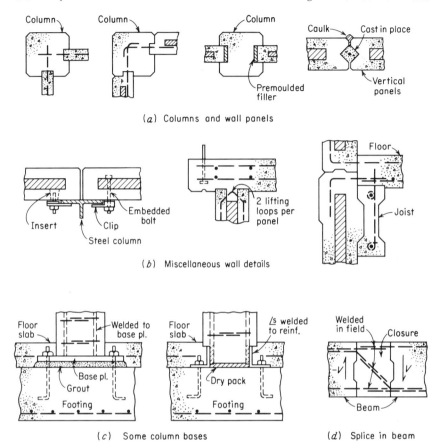

(*a*) Columns and wall panels

(*b*) Miscellaneous wall details

(*c*) Some column bases   (*d*) Splice in beam

Fig. **12-13**   Miscellaneous details.

Precast concrete blocks are used in making precast members of the general type shown in Fig. 12-10(*a*), one of which is the Dox system. These are useful in some cases for moderately light loads and short spans when simply supported and when not cut up by pipes and utilities. The blocks are machined, assembled in a row, bottom side up, and pulled tightly together. The reinforcement is mortared in the grooves, and the member is cured. It is shipped to the site and erected right side up. These units serve as forms to support the topping when it is poured. Together with the topping, they make a sort of T-beam construction.

Precast concrete boxes, such as those shown in Fig. 12-10(*b*),[1] have been used to form large roof slabs of waffle type or two-way pan construction. They serve as forms for the poured-in-place concrete ribs that

[1] Concrete Block Laid on Floor Slab Jacked Up on Columns to Form Roof, *Eng. News-Record*, Feb. 21, 1952, p. 30.

encase the bars, and their tops assist in resisting compressive stresses. They could also be used if erected directly in place by supporting them on a series of temporary bents.

Designers of precast-concrete structures should bear in mind the problem of stiffness. This may involve the knitting of a structure together by poured-in-place junction pieces as already shown. In other cases, it may be a question of the stiffness of individual members. For example, precast joists of relatively long span were used successfully when a poured-in-place floor slab was built on top of them, causing combined T-beam action. In another structure the same members were used except for the fact that the floor slab was made of precast-concrete panels laid on the joists but not rigidly connected to them. Because of the much smaller effective moment of inertia, the joists deflected sufficiently to cause serious cracking of the partitions erected above them. Each such problem should be studied in the light of the particular conditions that apply to it.

These are just a few of the applications of precasting. The illustrations here may give the reader some additional useful suggestions. Various persons will have various ideas about how best to do things. By no means are these illustrations complete or even expected to be the best examples of what ought to be done in precast-concrete construction.

By working together, engineers, architects, and the manufacturers of concrete products will undoubtedly develop precast-concrete construction to a point where much will be standardized. Also, engineering and architectural data regarding what has been and can be done will become available for all those who are in the design and construction business.

PRESTRESSED CONCRETE

13

13-1. Introduction. Although prestressed-concrete construction has come into prominence in this country only within the last few years, the basic idea is not very new. Its adaptation to reinforced concrete, however, is a relatively recent development that has grown considerably in Europe under the leadership of such men as Gustave Magnel and M. Freyssinet. Interest in this field is now growing rapidly in the United States.[1]

European leadership in prestressed-concrete construction seems to be due in large part to the relatively high cost and scarcity of materials, especially steel, and the low wage rates for and abundance of labor. In the United States, our mass-production methods and vast steel production have provided steel and many other materials in large volume at low cost, whereas labor has been both scarce and expensive. Design and construction methods have been developed so as to minimize the labor required and particularly to increase the use of mechanization. As time goes on, this situation will probably change, materials like steel may become more costly, and more thought will be given to economies in their use. It is therefore probable that the use of prestressed concrete will grow in popularity for construction to which it is adaptable. This does not mean that standard types and methods will be abandoned, but prestressing will probably become one of the alternatives to consider when concrete structures are planned.

Prestressed-concrete structures generally require less concrete than standard designs—perhaps 20 to 25 per cent less. The saving in steel is

---

[1] For details of precast, pretensioned framing, see C. W. Dunham, "Advanced Reinforced Concrete," chap. 8, McGraw-Hill Book Company, New York, 1964. Also, see Prestressed Concrete Institute's publications on joining methods for precast members.

likely to be even greater, but high-strength bars or wires are used.   When long spans, shallow depths, and light dead loads are desired, prestressing may be very advantageous.   For example, the depth-span ratio may be $\frac{1}{20}$ or $\frac{1}{30}$.   In general, such members are also relatively stiff until the pre-stressing is overcome.   After that, because of shallowness and small steel areas, they are likely to be quite flexible.

By no means is prestressed concrete to be considered as a substitute for all other reinforced-concrete construction.   It has disadvantages as well as advantages.   Among the former are the difficulty of securing continuity, the lack of adaptability for reversal of stress, and the effect of large moving live loads.

Space is not available for more than the basic principles of the analysis of prestressed-concrete construction.[1]   Much progress has been made in codifying the subject and in adapting it to engineering practice in the United States.   Because developments in this direction will probably be numerous and extensive, the author will try to stick to fundamentals.   A student of structural engineering should have a fair knowledge of these fundamentals on which he can later build whatever additional develop-ments his work may require.   Any detailed recommendations here are intended as suggestions for future guidance, and they apply to normal concrete only.

13-2.   Symbols.   There are many special symbols to be used in con-nection with prestressed concrete design.   They are grouped here for convenience.

$a = A_s f_{su}/0.85 f_c' b.$

A_b = bearing area of anchor plate of posttensioning steel.

A_b' = maximum area of the portion of the anchorage surface that is geometrically similar to and concentric with the area of the anchor plate of the posttensioning steel.

A_s = area of prestressed tendons.

A_{sf} = area of reinforcement to develop compressive strength of the over-hanging flanges of flanged members.

A_{sr} = area of tendon required to develop the web.

A_s' = area of unprestressed reinforcement.

A_v = area of web reinforcement placed perpendicular to the axis of the member.

b = width of compression face of flexural member.

b' = minimum width of web of a flanged member.

[1] See Gustave Magnel, "Prestressed Concrete," Concrete Publications, Ltd., London, 1954; P. W. Abeles, "Principles and Practice of Prestressed Concrete," Frederick Ungar Publishing Co., New York, 1949; August E. Komendant, "Pre-stressed Concrete Structures," McGraw-Hill Book Company, New York, 1952.

$d =$ distance from extreme compression fiber to centroid of the prestressing force.

$f'_c =$ compressive strength of concrete.

$f_{ci'} =$ compressive strength of concrete at time of initial prestress.

$f_{cp} =$ permissible compressive concrete stress on bearing area under anchor plate of posttensioning steel.

$f_d =$ stress due to dead load at the extreme fiber of a section at which tensile stresses are caused by applied loads.

$f_{pc} =$ compressive stress in the concrete, after all prestress losses have occurred, at the centroid of the cross section resisting the applied loads, or at the junction of the web and flange when the centroid is in the flange. (In a composite member f_{pc} will be the resultant compressive stress at the centroid of the composite section, or at the junction of the web and flange when the centroid is within the flange, due to both prestress and the bending moments resisted by the precast member acting alone.)

$f_{pe} =$ compressive stress in concrete due to prestress only, after all losses, at the extreme fiber of a section at which tensile stresses are caused by applied loads.

$f'_s =$ ultimate strength of prestressing steel.

$f_{se} =$ effective steel prestress after losses.

$f_{su} =$ calculated stress in prestressing steel at ultimate load.

$f_{sy} =$ nominal yield strength of prestressing steel.

$f_y =$ yield strength of unprestressed reinforcement.

$F_{sp} =$ ratio of splitting tensile strength to the square root of compressive strength.

$h =$ total depth of member.

$I =$ moment of inertia of section resisting externally applied loads.

$K =$ wobble friction coefficient per foot of prestressing steel.

$L =$ length of prestressing steel element from jacking end to any point x.

$M =$ bending moment due to externally applied loads.

$M_{cr} =$ net flexural cracking moment.

$M_u =$ ultimate resisting moment.

$p = A_s/bd$; ratio of prestressing steel.

$p' = A'_s/bd$; ratio of unprestressed steel.

$q = pf_{su}/f'_c$.

$s =$ longitudinal spacing of web reinforcement.

$T_o =$ steel force at jacking end.

$T_x =$ steel force at any point x.

$t =$ average thickness of the compression flange of a flanged member.

$v =$ unit shear at the plane being investigated.

$V =$ shear due to externally applied loads.

$V_c =$ shear carried by concrete.

V_{ci} = shear at diagonal cracking due to all loads, when such cracking is the result of combined shear and moment.

V_{cw} = shear force at diagonal cracking due to all loads when such cracking is the result of excessive principal tensile stresses in the web.

V_d = shear due to dead load.

V_p = vertical component of the effective prestress force at the section considered.

V_u = shear due to specified ultimate load.

y = distance from the centroidal axis of the section resisting the applied loads to the extreme fiber in tension.

α = total angular change of prestressing steel profile in radians from jacking end to any point x.

ϵ = base of Naperian logarithms.

μ = curvature-friction coefficient.

ϕ = capacity-reduction factor.

13-3. Fundamental Principles. In the preceding discussions of standard construction of reinforced concrete, considerable attention has been given to consideration of the effect of cracking due to the elongation of the reinforcement. The big difference in prestressed members is the use of a means of preventing these cracks. Thus the members more nearly approach solid monolithic ones.

What prestressing does is first to put the member under compression. Figure 13-1(a) shows an analogy of the condition. Here the springs squeeze the beam. Then, when a load is added, the beam deflects as shown to exaggerated scale in sketch (b), but it is not bent so much that tension in the bottom will overcome the compression that the springs caused initially. Therefore, the beam will not crack because there is

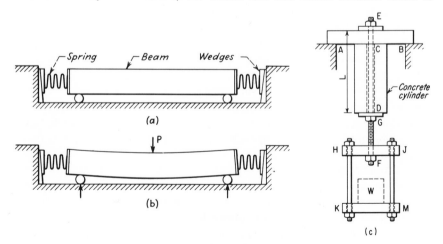

Fig. 13-1   Prestressing analogies.

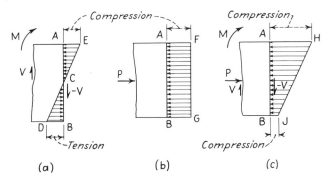

Fig. 13-2   Assumed stress conditions.

no tension to make it open up.   Of course, the top gets compressed more severely than it was in the first place.

As another illustration of prestressing, assume that a heavy frame spans an opening from A to B, which represent rigid abutments [Fig. 13-1(c)].   A threaded high-strength rod EF passes through a hole in the frame and is held by the washer and nut at the top.   The concrete cylinder CD has a hole in its center through which the rod passes.   This cylinder is held by the washer and nut at G.   A suspended pan KM is hung from the plate HJ, which is attached to the rod by nut F.   A load W is placed on KM.   Then rod EF stretches so that a space ΔL is opened at C below AB.   Now nut G is tightened until CD is again in contact with AB.   If W is now removed, the block CD is squeezed and it is compressed a little (or prestressed), but most of the elongation and resultant tension remain in rod EF, which continues to squeeze CD.   Any load less than W applied on KM hereafter will not relieve all of the compression or prestress in CD.   However, if a load greater than W is applied, the rod EF will stretch more than it did at first, so that a space (crack) will be created between AB and CD.

Let Fig. 13-2(a) represent a piece of a solid rectangular uncracked beam of elastic material that is in equilibrium as a free body.   The external bending moment M will be counteracted by the resisting moment of the compressive stresses and the tensile stresses pictured by the triangles ACE and DCB, respectively.   These act on face AB.   In sketch (b) the same piece is shown with a centrally applied direct load P which causes the compressive stresses illustrated in the diagram $AFGB$.   Now, if P is applied first and M is applied later, the condition may be as represented by the stress diagram $AHJB$ in sketch (c).   This means that $BG - DB$ results in a remaining compressive stress BJ, whereas $AF + AE$ produces a large compressive stress AH.   This in a broad sense is what prestressing tries to accomplish.

Compare Fig. 13-2 with Fig. 13-3.   In the latter, the bending effect of

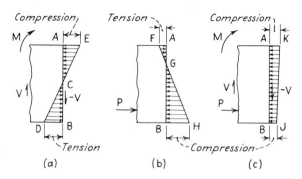

Fig. 13-3   Direct load applied eccentrically.

M is pictured in sketch (a) as before, but the load P in (b) is applied eccentrically so as to produce a combination of direct stress and bending. The large compression BH is at the bottom.   There may be little or no compressive stress at A, or even a small tension as pictured by FA, if the material can withstand it.   Now, when P is applied and M follows, the combined condition may be somewhat as pictured in sketch (c).   This is obviously much less severe in the resultant compressive stress AK at the top of the beam, and it is therefore more desirable.

Figure 13-4 shows some general arrangements of reinforcement to accomplish this eccentric pressure at the critical section near the middle. For the present, assume that the reinforcement has been tightened up in some way.   Also, assume that the transverse load on each beam is represented by its own weight w so that the bending-moment diagram due to these weights is a parabola and equal to $wL^2/8$ at the center.

In sketch (a), this tension in the steel will cause a direct compression P and a moment Pe throughout the entire length of the beam.   The compressive stress at the bottom of the beam may be largely offset by the moment $wL^2/8$ at the center of the span, but the dead-load moment will decrease toward the ends and will be unable to counteract the stresses caused there by the steel.

If the steel is curved in the form of a parabola as in Fig. 13-4(b), the eccentricity at the ends will be negligible and the remaining stress condition near the ends will be the compressive stresses due to P if the moments produced by P and w largely counteract each other.   Furthermore, the steel now acts somewhat like the cable of a self-anchored suspension bridge.   It tries to straighten out and therefore tends to produce an uplift on the central portion of the beam.   Sometimes the tensioning cables are made flat in the central portion, then sloped upward at the ends, or the cables may be straight and the beam built with a "hump" in the middle.   All these are for the same general purpose of producing a large eccentricity where it is most useful. However, these systems alone are not suitable when large moving live loads may be applied, especially near the ends of the beam.

In Fig. 13-4(c), part of the reinforcement is shown straight in the bottom and part is bent up above the center at the end so as to obtain part of each of the effects described in connection with (a) and (b). Sketch (d) is the same idea carried still further.

There are two general methods of prestressing. The word *prestressed* means that the steel is placed in tension before the main loads are applied to the structure. The words *pretensioned* and *pretensioning* will be used to denote members and processes in which the reinforcement is placed under tension before the concrete is cast around it. When the concrete has cured and the tensioning devices have been released, the bond on the surface of the steel prevents slippage so that the attempt of the steel to shorten under tensile stress produces a compression in the concrete that opposes such shortening. This pressure in the concrete makes it shorten more or less elastically, and this shortening relieves a corresponding amount of strain and stress in the steel. When equilibrium is attained, it is obvious that the steel is not stressed as highly as it was initially.

The terms *posttensioned* and *posttensioning* will be used to denote members and processes in which the concrete is first poured and cured and then the tension is applied to the reinforcement, which is not bonded to

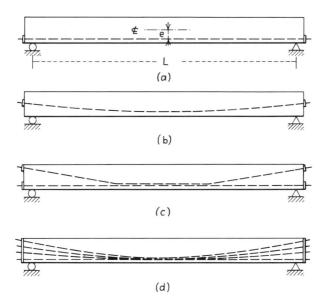

Fig. **13-4**   Various arrangements for prestressing systems in simply supported beams.

the concrete.   As the tension increases, the reaction of the tensioning devices against the concrete causes the concrete to shorten accordingly. When the bars are fully stressed, their ends are locked in place.   Thus compressive deformation of the concrete does not annul part of the effective stress-carrying capacity of the steel.   This is an obvious advantage of posttensioning.

In pretensioning, the shrinkage of the concrete causes, or tries to cause, some shortening of the member.   If the bond holds, this effects a corresponding shortening of the reinforcement or else it produces tension or hair cracks in the concrete that will be relieved or closed, respectively, when the tensioning devices are released from the steel.

Supposedly, most of the shrinkage has taken place before posttensioning is applied.   This is another advantage of posttensioning.

In both systems, plastic flow (creep) of the concrete may occur under the compressive stresses produced by the attempt of the steel to shorten under tensile stress.   This again causes a tendency to lose part of the tensile stress in the steel.   Furthermore, if the steel tends to yield (creep) under the effect of high stress, this also causes a loss of some of the tension in the reinforcement.

These ideas are illustrated to some extent in Fig. 13-5.   In (*a*) the pressure due to tension in the steel is centrally applied on a rectangular or symmetrical section.   It is supposed to cause a stress diagram *ACDB* in the concrete.   However, the actual pressure diagram is *AEFB* because of the shortening of the concrete due to shrinkage, plastic flow, compressive deformation, or any combination of them.   Sketches (*b*) and (*c*) illustrate the idea when the pressure and pretensioning are eccentric. Of course, the tensile deformation of the steel must be large or the deformation of the concrete will be sufficient to counteract it and cause complete loss of the prestress.   An approximation of the loss of prestress due to shrinkage, etc., is 20 to 25 per cent.   Magnel recommends the former figure.

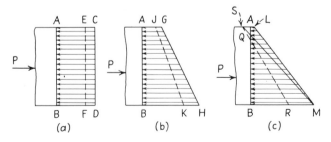

Fig. **13-5**   Effect of shrinkage and creep of concrete and steel.

Examine the situation for pretensioning particularly. Assume the following data (plastic flow and creep neglected for the present):

E_c for concrete = 4,000,000 psi
E_s for steel = 30,000,000 psi
Shrinkage = $\frac{3}{8}$ in. in 100 ft = 0.00031 in./in.

Average resultant compressive stress in concrete = 1,000 psi
The deformation per inch due to compression is

$$\delta_c = \frac{f_c}{E_c} = \frac{1,000}{4,000,000} = 0.00025 \text{ in./in.}$$

Add to this the shrinkage. Then the total unit deformation of the concrete is 0.00025 + 0.00031 = 0.00056 in./in. For the steel, this amount of deformation of the concrete corresponds to a tensile stress of

$$f_s = E_s \delta_s = 30,000,000 \times 0.00056 = 16,800 \text{ psi}$$

If the maximum pretensioning in the steel is 150,000 psi, this lost stress alone is approximately 11 per cent of the prestress. This shows that, if creep and plastic flow are included, ordinary reinforcing steel is not the thing to use as a prestressing medium because most of the pretensioning is likely to be lost when even the yield point cannot be more than 40,000 to 50,000 psi. Therefore, high-strength steels or cold-drawn wires with a high yield point are the appropriate materials to use for prestressed-concrete construction. In practice, high-strength bars having a yield point of 75,000 to 80,000 psi are to be used, or cold-drawn wires with an elastic limit of something like 160,000 to 180,000 psi are suitable. These wires are sometimes in the form of wire ropes called *tendons*.

As for the concrete, this too must generally be able to resist a much larger compressive unit stress than that required for ordinary reinforced concrete. Therefore, concretes with a 28-day compressive strength of 4,000 to 8,000 psi are necessary. Such strength can be obtained with rich mixes, good aggregates, proper water-cement ratios, excellent compaction, and careful curing. Good control is needed for such work.

Obviously the prestressing must cause sufficient compression in the concrete to annul any future tension that loads may develop. If not, the concrete is likely to crack and to behave similarly to ordinary reinforced concrete for the portions of the loads that are excessive—exceeding those which can be considered to overcome the prestressing. Since the cross-sectional area of high-strength steel is generally relatively small, the deformations beyond the prestressing limit are likely to increase rapidly with excess loading. Therefore, this condition is not desirable, and prestressing is only partially effective.

In prestressing, one must remember that what he wants is a tensioning medium that will take up large strains before the tension is eliminated. A large pressure on the ends of a member is not effective if it is destroyed by a tiny shortening of the member. Furthermore, if a bar sticks out of the ends of a member into which it is bonded, a jack applied at each end to put a tension in the bar and a compression in the concrete is primarily a pull-out test. It does not stretch the whole bar but only the end portions before bond transfers the stress into the concrete. The effect is local. If the beam is then bent, the part of the bar in the central portion will behave as usual and the concrete will crack as for any similar reinforced-concrete member. That is why the idea is pictured with springs in Fig. 13-1, since they can apply pressure to the concrete to a considerable extent in spite of the latter's deformation. Thus the compression is useful throughout the beam, and cracks cannot open up until the local tension from bending overcomes the compressive stress.

The beneficial effect of prestressing on the resistance of beams to shearing forces is especially important. In standard reinforced-concrete beams, the diagonal tension can become a serious problem when the shearing forces are relatively large. The situation is very different in prestressed members that always have compression on the entire cross-sectional area. Under such conditions, even though the members deflect and curve, there should be no cracks unless overloaded, and preferably not even tensile forces tending to cause cracks. Therefore, the shearing resistance approaches that of uncracked concrete subjected to transverse shearing stresses and to something approaching the longitudinal shearing stresses that exist in beams of homogeneous material.

Therefore, prestressed-concrete beams can be designed with relatively thin webs. Such sections as those shown in Fig. 13-6(a) may be used. The saving in weight is obviously considerable. This is likely to be more

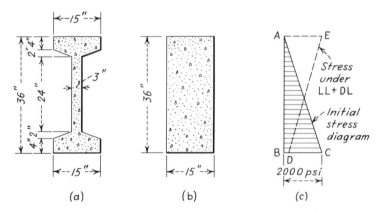

Fig. 13-6   Comparison of thin webs and thick sections.

important than the saving of the cost of the extra concrete, because prestressing is generally most helpful when long spans and shallow depths are needed and for which the minimizing of dead load is very important. The thin webs are also advantageous because if solid rectangular sections were used, the extra concrete near the center would not be needed for resisting flexure, but it would require more prestressing steel in most cases in order to secure the compressive stress (or compressive strain) that is needed in the member to counteract the expected or possible tensions for which the member is designed.

To illustrate this saving in prestressing reinforcement, assume the section in Fig. 13-6(*a*). Contrast it with that in (*b*) if the stress condition under the initial prestressing is to be as shown in (*c*) before the dead load is applied to the beam. Approximately, the total force to be applied to cause this pressure on the *I* section is 228,000 lb; for the rectangular member, it is 540,000 lb.

The elimination of cracking under diagonal tension is one of the great advantages of prestressing. Web reinforcement as ordinarily considered is not necessary, but some steel to tie across the member is usually desirable to knit the beam together, perhaps to help during handling, and to aid the web in case of unusually large loading. One source of information concerning the permissible proportions for the webs of *I* or channel shapes and box sections is the experience of others, especially the manufacturers, with structures that have been built successfully.

Partial prestressing denotes the application of some amount of prestressing force but not enough to avoid some, or even considerable, tension in the member under maximum loads. For example, a beam may be sufficiently prestressed to withstand all of the dead-load moments and perhaps one-half the specified live load without causing tension on, or cracking of, the section. This is done with the idea that the maximum live load will be applied very seldom. The beam can crack if necessary under that condition, but when this load is removed, the crack will close again. Perhaps such a scheme will result in a more economical member in first cost than will a fully prestressed design. However, the long-term service and the ultimate safety of the member may be open to question. Until more is known about this question, one will be wise to avoid partial prestressing in most cases because the live loads are often increased in spite of the designer's original intentions.

13-4. Methods for Pretensioning. Since pretensioning requires that large tensile forces in the steel must be resisted initially by the forms or by some strong anchorage, this method is more suitable for work done in the shop than it is for large jobs built in place in the field.

The manufacture of precast prestressed-concrete products seems to

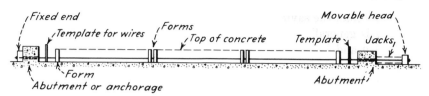

Fig. 13-7   Principles of equipment for multiple casting of pretensioned members, Hoyer method.

have tremendous possibilities.   It is possible for one to buy various standard sizes of members with predetermined dimensions and strengths just as he can purchase steel beams or heavy timbers.   He can then design his structures to use these standard sizes, thus taking advantage of both precasting and prestressing.

One illustration of shop methods of production is given in Fig. 13-7, which pictures an arrangement for casting prestressed slabs about 2 ft wide.   Concrete or other forms may be used as a base.   When multiple units are to be cast with several in a series, the anchorages or abutments may be made strong enough by·themselves to resist the pull without a complete intervening form to resist the compression.   The ends are designed to permit connection of the wires or tendons to some sort of head. One end may be stationary, and the stretching may be done by jacks at the other end with the reaction and bending resisted by the forms, or both ends may be equipped with jacks.   The movable end is then blocked to hold the tension.   Templates may be used to hold the wires in position beyond the end forms.   Bulkheads or intermediate forms are used to separate the pieces into the desired lengths.   The forms are oiled.   The concrete is cast, compacted, and cured.   Then the jacks are reapplied; the tension is released; the wires or tendons are cut, and the end and side forms are removed.   The slabs are then lifted off the bottom and stored.

By the use of high-early-strength cement, possibly by use of the vacuum process to remove excess water, and by steam curing, the attainment of proper strength in the concrete can be expedited so that reuse of the forms can be made fairly efficient.   Thus, even though the initial cost of the forms may be large, this is offset by their repeated use and long service. Vacuum lifters may be used for slabs and wall panels, but beams and girders are usually made with some sort of lifting loops or bolts for handling these products both at the plant and in the field.   With steam curing, the high-strength concrete will develop sufficient resistance to permit removal from the forms in a surprisingly short time.   The usual tendons made of twisted wires will develop enough bond resistance without mechanical anchorages—in fact, the latter are very troublesome, so that they are seldom used.   One can realize that excellent control of the manufacturing operations is essential for good results.

One advantage of pretensioning is the thorough encasement and protection of the reinforcement.

Since the bond is the primary means of the transfer of stress from the steel to the concrete, wires are preferable to bars because of their large surface area compared with their cross-sectional area. A diameter of 0.2 in. is a preferred maximum. The bond on the twisted wires of the usual tendons seems to be very effective.

The Code specifies that three or seven wire prestensioning strands should be bonded to the concrete from the cross section being considered for a minimum distance in inches of

$$(f_{su} - \tfrac{2}{3}f_{se})D \tag{13-1}$$

where D is the nominal diameter of the strand in inches and f_{su} and f_{se} are given in kips per square inch. This is of greatest importance near the ends of a member, and it should be investigated if an end is supposed to develop its ultimate strength under the action of the specified ultimate load. This action of the bond is called *prestress transfer*. The possible character of the rate of development of the tension in a bonded tendon might look something like the curve shown in Fig. 13-9(a). The bond required to develop flexural action is also important. It is likely to be most critical in short members with large loads and at interior supports of continuous members where large bending moments as well as shearing forces exist, with the possibility of tensile cracking.

Another method for getting prestress which seems to have considerable possibilities is the use of expansive cements.[1] A cement of the type described in Art. 1-10 but containing substantially more $C_4A_3\bar{S}$ is capable of being a self-stressing cement; one which expands sufficiently to tension suitable reinforcement to levels associated with mechanical prestressing. Steel stresses in excess of 100,000 psi have been achieved with this type of cement.[2] It is presently the subject of detailed experimental investigation and offers the promise of allowing prestressing to be applied to shapes and forms not now possible by mechanical means.

It seems that the general procedure for attaining prestress by this means is to erect the tendons in the forms with some medium to prevent bond to the forms. As the concrete hardens and expands, it bonds to the tendons and forces them to stretch, thus producing tension in the tendons and compression or prestress in the concrete. This is referred to as *self-stressing cement* or *chemical prestressing*. It has been used experimentally in some pavement slabs,[3] and it seems to have been successful. In this one case, the expansion apparently was 0.27 per cent of the original length

---

[1] See Art. 1-10 for general data about expansive cement.

[2] T. Y. Lin and A. Klein, Chemical Prestressing of Concrete Elements Using Expanding Cements, *J. ACI*, September, 1963.

[3] Data from C. W. Blakeslee & Sons, Inc., New Haven, Conn.

in spite of the stress in the tendons.   Much further development work is required, but the prospects for useful application seem to be good.

13-5.   Methods for Posttensioning.   The posttensioning of the steel has many advantages.   The concrete of the member can be used to resist the reaction from the tension in the steel.   The losses from shrinkage are largely eliminated so that less steel or less highly stressed steel can be employed to obtain a given compressive force on the concrete. Members can be cast in the shop, on the ground, or in place in the field, and the forms can be removed as soon as the strength of the concrete permits it.   The magnitude of the tension can be controlled without being influenced by shrinkage of uncertain extent.   Large tensile forces can be used without danger of slippage at or near the ends of the reinforcement.   The steel can also be curved as shown in Fig. 13-4(b), or inclined as in Fig. 13-13.

There are some disadvantages in posttensioning.   The reinforcement may be coated with some material of the nature of grease to allow it to slip if it is installed prior to pouring of the concrete.   This is troublesome and may not always be perfectly effective.   If holes are made in the concrete during casting so that the steel can be inserted later, some type of form is needed.   Cardboard ducts, thin metallic tubes, rubber "bars" that can be pulled out, and hoses that can be expanded by pressure and later released—these are some of the materials that may be used for forming these holes.   The tendons may be pulled through the holes.   Again individual wires may be enclosed as groups in light metallic or other sheaths before the concrete is poured.   This is a good way to do the work.   When the steel is finally installed and stressed, the clear space around and between the wires that constitute the reinforcement should be grouted for protection of the steel against corrosion.

The details of end anchorages and the spaces required for them are matters that need careful study.   Figure 13-8(a) shows an upset threaded bar of high-strength steel that is installed in the concrete and coated with grease.   The posttensioning is secured by tightening the nuts and stretching the rod inside the concrete.   This is not a practical arrangement for many cases.

Figure 13-8(b) illustrates the general idea of the sandwich-plate anchorages used by Magnel.   These can anchor several wires simultaneously, and they seem to be very effective.   The wires are placed in layers (four in a row in the illustration).   Some desired number is determined by the load requirements or the tensioning equipment.   These can be encased in a metal sheath or tube, the end of which may be flared to permit the wires a to spread apart so as to fit into the wedge-shaped slots in the plates b.   The wires are long enough to be gripped by the jack head.

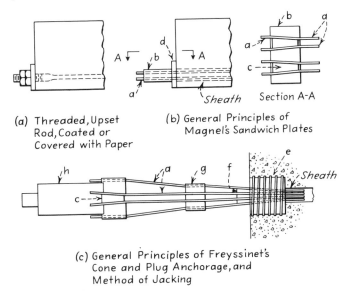

(a) Threaded, Upset
Rod, Coated or
Covered with Paper

(b) General Principles of
Magnel's Sandwich Plates

(c) General Principles of Freyssinet's
Cone and Plug Anchorage, and
Method of Jacking

***Fig.* 13-8**　Some methods of posttensioning.

The jacks stretch the wires as necessary, reacting through bearing plate *d* against the concrete of the member. The wedges *c* are driven or pressed in to grip the wires, and then the jacks are removed, and the excess length of the wires is cut off. The inside of the sheaths around the wires is grouted through holes left for that purpose. Generally the anchorages are finally encased in concrete for protection.

The conical anchorages used by Freyssinet are indicated in simplified form in Fig. 13-8(*c*). Such details as these require considerable study. The manufacturers of such materials should be consulted regarding their use. However, the principles are simple. Only the general features will be explained.

Freyssinet uses a concrete cylinder *e* with steel spirals in it to prevent bursting. It has a conical hole through which the parallel wires are splayed over a guide *g* on the jack *h*. The wires are anchored temporarily by wedges *c* to the body of the jack. The jack reacts against the concrete of the member. When the wires are tensioned, the jack holds them while a secondary jack plunger inside rams the conical plug *f* into the conical hole in the cylinder. The wires are then gripped between the two surfaces, one or more of which has the equivalent of "teeth" to help hold the wires. The jack is then released; the wires are cut, and the end may now be encased in concrete.

One device for tightening strands or tendons for posttensioned construction is shown in Fig. 13-18. The tendons are stressed by means of the jack which is calibrated so as to ensure the proper stress; then the nut is tightened to maintain the stress so that the jack can be removed.

13-6. General Data. Some general assumptions to be applied to prestressed-concrete design are as follows:

1. Strains vary linearly with depth for all loads.

2. Concrete will not resist tension at cracked sections.

3. In computing the properties of a section prior to posttensioning, the areas of open ducts or tubes are to be omitted. After the posttensioning tendons are stressed and grouted, the transformed area of the tendons may be included. In pretensioned members, the transformed section of the wires or strands may be included, but is often omitted when computing I.

4. The modulus of elasticity and the modular ratio of concrete should be taken the same as for standard reinforced concrete.

5. The modulus of elasticity of prestressing steel is to be that supplied by the manufacturer or ascertained by tests.

The following should be considered when estimating the loss of prestress to determine the effective prestress:

1. Slip at anchorage. This may be negligible in posttensioned members. In pretensioned ones, there may be some slipping close to the ends of bonded strands and wires, but the development of the tension in the tendons may be somewhat as indicated in Fig. 13-9(a),[1] or better. Of course, excessive rust, dirt, oil, or grease will be harmful because of weakening of the bond to the steel.

2. Elastic shortening of the concrete. The deformation to consider

[1] See J. R. Janney, Nature of Bond in Pre-tensioned Prestressed Concrete, *J. ACI*, May, 1954.

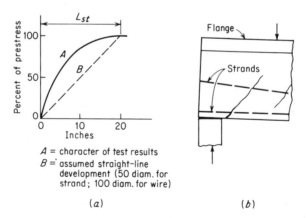

A = character of test results
B = assumed straight–line
     development (50 diam. for
     strand; 100 diam. for wire)

(a)   (b)

Fig. **13-9** Bond on strands.

is that of the concrete "along" the strands or wires. It may be different at the top and at the bottom of a beam because of curvature.

3. Creep of concrete. This is the plastic flow referred to in connection with Fig. 1-5.

4. Shrinkage of concrete. This is not only the chemical shrinkage but the shrinkage due to drying of the concrete.

5. Relaxation of steel stress—sometimes referred to as creep of the steel. This is usually small unless the stresses are very close to the elastic limit, which may not be very well defined, but it might come about if there is fatigue or temporary overstressing.

6. Frictional loss because of curvature. This applies to posttensioned members and will be discussed later on. These friction losses may be estimated as follows:

$$T_o = T_x \epsilon^{KL + \mu\alpha} \tag{13-2}$$

When $(KL + \mu\alpha)$ does not exceed 0.3, T_o may be assumed to be

$$T_o = T_x(1 + KL + \mu\alpha) \tag{13-3}$$

The total loss of prestress caused by the preceding actions may be about 20 per cent.[1]

The specified allowable unit stresses in concrete are as follows:[2]

1. Temporary stresses immediately after transfer of prestress force to the concrete but before creep and shrinkage losses:

a. Compression $= 0.60 f_{ci'}$

b. Tensile stresses in members without auxiliary reinforcement in the tension zone, whether prestressed or not, $= 3 \sqrt{f_{ci'}}$. If the computed tension exceeds this, provide steel to resist the total tension in the concrete, assuming an uncracked section.

2. Stresses at design loads after allowance for loss of prestress:

a. Compression $= 0.45 f_c'$.

---

[1] AASHO: pretensioned members $= 35,000$ psi; posttensioned $= 25,000$ psi.

[2] AASHO allowable unit stresses where different from Code:

  Concrete ($f_c' = 5,000$ psi):
    Temporary before losses:
      Compression:
        Pretensioned $= 0.60 f_{ci'}$
        Posttensioned $= 0.55 f_{ci'}$
      Tension: no nonprestressed steel:
        Single element $= 3 \sqrt{f_{ci'}}$
        Segmental element $= 0$
    Design load:
      Compression $= 0.40 f_c'$
      Tension $= 0$
    Cracking stress:
      From tests or $7.5 \sqrt{f_c'}$

b. Tension in the precompressed tension zone $= 6\sqrt{f_c'}$. This applies to members that are not exposed to freezing or a corrosive environment and that have bonded prestressed or unprestressed reinforcement located so as to control cracking.

c. All other members $= 0$.

3. Bearing stress f_{cp} under posttensioning anchorages with adequate end reinforcement:

$$f_{cp} = 0.6f_{ci'}\ \sqrt[3]{A_b'/A_b} \qquad (13\text{-}4)$$

but not over $f_{ci'}$.

The allowable unit stresses in the prestressing steel are:

1. Temporary stresses:

a. Due to temporary jacking force $= 0.80f_s'$ but not over that specified by the manufacturers of the steel and anchorages.

b. Pretensioned tendons immediately after transfer of stress to the concrete, or posttensioned tendons right after anchoring $= 0.7f_s'$.

2. Effective prestress $= 0.60f_s'$ or $0.80f_{sy}$, whichever is smaller.

13-7. Analysis of a Pretensioned Tee by WSD. Assume a Lin tee having the dimensions shown in Fig. 13-10. There are 12 strands ($A_s = 0.109$ in.² each) having their resultant 4.5 in. above the bottom of the stem or web. The allowable compressive stress in the concrete is 2,000 psi; the tensile stress, 600 psi. The tees are to be used in a roof

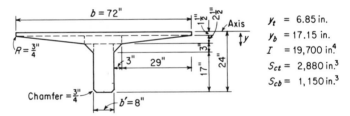

Part	A	y	Ay	Ay^2	I_0
$72''$ ‖$1\frac{1}{2}''$	$72 \times 1.5 = 108$	0.75	81	61	$\dfrac{72 \times 1.5^3}{12} = 20$
$29''\ 2\frac{1}{2}''\ 29''$	$29 \times 2.5 = 72$	2.33	168	392	$\dfrac{58 \times 2.5^3}{36} = 25$
$14''$ ‖$2\frac{1}{2}''$	$14 \times 2.5 = 35$	2.75	96	264	$\dfrac{14 \times 2.5^3}{12} = 18$
$3''\ 3''\ 3''$	$3 \times 3 = 9$	5.00	45	225	$\dfrac{6 \times 3^3}{36} = 5$
$8''$ ‖$20''$	$20 \times 8 = 160$	14.00	2,240	31,360	$\dfrac{8 \times 20^3}{12} = 5,333$
	$\Sigma = 384$		2,630	32,302	5,401

Fig. **13-10** Properties of plain section.

and spaced 6 ft c.c.   They have a span of 45 ft and are simply supported. The live load and roofing are to be 50 psf.   Determine the extreme fiber stresses at the time of transfer of stress, also with total load.   Assume that the strength of one strand at transfer is 18 kips and that the effective strength is 15 kips after all losses.   Also compute the over-all safety factor with respect to dead load and live load if the tensile stress in the concrete is not to exceed 600 psi.

The first step is to compute the properties of the concrete section alone as shown in Fig. 13-10, neglecting the strands.   The reference axis is at the top of the flange and y represents the lever arms of the parts.   Using these data, and letting y_t = the distance from the top to the center of gravity and y_b = the distance from the bottom to this center of gravity, find

$$y_t = \frac{\Sigma Ay}{\Sigma A} = \frac{2630}{384} = 6.85 \text{ in.} \qquad y_b = 24 - 6.85 = 17.15 \text{ in.}$$

The moment of inertia about the center of gravity will be

$$I = \Sigma I_o + \Sigma Ay^2 - Ay_t^2 = 5,401 + 32,302 - 384 \times 6.85^2 = 19,700 \text{ in.}^4$$

Then the section moduli in terms of concrete are

$$S_{ct} = \frac{19,700}{6.85} = 2,880 \text{ in.}^3 \text{ for the top}$$

$$S_{cb} = \frac{19,700}{17.15} = 1,150 \text{ in.}^3 \text{ for the bottom}$$

The weight of the tee alone is approximately

$$\frac{384 \times 150}{144 \times 1,000} = 0.4 \text{ kips/ft} \qquad \text{or} \qquad \frac{400}{6} = 67 \text{ psf}$$

This weight of the tee alone, with a span of 45 ft, would cause a bending moment of

$$M = \frac{0.4 \times 45^2 \times 12}{8} = 1,215 \text{ in.-k}$$

This will come into action when the prestressing jacks are released and the member tends to bow upward because of the compression of the bottom. The dead load will tend to make the tee sag down again.   If this dead load were acting alone, it would cause the following in the concrete if the latter were able to resist the stresses:

$$f_t = \frac{1,215,000}{2,880} = 442 \text{ psi compression in the top}$$

$$f_b = \frac{1,215,000}{1,150} = 1,056 \text{ psi tension in the bottom}$$

Now take the strands into account. Since their resultant is 4.5 in. above the bottom, the eccentricity e is $17.15 - 4.5 = 12.65$ in. At transfer, the total pull in the 12 strands is $P = 12 \times 18 = 216$ kips. Then the bending moment caused by the eccentricity of this pull is $Pe = 216 \times 12.65 = 2,730$ in.-k. Then, for this condition, the stresses are

$$f_t = -\frac{P}{A} + \frac{Pe}{S_{ct}} = -\frac{216,000}{384} + \frac{2,730,000}{2,880} = 386 \text{ psi tension in the top}$$

Similarly,

$$f_b = -\frac{P}{A} - \frac{Pe}{S_{cb}} = -\frac{216,000}{384} - \frac{2,730,000}{1,150}$$
$$= 2,936 \text{ psi compression in the bottom}$$

The dead-load stresses are shown in Fig. 13-11(*b*); those due to P, in (*c*). Combining these gives the results shown in sketch (*d*) for the temporary condition immediately after transfer.

After the various losses are assumed to have occurred, P will be less. It is $P = 12 \times 15 = 180$ kips. Then $Pe = 180 \times 12.65 = 2,280$ in.-k. Then the stresses which this reduced value would cause are

$$f_t = \frac{-P}{A} + \frac{2,280,000}{2,880} = 323 \text{ psi tension in the top}$$
$$f_b = \frac{-P}{A} - \frac{2,280,000}{1,150} = -2,452 \text{ psi compression in the bottom}$$

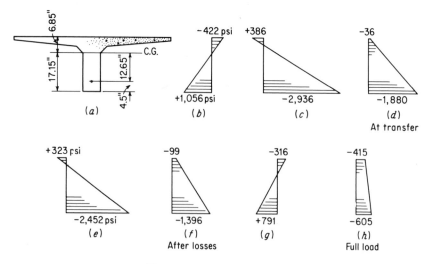

Fig. 13-11   Stresses in tee.

These values are shown in sketch (e). Combining (e) with (b) gives the values shown in (f). This represents the stress condition without live load and roofing after the losses have occurred.

Now consider the live load and roofing. The load per foot will be $50 \times 6 = 300$ plf. This will produce a bending moment of

$$M_{\text{LL}} = \frac{300 \times 45^2 \times 12}{8} = 910{,}000 \text{ in.-lb}$$

It will cause the following stresses:

$$f_t = \frac{910{,}000}{2{,}880} = 316 \text{ psi compression in the top}$$

$$f_b = \frac{910{,}000}{1{,}150} = 791 \text{ psi tension in the bottom}$$

These are shown in (g). Now, combining (g) with (f) gives the stresses shown in sketch (h). These are conservative and represent the ordinary service conditions.

The last step is to determine the over-all safety factor. To do this, the possible 600 psi tension in the bottom will be the critical stress. If this is added to the 605 psi compression in (h), it gives a possible effective stress of 1,205 psi tension. Then the accompanying moment would be

$$M = f_b \times S_{cb} = 1{,}205 \times 1{,}150 = 1{,}386{,}000 \text{ in.-lb}$$

Therefore, $w \times 45^2 \times \frac{12}{8} = 1{,}386{,}000$ or $w = 456$ plf or 76 psf. Since the total $\text{DL} + \text{LL} = 67 + 50 = 117$ psf, the ultimate allowable $w = 117 + 76 = 193$ psf. Hence, the safety factor $= 193/117 = 1.65$.

13-8. Analysis of a Pretensioned Tee by USD. Assume the same pretensioned tee as in Figs. 13-10 and 13-11(a), using the same span, loading, and materials as in Art. 13-7. Assume that the ultimate load is to be $U = 1.5\text{DL} + 1.8\text{LL}$. Let $f'_s = 250{,}000$ psi, $f_y = 40{,}000$ psi, $f'_c = 5{,}000$ psi, and $\phi = 0.9$. Is the proposed member safe in all respects?

$$U = 1.5 \times 67 + 1.8 \times 50 = 190 \text{ psf} \qquad \text{or} \qquad 190 \times 6 = 1{,}140 \text{ plf}$$

$$M_u \text{ required} = \frac{1{,}140 \times 45^2 \times 12}{8} = 3{,}470{,}000 \text{ in.-lb}$$

Figure 13-12(a) shows one way to start in this case. The tee is shown with an imaginary crack when the strands are fully stressed because their elongation will be considerable. With such a wide flange, it would seem that the resultant compressive force C_c not only is equal to T_s and is acting in the opposite direction but that it must be very near the top. Call it 1 in. from the top, as shown. The average thickness of the flange

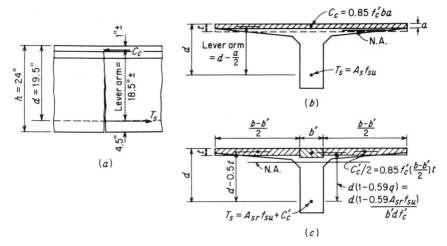

Fig. 13-12

can be taken as $(1.5 + 4)/2 = 2.75$ in.   Then it would seem that the neutral axis will be in the flange.   Incidentally, this neutral axis is not the same one as shown in Fig. 13-10 but is dependent upon the strains of the steel at the bottom and the concrete at the top.   As before, the center of gravity of the strands is 4.5 in. above the bottom.

In Chap. 2, the formula for the ultimate strength in flexure for rectangular sections (and approximately for flanged members with the neutral axis in the flange) was found to be

$$M_u = \phi[bd^2f'_c q(1 - 0.59q)] = \phi\left[A_s f_y\left(d - \frac{a}{2}\right)\right] \qquad (2\text{-}18)$$

where
$$a = \frac{A_s f_y}{(0.85f'_c b)} \qquad (2\text{-}16)$$

and
$$q = \frac{A_s f_y}{(bdf'_c)} \qquad (2\text{-}17)$$

If q is substituted in part of the left side of Eq. (2-18),

$$M_u = \phi\left[\frac{bd^2f'_c(A_s f_y)(1 - 0.59q)}{bdf'_c}\right] = \phi[A_s f_y d(1 - 0.59q)]$$

Then, with f_{su} for the strands replacing f_y, Eq. (2-18) can be written as

$$M_u = \phi[A_s f_{su} d(1 - 0.59q)] = \phi\left[A_s f_{su}\left(d - \frac{a}{2}\right)\right] \qquad (13\text{-}5)$$

This is the Code formula for this case.

If the neutral axis is outside of the flange of a flanged member, Eq. (2-37) gave

$$M_u = \phi \left[(A_s - A_{sf})f_y \left(d - \frac{a}{2} \right) + A_{sf}f_y(d - 0.5t) \right] \quad (2\text{-}37)$$

Substituting A_{sr} for $(A_s - A_{sf})$, f_{su} for f_y, the value of A_{sf} from Eq. (2-38), and the value of a from Eq. (2-39), Eq. (2-37) becomes

$$M_u = \phi \left[A_{sr}f_{su}d \left(1 - \frac{0.59A_{sr}f_{su}}{b'df_c'} \right) + 0.85f_c'(b - b')t(d - 0.5t) \right] \quad (13\text{-}6)$$

This is the Code formula for this second case.[1]

An estimate of A_{sr} can be made by using Eq. (2-38), which is

$$A_{sf} = \frac{0.85(b - b')tf_c'}{f_y} \quad (2\text{-}38)$$

and subtracting it from A_s after changing the symbols.   Then

$$A_{sr} = A_s - \frac{0.85f_c'(b - b')t}{f_{su}} \quad (13\text{-}6a)$$

Before this equation can be used, it is necessary to locate the neutral axis and to find f_{su}.

The Code states that the neutral axis will probably lie in the flange if the latter's average thickness is less than

$$\frac{1.4dpf_{su}}{f_c'} \quad (13\text{-}7)$$

In this particular problem,

$$p = \frac{A_s}{bd} = 12 \times \frac{0.109}{72 \times 19.5} = 0.00093$$

The Code states that, for bonded members for which f_{su} is not known, and provided that the effective steel prestress after losses, f_{se}, is not less than $0.5f_s'$, the following approximate values may be used:

$$\text{Bonded members:} \quad f_{su} = f_s' \left(1 - \frac{0.5pf_s'}{f_c'} \right) \quad (13\text{-}8)$$

$$\text{Unbonded members:} \quad f_{su} = f_{se} + 15{,}000 \text{ psi} \quad (13\text{-}9)$$

Therefore, using Eq. (13-8), $p = 0.00093$, and $f_s' = 250{,}000$ psi,

$$f_{su} = 250{,}000 \left(\frac{1 - 0.5 \times 0.00093 \times 250{,}000}{5{,}000} \right) = 244{,}000 \text{ psi}$$

[1] Figure 13-12(b) shows the conditions for Eq. (13-5) and how it applies. The stress block for the compression force C_c is shown hatched. Figure 13-12(c) is made similarly to illustrate the application of Eq. (13-6).

Now, substituting these numbers in Eq. (13-7) gives

$$\frac{1.4 \times 19.5 \times 0.00093 \times 244,000}{5,000} = 1.24 \text{ in.}$$

This is less than the average $t = 2.75$ in., so that Eq. (13-5) can be used. However, find a first from

$$a = \frac{A_s f_{su}}{0.85 f_c' b} \tag{13-10}$$

$$a = \frac{1.31 \times 244,000}{0.85 \times 5,000 \times 72} = 1.05 \text{ in.}$$

Then Eq. (13-5) gives

$$M_u = \phi \left[A_s f_{su} \left(d - \frac{a}{2} \right) \right] = 0.9 \left[1.31 \times 244,000 \left(19.5 - \frac{1.05}{2} \right) \right]$$
$$= 5,450,000 \text{ in.-lb}$$

This is considerably greater than the 3,470,000 in.-lb required by use of the assumed load factors. It is also far above the value of the maximum $M = 1,386,000$ in.-lb determined for the possible cracking point when this tee was analyzed by WSD in the preceding article. How does this come about?

It is partly due to the difference in the actual safety factors used in the two cases, but the principal cause seems to be the upward shifting of the neutral axis after severe cracking has occurred. With $a = 1.05$ in. as found from Eq. (13-10), this means that the neutral axis has shifted almost to the top of the tee. The lever arm of the tensile steel is thus greatly increased, and the member is acting pretty much like a conventional reinforced-concrete beam. The designer would be wise to consider the real measure of the safety factor of a pretensioned member to be the ratio of the load at severe cracking to the design working load. Severe cracking is that which would destroy the usefulness of the member.

Ordinary reinforcing bars are often used longitudinally in prestressed beams in addition to the strands. These bars may serve as ties for any web reinforcement, as bars to help resist initial tension near the top, or as an aid if cracking occurs. Any such bars which are not prestressed, when used with prestressed steel, may be assumed to contribute to the tensile resistance of the member at ultimate load, the amount being equal to the area of this steel times its yield-point stress, provided that $p f_{su}/f_c' + p' f_y/f_c'$ does not exceed 0.3.

The Code places the following limitations on the percentage of prestressing steel:

1. The ratio of prestressing steel used in computing M_u should be such that $p f_{su}/f_c'$ does not exceed 0.3. For flanged members, assume p is the

steel ratio found from only the portion of the total tensile steel area that is needed to develop the compressive strength of the web alone.

2. If the steel ratio used exceeds that specified in item 1, assume that M_u should not exceed the following:

a. Rectangular sections or flanged members with the neutral axis in the flange:

$$M_u = \phi[0.25f'_c bd^2] \tag{13-11}$$

b. Flanged sections with the neutral axis outside of the flange:

$$M_u = \phi[0.25f'_c b'd^2 + 0.85f'_c(b - b')t(d - 0.5t)] \tag{13-12}$$

3. The total prestressed and unprestressed steel should be enough to develop M_u at least 1.2 times the cracking moment computed on the basis of a modulus of rupture of 7.5 $\sqrt{f'_c}$.

This problem has purposely been solved by both WSD and USD in order to point out the difference in the procedures and answers, and to caution the reader to be sure that his structure will be safe no matter which method he uses.

In the WSD solution of the problem, the shear in the web was neglected. This feature will be investigated now because USD methods are used for computing shearing resistance.

It may be assumed that the portion of a member which resists shear is confined to the web and to its projection in the flange areas. For example, in Fig. 13-6(a) the resisting area may be assumed to equal a 3-in. strip 36 in. deep unless there are holes in the web that should be deducted or unless the holes for posttensioning reinforcement render the area beyond them ineffective.

The longitudinal shearing stress in a beam of homogeneous material is a maximum at the neutral axis. Its magnitude may be computed by the formula

$$v = \frac{VQ}{Ib} \quad \text{or} \quad \frac{VQ}{Ib'} \tag{13-13}$$

where Q = static moment of the area of the cross section one side of the plane being considered

I = moment of inertia of the entire cross section of the uncracked concrete

b = width of a rectangular member

b' = width of the web of a T or I member

However, the results of such computations may have little significance in a prestressed-concrete beam because of the great shearing resistance of uncracked concrete. If the member is stressed until it cracks, then an unreinforced thin web probably will be in difficulty.

The transfer length at the end of a pretensioned beam may be critical, and cracks might form as indicated in Fig. 13-9(*b*).   This danger of slippage does not exist at the end anchorages of posttensioned members because the tendons cannot slip.   However, one must be careful that the concrete under the bearing plates does not crush or spall off.   The webs at the ends are usually enlarged for this purpose.

If some of the prestressing tendons are curved or sloped upward, they will have a vertical component that assists in the resistance to transverse shearing.

Diagonal cracks due to the tensile resultant of the shearing and normal bending stresses usually precede a shear failure.   If the computations show that a shear failure will not occur before a flexural failure at ultimate load for a given case, the shearing and principal stresses at working loads need not be investigated.   It is obvious that compressive prestress will reduce the principal tensile stress and increase the load-carrying capacity before diagonal cracking will occur.

Generally, some web reinforcement is desirable and shear failure should not take place before the ultimate strength in flexure is reached.   The Code states that the area of web reinforcement placed perpendicular to the axis of a member should not be less than

$$A_v = \frac{(V_u - \phi V_c)s}{\phi\, df_y} \quad \text{or} \quad \phi A_v f_y = \frac{(V_u - \phi V_c)s}{d} \tag{13-14}$$

based on ultimate-strength principles, or

$$A_v = \frac{A_s}{80} \times \frac{f_s'}{f_y} \times \frac{s}{d} \sqrt{\frac{d}{b'}} \tag{13-15}$$

The similarity between Eq. (13-14) and Eq. (4-13) for conventional reinforced concrete should be noticed.   In the former $V_u - \phi V_c$ represents the force which the stirrups have to resist at a cross section of the beam.   Then $(V_u - \phi V_c)/d$ can be looked upon as the force per inch of web vertically and horizontally, and when multiplied by s, it gives the force per stirrup.   The principles used in the development of Eq. (4-12) can then be employed again, showing that $\phi A_v f_y$ will vary as s multiplied by the shear per inch.

The effective depth d in Eqs. (13-14) and (13-15) is as follows:

1. If the member has a constant over-all depth, d is the effective depth at the section having the maximum bending moment.   The length of the stirrups at the section being considered should be at least equal to that of the stirrups at the point of maximum bending.

2. In members having varying depth, $d = h(d_m/h_m)$, where d_m and h_m are the effective depth and total depth, respectively, at the section having maximum moment, and h is the over-all depth at the section being con-

sidered. The stirrups should extend into the member at least a distance d from the compression face.

Provide web reinforcement not less than that required by Eq. (13-15) at all sections at spacing not over three-fourths of the depth of the member or 24 in. maximum, using a yield strength not exceeding 60,000 psi.

The shear V_c carried by the member at the condition where diagonal cracking may occur should be taken as the smaller value V_{ci} or V_{cw}, as computed from Eqs. (13-16) and (13-18).

$$V_{ci} = 0.6b'd \sqrt{f'_c} + \frac{M_{cr}}{M/V - d/2} + V_d \qquad (13\text{-}16)$$

but not less than $1.7b'd \sqrt{f'_c}$, where d is the distance from the extreme compression edge to the centroid of the prestressing tendons and M/V is to be for the maximum moment conditions at the section due to external loads. Assume

$$M_{cr} = \frac{I}{y} (6 \sqrt{f'_c} + f_{pc} - f_d) \qquad (13\text{-}17)$$

Equation (13-17) is really a form of $M = fI/c$. The critical point in this tee is the bottom fiber because it will crack first. The intensity of stress at this point is that specified for the ultimate strength of concrete in shear or diagonal tension, $6 \sqrt{f'_c}$, with web reinforcement plus the compression produced by the prestress minus the tension at the bottom caused by the dead-load moment.

The formula for V_{cw} is

$$V_{cw} = b'd(3.5 \sqrt{f'_c} + 0.3f_{pc}) + V_p \qquad (13\text{-}18)$$

where d is the same as in Eq. (13-16) or 80 per cent of the over-all depth, whichever is greater. With this formula, the maximum v_c allowed by the Code without web reinforcement is $3.5\phi \sqrt{f'_c}$, which is the same limitation as for Eq. (4-11) for conventional reinforced-concrete beams. If ϕ is neglected, the shear value of the concrete may be called $3.5 \sqrt{f'_c}$, as shown in Eq. (13-18). The stress f_{pc} is caused by the prestress after losses. It will add to the resistance of the section to transverse shearing forces, but the coefficient 0.3 is a factor to be applied when considering the entire area $b'd$. Thus the first term of Eq. (13-18) gives a maximum value for plain concrete plus the extra from prestressing. Of course, the uplift V_p of sloping strands will help.

As an alternative for Eq. (13-18), V_{cw} may be assumed to be the DL plus LL shear which corresponds to the occurrance of a principal tensile stress of $4 \sqrt{f'_c}$ at the centroidal axis of the section resisting live load unless this axis lies in the flange, in which case the principal tensile stress should be computed at the junction of the web and flange.

In a pretensioned beam, one may assume that the transfer length is 50 diameters for strands and 100 diameters for single wires. If the transfer length is greater than $d/2$ from a support of the member, the reduced prestress in the transfer length should be considered when computing V_{cw} on the basis of a straight-line variation from zero at the end to a maximum at the end of the transfer length. In any case, determine the web reinforcement at $d/2$ from the end, then use this same reinforcement in the web from this section to the face of the support, anchoring both ends of the stirrups. Also, at least provide web reinforcement for a distance d beyond the point where it might otherwise be omitted.

In order to avoid the high compressive stress shown in Fig. 13-11(c), it is desirable to slope up some of the strands near the ends. Assume that the bend point is located at the quarter points of the span and that the eccentricity at the ends is 6 in., as indicated in Fig. 13-13(a). Then d at the extreme end is 12.85 in. from the top fibers to the center of gravity of the force in the tendons. Since the critical section for shear will be taken at about $d/2$ from the face of the support, call this point 12 in. from the center of the bearing or 18 in. from the end. Then e at this point $= 6 + 6.65 \times 1.5/11.75 = 6.85$ in. Therefore, let

$$d = 6.85 + 6.85 = 13.7 \text{ in.}$$

and

$$Pe = 12 \times 15,000 \times 6.85 = 1,230,000 \text{ in.-lb}$$

Then

$$f_b = -P/A - Pe/S_{cb} = -180,000/384 - 1,230,000/1,150 = -1,540 \text{ psi}$$

compression which is safe. Similarly,

$$f_t = -180,000/384 + 1,230,000/2,880 = -43 \text{ psi compression}$$

These are shown in Fig. 13-13(b).

The vertical component of the inclined pull in the tendons will help to resist the downward shears. This component after losses is

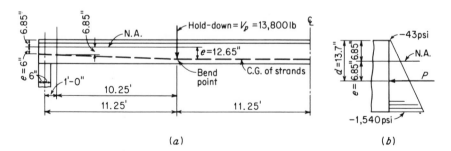

(a)   (b)

Fig. 13-13   Draped strands.

$V_p' = 12 \times 15{,}000 \times 6.65/(11.75 \times 12) = 8{,}500$ lb approximately, at working conditions. This should be multiplied by the safety factor to determine the value of V_p at ultimate loads. Since

$$DL + LL = 67 + 50 = 117 \text{ psf}$$

for WSD and $1.5DL + 1.8LL = 190$ psf for USD, call the general safety factor equal to S. F. $= 190/117 = 1.62$. Then

$$V_p = V_p' \times 1.62 = 8{,}500 \times 1.62 = 13{,}800 \text{ lb}$$

This is also the hold-down force needed at the bends.

The next problem is the design of the web reinforcement. In a case like this, some stirrups are to be used anyway at a spacing not over three-fourths of the over-all depth of the member, as required by the Code. As stated in connection with Eq. (13-16), the allowable shearing resistance of the concrete should be the smaller of the computed values for V_{ci} and V_{cw}, but not less than $1.7b'd\sqrt{f_c'}$. See what happens if this minimum is used together with the uplift due to the inclined tendons, considering both to be on the basis of USD. Test a section one foot from the center of the bearing, as shown in Fig. 13-13(a). Assume $d = 13.7$ in. from the compression edge at the top to the center of gravity of the prestressing force, as previously computed for this section. Then

$$V_c = 1.7b'd\sqrt{f_c'} + V_p = 1.7 \times 8 \times 13.7 \times 70.7 + 13{,}800 = 26{,}900 \text{ lb}$$

This is more than the actual shearing force at this section; that is, $V = 1{,}140 \times 21.5 = 24{,}500$ lb. One can see that the uplift of the tendons becomes very effective.

From the preceding computations, and from Eq. (13-14), no web reinforcing appears to be necessary. However, examine the requirements of Eq. (13-15). In item 2 of the explanation of this formula, it was stated that the effective depth to use when members have varying depths is $d = h(d_m/h_m)$. This member has a constant over-all depth but a varying effective depth. For this purpose, it will be called a member with constant depth. This is shown by the following, with $d_m = 19.5$ in. at the center and $h = h_m = 24$ in.:

$$d = 24(19.5/24) = 19.5 \text{ in.}$$

This value is apparently to be used, even though it gives a smaller value for A_v in Eq. (13-15) than would the effective depth of 13.7 in. for this section. Then, with an arbitrary spacing of 15 in. between stirrups,

$$A_v = \frac{A_s}{80} \times \frac{f_s'}{f_y} \times \frac{s}{d}\sqrt{\frac{d}{b'}} = \frac{1.31}{80} \times \frac{250{,}000}{40{,}000} \times \frac{15}{19.5}\sqrt{\frac{19.5}{8}} = 0.123 \text{ in.}^2$$

Therefore, use No. 3 U-shaped stirrups 15 in. c.c. throughout the member.

Theoretically, the full value of V_p used here should be reduced a little because the 18 in. beyond the section is not quite equal to 50 diameters of the $\frac{7}{16}$-in. strands.   However, the stirrups more than make up for this.

A comparison of the analysis of the bending resistance of this pretensioned tee of Fig. 13-10 with that for T beams by USD shown in Chap. 2 shows that both are similar.   This is chiefly because the wide flange provides so much area of concrete that, even in the pretensioned member, the bending resistance is limited by the strength of the steel.   However, when the matter of web reinforcement is examined, the two cases are very different.   Obviously, the avoidance of heavy web reinforcement in the pretensioned member is a great advantage.

13-9.   Design of a Pretensioned, Composite Tee.   Assume that a floor in a commercial building is to be made of pretensioned Lin tees having the dimensions shown in Fig. 13-10, with a 3-in. topping added to form a composite member for live load.   The tees support themselves and the topping.   Use working-stress design.   Assume the following data:

Type 6*H*24, 6 ft c.c.

Concrete for tee: $f'_c = 5,000$ psi; $E = 4,100,000$ psi; max $f_c = 2,000$ psi working stress or 2,500 psi temporary at time of transfer

Concrete for topping: $f'_c = 3,000$ psi; $E = 3,200,000$ psi; max $f_c = 1,350$ psi

Span: $L = 45$ ft c.c. bearings, simply supported

DL: tee = 0.4 klf; topping = 37.5 psf or 0.225 klf

LL: 50 psf or 0.3 klf with no tension in bottom

Strands: $\frac{7}{16}$ in. diam. with $f'_s = 250,000$ psi and area = 0.109 in.^{2} per strand

The computations will be made by slide rule and may be rounded off in some cases.

Assume the following forces in the strands:

Initial $P_i = 18.9$ kips per strand

Transfer $P_t = 18.0$ kips per strand

Final $P_f = 15.0$ kips per strand

The bending moments at mid-span are

$$\text{Tee} = \frac{400 \times 45^2 \times 12}{8} = 1,215,000 \text{ in.-lb}$$

$$\text{Topping} = \frac{225 \times 45^2 \times 12}{8} = 683,000 \text{ in.-lb}$$

$$\text{Live load} = \frac{300 \times 45^2 \times 12}{8} = 910,000 \text{ in.-lb}$$

The topping is weaker concrete than that of the tee itself. Therefore, in order to compute the properties of the transformed section of the composite member, it is desirable to convert the topping to an equivalent area of 5,000-lb concrete. This may be done by reducing the 6-ft width to make an equivalent amount of 5,000-lb concrete by multiplying the 72 in. by the ratio of $E_{3,000}/E_{5,000}$. Hence,

$$b_t = \frac{72 \times 3.2}{4.1} = 56.2 \text{ in. Call it 56 in.}$$

This is shown in Fig. 13-14. The properties of the composite member and the appropriate symbols are also shown in this figure. The data for the tee are taken from Fig. 13-10. The axis of reference for the table is the top of the topping. The moment of inertia and the section moduli or I/c are given below the table.

Use the bending moments previously computed and the section moduli for the tee ($S_{ct} = 2,880$ in.^{3} and $S_{cb} = 1,150$ in.^{3}); then compute the stresses in the extreme fibers of the tee when it is supporting its own weight and that of the topping before the latter has set.

Tee:

$f_t = 1,215,000/2,880 = -422$ psi compression in the top of the tee
$f_b = 1,215,000/1,150 = +1,056$ psi tension in the bottom of the tee

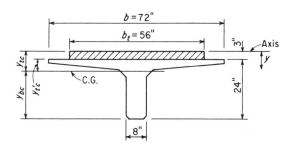

Part	A	y	Ay	Ay²	I_0
56 × 3	168	1.5	252	378	$\frac{56 \times 3^3}{12}$ = 126
Tee	384	9.85	3,780	37,300	19,700
Σ	552		4,032	37,678	19,826

$y_{tc} = 4,032/552 = 7.30$ in., $\quad y_{bc} = 27 - 7.30 = 19.70$ in., $\quad y'_{tc} = 4.30$ in.
$I_{C.G.} = \Sigma I_0 + \Sigma Ay^2 - (\Sigma A) y_{tc}^2 = 19,826 + 37,678 - 552 \times 7.30^2 = 28,100$ in.^{4}
$S'_{tc} = 28,100/7.30 = 3,850$ in.^{3}, $\quad S'_{bc} = 28,100/19.70 = 1,430$ in.^{3}

Fig. 13-14   Properties of composite section.

Topping:

$f_t = 683,000/2,880 = -237$ psi compression in top of the tee
$f_b = 683,000/1,150 = +594$ psi tension in bottom of the tee

Next, assume that the composite section of Fig. 13-14 is supporting the live loads. The tentative bending stresses at mid-span are

$$f_t = \frac{910,000}{3,850} = -236 \text{ psi compression at top of topping}$$

$$f_b = \frac{910,000}{1,430} = +636 \text{ psi tension at bottom of tee}$$

$$f_{tc} = -139 \text{ psi compression at top of tee (by proportion)}$$

These stresses are considered tentative because they have been calculated as though the concrete could withstand such stresses without cracking. Now add these three tensions in the bottom to represent the total tensile stress which should be overcome by the prestressing. Then

$$\Sigma f_b = 1,056 + 594 + 636 = 2,286 \text{ psi}$$

This stress is supposed to be counteracted by the prestressing force acting with some eccentricity e. The effect of P on the bottom fibers will be

$$f_b = \frac{P}{A} + \frac{Pey_b}{I} = \frac{P}{A}\left(1 + \frac{ey_b}{r^2}\right) \tag{13-19}$$

since $I = Ar^2$, where r is the radius of gyration.

The next step is to assume a value for P and then to see what eccentricity is needed to make $f_b = 0$ or a small compression. Since this stage is the final one after all losses have supposedly occurred, assume that each strand will have the specified pull of 15,000 lb. Then try 14 strands. The 636 psi included in the 2,286 psi total was based upon the transformed section of Fig. 13-14, whereas the remainder was based upon the tee in Fig. 13-10 acting alone. In order to be on the conservative side, assume the tee to act alone for all. Then, from the first form of Eq. (13-19), find

$$f_b = 2,286 = 14 \times 15,000/384 + 14 \times 15,000 \times e \times 17.15/19,700$$
$$e = 9.54 \text{ in.}$$

Now $y_b - e = 17.15 - 9.54 = 7.61$ in. from the center of gravity of the strands to the bottom of the stem. This places the steel farther from the bottom than necessary. The number of strands might be reduced, but in order to have some residual compression in the bottom, assume that the center of gravity of the strands at mid-span is 4.5 in. from the bottom as shown in Fig. 13-15(*a*). They will be sloped upward from hold-downs located at the quarter points of the span. At the ends of the members, the strands will be positioned as shown in Fig. 13-15(*b*).

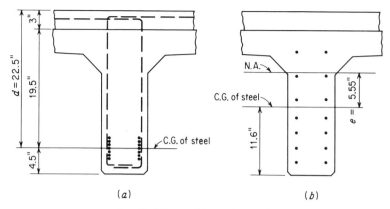

(a)   (b)

Fig. 13-15   Positions of strands.

The beam should now be analyzed to determine whether the stresses at mid-span are safe at all stages, using the selected 14 strands.

Transfer for tee alone:

$$M \text{ for DL} = 1,215,000 \text{ in.-lb} \qquad P = 14 \times 18,000 = 252,000 \text{ lb}$$

$e = 12.65$ in.; f_t for DL $= -422$ psi compression in top; and $f_b = +1,056$ psi tension in bottom.   These are shown in Fig. 13-16(a).

Prestress alone:

$$M = Pe = 14 \times 18,000 \times 12.65 = 3,190,000 \text{ in.-lb}$$

$$f_t = \frac{-P}{A} + \frac{M}{S_t} = -\frac{252,000}{384} + \frac{3,190,000}{2,880} = +452 \text{ psi tension in top}$$

$$f_b = \frac{-P}{A} - \frac{M}{S_b} = -656 - \frac{3,190,000}{1,150}$$

$$= -3,430 \text{ psi compression in bottom}$$

These stresses are not truly applicable because the tee will at least shorten as the prestress is applied.   After the prestress losses have occurred, which will soon happen, the strand loads are assumed to be 15,000 lb each.   This will reduce the stresses previously computed to the following:

$$P = 14 \times 15,000 = 210,000 \text{ lb}$$
$$Pe = 210,000 \times 12.65 = 2,660,000 \text{ in.-lb}$$
$$f_t = -210,000/384 + 2,660,000/2,880 = +377 \text{ psi tension in top}$$
$$f_b = -547 - 2,660,000/1,150 = -2,860 \text{ psi compression in bottom}$$

These are shown in Fig. 13-16(b).

Dead load combined with prestress:

$$f_t = -422 + 377 = -45 \text{ psi compression in top}$$
$$f_b = +1,056 - 2,860 = -1,804 \text{ psi compression in bottom}$$

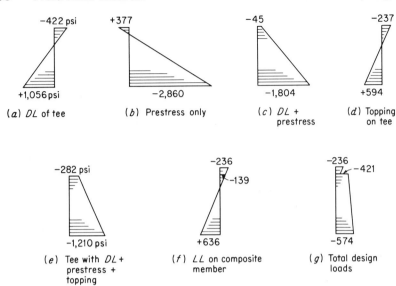

(a) *DL* of tee (b) Prestress only (c) *DL +* prestress (d) Topping on tee

(e) Tee with *DL +* prestress + topping (f) *LL* on composite member (g) Total design loads

Fig. 13-16 Stresses in composite section.

This stress f_b is less than the specified allowable 2,000 psi.   The preceding results are shown in Fig. 13-16(c).

Weight of topping on tee: From previous calculations, $f_t = -237$ psi compression in the top and $f_b = +594$ psi tension in the bottom of the tee. These are shown in sketch (d).   When these are combined with the prestress and dead load in (c), the results are as shown in (e).

Live load on composite member: The stresses due to the live load on the composite member were previously found to be as shown in Fig. 13-16(f). When these are combined with the stress in (e), the resultant stresses are as shown in (g).   These appear to be quite conservative.

Test of over-all safety factor: Using a load factor of

$$U = 1.5DL + 1.8LL$$

the ultimate load per square foot would be

1.5(67 for tee + 37.5 for topping) + 1.8(50 for LL) = 247 psf

Now, if the 574 psi residual compression in the bottom from Fig. 13-16(g) is added to the stated maximum possible tension permitted, which will be at the bottom, the possible change of stress is 574 + 600 = 1,174 psi. Then, using the properties of the composite member,

$$1,174 = \frac{(wL^2 \times 12)}{8 \times S'_{bc}} = \frac{(w \times 45^2 \times 12)}{8 \times 1,430} \qquad \text{or} \qquad w = 550 \text{ plf}$$

$$\text{or} \qquad \tfrac{550}{6} = 91.7 \text{ psf}$$

Therefore, this can be added to the actual

$$DL + LL = 67 + 37.5 + 50 = 154.5 \text{ psf}$$

giving a final total of 246.2 psf which can be withstood without exceeding the 600 psi tension. This is just a bit less than the 247 psf needed, but it is so close that it will be accepted. A review of these figures will show the reader what would have happened to the safety factor with reference to the cracking load if the number of strands were reduced. Remember that, once the tee has cracked, it will not again resist tension across the cracked area even if the removal of the load and the elastic recovery of the strands close the crack.

To get a scale on what cracks may mean in this case, assume that the loads on the composite member are just enough to cause a tensile stress of 600 psi in the bottom; whereupon the web cracks. This converts the member into something more like a conventional reinforced-concrete beam for resisting further loading. Since the steel area is so small, the theoretical neutral axis will rise much closer to the top. Even if this effect is neglected and the section modulus S'_{bc} is kept at 1,430 in.^{3}, as shown in Fig. 13-14, a loss of 600 psi in the tensile stress at the bottom means a loss of $M' = 600 \times 1{,}430 = 858{,}000$ in.-lb of resisting moment. This corresponds to a loss of

$$w = \frac{858{,}000 \times 8}{45^2 \times 12} = 282 \text{ plf} \qquad \text{or} \qquad \frac{282}{6} = 47 \text{ psf}$$

One might look upon the situation as though this amount of load-carrying capacity has been lost henceforth. Also, with the neutral axis of the cracked member way up in the flange, further loading will change the steel stresses—possibly rapidly and with considerable accompanying deflection.

The longitudinal shear between the tee and the topping should be checked. Use Eq. (13-13), with V equal to the end reaction under live load, because the ends are the points where the shearing will be at a maximum. Also, use Q = the static moment of the adjusted area of the topping about the neutral axis of the composite section as pictured in Fig. 13-14. Of course, I is also that of the composite member.

$$v = \frac{VQ}{Ib} = \frac{(300 \times 22.5)(56 \times 3 \times 5.8)}{28{,}100 \times 72} = 3.3 \text{ psi}$$

Even with the ultimate load of 247 psf minus the dead load and topping $(247 - 67 - 37.5 = 142.5 \text{ psf})$,

$$v = \frac{(6 \times 142.5 \times 22.5)(56 \times 3 \times 5.8)}{28{,}100 \times 72} = 9.3 \text{ psi}$$

This is very safe, since the Code allows 40 psi for a rough and clean surface without ties.   However, ties will be used as indicated in Fig. 13-15(*a*).

Now check the stresses at the ends due to the prestress, assuming the full 210,000 lb pretensioning force after losses if the strands are located as shown in Fig. 13-15(*b*).   The center of gravity of the strands is 11.6 in. above the bottom of the web, so that $e = 17.15 - 11.6 = 5.55$ in.   Then the moment due to prestress is

$M = 210,000 \times 5.55 = 1,165,000$ in.-lb
$f_t = -210,000/384 + 1,165,000/2,880 = -142$ psi compression at top
$f_b = -210,000/384 - 1,165,000/1,150$
$\qquad\qquad = -1,560$ psi compression at the bottom

These are satisfactory stresses.

The critical shear in the web will be for the composite member under full live load.   The shear V at $d/2$ from the support will therefore be

$$V = (400 + 225 + 300)(\tfrac{4.5}{2} - 1) = 20,000 \text{ lb (approx)}$$

The uplift produced by the strands, if the bends are at the quarter points, is again the vertical component of P.   The force P acts $11.6 - 4.5 = 7.1$ in. higher at the ends than at the bend points.   Let the horizontal distance of the slope $= 12 \times \tfrac{4.5}{4} + 6 = 141$ in. and $P = 210,000$ lb.   Therefore,

$$V_p = \frac{210,000 \times 7.1}{141} = 10,600 \text{ lb (approx)[1]}$$

Because of this large uplift, assume that the concrete is not capable of more resistance than the minimum specified as $1.7b'd \sqrt{f_c'}$, as in the preceding example.   Then, with $V_p = 10,600$ lb and $d = 22.5 - 7.1 = 15.4$ in., from Fig. 13-15,

$$V_c = 1.7 \times 8 \times 15.4 \times 70.7 + 10,600 = 14,800 + 10,600 = 25,400 \text{ lb}$$

This exceeds the 20,000 lb needed, but No. 3 stirrups like those shown in Fig. 13-15(*a*) at 15 in. c.c. will be used throughout.

Investigate the deflections in order to see how the effects of the prestress and the dead load of the tee oppose each other—it is sometimes said that "the dead load carries itself."   The computations are to be made by the use of the conjugate-beam method.   This uses the M/EI diagram as a load on a simply supported beam with the same span and moment of inertia.   For convenience, omit E until the end of the computations.   The end reaction calculated from this imaginary loading is the slope of the member's longitudinal axis at the support.   The bending moment calculated for any point is the deflection of that point from the chord joining

---

[1] The transfer length beyond this section is not quite the required 50 diameters, but the stirrups are more than adequate.

the end supports. In this case, to be conservative, use I for the concrete section, without including the steel, and $E = 4,100,000$ psi as a constant, without regard for any possible reduction because of plastic flow.

Dead load of tee: Since I of the tee $= 19,700$ in.[4] and the bending moment $= 1,215,000$ in.-lb, $M/I = 1,215,000/19,700 = 61.7$, which is the middle ordinate of the parabolic diagram shown in Fig. 13-17(a). The end reaction R_A will be one-half of the area of the M/I diagram or $0.667(61.7 \times 540/2) = 11,100$. The center of gravity of one-half of a parabola with the vertex at C is shown in sketch (c). Therefore, the fictitious bending moment at C is

$$M_C = \text{end reaction} \times AC - \text{area} \times 0.375 \times AC$$
$$M_C = 11,100 \times 270 - 11,100 \times 0.375 \times 270 = 1,873,000$$

Then the real deflection $= M_C/E$, or

$$\Delta_C = 1,873,000/4,100,000 = 0.46 \text{ in. downward}$$

Prestress after initial losses: The M/I diagram will be assumed to be as shown in Fig. 13-17(b), because of the sloping strands outside of the quarter points. Since the diagram is symmetrical, the imaginary end reaction will be

$$R_A = 135 \times 270/2 + 59 \times 135 + 0.5(135 - 59)135$$
$$R_A = 18,200 + 7,960 + 5,140 = 31,300$$

Then

$$M_C = 31,300 \times 270 - 18,200 \times 135/2 - 7,960(135 + 135/2)$$
$$- 5,140(135 + 0.333 \times 135) = 4,675,000$$

Therefore, $\Delta_C = 4,675,000/4,100,000 = 1.14$ in. upward

The net deflection under prestress and dead load is $1.14 - 0.46 = 0.68$ in. upward.

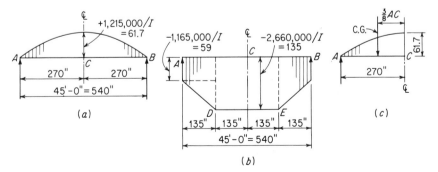

Fig. 13-17 Diagrams for estimating deflection by conjugate-beam method.

Topping: The deflection caused by the weight of the topping will be in proportion to the loading, or

$$\Delta \text{ for topping} = 0.46(225/400) = 0.26 \text{ in. downward}$$

After the topping is on, the net deflection for prestress and full dead load will be

$$\Delta \text{ net} = 0.68 - 0.26 = 0.42 \text{ in. upward (a camber)}$$

Live load: With the live load acting on the composite section for which $I = 28,100$ in.[4], the deflection will be that for the dead load of the tee alone corrected directly by the ratio of the loads and inversely by the ratio of the Is. Then

$$\Delta_{LL} \text{ for live load} = 0.46(300/400)(19,700/28,100) = 0.24 \text{ in. downward}$$

Therefore, with the live load on the member, the final deflection is

$$\Delta \text{ final} = 0.42 - 0.24 = 0.18 \text{ in. upward}$$

13-10.   Computations for a Posttensioned Girder.   Perhaps the best way to study posttensioning is to design a structure using it.   Therefore assume that Roebling prestressing cables, having end anchorages like those in Fig. 13-18, are to be used in a highway bridge of the character

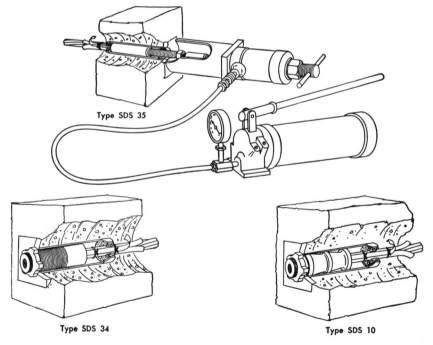

Type SDS 35

Type SDS 34

Type SDS 10

Fig. **13-18**   Jacking equipment for Roebling simplified posttensioning method.

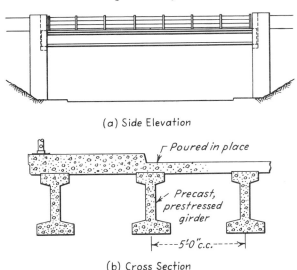

(a) Side Elevation

(b) Cross Section

Fig. **13-19** A highway overpass.

shown in Fig. 13-19.   The prestressed-concrete girders are to be I-shaped, with a concrete slab poured in place on top.   The girders are to be 5 ft c.c., simply supported, and are to carry the load of the wet concrete of the slab, any diaphragms, and the forms for the slab.   Thereafter, they will be part of a composite member.   Use AASHO specifications.   Assume that the span $= 56$ ft, $f_c' = 5{,}000$ psi, $f_c = 2{,}000$ psi, $n = 6$, $f_c' = 3{,}000$ psi for the slab, and the maximum tension allowed in the girder $= 7.5 \sqrt{f_c'} = 530$ psi at cracking load.   The strengths of various cables are shown in Table 13-1.   The cables are to be grouted after installation so that they will act with the concrete after the slab concrete has set.   The load factors are to be 1.5 for DL and 2.5 for LL at ultimate load, without exceeding the stated maximum tensile stress.   The live load plus impact will be taken as 80 psf uniform load and as a concentrated load of 2,300 lb/ft of width of lane placed so as to produce maximum bending or shear.   This corresponds to H20 loading with 0.275 for impact.   Use WSD in general.

The first step is to get a reasonably good trial section to start with before making any detailed computations.   One way to do this is to make an approximation for preliminary design, then to analyze it and to modify it as necessary.

Assume an interior girder like the one at the right in Fig. 13-19(b). The LL bending moment for full uniform live load and the concentrated load at the center will be

$$M_{\text{LL}} = \frac{(5 \times 80)56^2}{8} + \frac{5 \times 2{,}300 \times 56}{4} = 318{,}000 \text{ ft-lb}$$

$$\text{or } 3{,}820{,}000 \text{ in.-lb}$$

Table 13-1.   Roebling Galvanized Prestressed Concrete
Strand for Posttensioned Design

Diameter, in.	Weight per ft, lb	Area, in.^{2}	Minimum guaranteed ultimate strength, lb	Design load, lb
0.600	0.737	0.215	46,000	26,000
0.835	1.412	0.409	86,000	49,000
1	2.00	0.577	122,000	69,000
$1\frac{1}{8}$	2.61	0.751	156,000	90,000
$1\frac{1}{4}$	3.22	0.931	192,000	112,000
$1\frac{3}{8}$	3.89	1.12	232,000	134,000
$1\frac{1}{2}$	4.70	1.36	276,000	163,000
$1\frac{9}{16}$	5.11	1.48	300,000	177,000
$1\frac{5}{8}$	5.52	1.60	324,000	192,000
$1\frac{11}{16}$	5.98	1.73	352,000	208,000

These strands are fabricated from hot-dip galvanized wire, which assures complete protection against corrosion without further treatment.

The slab is to be 7 in. thick, weighing 88 psf or 88 $\times$ 5 = 440 plf of girder. With a 56-ft span, it would seem that the girder should be 3 or 4 ft deep, with a web of 5- or 6-in. thickness and with fairly large flanges. This may weigh more than the slab, so call it 500 plf. Then a trial M for DL is

$$M_{\text{DL}} = \frac{(440 + 500)56^2}{8} = 368,000 \text{ ft-lb}$$

The girder must resist this bending moment by itself, and it must also act as part of the composite member in supporting the live load. To start with, assume that the girder withstands the total moment because the bottom flange is likely to be stressed highly in both cases and because the prestressing is likely to make this flange the critical one. Then, for design,

$$M_G = 318,000 + 368,000 = 686,000 \text{ ft-lb or } 8,230,000 \text{ in.-lb}$$

Try the depth of section shown in Fig. 13-20(a). Then

$$C = T = \frac{M_G}{32} = \frac{8,230,000}{32} = 257,000 \text{ lb}$$

For the bottom flange to withstand the compression from a prestressing force at least equal to T, with an assumed average stress of 1,600 psi, the area needed is $A_B = 257,000/1,600 = 160$ in.^{2} This should be increased a little because the prestressing force may be larger, and a few square inches will be lost because of the metal ducts that form the spaces around the cables.

Try the section shown in Fig. 13-20(*b*), with a gross area of 189 in.² in the bottom flange below a line 9 in. above the bottom. Make the top flange of the precast girder a little smaller than the bottom one. Find the center of gravity of this girder, using the axis *X-X* at mid-depth. Deduct 6 in.² for the area of the assumed three ducts in the bottom. The net area of the section is then 441 in.² The center of gravity is 1.5 in. below *X-X*. Then the moment of inertia of the girder about the center of gravity is approximately 77,500 in.⁴ From these data,

$$r^2 = I/A = 77,500/441 = 176 \text{ in.}^2$$

Figure 13-21 shows the eccentricity of the strands at mid-span to be 14.5 in. below the center of gravity of the section, this being the desired location. The posttensioning force *P* in the strands will therefore cause the girder to curve upward and to rest on its ends. The dead load will then tend to counteract this curvature. Since the girder weighs $441 \times \frac{150}{144} = 460$ plf, the dead-load bending moment at the center will be

$$M_B = \frac{460 \times 56^2 \times 12}{8} = 2,160,000 \text{ in.-lb}$$

Now assume that the bottom fiber stress f_b is to be a maximum and that the stress at the top is to be zero under the action of the prestress force *P* and the dead-load moment. Then, for the top fibers, as shown in Fig. 13-21(*b*), the stress in them will be

$$f_t = 0 = \frac{P}{A} - \frac{(Pe)y_t}{I} + \frac{M_B y_t}{I}$$

or
$$\frac{P}{A} = \frac{(Pe - M_B)y_t}{I} \tag{13-20}$$

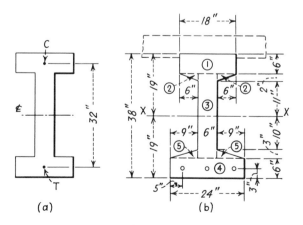

(a)   (b)

Fig. 13-20

Substituting Ar^2 for I gives

$$P = \frac{M_B}{e - (r^2/y_t)} \tag{13-21}$$

Since the desired eccentricity of P at the center of the girder is 14.5 in., as shown in Fig. 13-21(a), substitution of the numerical values in Eq. (13-21) gives

$$P = \frac{2,160,000}{14.5 - 176/20.5} = 366,000 \text{ lb}$$

This force does not need to be modified for shrinkage to any great extent because the girder is supposed to be well cured before the prestressing force is applied.   Furthermore, compressive shortening of the member will occur as the strands are tightened.   However, the magnitude of P might be increased to allow for creep and some small subsequent shrinkage. In this case, let $P = 420,000$ lb in order to have some reserve strength. If P is equally divided between three Roebling cables, $P_1 = 140,000$ lb. For this, use a $1\frac{1}{2}$-in. strand which, as shown in Table 13-1, has a safe design strength of 163,000 lb and an area of 1.36 in.^{2}

Using three of these strands at 140,000 lb each, the stress in the top fibers for dead load and prestressing is

$$f_t = \frac{P}{A} - \frac{Pey_t}{I} + \frac{M_By_t}{I}$$

$$= \frac{420,000}{441} - 420,000 \times 14.5 \times 20.5/77,500 + 2,160,000$$

$$\times 20.5/77,500 = 85 \text{ psi tension (satisfactory)}$$

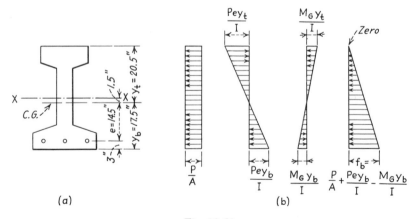

Fig. 13-21

For the bottom fibers,

$f_b = 420{,}000/441 + 420{,}000 \times 14.5 \times 17.5/77{,}500 - 2{,}160{,}000$
$\times 17.5/77{,}500 = 1{,}840$ psi compression (safe)

The bending moment for the weight of the deck slab alone is

$$M_D = \frac{12 \times (88 \times 5)56^2}{8} = 2{,}070{,}000 \text{ in.-lb}$$

The resultant bending stresses in the girder, neglecting diaphragms and forms, are

$$f_t = \frac{2{,}070{,}000 \times 20.5}{77{,}500} = 550 \text{ psi compression}$$

$$f_b = \frac{2{,}070{,}000 \times 17.5}{77{,}500} = 470 \text{ psi tension}$$

The resultant stress at the bottom with the girder thus loaded is $1{,}840 - 470 = 1{,}370$ psi compression, and that at the top is $550 - 85 = 465$ psi compression, as shown in Fig. 13-23(a). Obviously, there is plenty of reserve strength for the forms and other construction loads.

After the slab has cured, the serrated top and the ties shown in Fig. 13-22(a) will enable the slab to participate in resisting compression. However, how much of the slab can be relied upon? The width of 18 in. plus $4t$ each side would give $b = 18 + 2 \times 4 \times 7 = 74$ in., but 60 in. c.c. of girders is naturally an upper limit. However, the slab is made of 3,000-lb concrete, whereas the girder has 5,000-lb concrete. Therefore, reduce the 60-in. width by the ratio of the moduli of elasticity to get an equivalent section. Then $b_t = 60 \times 3.2/4.1 = 47$ in., neglecting the effect of any longitudinal bars in the slab. On the other hand, the 6 in.^{2} previously deducted for the space taken up by the tubes around the strands will now be included in the section because they will have been grouted in. The transformed area of the strands themselves will also be included but that of the small longitudinal ties shown in Fig. 13-22(a) will be neglected. The results of the computations are shown in (a).

The live-load bending moment for design was previously computed as 3,820,000 in-lb. Therefore,

$$f'_{tL} = \frac{3{,}820{,}000 \times 17.7}{190{,}000} = 356 \text{ psi compression at top of slab}$$

$$f_{bL} = \frac{3{,}820{,}000 \times 26.3}{190{,}000} = 530 \text{ psi tension at bottom of girder}$$

$$f_{tL} = \frac{3{,}820{,}000 \times 11.7}{190{,}000} = 235 \text{ psi compression at top of girder}$$

Therefore, from Fig. 13-23(a), the total stresses in the girder are

$$f_t = 465 + 235 = 700 \text{ psi compression at top}$$
$$f_b = 1{,}370 - 530 = 840 \text{ psi compression in bottom}$$

These are shown in Fig. 13-23(b).

Now check the design to determine whether or not the required load factors can be attained without cracking the bottom.

Check of precast girder: It has been shown that a dead-load bending moment of $2{,}160{,}000 + 2{,}070{,}000 = 4{,}230{,}000$ in.-lb is already being resisted by the girder in supporting itself and the slab, causing in it the stresses shown in Fig. 13-23(a). Therefore, the added bending moment for 50 per cent increase of dead load is

$$\text{DL:} \quad 4{,}230{,}000 \times 0.5 = 2{,}115{,}000 \text{ in.-lb}$$

The specified load factor of 2.5 for live load is to be applied to the live-load bending moment previously computed. Then

$$\text{LL:} \quad 3{,}820{,}000 \times 2.5 = 9{,}550{,}000 \text{ in.-lb}$$

This makes a total of $2{,}115{,}000 + 9{,}550{,}000 = 11{,}665{,}000$ in.-lb for design. The stresses already in the girder due to dead load are locked in. For the theoretical ultimate load, the member will act as a composite section. Therefore, the additional bending stresses will be determined by the use of the data shown in Fig. 13-22(a). Hence, for the top and bottom

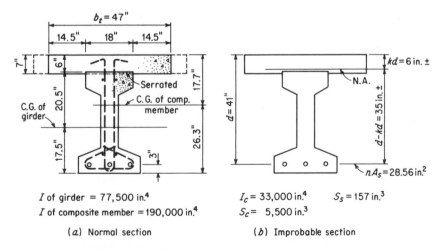

I of girder = 77,500 in.⁴ I_c = 33,000 in.⁴ S_s = 157 in.³
I of composite member = 190,000 in.⁴ S_c = 5,500 in.³

(*a*) Normal section   (*b*) Improbable section

Fig. 13-22   Posttensioned composite construction.

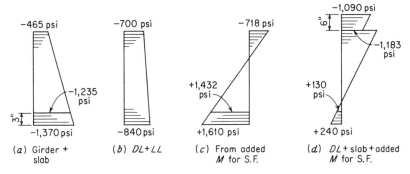

(a) Girder + slab   (b) *DL+LL*   (c) From added *M* for S.F.   (d) *DL*+slab+added *M* for S.F.

Fig. 13-23

of the girder,

$$\text{Added } f_t = \frac{11,665,000 \times 11.7}{190,000} = 718 \text{ psi compression}$$

$$\text{Added } f_b = \frac{11,665,000 \times 26.3}{190,000} = 1,610 \text{ psi tension}$$

These are shown in Fig. 13-23(c). When combined with those shown in (a), the resultant stresses are as shown in (d). The tension in the bottom is below the 530 psi specified for the assumed ultimate or cracking load, giving a reserve of 290 psi, which is satisfactory.

Check of slab of composite section: The total bending moment applied to the composite member will be the same 11,665,000 in.-lb which was used for testing the girder. The stress in the top fibers will be

$$f_t' = \frac{11,665,000 \times 17.7}{190,000} = 1,090 \text{ psi compression}$$

This is safe since it is even less than the allowable working stress for 3,000-lb concrete, and it is shown in Fig. 13-23(d).

The cables are supposed to have a stress of $140,000/1.36 = 103,000$ psi (approx) after first being stressed and supporting the weight of the slab, causing a compressive stress of 1,370 psi, as shown in Fig. 13-23(a), in the bottom or about 1,235 psi at the level of the cables when grouting locked them in. With a change of stress in the concrete alongside from 1,235 psi compression to about 130 psi tension, as shown in Fig. 13-23(a) and (d), the change of stress in the concrete is 1,365 psi. If the change in the steel stress is n times that in the adjacent concrete, the additional tension in the cables at the center of the span may be $6 \times 1,365 = 8,200$ psi, resulting in a total of about $103,000 + 8,200 = 111,200$ psi in the cables. This is very safe.

If the girder cracks considerably, what might happen?   Figure 13-22(*b*) shows the properties of the cracked section if it were a conventional reinforced-concrete beam.   This would mean that the tension in the concrete of the bottom would vanish.   As the cracking became severe, the member would approach a condition somewhat like that of Fig. 13-22(*b*), provided that the shearing resistance between the girder and the slab is sufficient.   The slab itself is so large that one need not worry about a compression failure of the concrete.   The chief question is whether or not the cables would fail.   Now imagine that the overload caused the assumed cracking but that the elastic recovery of the cables closed the cracks when the load was removed.   The worst that could happen thereafter seems to be that the entire bending moment might have to be withstood by the cables acting with a lever arm something like that shown in (*b*).

Remember that the overcoming of a compression is equivalent in a way to a tension in this case.   In other words, the strands are in tension when the dead load is acting.   Cracks will not open up until the prestressing force is overcome, and it requires a lot of bending moment to do this. If the DL moment of 4,230,000 in.-lb and the overload moment of 11,665,000 in.-lb were combined and the cables were acting with a lever arm of something like 35 in., an estimate of the stress in the cables at mid-span might be

$$f_s = \frac{15,895,000}{35 \times 3 \times 1.36} = 111,000 \text{ psi}$$

This is not even an increase above the 111,200 psi estimated from prestress plus the strain effect previously computed.   This shows that, before the structure would fall down, it might behave something like this extreme assumption.   At any rate, the raising of the neutral axis (if it could be called such) greatly increases the moment resistance of the cables.   In any event, the stress in the cables need not be more than enough to produce equilibrium.

At the ends of the girder, the prestressing force will be acting alone because the bending moment caused by the transverse loads is zero.   It is therefore desirable to reduce the eccentricity of the cables here.   Assume that, as shown in Fig. 13-24, the two outer cables are straight because they are in the flange but that the middle one is curved upward 18 in. in a more or less circular curve.   As an approximation, assume that the curve is circular and that the midordinate $m = c^2/8R$, where c is the chord length or span and R is the radius of curvature.   Then $R = 56^2/(8 \times 1.5) = 262$ ft.   Next, assume that the tension in the cable will have to be increased in order to make the cable slide inside the duct because of frictional resistance.   The original tension of 140,000 lb in the cable will produce a pressure against the inside of the curve.   Let $T = pr$, where p is the

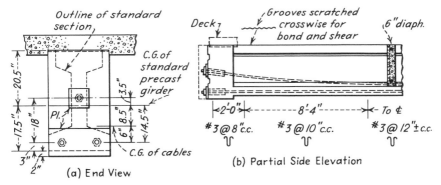

Fig. 13-24   End anchorage details.

pressure per foot of cable length. Then $p = T/R = 140,000/262 = 535$ plf. Then, assuming that one end is jacked and the other fixed, the total pressure on a length of 56 ft is $56 \times 535 = 30,000$ lb. With a coefficient of friction of cable against galvanized duct $f = 0.25$, the total frictional resistance is

$$F = fN = 0.25 \times 30,000 = 7,500 \text{ lb}$$

This is a minimum additional tension in the curved cable. Any lateral curvature or "wobble" will increase the resistance. This might be estimated as an extra increase of 10 to 15 per cent above the initial cable pull, depending upon the accuracy of the field work. However, perfect alignment is difficult to attain.

The Code formula for the total tension in a curved strand is

$$T_o = T_x(1 + KL + \mu\alpha) \qquad (13\text{-}3)$$

if $KL + \mu\alpha$ does not exceed 0.3. According to the Code, reasonable values for galvanized cables and ducts are $K = 0.0015$ and $\mu = 0.25$. Let $L = 56$ ft. The approximate angle of slope at the end of the curved cable is $\tan^{-1} = 2 \times 1.5/28 = 0.107$ or $6.1°$. Then $\alpha = 6.1/57.3 = 0.107$. radians. Then $KL + \mu\alpha = 0.0015 \times 56 + 0.25 \times 0.107 = 0.111$. Therefore, Eq. (13-3) can be used, and it gives

$$T_o = 140,000(1 + 0.0015 \times 56 + 0.25 \times 0.107) = 156,000 \text{ lb}$$

This is even below the limit of 163,000 lb allowed for the design load in this particular cable.

The two straight cables at 140 kips each and the central one at 156 kips produce a total P at the end equal to 436 kips. The centroid of this cable pull will therefore be $156 \times \frac{18}{436} = 6.4$ in. above the plane of the bottom cables. Then $e = 14.5 - 6.4 = 8.1$ in. below the center of gravity of the precast girder.   Therefore, the stresses at the end are

$$f_t = -\frac{436,000}{441} + \frac{(436,000 \times 8.1)20.5}{77,500} = -55 \text{ psi compression}$$

$$f_b = -\frac{436,000}{441} - \frac{(436,000 \times 8.1)17.5}{77,500} = -1,790 \text{ psi compression}$$

These stresses are satisfactory.

The ends of the girder should be thickened to accommodate the anchorage details and to prevent crushing, buckling of the web, and excessive vertical or longitudinal shearing stresses.   It is important to reinforce the ends under and around the cable anchorages so as to prevent local bursting or spalling of the concrete.   This reinforcement may be short spirals or hoops in the concrete under the bearing plates together with a cage of small bars to knit the entire end together.

The pressure under the bearing plate of a posttensioned cable, when the concrete is adequately reinforced, should not exceed

$$f_{cp} = 0.6f_{ci'} \sqrt[3]{A_b'/A_b} \tag{13-22}$$

but not greater than $f_{ci'}$.   In this case, the ends are to be thickened to equal the width of the bottom flange, also deepened 2 in.   The area A_b' for one bottom cable will be assumed to be 10×10 in. $= 100$ in.^{2} Assume that the area of the bearing plate A_b is $7 \times 7 = 49$ in.^{2}, but deduct 3 in.^{2} for the hole in it.   Then

$$f_{cp} = 0.6 \times 4,000 \sqrt[3]{\tfrac{100}{46}} = 3,100 \text{ psi}$$

if $f_{ci'}$ is assumed to be as low as 4,000 psi.   The pressure under a bearing plate will be

$$140,000/46 = 3,040 \text{ psi (safe)}$$

The middle cable will also be safe because of the larger area of concrete under and near the bearing plate.

After prestressing of the cables, the space between the inside of the metallic ducts and the cables must be grouted.   Holes or connecting tubes are to be provided for this purpose, and the work must be well done.   As indicated in Fig. 13-24(b), the ends of the anchorage details are to be encased in concrete, which must be tied to the end of the girder.

The curved cable actually helps to support transverse loads.   In this case, assume that the effective pull equals only the initial 140,000 lb, neglecting the effect of friction on the curved duct when the cable is tightened.   With a midordinate of 1.5 ft and the end slope previously computed, the approximate vertical component of the cable pull is

$$P_v = 140,000 \times 0.107 = 15,000 \text{ lb}$$

With the concentrated live load 3 ft from the end of the member, and with uniform load over the entire span, the maximum end shear is

Dead load: $(440 + 460)28$ $= 25{,}200$ lb
Live load: $5 \times 80 \times 28 + 5 \times 2{,}300 \times \frac{53}{56} = 22{,}100$ lb

$$\Sigma V = 47{,}300 \text{ lb}$$

Deducting an uplift of 15,000 lb gives a net $V = 32{,}300$ lb to be taken by the web for WSD. The corresponding shear, computed rather arbitrarily for USD, is

Dead load: $1.5 \times 25{,}200 = 37{,}800$ lb
Live load: $2.5 \times 22{,}100 = 55{,}200$ lb

$$\Sigma V_u = 93{,}000 \text{ lb}$$

Deduct only the regular strand uplift of 15,000 lb, giving a net

$$\Sigma V_u = 78{,}000 \text{ lb}$$

AASHO specifies that stirrups shall be perpendicular to the member's axis and that

$$A_v = \frac{(V_u - V_c)s}{2f_y jd} \text{ or not less than } 0.0025b's$$

The question is what to use for V_c. For beams without web reinforcement, the maximum v_c specified is $0.03f'_c$ (but not over 90 psi) for WSD. For USD here, assume that the maximum $v_c = 2 \times 90 = 180$ psi, doing so in order to be conservative, since this neglects prestress. Then the concrete of the web might be allotted an amount

$$V_c = 180b'd = 180 \times 6 \times 35 = 38{,}000 \text{ lb}$$

Using this with $j = 0.85$, $d = 35$ in., and $s = 10$ in.,

$$A_v = \frac{(78{,}000 - 38{,}000)10}{2 \times 40{,}000 \times 0.85 \times 35} = 0.17 \text{ in.}^2$$

This is greater than $0.0025b's = 0.15$ in.^{2} Therefore, No. 3 U-shaped stirrups will be enough, spaced as shown in Fig. 13-24(b). Notice that this design does not allow any vertical shearing resistance for the roadway slab or the wings of the flanges. Neither does it allow for the uplift effect of the central cable. These stirrups are to extend into the slab, and they will fulfill the requirement that the ties shall be not less than two No. 3 bars 12 in. c.c.

As a last step, check the longitudinal shearing stress between the slab and the girder by Eq. (13-13) for the live-load end shear 3 ft from the center of the support (or 3 ft from its edge if desired), since the dead load is already carried by the girder alone.

$$v = \frac{VQ}{Ib'} \tag{13-13}$$

For WSD, $V = 22{,}100$ lb and Q from Fig. 13-22(a) $= 29 \times 7 \times 14.2 + 18 \times 6 \times 14.7 = 4{,}470$ in.[3]   Then

$$v = \frac{22{,}100 \times 4{,}470}{190{,}000 \times 18} = 29 \text{ psi}$$

This is safe, since AASHO allows 75 psi when minimum ties are used and 150 psi when the contact surface is artificially roughened as it is here. Of course, the above applies when the girder is made of concrete.

The plans for casting the girders are to be made in accordance with the requirements of the particular situation.   The girders may be cast on one approach, then skidded or lifted into position.   They may be cast below the structure and lifted into place.   They may even be manufactured elsewhere and transported to the site.   If so, care must be exercised to make sure that they do not tip over because the dead load is relied upon to counteract some of the prestress bending.   In any case, the stresses caused by the erection procedures should be investigated and provided for.

As with ordinary reinforced-concrete construction, many beams of varying dimensions and make-up might be used for this particular service. The one designed here is shallow but not unduly so.   It might be made a little lighter, but the saving, if any, would not be great.

13-11. Miscellaneous Construction.   Many   adaptations   of   the principles of prestressing can be made.   Some will be explained briefly.

Beams of considerable length and strength can be made by using a series of accurately faced concrete blocks with prestressing wires or strands somewhat as shown in principle in Fig. 13-25.   The blocks are assembled and then posttensioned.   Here the blocks are of three kinds.   The end ones a have some kind of detail that will accommodate the anchorages, and they are strong enough to resist the local pressures.   The intermediate

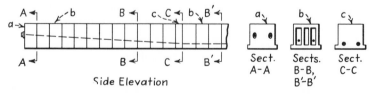

Fig. **13-25**   One arrangement for a prestressed block beam.

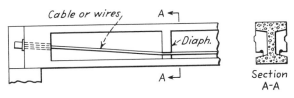

Fig. 13-26　Precast posttensioned unit with unencased steel.

blocks *b* are hollow so that the strands can slope down to the "kink point" at block *c*, from which they extend straight across to a corresponding block beyond the center. Generally the joints between the blocks of the precast beams are not mortared. Friction due to the pressure resists most of the transverse shear. In this case, the strands are bare inside the blocks. The lips on the bottoms of the blocks provide for mortaring or concreting between adjacent beams when a concrete topping with wire mesh is placed over the erected units.

A different arrangement for a precast prestressed-concrete beam is pictured in Fig. 13-26. This is a posttensioned unit. The strands are not encased but are alongside the web, as shown in section *A-A*. The units are to be set side by side, with the diaphragms interlocked; then a 2-in. layer of concrete with wire mesh is to be placed across the top. If desired, the diaphragms and ends could have a hole in them transversely with respect to the beam so that, after erection, rods or strands could be threaded through them and all could be tensioned so as to pull the beams tightly together in order to make them act more like a monolithic unit.

Whether or not the use of bare strands, bars, or wires is desirable will depend largely upon the conditions applying to the particular case in which they are to be incorporated. Is there danger of serious corrosion?

Poles, piles, roof slabs, wall panels, and many similar small units are especially well adapted for manufacture as precast prestressed units, especially with bonded wires or strands. Although a faint tinge of rust may be beneficial for the bond on the tendons in such units, be careful that no serious rusting has occurred before they are used and that they are clean.

Prestressed-concrete tanks, and even prestressed pipes, are very useful. The tendons are generally posttensioned and then covered with gunite, mortar, or concrete protection. The purpose is not to resist bending but to squeeze the concrete so that shrinkage, temperature changes, and ring tension cannot cause even minute cracks that would permit leakage through the concrete.

Some other sections that are practicable as prestressed-concrete beams are shown in Fig. 13-27.

Continuous beams and girders are more difficult to prestress than

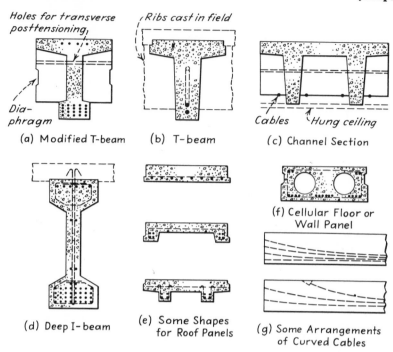

Fig. 13-27   Various types of precast prestressed members.

simply supported ones. In general, they are posttensioned. Some principles that may be used are indicated very sketchily in Fig. 13-28. It is probable that continuous members should be poured in place and the strands subsequently tightened, but this may not always be necessary. However, since reversal of stress, fatigue, and rapid variations of stress are

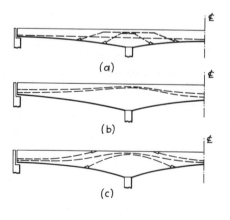

Fig. 13-28   Some studies of schemes for posttensioning a three-span girder.

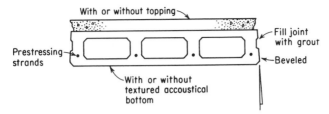

Fig. **13-29**   Hollow prestressed floor units.

likely to occur in continuous structures, prestressing of the particular kind of structure desired, and the details and procedures to be used in building it, should be studied very carefully when the structure is planned.

PRACTICE PROBLEMS

For the following problems, assume the following data: $f_c' = 5,000$ psi and $f_c = 2,000$ psi for prestressed members; $f_c' = 3,000$ psi for topping; $f_s' = 250,000$ psi; $f_y = 40,000$ psi for web reinforcement; max tension in prestressed members $= 600$ psi; strengths of strands as in Table 13-1; and all members simply supported. Design for the condition of no tension in the bottom flange for working loads.

13-1   Design a Lin tee like that of Fig. 13-10 except that the depth is 28 in. instead of 24 in. to support a live load of 60 psf when the span is 50 ft.

13-2   Assume that a Lin tee is 8 ft wide and that the flange thickness is increased by $\frac{1}{2}$ in. over that in Fig. 13-10. The web is 8 in. wide and the total depth is 36 in. Design the member to carry a live load of 50 psf without a topping. The span is 56 ft.

13-3   Assume that the Lin tee shown in Fig. 13-10 is to have a topping 3 in. thick and is to support a live load of 50 psf. The span is 42 ft. Design the member.

13-4   Design a posttensioned bridge girder like that of Fig. 13-20(*b*), but 42 in. deep. The deck slab is to be 6 in. thick and its bottom is to be 1 in. below the top of the top flange. The girders are to be 6 ft c.c. The live load is 75 psf and a concentrated load of 2,500 lb/ft of width of lane. The span is 60 ft.

ARCHITECTURAL AND
MISCELLANEOUS DETAILS

14

14-1. Basic Principles Underlying Architectural Treatment. The design of the architectural features of any reinforced-concrete structure must be based upon the nature of concrete itself and upon the processes that are to be used in the building of that structure. In other words, concrete is a substance that, in a plastic state, is placed against forms that give to it whatever shape the forms possess so that, after the concrete has set, these outlines and shapes are permanently maintained by the concrete. Therefore, the details of the whole subject must be thought about in terms of forms, simplicity, pouring schedules, and other practical matters of construction so that the finished structure will become one harmonious entity whose parts automatically blend to produce the desired architectural ensemble. Figure 14-1 is an example of this.

Furthermore, the general architectural conception of the structure should also be based upon considerations of beauty, proportion, surface texture, permanence, color, economy, and functional effect. The designer should abandon many of the ideas that have been developed through the past in using stone and brick masonry which emphasize moldings, cornices, decorative carvings, and other features for which concrete is not suited. Concrete is a special material, and the architecture of a concrete structure should be adapted to it.

There is considerable difference in the planning of a structure which is made of poured-in-place concrete and one that is composed in whole or in part of pretensioned-concrete units. Figure 14-2 shows one of the latter type. Here the structure was designed to utilize available prestressed-concrete elements. The exterior columns are modified pretensioned tees with the stems turned outward. The floors are composed of pretensioned

518

Fig. 14-1 Los Angeles County Hospital, Los Angeles, California. (*Courtesy of Portland Cement Association.*)

Fig. 14-2 Doctor's Building, New Haven, Connecticut, Westcott and Mapes, Inc., New Haven, Connecticut, architects and engineers. Prestressed concrete made and erected by C. W. Blakeslee and Sons, Inc., New Haven, Connecticut.

units with a poured-in-place topping.   Even the central utility core is built with precast panels which are joined together by means of welded bars or other steel details which were later encased in narrow poured-in-place strips.

Much of the architectural detail of the exterior surface of a concrete structure should be based upon a consideration of light and shadows. Excessively large flat surfaces should be avoided because any irregularities are so noticeable; these large areas should be broken up by horizontal or vertical markings or moldings, by moldings in both directions, by offsets, or even by the use of various finishes.   Such lines or markings should be straight and distinct, avoiding intersections at large obtuse angles.   The special features should be coordinated with the necessities of the building operations so that the construction joints can be placed at markings or offsets, the joints themselves thus being indistinguishable.

An interesting treatment of large surfaces is shown in Fig. 14-3.   With the base and panel treatments, one hardly realizes that this might have been an unattractive retaining wall.

In planning details to accommodate formwork, one should remember that plywood is a common lining for the forms for architectural concrete. However, the greatest reasonable length of plywood is about 8 ft (maximum 10 ft).   If this is the limit between moldings, offsets, or other details, it will avoid the marks that might otherwise be visible in the con-

Fig. 14-3   Retaining wall at portal of Figueroa Tunnel, Los Angeles, California. (*Courtesy of Portland Cement Association.*)

crete at the junctions of the ends of abutting pieces.   Also, with the panels of plywood laid with their long dimension horizontally, joints must often occur between the upper and lower pieces, but these joints can be braced thoroughly.   If slight marks occur in the concrete at such points, they are not usually objectionable because of their horizontal position.   However, care should be used to line up these joints across or around the structure.   A patchy effect should be avoided.

14-2.   Moldings and Minor Details of Forms.   The moldings or markings that are used to break up flat surfaces should be detailed properly.   It is very easy to nail strips on the insides of the forms, thereby causing recessed moldings or markings in the finished concrete, but it is rather difficult to make recesses in the forms themselves.   It is also better to recess these cuts in the concrete, because small projections are likely to be damaged.

The following points, some of which are illustrated in Fig. 14-4, should also be considered:

1. Wide, shallow recesses do not cast sharp shadows.

2. Horizontal recessed moldings or markings should be V-shaped, beveled outwardly at the top and bottom, or horizontal at the top and beveled at the bottom.   Horizontal ledges are likely to collect and hold dirt; they are also more difficult to fill so as to produce sharp edges, unless the hydrostatic head of the concrete is considerable.

3. Sharp acute angles in moldings are difficult to fill properly and may spall off easily.

4. Chamfering of projecting corners is desirable, but filleting is difficult at reentrant angles.

5. Small, recessed V strips or half rounds ($\frac{3}{4}$ to $1\frac{1}{2}$ in. deep) should be used at construction joints.   If this is not permissible, a straight finish strip should be attached to the forms at the line of the joint so as to avoid

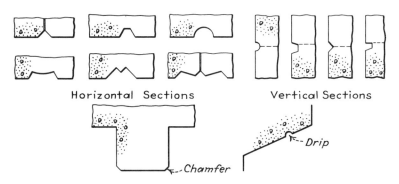

Horizontal Sections   Vertical Sections

Fig. 14-4   Details of moldings.

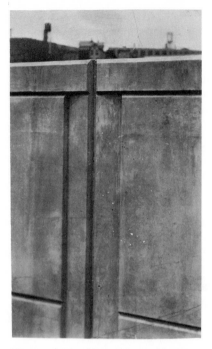

Fig. 14-5   Details at expansion joint.

a ragged line in the concrete. Sometimes the top of the forms can be a straight edge along which the concrete can be finished with a trowel for a depth of about 1 in.

6. Lining up of moldings in successive pours is very important; so is proper mitering of the form strips at intersections.

7. Vertical keyways should be made, if possible, by fastening the key strip on the inside of the forms of the first pour.

8. Moldings must be sloped along their edges sufficiently to facilitate the removal of the forms.

9. The depths of moldings and V cuts automatically affect the location of the reinforcement because the bars should have $1\frac{1}{2}$ to 2 in. of cover at the deepest part of the recess. When the cuts are too deep, the reinforcement that crosses behind them has to be bent in order to keep the bars near enough to the main surface, or they have too small a cover at the cut.

A very simple but effective treatment at a contraction joint in a retaining wall is shown in Fig. 14-5. The grooves are only $1\frac{1}{2}$ in. deep, but they are sufficient to cast sharp shadows.

14-3. Surface Finish and Appearance. It is ordinarily desirable to have a smooth finish on the surface of the concrete. Plywood or similar materials generally serve this purpose very well. However, in the case of buildings or other structures that will be seen at close range, it is often desirable to rub the concrete with a carborundum stone and water so as to remove all objectionable irregularities and variations in the pattern or the texture of the surface.

Various linings for forms are manufactured which give to the concrete surface a specific pattern or texture. The use of steel forms generally causes a glassy surface or texture unless they are lined with other material. It is possible to coat the forms with a special material which retards the set of the concrete at the surface so that, after removal of the forms, the surface can be brushed or rubbed to produce a rough texture and to expose some of the coarse aggregate. In some cases, the surfaces have been made with projecting fins that crack off irregularly when the forms are removed,

or which are purposely broken later, so
as to give the structure a rough surface
which, from a distance, may look "aged."
Of course, with precast members, it is
possible to obtain a wide variety of
finishes.

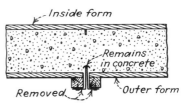

Fig. 14-6   Dummy joint.

Bushhammering or other treatment
that destroys the mortar coating over
the aggregate is sometimes used on massive structures, but the effect
upon durability is open to question.   However, this type of finish has the
advantage of being very difficult to deface with chalk and similar materials; it also can be used in securing paneled effects, but the aggregate
should be suitable in color, about $\frac{3}{4}$ to 1 in. in size, and a good gravel or
crushed stone.   The general layout must be simple, and the bushhammering must be done thoroughly if the effect is to be satisfactory.

When any painting of exterior surfaces is desirable, a wash or paint of
cement is recommended by the Portland Cement Association.   Consult
the manufacturers for other suitable paints.

The finishing of horizontal surfaces should be done by screeding, using
a wood float, or troweling moderately.   Excessive troweling or other
working that flushes too much of the mortar to the surface will be likely
to cause "crazing"—very fine cracking of the surface.

The builder should be very careful to use aggregates that are properly
graded and that are uniform in color.   Otherwise, variations between
pours will be distinguishable.   Also, he should be careful to avoid segregation of the aggregates during the placing of the concrete, or else the
bottom of a pour will appear coarse while there is an accumulation of
finer material at the top of the previous one.   This requires careful
placing, the use of a concrete that is not too dry or too wet, and careful
spading near the surface.   Generally, the use of vibrators which are
applied to the forms or in the body of the wet concrete is not sufficient to
guarantee a good surface regardless of all other conditions.   Great care
must be exercised when using vibrators to make sure that the forms do not
become bulged, displaced, or leaky at joints.

Another thing to be avoided in the surface of concrete is cracking due
to shrinkage or variations in temperature.   Adequate reinforcement will
prevent this usually if the pours are not too large and if the details are
such that the necessary deformation can occur at vertical and horizontal
contraction joints along the lines of the moldings.   Doors, windows,
and other openings in walls are likely to cause such cracking of the concrete because they are points of relative weakness.   In these cases,
"dummy joints" such as that of Fig. 14-6[1] may be used together with a
decrease in the reinforcement across the joint, thus inviting the crack

[1] A. M. Young, Crack Control in Concrete Walls, *Eng. News-Record,* Aug. 11, 1938.

to occur at a predetermined concealed point.   In very long buildings, it is even desirable to divide the structure into units about 200 ft or less in length by using expansion joints which completely isolate adjacent units.

Other features that are desirable in order to maintain a satisfactory appearance of concrete surfaces are as follows:

1. Waterproofing of the back of the concrete when it is subjected to the penetration of moisture—as in retaining walls.

2. Avoidance of metallic accessories which are almost certain to cause streaking.

3. Provision of a wash (or slope) on the tops of parapets so that drainage will carry dirt back away from the surface which one wishes to maintain in a satisfactory condition.

4. Provision of a "drip," or groove, under cantilevered construction as shown in Fig. 14-4.   This is very important to prevent streaking of the surface due to rains.

Designers often believe that the concrete should have some reinforcement near the surface even when not required by stress.   When the structure is exposed to sea water, salt spray, and even ordinary weathering, the presence of steel near the surface may eventually cause spalling of the concrete because of corrosion of the bars.   Some structures for which one might consider the avoidance of "skin reinforcement" on the outside or top are sidewalks, pavements, bridge decks, pier shafts, abutments, retaining walls, sea walls, and water-front structures in general.   However, sufficient steel should be provided to prevent structural weakness, and enough contraction joints should be used to avoid temperature and shrinkage cracking.   Many a structure has had its appearance marred by spalling because of this corrosion of the steel and then by rusting of the exposed bars and staining of the concrete.

14-4.   Temperature Reinforcement.   If a piece of concrete could be rigidly connected to immovable supports, and if this piece could be subjected to a fall in temperature of 1°F, the resultant tension in the concrete due to the latter's attempt to contract would be

$$f_c = E_c \frac{\Delta L}{L} = E_c \omega$$

where ω is the coefficient of thermal expansion or contraction.   Then, if $E_c = 3,000,000$ psi and $\omega = 0.000006$, $f_c = 18$ psi.   However, since the range of temperature variations is frequently $\pm 60°F$, the corresponding maximum stresses in this concrete would be $f_c = \pm 18 \times 60 = \pm 1,080$ psi.   From this, it is clear that the compressive stresses alone would not be critical, but the concrete would surely crack if it were subjected to such a tensile stress. The amount of steel that would be needed to carry this tension is clearly excessive.

In practice, structures are not often fully restrained this way, unless they are keyed to rock foundations. However, if they are very long, there may be sufficient restraint or frictional resistance to cause them to crack in tension. Therefore, the engineer should build his structures in short units so that these cracks cannot develop.

When one considers that ω for steel is practically the same as for concrete, he realizes that the use of bars to prevent cracks due to temperature variations is not for the purpose of stopping the expansion and contraction but merely to knit the structure together and to avoid localized cracks. The amount of steel that is required to do this is not definite, but it may be assumed to be about 0.0025 times the cross-sectional area of the concrete. However, parapets and similar relatively thin parts which are attached to more massive structures should be reinforced more heavily.

The temperature reinforcement should be placed somewhere near the surfaces of concrete structures, and especially near edges where cracks may start. When only one side is exposed, about 60 to 70 per cent of the steel should be near the exposed side, unless the walls are thin. Even when the foundations are on rock, the parts of the structure that are above the footings should be provided with adequate joints and reinforcement.

Special attention should be given to concrete aprons at doorways; pavements of areaways; sidewalks that adjoin foundation walls; basement walls of poured concrete that are largely or entirely exposed to the atmosphere, as at the back of a structure on a lot that slopes down and toward the rear; and long monolithic walls like that of Fig. 14-7(a), particularly when they have southern exposure in the northern temperate zone.

An apron should be seated on the foundation wall, and perhaps doweled to it, when heavy loads will pass over it. The apron should be cut by contraction joints into sections about 10 to 15 ft wide. Pavements of areaways on earth should be free from main structures, and they should be cut into sections approximately 10 by 20 ft in size. Adjoining sidewalks should be cut free from foundation walls, and they should have contraction joints 10 to 20 ft c.c. In cold climates, retaining walls, long monolithic walls of basements, and such walls of first stories without basements, theoretically should have contraction joints 50 to 60 ft c.c., even though this is troublesome in construction. Such joints should also be near corners, at offsets, and at points of weakness where cracks may occur.

Even with prestressed-concrete construction, the thermal deformation of large, flat roofs may cause troublesome stresses and cracking in spite of insulation. The provision of joints is desirable if a width or length exceeds

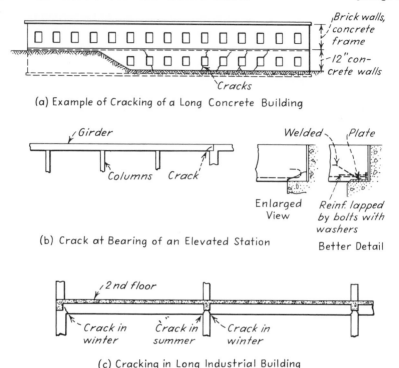

(a) Example of Cracking of a Long Concrete Building

(b) Crack at Bearing of an Elevated Station

(c) Cracking in Long Industrial Building

Fig. 14-7   Examples of cracking apparently caused by changes in temperature.

about 100 to 125 ft, especially if the walls are masonry or panels that may crack when the roof expands more than does the floor below it.   Also, the columns should be able to bend a little to permit the deformations of the roof because they will seldom be strong enough to force heavy roof members to slide on seats or to compress or stretch.   Rigid stairwells, elevator shafts, and concrete ducts, when near the ends or corners of a structure, will tend to resist the deformation of the roof, and their walls may crack.

Structures that are narrow and high, such as piers, do not need steel to resist cracking due to temperature to the same extent as do long ones. They are free to expand or contract vertically.   Their dead load alone will not let cracks open up.

When concrete is used in places that are subjected to very high temperatures,[1] there is a far different problem.   Ordinary concrete may be weakened by dehydration above 500 to 600°F; at 1200°F it may be almost

[1] Alfred L. Miller and Herbert F. Faulkner, A Comparison of the Effect of High Temperatures on Concretes of High Alumina and Ordinary Portland Cements, *Washington Univ. Expt. Sta., Bull.*, Series 43, Sept. 15, 1927; also Concrete Subjected to High Temperatures, *ACI Proc.*, Vol. 35, p. 417, 1938–1939.

worthless. No amount of reinforcement will stop this action. On the other hand, concrete made with Lumnite and low-silica aggregate may withstand about 1000°F. If ladles of molten metal are to be placed on or alongside concrete surfaces, these should be protected by replaceable materials such as bricks. Unless ventilation can remove the heat, even these are inadequate; e.g., an uncooled concrete foundation slab for a reverberatory furnace, if placed on earth, will have its temperature gradually increased to somewhere near that of the furnace in spite of the latter's lining because the heat has no way of being dissipated. Surface spalling may occur when one side of a section is subjected to sudden high temperature locally applied.

14-5. Shrinkage. The shrinkage of concrete has been discussed briefly in Art. 1-10. Other than with expansive cements, it is a characteristic which should be guarded against when portland-cement concrete is used.

According to Lyse,[1] the shrinkage of concrete is approximately proportional to the amount of cement paste in the concrete regardless of the composition of the paste. High relative humidity reduces the amount of shrinkage; saturation usually eliminates shrinkage or may cause slight expansion.

It seems that shrinkage of the mortar in a concrete, whereas the coarse aggregate particles do not shrink, causes microcracks[2] which are more or less dispersed throughout the mass. This not only causes a weakening of the strength but may aggravate disintegration. A 1:2 mortar itself may have a tensile strength of 500 to 600 psi at an age of 30 days; a 1:3

[1] I. Lyse, Shrinkage and Creep of Concrete, *J. ACI*, February, 1960.

[2] T. T. C. Hsu, Mathematical Analysis of Shrinkage Stresses in a Model of Hardened Concrete, *J. ACI*, March, 1963; T. T. C. Hsu and F. O. Slate, Tensile Bond Strength between Aggregate and Cement Paste or Mortar, *J. ACI*, April, 1963.

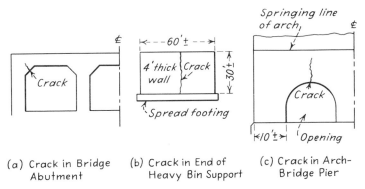

(a) Crack in Bridge   (b) Crack in End of   (c) Crack in Arch-
    Abutment   Heavy Bin Support   Bridge Pier

Fig. 14-8   Examples of shrinkage cracking of heavy structures.

mortar, 400 psi; and a concrete made with these mortars, about 50 per cent of that of the mortar itself.   This may indicate that the resistance of the concrete to compression does not constitute the same kind of thing as its resistance to tension.   Thus, the latter may not be related directly to the former.

Shrinkage may cause warping of reinforced concrete.   Plain concrete is free to shrink equally across the section of a member.   However, when heavy bars are placed close to only one side of the section, these tend to resist the shrinkage.   This means that the plain side shortens more than the reinforced side, causing curvature.   Also, the resistance of the bars causes a moment in the section—a tension near the bars and a compression in the other side.   This also may contribute to the curvature.

Shrinkage of concrete may be looked upon as somewhat like a drop in temperature of 30 to 80°F (depending upon the "richness" of the concrete). Since shrinkage actually tends to produce compressive stresses in the reinforcement, the bars, if the area of the steel is great enough, may be stronger in compression than the concrete is in tension, thereby producing tiny cracks instead of eliminating them.

The best way to eliminate harmful effects from shrinkage of a structure is to try to design it so that the concrete can shrink without causing trouble.   With proper planning, the work may be arranged so as to build long structures in alternate sections, pouring rather long portions first, then filling in the shorter, intermediate sections later—preferably allowing the first ones to set for 1 or 2 weeks.[1]

In multistory structures with heavy, rather solid walls and intermediate floors or in massive ones poured in complete horizontal lifts, the lower portions necessarily shrink first.   When a higher lift or a floor is poured, its shrinkage is restrained so that it may develop cracks.   Proper jointing is the best remedy, but full-height wall reinforcement inclined upward toward the center of shrinkage may minimize cracking.   However, this is seldom practicable.   Buildings with reinforced-concrete framework and filled-in walls do not have this difficulty because the columns can safely deform if full-height expansion joints are used about 300 ft apart.

Cooling of concrete by using ice in the mixing water and by means of refrigeration during the period of setting is sometimes thought to be desirable for massive structures such as dams.   The purpose is to prevent too high a temperature resulting from the chemical reaction of the cement, and thereby to avoid cracking from subsequent cooling.   Such means are costly and may not eliminate the trouble entirely.   Proper

---

[1] AASHO specifies the coefficient of shrinkage as 0.0002, and that of thermal deformations as 0.000006.

planning of pours and joints, construction of alternate sections, and provisions for future grouting may secure satisfactory results.

When shrinkage causes bending moments in such structures as arches and rigid frames, the resulting stresses should be considered in the design.

14-6. Construction Joints. Construction joints must be located so as to cause no serious weakness in the structure. It is therefore desirable to place them in regions where the shearing stresses and the bending moments are small or where the joints will be supported by other members. However, they must be located so as to facilitate construction.

Such joints must be adequately keyed in order to transfer the necessary shearing forces. In general, joints in floors should be located near the centers of the spans of beams, slabs, and girders because the shear is usually small at such points. If a beam intersects at the center of a girder, it is obvious that the joint has to be offset. An offset equal to twice the width of the beam is generally sufficient. Such a joint which must resist a substantial shear might well have at least one bar or stirrup passing through its center at an angle of 45° to take tension when resisting the shear. When the shear is reversible, inclined bars might be placed at 45° on each side to form an X centered on the joint.[1]

Figure 14-9 shows various arrangements for keys. The numbers in the circles denote the sequence of the pours.

The following comments should be noted, the letters referring to the various sketches in Fig. 14-9:

(*a*) This key is easy to form, but it holds water when it is horizontal. The water should be removed. When a keyway is vertical, the form for it should be attached on the inside of the forms for the first pour.

(*b*) The raised key causes objectionable formwork.

(*c*) These intermittent precast blocks set by hand in the wet concrete are not very strong. They are easily forgotten, and they require extra hand operations. This intermittent key idea can be used to advantage in vertical joints that must withstand vertical and horizontal shears by placing pieces of 2 by 6 planks about 8 in. long and 18 in. c.c. on the inside of the end forms of the first pour.

(*d*) This type of key can be used only when the shearing forces are as shown, but it is good for such cases.

(*e*) This is theoretically better than (*d*), but when the bars are close together, it is difficult to finish properly. It is also easily forgotten by the workmen.

(*f*) This V can be made by hand after the concrete is poured, thus

[1] R. J. Hansen, E. G. Nawy, and J. M. Shah, Response of Concrete Shear Keys to Dynamic Loading, *J. ACI*, May, 1961.

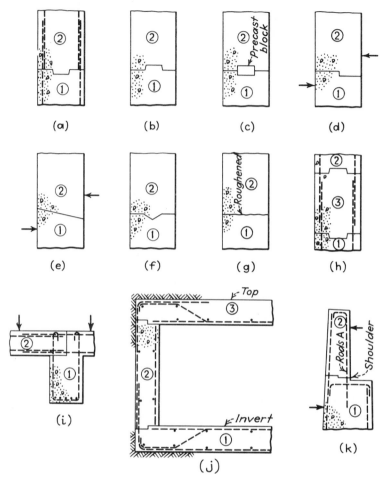

Fig. 14-9 Illustrations of construction joints.

eliminating form strips which would interfere with the pouring of the concrete.   It is good for thin walls and often for others (if it is not forgotten).

(*g*) If the shears are small or the direct compressive loads are large, this hand-roughened surface is often sufficient.   Trowel the edge to get a straight line for appearance.

(*h*) This type of key is good for arches or other structures that it is desirable to pour in sections so as to minimize shrinkage stresses.

(*i*) This is a possible arrangement.   There is no need for a true key when the joint is supported.   This may weaken the T beam.

(*j*) This shows suggested keyways at corners—such as for box culverts where the invert must be troweled or screeded.   The wall forms can be braced on the inside. It is assumed that the shears in the walls are small compared with those which are in the top and bottom slabs. The vertical pressure on the walls causes friction that helps the keying action, whereas if the full width of the wall were recessed into the horizontal slabs, the effective thickness of the latter for resisting shears would be seriously decreased.

(*k*) This illustrates the provision of a shoulder for use in setting forms.

A key should be designed so that its width w [Fig. 14-10(*a*)] is sufficient to transfer the shearing force, using ordinarily a shearing unit stress of about $0.1f_c'$. When the shearing force is reversible, the width of the key should be $t/3$ or slightly smaller, so that all three parts of the joint will have approximately equal strength. The thickness t' should provide sufficient area in bearing along the edge at a unit stress of not over about $0.2f_c'$. A thick but narrow key (large t') is likely to break off.

Figure 14-10(*b*) shows why a key at a point of large bending stresses may have very little value. The tensile forces may cause cracks which destroy the ability of a piece like A in the figure to resist a pressure applied at B. In such cases, it is advantageous to use vertical or inclined stirrups on both sides of the joint, as shown in the figure. Diagonal ones crossing the joint are best.

Sometimes keyways are called for in places where they are not necessary; e.g., at the junction of a column with a floor or the junction of a tall bridge pier with its footing. When no large shearing forces exist, there is no sliding action that requires a key. Bond and friction are sufficient to hold the members in position even without the dowel action of the reinforcement.

It is usually necessary to have reinforcement which passes through construction joints. If so, the bars from the first pour should project through the joint enough to secure the desired bond, as in Figs. 14-9(*a*), (*h*), (*i*), (*j*), and (*k*). In this way, the bars in the later pour can rest upon or bear against the concrete of the previous pour. In some special cases, separate dowels like bars A of sketch (*k*) can be set by hand in the wet concrete of the first pour. However, bars or dowels should not be

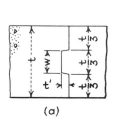

(a)

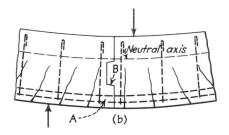

A   (b)

Fig. **14-10**

relied upon too much for resisting the shear at the joint, because they will crush the concrete locally before they will withstand much shearing force unless special provision is made to avoid this action.

In all cases, construction joints must be cleaned thoroughly before the next pour is made. All laitance must be removed, using wire brushes, water under high pressure, or other means. It is often desirable to coat the joint with a little mortar just prior to the placing of the concrete upon it. The exposed edges of the joint should be finished straight, or they should have a small V strip on the forms ($\frac{1}{2}$ to 1 in. deep).

The volume of concrete that can be deposited in one continuous pour will influence the locations of construction joints in massive structures. These should be planned far in advance. The other extreme occurs in the case of very thin walls—4 to 6 in. thick. These cause difficult pouring if they have any great height. They must be built in short lifts, by the use of pumping or by depositing through a small spout and hopper called an "elephant's trunk."

The maximum number of cubic yards of concrete for one continuous pour depends upon the equipment available, the distances that the concrete has to be transported, and the character of the structure. In general, 300 yd³ will be enough for one pour in 8 hr on even a big job.

Needless construction joints should be avoided, especially in retaining walls and other structures in which such joints may be a cause of the seepage of water.

14-7. Expansion and Contraction Joints. An expansion joint is used, generally with a premolded mastic or cork filler, when an elongation of adjacent parts and a closing of the joint are expected. A contraction joint is usually made without a filler except for a paint coat of asphalt, paraffin, oil, or some other material to break the bond. Its purpose is to prevent cracking. In general, the contraction caused by shrinkage will offset a large part of the subsequent expansion due to a rise of temperature, and a little pressure on the joint is seldom harmful anyway.

One of the first things to consider about expansion and contraction joints is that of the best locations for them from the standpoint of their proper functioning. They should be at points of change in thickness, at offsets, and at other points where the concrete will tend to crack if shrinkage and temperature deformations are restrained or prevented. The engineer must study a structure carefully in order to discover these points. Ordinarily, joints should be about 30 to 50 ft c.c. in exposed structures.

The second consideration should be that of coordination with the pouring schedule and avoidance of extra construction joints.

The third matter is that of satisfactory details. For these, the following points are mentioned:

1. The keyways should be of the types shown in Figs. 14-9(a), (b), and (d) if there is a considerable shearing force at the joint. Figure 14-11(a) shows a vertical expansion joint which has been used in some of the retaining walls of the approaches to the Lincoln Tunnel. The compressible material can be fastened to the first pour by tacking it to the forms and by having nails protruding so as to bond into the concrete. However, one must be careful to use materials that will not squeeze out, slump when heated by the sun, or stain the surface of the concrete. In the space outside the key itself, beveled strips may be used (instead of fillers) and later withdrawn, but this is difficult when the walls are thick.

2. The edges of the keyways should be beveled slightly; they should be coated with mastic paint or with some material that will break the bond but that is not thick enough to destroy the bearing value of the key.

3. When the joints are likely to leak, they should be sealed in some way. Copper flashing is sometimes used as in Fig. 14-11(a). This copper should be folded into the joint so as to permit it to open slightly without rupturing the flashing; it must also be strong enough to hold its position during the placing of the concrete—an operation that must be done very carefully.

A second method of flashing, when the back of the joint is accessible, is pictured in Fig. 14-11(b). This is expensive because it is a sort of roofing job, but it is sometimes more reliable.

Copper water stops are likely to become crumpled during the con-

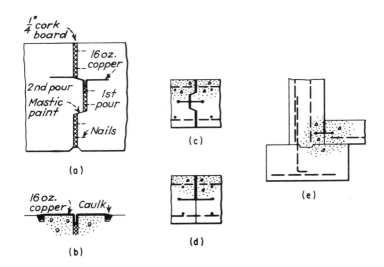

Fig. 14-11 Some types of water stops.

creting.   The depositing must be done carefully.   Steel plates $\frac{1}{8}$ to $\frac{1}{4}$ in. thick, with bolted and gasketed joints, can be used at construction joints, but the accordion action of the fold shown in the sketches is needed whenever movement can occur.

Rubber "dumbbell" water stops like that indicated in Fig. 14-11(c) are available.   These are made to fit corners and intersections of various kinds, both vertically and horizontally.   Still another type of rubber water stop is illustrated in (d) and (e).

4. When it is possible to do so, expansion joints should be entirely open, with an air space of 1 to 2 in. between the concrete sections.   This is especially desirable in bridges where considerable motion occurs.   However, to avoid visibility through parapets and similar parts, the joints may be offset in plan instead of being straight.   In any case, one should be able to clean out such joints.

5. V cuts should be used at the joints, or the joints should be placed at moldings, reentrant corners, and other suitable points.   The V cuts should be 1 to 2 in. deep or large enough (and the joint fillers thick enough) to guard against spalling of the edges because of the compressive forces which might be caused at the joint by expansion.

When a keyway is used at a construction joint AF to resist shearing forces as pictured in Fig. 14-12(a), the bearing on the surface DE is confined, and the unit pressure can safely be large because the concrete cannot get away.   The key may shear along BE or DG, but since this is ordinary shearing action without much diagonal tension, the unit resistance will probably be great.

On the other hand, assume a contraction or expansion joint like that of Fig. 14-12(b).   Here the faces AB, CD, and EF are not in direct bearing.   The pressure on DE may cause a crack along DG because of combined shear and bending.   The reinforcement is not close enough to resist this diagonal tension.   The key might crack along EH but this is less likely.   The section should then be proportioned so that the shearing stress across BE and DF does not exceed $0.03f'_c$ to $0.05f'_c$.

In comparing Figs. 14-12(a) and (b), notice that $GDEF$ of the former

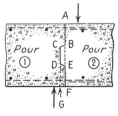

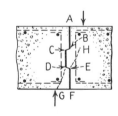

Fig. 14-12   Action of keys in shear.

(a) Key at Construction
Joint

(b) Key at Contraction
or Expansion Joint

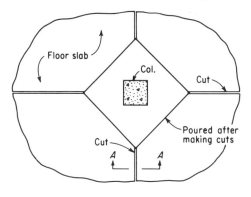
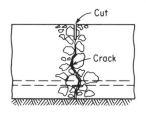

Fig. 14-12*A*

cannot rotate clockwise under the load on *DE* because the adjoining concrete will not let it do so, whereas *GDEF* of sketch (*b*) may do this, since
an opening or yieldable material exists along *EF*.

"Dummy" joints like that which is shown in Fig. 14-6 are useful in
some cases as contraction joints that will avoid unsightly shrinkage
cracks. The reinforcement should be weaker at these joints than it is
elsewhere.

One way to handle the problem of contraction joints in floors which are
poured on ground (and pavements also) is to pour and finish a large area
in one day, then on the next day to cut the slab into sections by making
1- to 1½-in. deep cuts by means of a diamond saw so that tension due to
shrinkage and temperature will cause the lower part of the slab to pull
apart below the cuts. This deformation will usually offset any future
expansion of the slab. The jagged vertical cracks below the cuts will
generally provide a sufficient keying effect for the transfer of any necessary
shearing forces. Of course, the steel which extends through the cracks
may be locally overstressed. One pattern for such cuts is that shown in
Fig. 14-12*A*. The diamond piece of slab is formed at first around the
column location and is not poured until after the main slab has set and the
cuts have been made.

Another problem is the movement at seats. In one case, a series of
beams was made so that they were continuous over some central piers and
simply supported at the ends, one of which was anchored and one of which
was labeled a "contraction joint." The notes on the drawing stated that
contraction joints were to be painted with asphalt. The workmen painted
the vertical portion of the joints but did not bother with the horizontal
seats. When the concrete of the beams set, it bonded to these supports.
Shrinkage later caused cracks close to the end in spite of the use of hooked
bars. If the seats are rough (not troweled smooth), it is advisable to use

a layer of tar paper or plastic on top of the asphalt paint so as to let the shrinkage and temperature changes cause sliding easily.

14-8. Waterproofing. Retaining walls, basement walls, spandrels of earth-filled arches, subways, tunnels, and similar structures must be watertight if they are to present a pleasing appearance. The denser the concrete itself is, the more impervious it becomes, so that too lean a mix is likely to facilitate leakage. However, it is exceedingly difficult—or almost impossible—to keep construction joints from leaking when there is an appreciable pressure of water behind them.

There are three ways in which this problem of seepage may be attacked, viz., integral waterproofing, inside surface coatings, and outside surface coatings.

Integral waterproofing denotes materials added to the concrete when it is mixed, in order to make the concrete itself impervious. The admixture supposedly fills all the voids through which the water might pass. In any event, complete reliance upon integral waterproofing is dangerous, especially at the construction joints.

Coatings on the insides of the walls are used sometimes as a means of stopping leaks, but they are very likely to be unsatisfactory. Mortars with impervious mixtures in them, paints with water glass or other chemicals which evaporate or congeal and leave crystals or chemicals in the pores of the surface of the concrete, and asphaltic paints—these are some of the coatings used. However, these are expensive, and it is unreasonable to expect them to stop the water at the last line of defense —the inside surface—especially when the structures are subject to temperatures that cause the water to freeze behind the surfacing. If used, inside applications should be chosen and applied by experts in that line.

The best place to stop the leakage is at the outside surface—the point of entrance of the water. There are two customary ways of doing this. The first, and the most effective, is the use of membrane waterproofing which forms a continuous watertight sheet outside the structure; the second is the use of asphaltic emulsions or similar bituminous coatings, or some plastics, forming waterproofing without membrane.

Membrane waterproofing is usually built up by coating the surface that is to be waterproofed with hot asphalt or coal-tar pitch, laying thereon successive layers of special fabric—placed shingle fashion and each layer coated with the mastic—until the desired number of layers or "plies" is obtained (two-ply, three-ply, etc.). Figure 14-13(a) shows this principle.

It is necessary to protect the membrane waterproofing against damage during backfilling; against penetration of oils, gasoline, or other solvents; and against the cutting tendency of sharp stones in the backfill. The first of these may be accomplished by the use of plywood which is laid

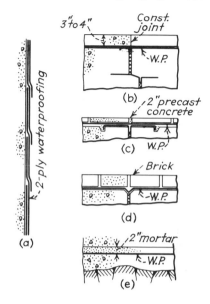

Fig. 14-13   Waterproofing details.

against the membrane, but this is only a temporary material.   Better
protective coatings are poured concrete 3 or 4 in. thick and precast-concrete blocks or bricks set in mortar as shown in Fig. 14-13(*b*) to (*d*), which
picture details at expansion joints—always troublesome points.

It is best to place membrane waterproofing directly upon the outside
of the wall, but this is not always economical, desirable, or possible.
When one is waterproofing subways, tunnels, and other structures which
are not accessible from the outside, it is necessary to provide a surface
on which the membrane may be placed before the concrete of the main
structure is poured.   A sample of this work is shown in Fig. 14-14, which
pictures part of the bottom of the New Jersey shaft of the Lincoln Tunnel.
A concrete lining, or "sand wall," is poured against the rock as shown
above the line of black waterproofing.   The membrane is then applied as
shown, using a 2-in. coating of mortar on the invert or bottom to protect
the waterproofing when the reinforcement and concrete are placed, as
shown in Fig. 14-13(*e*).   Of course, all roughness, projecting form ties,
and other sources of damage are removed before the membrane is applied.
It is necessary to keep the steel far enough away so that the subsequent
spading of the concrete will not cause damage to the membrane.   It is also
obvious that such waterproofing work should not be attempted when any
reinforcement is in the way.

The concrete must be dry when membrane waterproofing is applied.
It is also necessary to keep water pressure from building up behind the
sand walls, which are shown in Fig. 14-14, in order to avoid bulges in the

Fig. 14-14   Construction of invert of New Jersey shaft of the Lincoln Tunnel at New York City.

membrane or displacement of the concrete.   There is a shallow gutter-like drainage system behind the walls in the picture.   It leads to a temporary sump.

Waterproofing without membrane—sometimes called *dampproofing* —is merely the application of bituminous materials to the concrete as a paint coat or as multiple coats.   It is beneficial, but it can be applied only from the outside; it is easily damaged; it has no particular elastic properties; and it is likely to be ineffective after a number of years.   However, it is far better than no waterproofing at all.

It is less detrimental if waterproofing is omitted from stone-faced walls than from plain concrete-surfaced ones if the stones and the concrete are placed monolithically, because slight staining will not show very much, and the leakage is not likely to be serious.   However, it is hardly worthwhile to take such chances.   At least, one should apply waterproofing without membrane.

When a concrete slab is poured directly on the ground inside a building, it may be well protected against weather, but something should be done to prevent the penetration of moisture through the concrete from below. This is usually accomplished (1) by having a permeable base of stone or gravel below the concrete, with a positive drainage system so that neither standing nor flowing water can reach the concrete and so that capillary action of the moisture will be prevented from causing dampness in the concrete, or (2) by placing some kind of vapor seal on the ground immediately below the slab in order to keep the capillary action of the moisture from reaching the concrete, and by having a drainage system also to prevent the proximity of standing or flowing water. Various materials are used for this vapor seal, but they should be both waterproof and tough enough so that construction operations will not rupture them.

14-9. Drainage. When the conditions are such that natural drainage can be provided readily, it is advisable to construct a drainage system whose function is to remove the ground water behind or outside the structure. This is desirable even with waterproofed structures, especially when they rest upon rock. Such installations vary from simple weep holes through the walls to elaborate interconnected systems.

Any such drainage system is subject to clogging by silt and to freezing. It should therefore be laid out so that it can be cleaned by rodding or flushing and so that it is below or behind the frost line, if possible. This can be done by a system of manholes, Y connections at intervals, or even galleries such as that in Fig. 14-15(c) which can be inspected.

A few means of drainage are shown in Fig. 14-15. Sketch (a) shows a simple "blind," or stone, drain behind a wall. The water passes through the ground to this drain, thence to outlets through the wall, and finally through another system to a sewer or outlet. The addition of a vitrified-

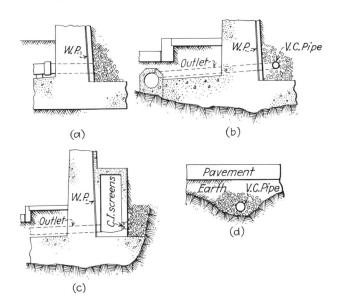

Fig. 14-15　Examples of drainage details.

clay pipe line which has open joints wrapped with burlap or tar paper, as shown in sketch (*b*), facilitates the removal of the water. The gallery drain in sketch (*c*), which can be entered at manholes, has been used behind some important walls along the approaches of the Lincoln Tunnel. The adaptation of the same principles for use under pavements—called a "French drain"—is pictured in sketch (*d*).

14-10. Deflections. The magnitude of the deflection of a reinforced-concrete beam is difficult to determine with accuracy. If the member is highly stressed, it will have many hair cracks; if lightly loaded, it may have very few. The former case is the one that usually concerns the engineer. The importance of the deflection is not so much a matter of its effect on the beam itself as it is of the effect upon the use and appearance of the structure. Satisfactory service is an important element of design.

It is natural to expect that high stresses in the tensile reinforcement will aggravate the cracking and the accompanying deflection; so will high stresses in the concrete. Shrinkage may also cause early deflection because of the shortening of the concrete on the compression side, but a temperature change will usually affect both steel and concrete similarly. If a beam has been loaded and cracks have occurred before shrinkage has been completed, further shrinkage alone will contribute to the deflection because of the shortening of the compression side, whereas the cracks isolate the tension side into short pieces. Creep of the concrete may also have considerable effect in increasing deflection caused by dead load and by long-term live load, as in warehouses. However, compression reinforcement will reduce the deflection because it opposes the compressive forces, the shrinkage, and the creep, but it may be a costly way of reducing deflections. Not only will shrinkage and creep affect deflection, once a beam has cracked, but they may also cause a redistribution of the bending moments in statically indeterminate structures.[1] This may be especially so when the live loads are light and the shrinkage is large.

It is particularly important for the designer to investigate the deflection of thin slabs and shallow beams, especially when the ratio of L/d equals 20 to 25 or more. This may apply to precast members as well as to those that are poured in place. The Code limits for the over-all depth of members are shown in Table 14-1. Smaller depths may be used if the deflections are computed and are found to be satisfactory.[2]

The Code differentiates between long-term and immediate or initial deflections. The big question in computing deflections is generally what

---

[1] H. Gesund, Shrinkage and Creep Influence on Deflections and Moments of Reinforced Concrete Beams, *J. ACI*, May, 1962.

[2] Y. Wei-Wen and G. Winter, Instantaneous and Long-time Deflections of Reinforced Concrete Beams under Working Loads, *J. ACI*, July, 1960.

Table 14-1   **Suggested Minimum Thicknesses**

Member	Minimum thickness or depth t			
	Simply supported	One end continuous	Both ends continuous	Cantilever
One-way slabs	$L/25$	$L/30$	$L/35$	$L/12$
Beams	$L/20$	$L/23$	$L/26$	$L/10$

to use for the moment of inertia I in the customary formulas and procedures for determining deflections. The Code suggests that I_c for immediate and short-term deflections should be based upon the gross section of the concrete alone when pf_y is equal to or less than 500, and upon the transformed cracked cross section when pf_y exceeds 500. For $f_y = 40,000$ psi, this limit corresponds to a value of $p = 0.0125$—very lightly reinforced members. The sudden change may seem strange because it means that the acceptable value of I_c for the uncracked section changes to approximately $I_{ct} = 0.5I_c$ when pf_y exceeds 500, as it would in much ordinary construction. Nevertheless, the general principle seems to be sound. It might be satisfactory to make the change more gradually, as suggested in Fig. 14-16(a). For continuous spans, the moment of inertia may be taken as the average of the values for the positive and negative moment regions. For members having moments of inertia which vary considerably throughout the span, the differences may be taken into account in the computations, or a sort of weighted average may be assumed.[1]

The justification for using I_c for lightly reinforced beams is the fact that the area of the concrete bd is relatively large, that the concrete is not likely to be highly stressed in compression, and that there may be less cracking of the tensile side, resulting in a small deflection. However, heavily reinforced beams will be relatively more shallow, more highly stressed, and more severely cracked, with more accompanying deflection. The use of I_{ct} for these beams is one way of taking this severely cracked and stressed condition into account. The more gradual transition from I_c to I_{ct} shown in Fig. 14-16(a) for use in the computations instead of the sudden change specified in the Code may be a better way of computing a reasonable approximation of the actual deflection.

In these computations for deflections, the modulus of elasticity E_c is assumed to be that given in Table 1-8 for the strength of concrete used.

---

[1] AASHO specifies that I_c of the uncracked concrete, without considering the steel, is to be used when computing deflections of beams and slabs but that the modulus of elasticity of the concrete is to be taken as $\frac{1}{30}$ times that of steel, or approximately $E_c = 1,000,000$ psi. This is their method of allowing for creep and shrinkage.

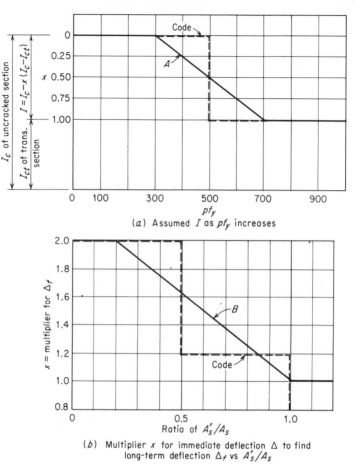

Fig. **14-16**   Data for estimating deflections of beams.

Of course, live-load deflections are classed as "immediate" or temporary in most cases.

The long-term deflection will usually be greater than the initial or immediate deflection when the loads are first applied. The Code states that the former may be estimated by multiplying the latter by 2.0 when $A_s' = 0$; by 1.2 when $A_s' = 0.5A_s$; and by 0.8 when $A_s' = A_s$. Perhaps this multiplier might be made to change more gradually by using the suggested line B in Fig. 14-16(b). The Code factor of 0.8 apparently is supposed to reflect the effect of the heavy compressive reinforcement on f_c, I, shrinkage, and creep. It is obvious that compressive reinforcement will stiffen a beam, and its effect should be somewhat in proportion to the amount of steel used in the compression side. Looking at the situation from the standpoint of USD, if $A_s' = A_s$ and both are stressed to f_y, the compressive force C_s is the same as the tensile force T, so that the concrete might seem to be unnecessary for resistance. However, the concrete will have some compressive stress when the steel in it is shortened.

In Fig. 14-16(a), notice that x is a reduction factor to be applied to the difference between I_c of the uncracked section and I_{ct} of the transformed cracked section, using the values found from line A. In sketch (b), the line B is to be used to find the factor x by which the immediate or short-term deflection Δ is to be multiplied to estimate the long-term deflection Δf. Notice that line B is purposely made to produce a minimum limit of $x = 1.0$. This is in order to be a bit conservative because once the structure has been built and found to be too flexible, nothing can be done about it without considerable cost—if at all. Furthermore, shrinkage of the concrete may reduce the effective participation of the concrete in resisting compression when A_s' is so relatively strong.

A usual limit for the immediate deflection due to live load is $L/180$ for roofs not supporting plastered ceilings and $L/360$ for roofs which do support such ceilings and for floors not supporting partitions. For roofs and floors supporting or attached to partitions or other parts harmfully[1] affected by deflection, $L/360$ is a maximum limit for the *total* deflection, but in many cases, this deflection is too large to be satisfactory.

The determination of what will be a satisfactory deflection is a matter requiring good engineering judgment. For example, a floor in a multistory parking garage may deflect without anyone noticing it. However, the floor of a ballroom in the third story of a building should be stiff enough so that dancing will not cause the lights in the ceiling below to vibrate objectionably.

In this problem of deflection, one should remember that, once a beam has been heavily loaded and the section has cracked considerably, the cracks will not heal when the load is removed, so that the cracked section will probably control thereafter.

The use of the conjugate-beam method of computing deflections for a prestressed tee was illustrated in Art. 13-10. To illustrate this method further, assume the bending-moment diagram and other data shown in Fig. 14-17. For this case, compute the total deflection at the center of the span.

The hatched area in Fig. 14-17 represents a typical bending-moment diagram for the end spans of a floor consisting of wide, shallow beams along the column lines with square panels of two-way waffles between them. This diagram may be considered to be made up of that for the

---

[1] Deflection of a floor above a partition may crush it; deflection of a floor under a partition may cause the floor to make a crack along the junction of the two, or it may cause shear cracks in the partition.

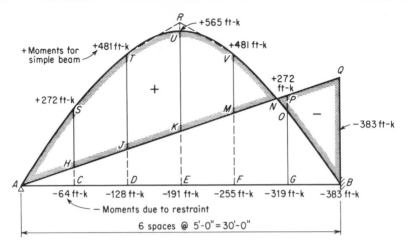

Fig. 14-17   Diagram for computing deflection.

positive bending moment alone as a simply supported beam represented by $ASTUVNOB$ and the triangular negative bending-moment diagram AQB acting on the simply supported span AB.   The deflection will be the algebraic sum of those caused by these two loading diagrams acting separately.   Compute the deflection at the center line E even though this is not the point of maximum deflection.

Assume that $f'_c = 4,000$ psi; $f_y = 50,000$ psi; for the uncracked section, $I_c = 0.86$ ft^4; for the transformed section, $I_{ct} = 0.49$ ft^4; $A'_s = 0.33A_s$; dead load $= 0.6$ times total load; $p = 0.0225$; and $E_c = 3,650,000$ psi or 525,000 ksf.

1. *Positive-moment diagram.*   It will be sufficient to assume that this is bounded by a series of straight-line chords except near the top, where the lines TR and RV will be substituted for the curves, with point R representing a moment of 590 ft-k.   Both E_c and I will be considered to be constant, and they will be introduced into the computations later.

In the conjugate-beam method, each span is considered to be simply supported and loaded with the M/EI diagram.   The bending moment at any point produced by this M/EI diagram loading gives the deflection at that point from the straight line or axis of the beam between the supports. In this case, the M diagram will be used directly at first.

Since the positive bending-moment diagram ARB is symmetrical about the center line, the end reaction will be equal to the area of one-half of the diagram.   Divide the diagram into triangles and rectangles, with the horizontal dimensions shown and with vertical ordinates equal to the positive moments shown above the diagram in Fig. 14-17.   Then find the following:

$$R_A = 272 \times 2.5 + 272 \times 5 + (481 - 272)2.5 + 481 \times 5$$
$$+ (590 - 481)2.5 = 5{,}240$$

$$M_E = 5{,}240 \times 15 - [272 \times 2.5 \times 11.67 + 272 \times 5 \times 7.5$$
$$+ (481 - 272)2.5 \times 6.67 + 481 \times 5 \times 2.5$$
$$+ (590 - 481)2.5 \times 1.67] = 50{,}300$$

Using I_c for the uncracked section, the immediate deflection is

$$\Delta_E = \frac{M}{EI_c} = \frac{50{,}300}{0.86 \times 525{,}000} = 0.1113 \text{ ft or } 1.34 \text{ in. downward}$$

2. Negative-moment diagram. Since this diagram is triangular,

$$R_A = \frac{383 \times 15 \times 10}{30} = 1{,}915$$

$$M_E = 1{,}915 \times 15 - 191 \times 7.5 \times 5 = 21{,}500$$

Since this is negative,

$$\Delta_E = \frac{M}{EI_c} = \frac{21{,}500}{0.86 \times 525{,}000} = 0.0478 \text{ ft or } 0.57 \text{ in. upward}$$

3. Final deflection. The net deflection Δ_E is $1.34 - 0.57 = 0.77$ in. downward for the total load. This is the immediate deflection without correction. Now assume that the dead-load deflection has to be modified by the data from Fig. 14-16(a) and (b) but that the live-load deflection is to be modified by Fig. 14-16(a) only, neglecting the effect of plastic flow, etc. Then

$$\text{Dead load } \Delta_E = 0.77 \times 0.6 = 0.462 \text{ in. downward}$$

This has to be adjusted as follows:

1. The deflection 0.462 in. downward is the immediate or short-term deflection based upon I_c of the uncracked concrete. This may have to be adjusted for the cracked condition.

2. The value of $pf_y = 0.0225 \times 50{,}000 = 1{,}125$. In Fig. 14-16(a), this value falls beyond the diagram. Following line A, this means that I_{ct} of the transformed section should be used in the calculation of deflections. Therefore, the immediate deflection should be increased as follows:

$$\Delta_E = \frac{0.462\,I_c}{I_{ct}} = \frac{0.462 \times 0.86}{0.49} = \frac{0.462}{0.57} = 0.81 \text{ in. downward}$$

3. The next step is to modify this adjusted immediate deflection to allow for creep. In Fig. 14-16(b), with the stated ratio of $A_s'/A_s = 0.33$, project upward from this value to line B, then to the left to determine x. Here x is about 1.82. Therefore, multiply the adjusted $\Delta_E = 0.81$ in. by this figure, giving

$$\Delta_E = 0.81 \times 1.82 = 1.47 \text{ in. downward}$$

Next, find the live-load deflection at E. Using I_{ct} instead of I_c, with $I_{ct}/I_c = 0.57$,

$$\Delta_E = \frac{0.77 - 0.462}{0.57} = 0.54 \text{ in. downward}$$

Therefore, the total deflection under full load is

$$\Sigma\Delta_E = 1.47 \text{ for DL} + 0.54 \text{ for LL} = 2.01 \text{ in. downward}$$

Notice that creep is not considered in the live-load deflection unless it is of long duration, as in a warehouse.

Is this total deflection satisfactory? The Code limit is

$$L/360 = 30 \times \tfrac{12}{360} = 1 \text{ in.}$$

Then the deflection is not satisfactory if there are partitions. The dead-load deflection alone would probably be so apparent that it would be objectionable. Cambering the forms might counteract it, but this would be troublesome and costly.

Although the magnitude of the deflection is questionable, a consideration of it qualitatively is very important. Otherwise, the designer may receive some rude shocks when his structures crack badly. Just three cases will be cited to illustrate this point and its consideration in the detail planning of structures, and in design matters.

Figure 14-18(a) shows one construction that was designed to hold a pair of heavy machines that would cause considerable shock. One machine was centered over the opening $KLOP$. Beam $ABCD$ was continuous across the wall, whereas beam $EFGH$ was simply supported at the wall. An 8-in. floor slab was a common part of both T beams. Even with the dead load alone, these beams will deflect differently—somewhat as shown in sketch (c). This, as well as the load of the

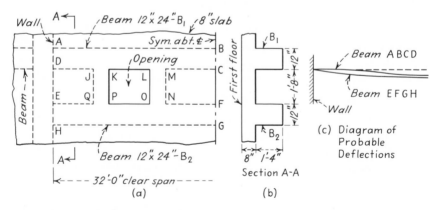

Fig. **14-18** Study of difference in deflections of beams under a machine.

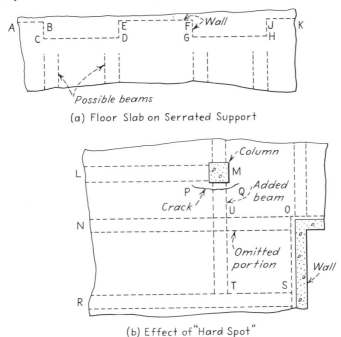

(a) Floor Slab on Serrated Support

(b) Effect of "Hard Spot"

Fig. **14-19**   Illustrations of effects of deflections.

machines and their vibrations, will tend to crack the slab and the diaphragms under the machines. When the situation was discovered, it was too late to redesign the structure. Therefore, posts were put under the centers of both beams so that deflections would be negligible.

If a long-span floor slab rests upon a wall that is serrated in plan, as in Fig. 14-19(a), the deflection of the slab tends to make it "ride" the edges CD and GH so that the load does not reach AB, EF, and JK satisfactorily. It may be desirable to use beams extending from the corners C, D, G, and H, then to span the slab across them.

If there is a jog in the supports, as pictured in Fig. 14-19(b), cracks may occur near the region PQ because beam NO may deflect enough with respect to the beam LM and the column to overstress the slab. It may be desirable to use a beam from M to T, make RS strong enough to hold it, and stop the first beam at point U.

14-11.   Encasement of Structural Steel.   Concrete is often used as an encasement around steel beams, girders, and columns in order to fireproof them or to protect them from corrosion. Gunite is used for the same purpose. However, it is necessary to do this work correctly so as to avoid unsightly cracking.

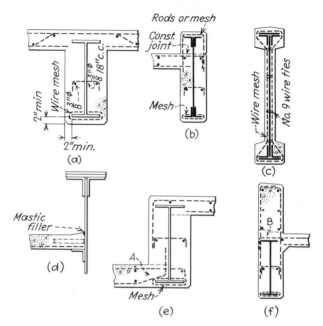

Fig. 14-20   Encasement of structural steel.

In the first place, if this encasement is placed on beams and girders before the major part of the dead load is applied, there is a probability that cracking of the encasement will occur, especially if the member has smooth flanges. For this reason, and in order to avoid spalling, the bottom flanges should be covered with wire mesh or beam wrapper, using 4- by 4- or 6- by 6-in. mesh and No. 8 or 9 gauge wires—preferably welded. The minimum encasement should be about 2 in. for smooth flanges or $2\frac{1}{2}$ in. when there are rivet heads to be covered. The concrete on the sides of the webs should be fastened together by bars or wires which pass through the webs. These details are pictured in Fig. 14-20.

Stiffeners on the webs of encased girders are likely to cause cracks. Thin encasement is relatively light, but it must be carried out around these stiffeners, and that causes expensive formwork. With heavy encasement, the longitudinal bars along the web should be carried through holes in the stiffeners. However, unnecessary stiffeners and other details that complicate the work of encasement should be eliminated.

When a bridge deck is poured against exposed steel, as in Fig. 14-20(d), the load of the slab should be carried on a shelf angle; welded ties should bond it to the metal; the surface should be pitched slightly away from the steel for drainage; and some good expansion joint cement or mastic should be placed in a small tooled groove at the junction with the metal because the joint must be sealed to prevent rusting of the steel.   Similar treatment is generally advisable in any case where bare steel enters concrete if it is exposed to the weather.   In important cases, it is often advisable to keep the concrete entirely free from the steel so that the latter can be painted.   It is obvious that channels with the flanges turned down, beams with horizontal webs, and similar steelwork causing air pockets are undesirable.   They should have at least 3- or 4-in. holes about 18 in. c.c. in the webs to permit concreting or grouting the hollow space.

Figure 14-20(e) shows another problem which arises sometimes when an offset in levels occurs at an encased beam.   The lower slab must be thick enough at A to deliver its reaction on to the top surface of the bottom flange of the I beam, because the part of the slab that is below this flange is almost valueless for transmitting shear.

When a parapet or projection is supported as shown in Fig. 14-20(f), the tendency of the steel beam to act by itself is likely to cause cracks at B due to longitudinal shear.   The parapet should have joints at close intervals, or else the member should be designed so as to act as a composite beam.   Furthermore, special care must be exercised to tie the parapet down to the encasement of the beam itself.

14-12.   Reinforcement around Openings.

When a rectangular opening must be made in a wall or some other member that is not subjected to bending, it is advisable to add special reinforcement, such as bars A in Fig. 14-21(a), in order to avoid the formation of cracks due to shrinkage, settlement, and changes in temperature.   Such corners are points of weakness.   However, the number of bars and the size to use are matters of judgment. In general, bars A should be at least sufficient to replace the temperature steel that the opening has eliminated; in other words, the area of two sets of bars should be 0.0025 times the area of the cross section of the cut —or more.

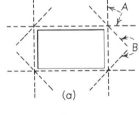

(a)

In cases where there are large shearing forces acting in the plane of the opening that is shown in Fig. 14-21(a), it is advisable to use additional diagonal bars like those marked B.   If there is a series of openings in a long structure, it may be advisable to reinforce the solid portions above and below the openings as girders with special full-height bars designed as stirrups.   The

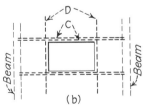

(b)

Fig. **14-21**   Reinforcement at openings.

area of the steel must be determined by one's judgment of the seriousness of the particular situation.

When the opening occurs in a slab that carries bending, as in Fig. 14-21(b), bars C should be added to make up for the bars that have been eliminated by the opening—or these bars should be sufficient to carry all the loads. Bars D are used to secure lateral spreading. The area of metal in them should equal that of the missing lateral bars or 0.0025 times the cross section of the opening. When the opening is small, the slab may not need to be thickened; but if it is large, small reinforced-concrete beams should be used instead of bars C, extending the beams across to the nearest main supports. Of course, when the conditions are very severe, bars D should also be replaced by beams.

14-13. Torsion in Concrete. A column that is subjected to torsion as well as to compression will have its entire cross section available to resist the torsion. When such a member is rectangular and is twisted enough to cause failure, it may start to crack near the middle of a long side, the crack tending to follow a helix at approximately 45° with respect to the longitudinal axis of the member. The failure is primarily due to tension in the concrete, otherwise the cracks cannot open up. The longitudinal reinforcement will try to act as dowels across any cracks and to resist any shearing movement of the two surfaces forming the crack as well as to resist their tendency to open up. However, the bars are relatively slender, and they depend upon the concrete to hold them in position laterally. Hoops and stirrups are useful in resisting the torsion if they are not too far apart. The most effective reinforcement for such a case is a spiral having its bars at an angle of 45° to the member's axis and arranged so as to cross the possible cracks.[1]

The maximum shearing stresses[2] in sections of isotropic material may be assumed to be the following:

Edge of circular section:

$$v_e = 0.637 \frac{M}{r_e{}^3} \tag{14-1}$$

---

[1] H. Gesund and L. A. Boston, Ultimate Strength in Combined Bending and Torsion of Concrete Beams Containing Only Longitudinal Reinforcement, *J. ACI*, November, 1964; H. J. Cowan, Design of Beams Subject to Torsion Related to the New Australian Code, *J. ACI*, January, 1960; G. P. Fisher and P. Zia, Review of Code Requirements for Torsion Design, *J. ACI*, January, 1964.

[2] See Paul Andersen, Rectangular Concrete Sections under Torsion, *J. ACI*, September–October, 1937. Equations (14-1) to (14-5), inclusive, are taken from this publication with minor changes in symbols.

Middle of long side of rectangular section:

$$v_m = \left(3 + \frac{2.6}{0.45 + \dfrac{B}{a}}\right) \frac{M}{8Ba^2} \qquad (14\text{-}2)$$

Middle of short side of rectangular section:

$$v'_m = \left(3 + \frac{2.6}{0.45 + \dfrac{B}{a}}\right) \frac{M}{8aB^2} \qquad (14\text{-}3)$$

where M = twisting moment
  r_e = radius of concrete section
  B = one-half of the long side
  a = one-half of the short side

In general, B should not exceed $1.5a$. The diagonal tensions and compressions that result from these shearing stresses may be considered to be equal to the intensities of the shearing stresses themselves.

The designer should endeavor to avoid torsional stresses that, when computed by the preceding formulas, will cause in the concrete a net tension that exceeds the allowable tensile stress in the concrete (after overcoming the compressive stresses due to longitudinal loads, if any).

When the torsional moments cause excessive tensile stresses in a circular member with a 45° spiral, Andersen states that the following relation exists:

$$6Kv_e{}^3 = (v_e - t_c)^2(3v_e{}^2 + 2v_e t_c + t_c{}^2) \qquad (14\text{-}4)$$

where

$$K = \sqrt{2} \times N \times t_s \times A_s \times \frac{r_s}{\pi r_e{}^3} \qquad (14\text{-}5)$$

  t_c = allowable tension in concrete
  N = number of spiral bars 45° to the axis that are cut by a horizontal plane
  t_s = allowable tensile stress in the steel
  A_s = cross-sectional area of one spiral bar
  r_s = radius to the line of the spiral

Even when the member is square, it may be considered to be approximately the same as one whose cross section is the circle that may be inscribed in the square.

Andersen's experiments also indicate that the ultimate torsional shearing unit stress in specimens that have no 45° spirals is approximately $v_m = 0.1f'_c$; spirals generally increase the magnitude of this ultimate stress; and the modulus of elasticity in torsion is about $0.45E_c$. The safe working stress in torsional shear should be about $0.03f'_c$ when no spirals are used.

The angular rotation of an uncracked rectangular member with a depth h which is not over 1.5 times the width b may be approximated as follows:[1]

$$\theta_T = \frac{3.33ML}{E} \times \frac{b^2 + h^2}{b^3 h^3} \tag{14-6}$$

where M = twisting moment
E = modulus of elasticity in torsion
L = length of the part of the member through which the torsion acts

Here L represents the distance from the point of application of the torque to the reaction point.   This applies when the torque is applied as a concentration; the twist caused by a torque uniformly distributed over the length of a member might be estimated by assuming that the total torque acts through $0.5L$ if the member is supported at one end or through $0.25L$ if it is supported against twist at both ends.

A beam which supports large transverse bending moments and is also subjected to torsion may be seriously affected by the torsion because of the probability that there are hair cracks in the tensile side of the member.

Assume a simply supported rectangular beam AB as shown in plan in Fig. 14-22(a) with the ends restrained laterally.   Let the projection CD have a downward force applied at C merely to picture something which applies a torque to the beam, but this is not intended to represent practicable construction.   A cross section of the beam is shown in (b).

It seems that a torque applied at the center of this beam and increased to failure might cause a cracking somewhat as pictured in Fig. 14-22(a) if the bracket is strong enough.   Plain concrete might fail suddenly.   It is obvious that closed double U-shaped stirrups or hoops, as in (b), will cross such cracks near any surface and that they will resist the cracking.

The German code[2] states that the torsional shearing stress at the edge of a circular or rectangular section without reentrant angles may be computed as

$$v_T = \frac{\eta t}{AD} \tag{14-7}$$

where A = cross-sectional area
D = diameter of the circular section or the width of the rectangular section at the mid-point of the long side
η = a coefficient related to the shape of the cross section

[1] Taken from Mauer and Withey, "Strength of Materials," John Wiley & Sons, Inc., New York, with changes in symbols.

[2] G. P. Fisher and P. Zia, Review of Code Requirements for Torsion Design, *J. ACI*, January, 1964.

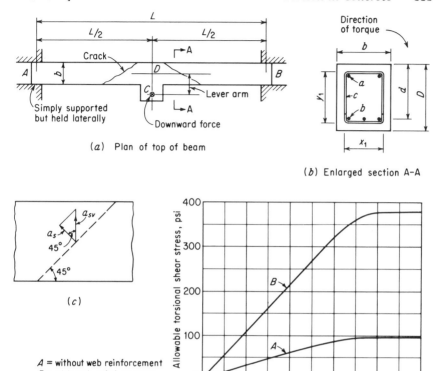

Fig. **14-22**   Schematic illustration of torsion applied to a beam.

and
$$\eta = 3 + \frac{2.6}{h/b + 0.45} \tag{14-8}$$

An approximate value for η is 4.5.

Cowan[1] showed that a_s, the cross-sectional area of one bar of a spiral reinforcing cage in a rectangular member that is subjected to torsion, is

$$a_s = \frac{\sqrt{2}\, M_{ts}s}{4\lambda f_s x_1 y_1} \tag{14-9}$$

[1] H. J. Cowan, Torsion of a Rectangular Elastic Isotropic Beam Reinforced with Rectangular Helices of Another Material, *Appl. Sci. Res.*, Vol. 3A, pp. 344–348 1953.

where M_{ts} = additional twisting moment due to use of the continuous
        spirals
     s = spacing between spiral bars
    x_1 = short side of the hoop
    y_1 = long side of the hoop
    λ = a function of y_1/x_1

For vertical hoops, as seen in Fig. 14-22(c),

$$a_{sv} = \frac{a_s}{\sin 45°} = \sqrt{2}\, a_s = \frac{M_{ts}s}{2\lambda f_s x_1 y_1}$$

For the area of *both* sides of the hoop, the total required area is

$$A_{sv} = 2a_{sv} = \frac{M_{ts}s}{\lambda f_s x_1 y_1} \qquad (14\text{-}10)$$

Cowan[1] states that the use of $\lambda = 0.8$ is reasonably accurate. Therefore,

$$A_{sv} = \frac{M_{ts}s}{0.8 f_s x_1 y_1} \qquad (14\text{-}11)$$

or
$$A_{sv} = \frac{M_{TN}s}{0.8 f_s x_1 y_1} \qquad (14\text{-}12)$$

where M_{TN} is the net torque after subtracting that taken by concrete.
    A piece of beam s long will contain one hoop, the volume of which is

$$A_{sv}(x_1 + y_1) \qquad (14\text{-}13)$$

An equal total volume of longitudinal steel is needed in the same piece of
beam if $f_v = f_s$, as it usually does.   This can be seen in Fig. 14-22(c).
Then

$$A_{sl}s = A_{sv}(x_1 + y_1) \qquad \text{or} \qquad A_{sl} = \frac{A_{sv}(x_1 + y_1)}{s} \qquad (14\text{-}14)$$

This can be expressed as

$$A_{sl} = \frac{M_{ts}(x_1 + y_1)}{0.8 x_1 y_1 f_s} \qquad (14\text{-}15)$$

This is in accordance with the Australian code.
    The stresses in the hoops from the torque and vertical loads may be
considered to be directly additive.   If part of the shear is assumed to be
resisted by the concrete alone at an allowable stress v_c for plain concrete,
this v_c should be deducted from the computed shearing stress caused by the
loads producing flexure *or* from those caused by the torque—not from both.
However, there is considerable question as to how much, if any, of the
shear should be taken by the concrete when reinforcement is provided.

---

[1] H. J. Cowan, Design of Beams Subject to Torsion Related to the New Australian
Code, *J. ACI*, January, 1960.

If part of the torque is allocated to the concrete, assume M_{TN} in Eq. (14-12) to be the remaining part to be resisted by the hoops. Taking all of the torque and vertical shears in the hoops is the safest, but some increase in the allowable unit stress might be permitted.

For T beams, the torsional resistance may be somewhat greater than that of the area $b'D$, but when the action is uncertain and safety is vitally important, the assumption that the member is the rectangular beam $b'D$ is the safest procedure.

Figure 14-22(d) is a diagram in which line A shows suggested allowable torsional shearing stresses for members without web reinforcement and curve B shows those allowable with web reinforcement. These curves are approximations of those allowed by the Australian code as reported by Fisher and Zia, but slightly more conservative. These are for WSD only.

In general, small-diameter hoops near together are better than large ones farther apart when needed as torsional web reinforcement.

To illustrate the procedures given herein, assume the beam shown in Fig. 14-22(a) and (b). Let $L = 16$ ft, $b = 15$ in., $D = 20$ in., $d = 17$ in., $x_1 = 10.5$ in., $y_1 = 15.5$ in., $a =$ two No. 8 bars, $b =$ three No. 11 bars, $c =$ No. 4 at 5 in. c.c., $f_s = f_v = 18,000$ psi, and $f'_c = 3,000$ psi. A downward force of 15,000 lb is applied at C with $CD = 16$ in. Are the hoops and the longitudinal bars safe for the vertical loads and the torque, not relying on the concrete for shear?

$$w \text{ or DL} = 1.25 \times 1.67 \times 150 = 312 \text{ plf}$$

$$M_D = 312 \times 16^2 \times \tfrac{12}{8} + 15,000 \times 16 \times \tfrac{12}{4} = 840,000 \text{ in.-lb}$$

For flexure,

$$A_s = \frac{M}{f_s j d} = \frac{840,000}{18,000 \times 0.9 \times 17} = 3.05 \text{ in.}^2 \text{ (approx)}$$
$$V_{\max} = 312 \times 8 + 7,500 = 10,000 \text{ lb}$$

Since torsion complicates the problem, compute v at A, not at d therefrom.

$$v = \frac{V}{bd} = \frac{10,000}{15 \times 17} = 39 \text{ psi}$$

From Eq. (14-8),

$$\eta = 3 + \frac{2.6}{20/15 + 0.45} = 4.46$$

Then, from Eq. (14-7), using one-half of the total torque because both end sections will resist the twisting, and allowing nothing for the concrete,

$$v_T = \frac{4.46 \times 120,000}{15 \times 20 \times 15} = 119 \text{ psi}$$

This is greater than what curve A of Fig. 14-22(d) allows, so that web reinforcement is needed. With $M_{TN} = 120,000$ in.-lb, Eq. (14-12) gives

$$A_{sv} = \frac{120,000 \times 5}{0.8 \times 10.5 \times 15.5 \times 18,000} = 0.255 \text{ in.}^2$$

For vertical loads, the area of hoops required is

$$A_v = \frac{vbs}{f_v} = \frac{39 \times 15 \times 5}{18,000} = 0.162 \text{ in.}^2$$

Combining A_{sv} and A_v gives the needed area of one complete hoop = $0.255 + 0.162 = 0.417$ in.^{2} The two sides of a No. 4 hoop provide only 0.4 in.^{2} This could be accepted, or the hoops could be placed closer together.

The required extra longitudinal steel in the bottom of the beam, from Eq. (14-14), is

$$A_{sl} = \frac{0.255(10.5 + 15.5)}{5} = 1.33 \text{ in.}^2$$

Adding this to the 3.05 in.^{2} needed for flexural resistance gives the required $A_s = 3.05 + 1.33 = 4.38$ in.^{2} This is less than the 4.68 in.^{2} provided, and it is safe.

Although the top bars a in the beam are under some compression, it seems that their area should equal A_{sl} because the torque effect goes clear to the ends, whereas the bending moment from vertical loads approaches zero at the supports. This area is provided.

In any rectangular beam that is subjected to a large torque, there should be at least one No. 4 bar near each corner to provide something to which the stirrups can be attached during concreting and to provide some dowel action. The stirrups should be continuous or double U-shaped ones lapped on the sides. The spacing of the stirrups probably should not exceed $0.33d$ for beams and 0.5 times the least lateral dimension for columns when the torque is sufficient to be considered in the design.

When a beam is subjected to a large torque, and when the beam is relatively deep with respect to its width, it may be sufficient to make an approximate design. This will be illustrated qualitatively. For example, assume the canopy pictured in Fig. 14-23(a). The cantilevered slab $CDFE$ tends to twist as well as to bend the spandrel beam. The top of the spandrel beam is restrained laterally by the floor slab, but the floor is not supposed to resist the torsion caused by the canopy. It would be possible to use intermediate haunched beams in the floor to resist the torsion, as indicated by the dotted lines in sketch (b).

Figure 14-23(c) shows the general character of the details of the construction. The moment M per foot of spandrel is the weight W of 1-ft

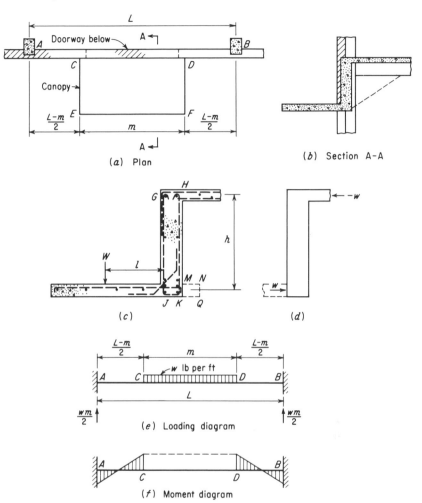

Fig. 14-23   Example of torsion caused by a canopy.

width of the canopy and its live load times the distance to the bars GJ.
If M is divided by some approximated depth h, this gives the force w
shown in (d), and it may be considered as a load applied horizontally to
the bottom portion of the beam.   This gives a loading somewhat as
shown in (e).   Since the spandrel is continuous here, and since the canopy
is very stiff laterally, the bending moment due to w may be assumed to be
the reaction $wm/2 \times AC/2$ at A and B and of the same magnitude and
opposite sign at C and D.   This bending moment should be assumed to be
resisted by the bottom portion of the spandrel, but the limits of the
assumed effective member are indefinite.   The resulting tension in the

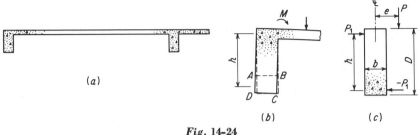

(a)

(b)   (c)

Fig. 14-24

bars near J and K in sketch (c) should be added to those produced in these bars by vertical loads, and the sum should not exceed the safe value.

The shearing forces must be considered also.   The effect of the vertical shear in the spandrel $GHKJ$ will be small because of its large depth. However, shearing stresses horizontally in the bottom of this member in the portions AC and DB in sketch (e) may be large and difficult to resist.

If the bottom of the beam is too flimsy, the width of the spandrel may be increased for the full height, or a projection $MNQK$ [Fig. 14-23(c)] may be made inside from column to column, with a construction joint at the level MN and the top of the canopy.   If the canopy is long enough so that it exceeds L, the canopy itself will act as a horizontal beam that will deliver the reactions to the columns.   Haunches on the floor beams may be needed to assist the columns in resisting the horizontal forces.

The tension in the bars GJ in Fig. 14-23(c) should be computed by taking moments of the load W about the inner bars HK.

A spandrel beam used to support the end of a large slab has a torque applied at its top due to the spandrel's attempt to restrain the slab.   The construction in Fig. 14-24(a) is shown greatly distorted in (b).   The bottom portion $ABCD$ may be designed to resist a horizontal load $w = M/h$ for its full length.   However, the lateral deflection or twisting of the beam tends to reduce the moment M.   Therefore, some attempt should be made to estimate the relative stiffness of the spandrel compared with that of the slab.   If the beam can safely withstand the full load w, a smaller load certainly will not hurt it.

Again, assume a deep beam like that shown in Fig. 14-24(c) for which D greatly exceeds $1.5b$.   The beam is acted upon by a force P with a lever arm of e, or by an equivalent moment M which can be called Pe.   Assume a depth $h = D - 12$ in., approximately.   Then a moment P_1h can be substituted for Pe.   Consider the upper portion of the beam to be loaded by a force $P_1 = Pe/h$ acting horizontally, and applied in the same longitudinal position as P.   The bottom of the beam is to be designed for P_1 acting in the opposite direction, as shown.   It is usually sufficient to add the stresses caused by P_1 directly to those produced by the load P acting vertically on the member.

If a heavily loaded beam is subjected to large torsional action such that the safety of the member is uncertain, it probably is good engineering to give thought to changing the system of framing to remove all doubts regarding its action and safety.

14-14. Reinforced-concrete Brackets and Seats. When a load is applied upon a bracket of reinforced concrete as pictured in Fig. 14-25, the tendency is to cause tension in bars A, also to open up the column at E and G so as to produce tensile stresses (or reduced compressive ones) in bars B and C at these two points, with compressive stresses at F and H.

The eccentricity of the load, as far as the bracket is concerned, may be assumed to be e, the distance from the nearer row of steel to the load, because the point of compressive resistance is near F at the bottom. When e is less than d, it is sufficient to assume that the cracks will be about as shown in the figure so that the compressive stresses will be small and the bars A can be designed by the simple formula

$$M = Pe = A_s f_s jd = 0.87 A_s f_s d$$

When e exceeds d, the bracket should be analyzed, as a cantilever beam, for compressive, tensile, and shearing stresses. The applied moment may still be called Pe.

When the shearing stress in a rectangular reinforced-concrete bracket is computed, the critical section is to be taken at the face of the support, not a distance d therefrom. If the bracket is trapezoidal, as in Fig. 14-25, the critical shearing stress is to be computed for the affected section providing the least resistance.

For the column itself, the effect of the load P should be determined upon the basis of combined bending and direct stress (Chap. 7).

The cracking of brackets and shelves holding heavy concrete members (such as small bridges and beams which are supposed to have simply supported ends) sometimes seems to arise from a combination of the effects of vertical loads, shrinkage, temperature drop, and release of live-load deflections. Even when steel plates have been used as bearings, the frictional resistance (or "freezing") has been too great and caused a horizontal resistance to movement of the member resting on the bracket or shelf. The tieback steel in the top of the support should be designed to resist this combination of forces.

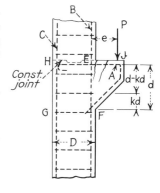

Fig. **14-25**

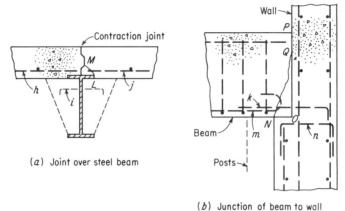

(*a*) Joint over steel beam   (*b*) Junction of beam to wall

Fig. 14-26   Cracking caused by inadequate bearing.

Figure 14-26(*a*) illustrates a case of inadequate bearing.   As shown, a contraction joint was made over the steel beam, and the bars *h* and *j* were stopped short.   Of course, the keyway here was of no use, but was actually harmful.   The concrete cracked as illustrated by *LM*.   By drilling holes in the steel beam, inserting bars *i*, and packing in the concrete haunches, as shown by the dotted lines, it was possible to make a seat which at least would prevent complete failure.

Another failure is shown in Fig. 14-26(*b*).   Here a continuous shelf was left for supporting a series of beams on a basement wall.   Small ties *k* were installed, and the main bars *m* were hooked.   Ties *n* were able to keep the seat from failing, but some of the beams cracked as shown by *NQ*. The hooks could not do much good and the ties were ineffective—probably because the pull caused by shrinkage and temperature was too great and therefore caused the concrete to have tensile stress as well as shearing stress at this end section.   The construction of posts as shown by the dotted line was the most practicable remedy.

14-15.   Planning Reinforcement.   As an example of some of the things that should be considered when planning reinforcement, examine the case illustrated in Fig. 14-27.   Sketch (*a*) shows the cross section at one portion of a twin intake tunnel for condenser water that is proposed for a power plant.   It rests upon sand, and it supports a screen and pump house above it.   The downward pressure on the bottom slab caused by the weight of the structure exceeds the hydrostatic uplift, even when the gates are closed and the tunnels are empty for cleaning.   It is assumed that water can seep between the tremie seal and the slab.   This empty state is a critical load condition for the invert.   In such a case, the bottom acts as a continuous slab with tension in the lower reinforcement near E, F, and G, and in the top bars near H and J. When one tunnel is emptied, the pressure of the water in the other one tends to cause tension near B or C in the center wall.

Assume that the bending moments and shears have been computed for design, that tentative thicknesses have been selected, and that a section requiring no web reinforcement is to be used. The reinforcement may be planned in various ways, but it is desired to find the best arrangement. The following comments refer to the arrangements shown in the various sketches in Fig. 14-27:

(*b*) Bars a extend clear across the bottom and are bent up to resist the moments at the bottoms of the outer walls. Extra straight bars b are added under the center wall to resist the greater tension there. Bars c are extended clear across the top of the slab, with d added to reinforce the centers of the spans. To resist the unbalanced tension in the center wall, dowels e are used. Then a few light dowels f are added to hold the inner reinforcement of the outer walls during concreting.

(*c*) Here bars g in the bottom are lapped under the center wall to

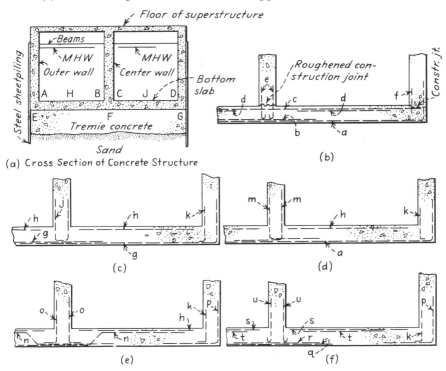

Fig. 14-27 A study of arrangements of reinforcement for a heavy slab subjected to hydrostatic uplift.

provide for the greater bending moment there.   Bars h are made only as long as the tension in the top of the slab requires them to be.   Single U-shaped dowels j are used for the center wall.   They can then be rested on the lower mat during concreting.   The bend at the bottom of k is for the same purpose.

(d) Bars a are used clear across the bottom, but dowels m are bent alternately left and right to reinforce both the wall and the slab.   However, the bond stresses may be undesirably high between the bends when the bars act in resisting tension in the bottom of the mat.   When the bars are close together, this tends to cause a screen effect during concreting.

(e) Bars n are bent up to serve as top reinforcement after they are not needed in the bottoms of the slab, thus helping bars h.   When the slab is deep, this arrangement is not usually desirable.   Dowels o are made with the bends turned outward.   This is undesirable and is not the most effective anchorage against bending in the center wall.   Bars p are extended to take care of the tension in the outer corners.

(f) This shows bars q and r of minimum length to provide for the bottom tension under the center wall and bars p to reinforce the outer corners. Bars s and t are the minimum for the tension in the top.   Dowels u are obviously turned the wrong way.

Conclusions.   Such a heavy important structure should be "knitted" together well.   The author prefers bars g in sketch (c) for bottom reinforcement, with extra bars p in sketch (e) at the corners if needed.   All are of moderate length and are easy to place, and the bottom is tied clear across. For the top of the slab, bars c and d of sketch (b) seem to be the best since the slab is tied together well, and there is some reinforcement crossing the center wall to resist the effect of tension near the corners.   For the center, dowels j in sketch (c) are preferable when the U is broad enough; otherwise m in sketch (d) is preferable.   A few light dowels k in sketch (c) placed perhaps 3 ft c.c. are generally worth their cost.

The detailing of reinforcement[1]—the making of the shop drawings—is an important job because they are used for the cutting and bending of the bars as well as for erection in the field.

14-16.   Heavy Concrete.   Inasmuch as shielding persons from radiation has become an important problem, and since the effectiveness of the shielding depends, to the first approximation, upon the mass of the interposed materials, there has recently been considerable attention given to the development of heavy concrete.   This shielding may have to be fixed in position, or it may be movable in the form of precast units.   The Brook-

---

[1] See Manual of Standard Practice for Detailing Reinforced Concrete Structures (ACI 315-57) and revisions thereof as discussed in *J. ACI*, September, 1964.

haven National Laboratory, Associated Universities, Inc., Upton, Long Island, New York has given this matter considerable study.   The following paragraphs are based upon their experience.

It is possible to use ferrous metal pieces as aggregates, producing concretes weighing perhaps 300 pcf.   However, these are very costly.   One can realize the difficulties of mixing such concretes with conventional equipment and of preventing segregation when the concrete is placed in the forms and is vibrated for consolidation.   The coarse aggregate might be prepacked in the forms and then pressure-grouted, but this is also expensive.

Apparently, it is practicable to obtain concretes weighing 230 to 240 pcf by the use of Ilmenite coarse and fine aggregates from Canada or magnetite ores from the United States.   These concretes have been tested and found to yield strengths of 3,000 to 4,000 psi at an age of 28 days. The cost of concrete heavier than about 240 pcf rises rapidly with an increase in the unit weight.

Such concretes may be produced at the site or transit-mix trucks may be used.   Of course, the volume to be mixed at one time by standard equipment must be reduced to yield about the normal total weight of the charge.   Even then, extra mixing blades may have to be installed, and extensive repairs and replacement of equipment must be expected because of severe wear.

In general, it seems to be desirable to design movable shielding in the form of strongly reinforced precast units with recessed heavy U-bolts placed over the center of gravity of each unit so that it can be lifted by a crane and readily turned in any direction.   This lifting device may be in one side and in one end so that a given piece may be stacked on its side or stood on end.   It is also desirable to protect the corners of such blocks by steel angles anchored into the units.   Furthermore, the workmanship may be better if the blocks are cast at the plant of an experienced contractor and shipped to the site rather than to attempt to set up special equipment in the field.[1]

14-17.   Miscellaneous Details.   There are many practical and theoretical points which the engineer learns by experience.   A few of these are pointed out here because they sometimes cause trouble.   Referring to the sketches in Fig. 14-28, note the following:

(*a*) When a beam parallel to a wall is poured against it, when it is keyed or bonded to it, or when it rests upon the edge of the wall, it will try to shift its load to the wall because its deflection is prevented.   The result

[1] See K. Shirayama, Properties of Radiation Shielding Concrete, *J. ACI*, February, 1963.

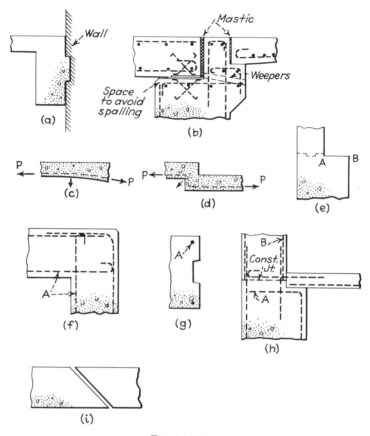

Fig. 14-28

is a breaking of the junction or an eccentric load on the wall.   The beam
and the wall should be separated by a deflection joint (tar paper or similar
isolating material), or the edge should be made wall-bearing.

(*b*) When an expansion bearing is provided as shown in the sketch, the
frictional resistance to motion will set up tensile stresses (at both fixed
and expansion ends) which require special hooks in the bars.

(*c*) and (*d*).   When bars are curved or offset, tensile stresses in them
will tend to straighten them out and to spall the concrete unless they are
tied back or otherwise detailed properly.

(*e*) When horizontal shelves, offsets, or brackets occur, like *AB* in the
sketch, it is not advisable to try to pour the top surface against forms
because of the uplift and the difficulty of filling the forms properly.   When
AB is sloped appreciably, forms can be used, but uplift must be guarded
against.

(*f*) Corner reinforcement should be made so that one set of outside bars is bent around the corner. When there is a tendency to open up the inside corner, the inner bars *A* should be hooked near the outside of the wall so as to reinforce this corner. Without these hooks, bars *A* have very little effective anchorage. If the walls are thin, the continuity will be poor. There is insufficient room for the hooks, so that it may be best to use a single bar for *A* and to bend it around 270° to the right so as to form an inside loop, the other end extending to the left.

(*g*) Bars at corners and in projecting parts—like *A* in the sketch—may cause trouble unless they are tied in properly. They should be placed so as not to interfere with the placing of the concrete.

(*h*) In cases of shelves where bars *A* are bent back, one must be careful to space them far enough apart to avoid forming a screen effect that will interfere with the placing of the concrete. Small U-shaped bars which are placed by hand are often more advantageous than the bending of the main bars. The vertical bars *B* should rest on the construction joint, with dowels to tie them into the lower pour. Any horizontal ties between the slab and the wall should be above the construction joint if the floor is poured last.

(*i*) Sharp corners such as those at this expansion joint should be avoided because they may break off.

In planning all layouts of reinforcement, one must consider that wire chairs or mortar pads must be used to support bars that are above horizontal forms; vertical bars cannot hang in the air but must rest upon a support, such as the previous pour, or they must be wired to other bars that are so supported; horizontal bars must be wired to the vertical ones which act as small columns in supporting them; intersecting bars must be tied together thoroughly; tie bars or spacer bars must be used to hold the main reinforcement in line; and multiple layers of bars must be held by ties and separators so as to make sure that their proper relative positions are maintained.

When surfaces have to be screeded, it is important to avoid projecting reinforcement that will impede the screeding. Small dowels or other reinforcement may be inserted after the finishing is complete. This often applies to bridge decks.

Excessively large groups of closely spaced bars may look well on a drawing, but they are very difficult to place accurately and to hold in position in the field. If they get out of line, if the upper layers sag too close to the lower ones, or if one of the bars is bent sideways so as to get too close to the forms or to an adjacent bar, they tend to act as a screen which makes the placing of the concrete very difficult—and the development of adequate bond very questionable. Honeycombing, segregation of the aggregate, and air pockets are likely to be the result. When such a group of bars is at the top of a large member, it is desirable to arrange the bars so that there will be one or two strips of clear space (about 5 or 6 in. wide) through which the concrete can be deposited and compacted.

When heavy mats are poured on soils, it may be desirable to use a 1-in. mortar or concrete protection coat over the ground so that the supports for the heavy reinforcement will not sink into the soil and reduce the cover over the steel, and so that the ground will not get mixed with the bottom of the concrete.

There may be fallacies in the use of fillets or haunches near the ends of beams and slabs with the intent of increasing the depth and securing greater resistance to diagonal tension. For example, take the case of Fig. 14-29, which pictures construction for the roof of a tunnel. In sketch (a) this roof is designed to be restrained at the ends and continuous with the side walls. The compression side is therefore the inside portion, and the effective depth d_2 near each end is increased by the fillet. In (b), the roof is designed as a simply supported slab, and the fillet does little or no good because the effective depth d is still measured from the top to bars c. The structure might as well be made as in (c). Surprising as it may seem, the simply supported slab in sketch (c) may be more economical in over-all cost than is that in (a) with the latter's heavier walls, corner reinforcement, and fillets. Again, the simple-span roof might be sloped as in (d) so as to reduce the size of bars f, the depth d being measured from the top.

Welded wire mesh and mats made of welded small bars are types of reinforcement which are very useful in pavement slabs, lightly loaded floors, roofs, and joist or waffle construction. If used, the sides and edges should be lapped properly.

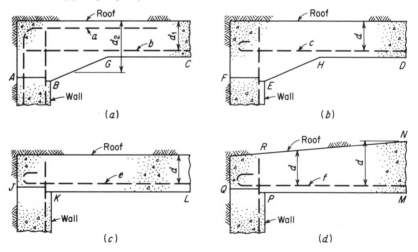

Fig. 14-29   Alternate schemes for construction for tunnel roof.

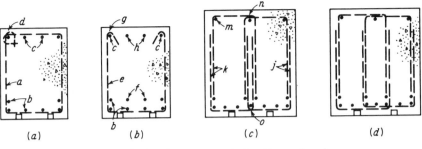

Fig. 14-30 Study of arrangements for stirrups in a beam.

There are likely to be troublesome details required in connection with the attachment of wood, steel, and other materials to concrete that is exposed to weathering. The initial construction may be easy but maintenance and replacement may be difficult. Anchor bolts embedded in the concrete may rust off. Furthermore, if the bolts become damaged, they cannot be repaired easily. One possible arrangement is the use of pipe sleeves through the concrete or to recesses so that the bolts can be replaced. When the bolts have to resist shear, the use of double nuts will hold the anchor bolts to the forms or templates during the concreting, and the inner nut will serve to help spread the shearing force to the concrete.

Figure 14-30 shows some types of reinforcing arrangements for rectangular beams (and T beams also). Notice the following:

(*a*) Stirrups *a* support the longitudinal bars satisfactorily, but the hooks *d* should be replaced by 90° bends so that they can be prefabricated and closed after the main steel is inside.

(*b*) Bars *f* and *h* are not supported by the stirrups but will require other means to hold them.

(*c*) The main bars are supported, but the double stirrups *k* and *j* overlap too little at *n* and *o*.

(*d*) These stirrups are properly placed, and all longitudinal bars are supported.

These matters that have been discussed here may seem to be minor details, but careful attention to such practical things often makes the difference between a good job and an unsatisfactory one. Furthermore, good judgment, common sense, and the ability to supervise work carefully will always be among the greatest assets of the designer and the builder of reinforced-concrete structures.

These are some of the things that one should bear in mind when he is designing a concrete structure. Continued study by architects and engineers, more extensive experience on the part of building contractors, and greater skill and care on the part of the men in the field—all these things together will bring still further advances in this great field of construction.

DESIGN PROBLEMS

15

15-1. Introduction. It is the purpose of this chapter to present several general layouts or plans of structures that are to be built of concrete. The plans will show the barest outline as it might be prepared by an architect, an owner, a mechanical-layout man, or someone else who establishes the general dimensions required and the facilities to be housed. The student, as a structural engineer, is to develop the framework and details of the structure built of concrete.

This method of approach is the one required in practice. It is essential that the student learn to tackle such creative problems, not just to determine stresses in or sizes of particular isolated members. This is because an engineer does not design a bunch of beams and columns first, then try to fit them together to form a structure. He has in mind first a structure that he wants; then he determines how to frame it and to make it strong enough. Practice with detail problems has been given with most of the preceding chapters in order to train the reader in the use of the tools of analysis and detailed designing. Now he should use these tools and practice building something with them.

The problems given here vary from small structures that are relatively simple to some that involve considerable difficulty. Thus the reader can practice upon whatever his abilities and the available time permit.

15-2. Assumptions. It is assumed that the student is not familiar with the analysis of statically indeterminate structures. Therefore, when continuity is involved, as it so frequently is in structures built of concrete, he has to make approximations. He may assume that the bending moments in ordinary continuous beams are the same as they would be if their ends were fixed, or if one end were fixed and the other were simply supported. Then he can use the data given in Figs. 1 and 2 of the

Table 15-1   **Approximate Bending Moments and Shearing Forces**
in Continuous Beams with Uniform Loads*

End spans:
  Positive moments:
    Discontinuous and unrestrained................................... $\frac{1}{11}wL'^2$
    Discontinuous but integral with supports......................... $\frac{1}{14}wL'^2$
  Negative moments:
    At exterior face of first interior support:
      Two spans.. $\frac{1}{9}wL'^2$
      More than two spans.. $\frac{1}{10}wL'^2$
    At face of exterior support for members built integrally with their
      supports:
      Spandrel beam or girder.................................... $\frac{1}{24}wL'^2$
      Column... $\frac{1}{16}wL'^2$
Interior spans:
  Positive moments.. $\frac{1}{16}wL'^2$
  Negative moments:
    Other faces of interior supports............................... $\frac{1}{11}wL'^2$
Negative moment at face of all supports for:
  Slabs with spans not exceeding 10 ft............................. $\frac{1}{12}wL'^2$
  Beams and girders where ratio of sum of column stiffnesses to beam stiff-
    ness exceeds eight at each end of the span..................... $\frac{1}{12}wL'^2$
Shearing forces:
  In end members at first interior support........................ $1.15\dfrac{wL'}{2}$

  At all other supports.. $\dfrac{wL'}{2}$

* Prestressed-concrete members are excluded.   The longer of two adjacent spans is not to differ from the shorter by more than 20 per cent.   Unit live load is not to exceed 3 times the unit dead load. L' is the clear span for positive bending moments and shears, and the average of two adjacent clear spans for negative bending moments.

Appendix.   On the other hand, he may use the data given in Table 15-1, these equations being based upon the ACI Code (318-63).

Obviously, both of the preceding procedures for estimating bending moments in beams are arbitrary approximations and are limited in their application.   Therefore, the author has prepared the influence-line diagrams shown in Figs. 15-1 to 15-4, inclusive, to assist the reader when there are special loading conditions on continuous beams.   These diagrams are for use in estimating reactions and shearing forces as well as bending moments.   The curves give the magnitudes of the coefficients to be used in the equations shown under the diagrams.   Remember that an influence line gives the value of a function at one given point for a load placed anywhere on the beam—in this case, the three-span beam shown under the drawings.

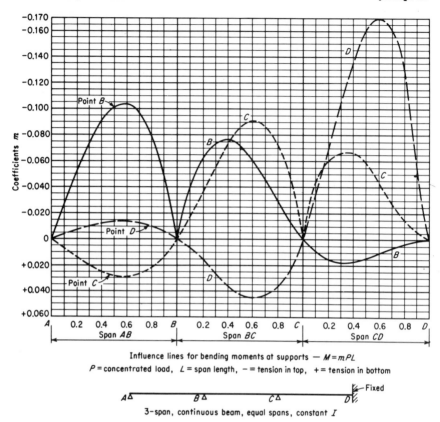

Influence lines for bending moments at supports — $M = mPL$

P = concentrated load, L = span length,   − = tension in top,   + = tension in bottom

3-span, continuous beam, equal spans, constant I

Fig. 15-1   Influence lines for negative bending moments at supports of continuous beams.

This particular beam has been chosen for the following reasons:

1. Spans AB and BC are fairly representative of the end two spans of a beam which is continuous over a series of supports and simply supported at one end.   The bending moments at the middle of BC and at point C can be used as typical values for all interior spans.

2. Spans BC and CD can be used for cases where the end of a continuous beam is fixed, as shown at D.   Then the bending moments at C and at the middle of span CD can be assumed to be typical for all interior spans.

3. The spans assumed here are all equal and the member has a constant moment of inertia throughout.   Spans and members which differ widely from these assumptions will naturally be beyond the scope of the diagrams.

Figure 15-1 contains three influence lines for bending moments at the various interior supports.   The solid line B is to be used to find the bending moment at point B; the dotted one, for C; and the dashed one, for D. Of course, the bending moment at A is zero.

To illustrate the procedure, assume that one wishes to find the maximum bending moment at B for two concentrated loads equal to 5 kips each spaced 5 ft apart on a continuous beam having spans equal to 20 ft. One way to do this is to draw two vertical lines representing the loads on a small piece of thin tracing paper, making the spacing equal to $0.25L$ to the same scale as the diagram. Slide this paper over the influence-line diagram for B, read the intercepts from the base to the curve, and add the two values. Try various positions until the greatest sum for the coefficient m is found. Then $M = mPL$. In this case, for example, try one load at $0.45L$ and the other at $0.7L$. Then Σm equals approximately $0.097 + 0.097 = 0.194$. Therefore, $M = 0.194 \times 5 \times 20 = 19.4$ ft-k.

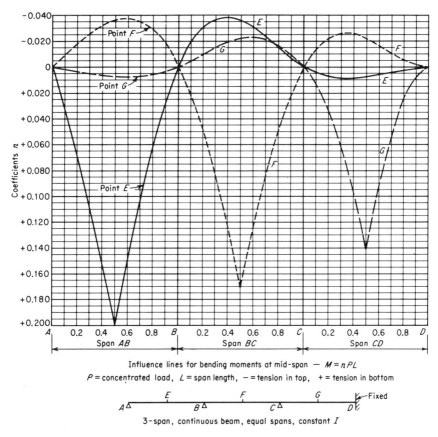

Influence lines for bending moments at mid-span — $M = nPL$

P = concentrated load, L = span length, — = tension in top, + = tension in bottom

3-span, continuous beam, equal spans, constant I

Fig. **15-2** Influence lines for positive bending moments at midspan of continuous beams.

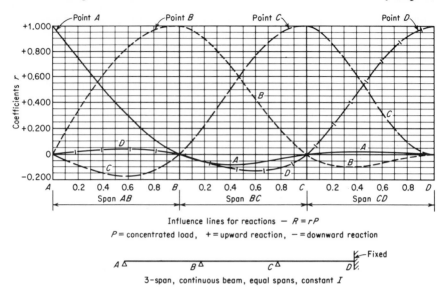

Fig. 15-3   Influence lines for reactions under continuous beams.

If these two loads can be in any position but not closer than 10 ft on centers, then one placed at 0.6*L* on *AB* and one at 0.4*L* on *BC* would obviously produce the greatest bending moment at *B*.   The other curves can be used similarly.

Figure 15-2 contains influence-line diagrams for the coefficient *n* for bending moments at the centers of the three spans.   The curves are

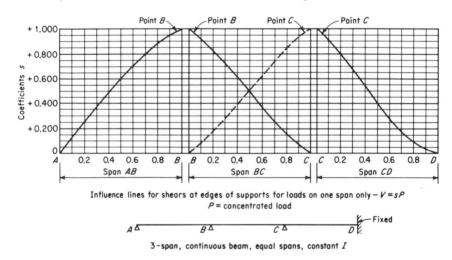

Fig. 15-4   Influence lines for shears in continuous beams.

labeled to correspond with the points to which they apply. Their use is similar to that explained for those shown in Fig. 15-1.

Figure 15-3 shows influence-line diagrams for the coefficients to be used in finding the reactions at all four supports as labeled. On the other hand, Fig. 15-4 shows influence-line diagrams for finding the coefficients to be used in connection with computing the shearing forces at the assumed edges of narrow supports for each of the three spans separately, each curve being labeled according to the point to which it applies. These particular curves do not consider the effect of loads on other spans because these latter effects are not great. It will be sufficiently accurate to call the shear at A equal to the reaction at A found from Fig. 15-3, and the shearing force at D equal to the reaction there as found from curve D of Fig. 15-3.

Careful study of these influence-line diagrams will help the reader to understand the behaviour of continuous beams under the action of loads.

15-3. Deformed Structures. When dimensioning structural members, especially when determining and locating the reinforcement, it is very helpful if the engineer will visualize the action of the structure under loads. This picturing of deformed structures to exaggerated scale is one of the things that Hardy Cross emphasized so effectively. A good engineer will do it mentally even if he does not take time to put it on paper. The beginner cannot afford to neglect this simple but effective aid to his designing.

Just a few illustrations are given to show the kind of thinking and the method of picturization. Figure 15-5 shows the deformations of a two-span beam. Figure 15-6 pictures a single frame under the action of two different loading conditions. In Fig. 15-7(a) is shown a loaded bin on columns, whereas (b) illustrates the action of a deep trench in the earth. The frames in Fig. 15-8 are given in order to picture the effects of some loads on certain continuous structures. In all the illustrations, the dotted lines show the location of the principal tensile reinforcement for the conditions pictured.

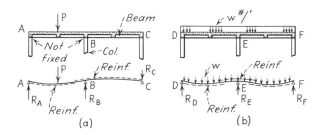

Fig. **15-5** Deformations of a two-span continuous beam.

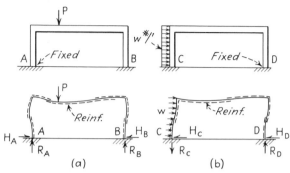

Fig. 15-6 Deformations of a rigid frame.

The reader should practice making such pictures until he develops a good sense of structural action. These problems will give him a chance to test his ability.

15-4. Concept of Design. The student probably does not realize how greatly the design of a structure is influenced by what the owner wants, by what the engineer wants, by what the architect wants, by what is good practice, by what is known to be satisfactory construction, by what is most suitable for foundation conditions, by what will minimize operating and maintenance costs, and by what is economical construction in a particular region. In general, a structure is planned in considerable detail before the strength of its component parts is determined. These parts are then made strong enough to serve their purposes. Not always are they the minimum sizes that will support the loads; they are what seem to

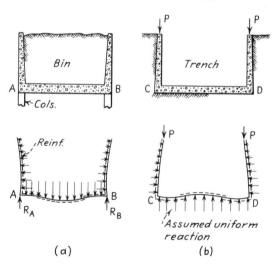

Fig. 15-7 Deformations of bin and trench.

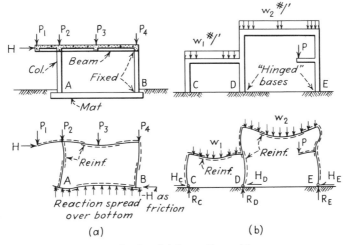

Fig. 15-8 Study of deformations of frames.

be suitable and appropriate members for that particular structure.   Thus good engineering judgment is rightly one of the assets of the planner of such structures.

In a broad sense, *design* means the planning, shaping, styling, and general proportioning of a structure.   By custom, design is also used to denote *dimensioning*, which is the assignment of depths, widths, thicknesses, details, and reinforcement.   These last have little effect upon economy after the general design has been determined.   The student should realize this.   Far too often, he seems to think that a structure is or should be an assembly of members each having just the theoretical balanced design of steel vs. concrete, and that the structure must be made to use these members.

Stiffness is often one of the desired qualities of a structure.   Admittedly it may be difficult or almost impossible to define or to state just what satisfactory stiffness means.   It varies with different structures and their uses.   Here again judgment and experience are of value.

As examples, two cases will be used to illustrate some of these ideas in an effort to show how that quality of judgment enters the picture.

A school was planned with a central corridor in each wing.   The adjacent classrooms were 24 ft wide.   The architect wanted to have a flat ceiling without beam haunches, yet he wanted it to be thin.   He planned to use a lightweight floor spanning the 24 ft.   Although it was only 9 in. deep, it was theoretically strong enough to resist the shears and bending moments.   However, his engineer changed the depth to 15 in., used T beam (tin-pan) construction with a hung ceiling, and deliberately used overreinforced members.   The purpose was to reduce the deflection under live load. Any apparent springiness of the floor, vibration of lights attached to the ceiling below, cracking of the junctions of floor and partitions above it, or crushing of partitions below were to be avoided. Hence the conservatism.

A beam-and-slab highway overpass was to be built at a metallurgical plant. Over the bridge was to pass a more or less continuous stream of ore trucks weighing 60 tons each when loaded. The bridge had been designed for a 45-ft span. Then it was decided to reduce the span to 35 ft by narrowing the roadway below. The designer reduced the sizes of the beams and widened their spacing to get what seemed to be a very efficient design in terms of the use of concrete and reinforcement. The supervisor ordered that the beams be spaced as before and that diaphragms be used between them in order to be sure that differential deflection of adjacent beams would not rupture the stiff floor slab. He also ordered that the T beam stems be deepened and widened, that the steel stress be limited to 18,000 psi in order to reduce cracking and to provide an allowance for impact of unknown magnitude, that the stirrups be designed to withstand all the shear, and that part of the longitudinal reinforcement be bent up to give a cradled effect that would help to tie the simply supported T-beam construction together. Was this wasteful? Probably it was wise action. How much did it cost to make sure that those heavy trucks would not cause fatigue and disintegration of the structure? The extra cost was really small since it merely added a little more steel and concrete to the deck but did not materially change the costly abutments, forms, finishing, and many other items of expense.

What is the most economical type of structure? That question is difficult to answer. In the case of small structures, any sensible type will probably be satisfactory, and the savings due to framing systems of various sorts will not be important. However, when one plans a large structure, and especially a multistory one, where considerable duplication exists, it may be wise to make comparative studies to determine what will be best for that particular case.

Again it should be called to the reader's attention that the cost of a concrete structure does not vary directly with the quantity of concrete used in it. Skimpy members may not be worthwhile even if they do yield small savings because a structure should be satisfactory as well as safe. Unsatisfactory structures mean money wasted or, at least, money spent unwisely.

For example, assume a beam-and-girder warehouse floor to hold a live load of 300 psf. The columns are 20 by 24 ft on centers. The beams are 24 ft long and 10 ft c.c. The girders are 20 ft long, with one concentrated load in the middle. The quantities are estimated for a sturdy design and for another using skimpy sections. The same materials and allowable

stresses are used for both designs.   The estimated quantities are multiplied by the following assumed unit prices:

1. Transit-mixed concrete delivered and placed $= \$25$ per yd^3
2. Reinforcing in place $= \$0.15$ per lb
3. Forms $= \$0.75$ per ft^2
4. Finishing of floor surface $= \$0.15$ per ft^2

The results are the following:

Material	Sturdy design		Skimpy design	
Concrete	$18.6 \times 25 =$	\$ 465.00	$12.8 \times 25 =$	\$ 320.00
Reinforcement	$3{,}460 \times 0.15 =$	519.00	$4{,}120 \times 0.15 =$	618.00
Forms	$736 \times 0.75 =$	552.00	$722 \times 0.75 =$	541.50
Finishing	$480 \times 0.15 =$	72.00	$480 \times 0.15 =$	72.00
		$\overline{\$1{,}608.00}$		$\overline{\$1{,}551.50}$

Thus the costs are not much different.   In the sturdy design, the slab is
8 in. thick; in the other, 5 in.   These figures are given only for the purpose
of showing some scale of how costs may be affected by different features.

For the work of this chapter, the student is urged to invent a structural system that is as simple as possible, one that transmits the loads to
the foundation as directly and efficiently as it can be done, and one that
is structurally reliable.   It should also be easy to build.

It is natural for one to ask, "Why do we want this thing?" and "Why
not make it some other way?"   In general, the owner wants a structure
to serve a special purpose, and he wants the engineers and contractors to
design and build it for him.   They should do so, but if better and more
economical things can obviously be done, they very properly should discuss the matter with him.   On the other hand, difficulties in the engineering work are seldom sufficient reason for not giving the owner what
he wants when operating economies and special service are to be secured
for him.   Structural safety is naturally essential because an owner should
not have to worry about the safety of his structure.   If he is willing to
pay for it, and if he prefers it to some suggested alternate, he should be
given what he wants.   Therefore, changing the general scheme is not the
solution for any of the problems here.

Where not shown specifically in the problems, the dimensions, locations, and details are to be assumed for doors, windows, elevators, electric lighting, heating, etc.   Sweet's catalogues, books on architectural
standards, and the publications of manufacturers are suitable sources
for such information.   In practice, all these matters have to be worked
out and provided for in the design.   However, in these problems, the
principal emphasis must necessarily be upon the structural part of each
one.

15-5.   Problems.   In order to have uniformity in the qualities of the materials used in the problems in this chapter, assume the following unless stated otherwise in a particular case: $f_c' = 3{,}000$ psi; reinforcement is intermediate-grade A 305 bars; and the allowable unit stresses are to be as given in the Code, Tables 1-7 and 1-8.

In many cases, the drawings show the live loads and other pertinent data needed for the design.   Graphical scales are given when they seem to be necessary.

The problems purposely cover a wide range of ordinary concrete construction.   They are all based upon practical cases, but many have been simplified in some minor details.   It is believed that enough problems are provided to suit the skill and available time of individual students and of groups working as engineering squads.

Problem 15-1.   Figure 15-9 shows the floor plan and some other pertinent data for a pump house at an industrial plant.   The entire structure is to be built of concrete.   The crane rails are to be supported upon concrete beams that rest upon pilasters.   The crane has two wheels 8 ft apart at each end.   The maximum wheel loads are 15,000 lb each.

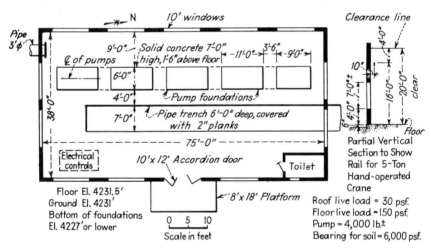

Fig. 15-9   Plan for a pump house.

Problem 15-2.   Figure 15-10 shows the general plans for a small warehouse to be supported upon 30-ton piles.   The floor is to be poured on the ground, but it must be self-supporting.   The structure is to have a concrete roof, concrete columns, and spandrel beams.   The walls are to be 4 in. of brick veneer with 8 in. of cinder blocks as backing.

Problem 15-3.   Figure 15-11 shows the floor plan and other data for a small office building at an industrial plant.   Roof, floors, and skeleton framework are to be of concrete.   Use hung ceilings except in the basement.   Other materials are optional.

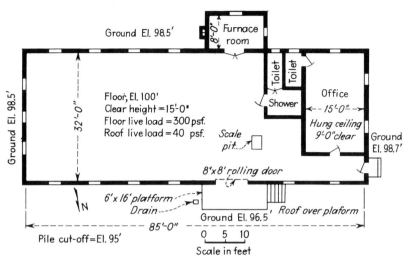

Fig. 15-10   A small warehouse on poured-in-place concrete piles.

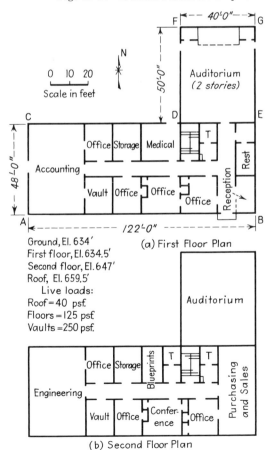

Fig. 15-11 Plan of an office building at an industrial plant.

Problem 15-4.   Figure 15-12 pictures the plan for a multistory warehouse.   Type and materials are optional except that the structure is to be built with concrete framing or flat-slab construction.   The layout is symmetrical about the center line of the building except for the offices, vault, and furnace room in the southeast corner.   There are no similar facilities in the northeast corner, but the toilets (*T*) are duplicated.

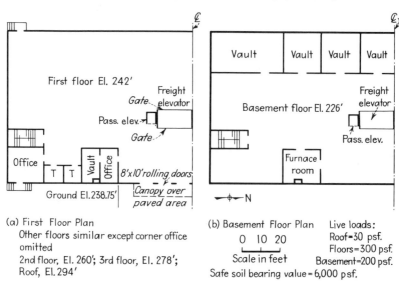

(a) First Floor Plan
Other floors similar except corner office omitted
2nd floor, El. 260'; 3rd floor, El. 278';
Roof, El. 294'

(b) Basement Floor Plan   Live loads:
0   10   20   Roof=30 psf.
Scale in feet   Floors=300 psf.
                                  Basement=200 psf.
Safe soil bearing value=6,000 psf.

Fig. 15-12   Partial plans of a large warehouse approximately 80 by 200 ft.

Problem 15-5.   The highway embankment in Fig. 15-13 is to have a box culvert extending through it as shown.   Design the culvert, including its floor and end walls.   Assume that the culvert supports all the fill above it plus a live load of 200 psf.   The fill weighs 110 pcf.   The soil under the culvert is sand of good bearing value.

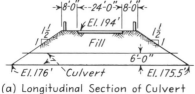

(a) Longitudinal Section of Culvert

Fig. 15-13   A culvert under a highway embankment.

(b) Transverse Section of Culvert

Problem 15-6.   Design the tunnel shown in Fig. 15-14.   Consider the varying depths of earth and ore.   Assume $w = 110$ pcf for ore and earth, $\phi = 35°, f'_c = 4{,}000$ psi, $f_s = 24{,}000$ psi, and p for soil $= 12{,}000$ psf.

This problem seems to be simple, but the two sections under the ore pile must support very large loads.   Assume that the tunnel is built in open cut and that it must support all the material over it.   The structure is built in short units in order to avoid cracks if unequal settlements occur.

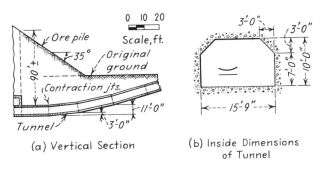

(a) Vertical Section   (b) Inside Dimensions
                                    of Tunnel

Fig. 15-14   A conveyor tunnel under an ore pile.

Problem 15-7.   Figure 15-15 pictures the general dimensions for a three-span highway overpass.   Assume a uniform live load of 250 psf, including impact.   The lower roadway is in a cut.   The soil is sandy, and it will support a pressure of $3\frac{1}{2}$ tons/ft^2 safely.

Design the superstructure and draw pictures of the piers and abutments.   The student may invent his own architectural and engineering features for the project.

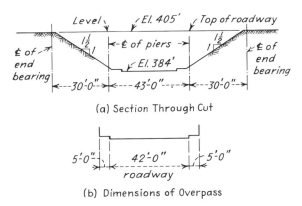

(a) Section Through Cut

(b) Dimensions of Overpass

Fig. 15-15   A highway grade-crossing elimination.

Problem 15-8.   Figure 15-16(*a*) shows a preliminary study of a highway bridge or trestle; (*b*) shows the dimensions of the proposed roadway and sidewalks.   The superstructure is assumed to be a pair of two-span continuous sections of beam-and-slab type supported upon abutments at *A* and *F* and upon piles and a cap or cross girder at *B*, *C*, *D*, and *E*. The abutments are 40 ft wide at the top, and they have wing walls parallel to the main abutment and sloping at $1\frac{1}{2}$:1 on top for a distance of 6 ft each side.   Beyond that point, the embankment may extend around the ends of the wings.   The piles are precast concrete.   For the abutments, they are 12 in. square and good for 18 tons each.   For the piers or pile bents, they are 16 in. octagonal and can safely support 25 tons apiece.

Assume a uniform live load of 250 psf on the roadway and 100 psf on the sidewalks.   This includes impact.   Design the superstructure and the bents, and at least make a sketch of one abutment.

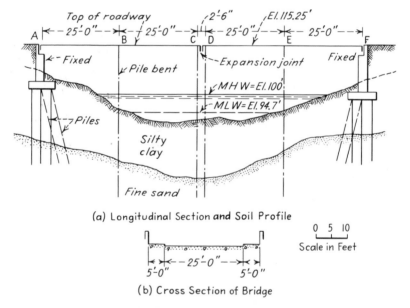

(a) Longitudinal Section and Soil Profile

(b) Cross Section of Bridge

Fig. 15-16   Highway crossing of a small tidal stream.

Problem 15-9.   Figure 15-17 pictures an elevated concrete bin for a granular product weighing 110 pcf.   Assume $\phi = 30°$.   Design the bin, assuming it to be filled level with the top.   Assume that the lateral pressures are the same as they would be if the sides were retaining walls. Also design the supporting structure, assuming that a width of 15 ft and a height of 15 ft 6 in. must be maintained for trucks and gates under the openings.   The soil is good for a pressure of 5 tons/ft².

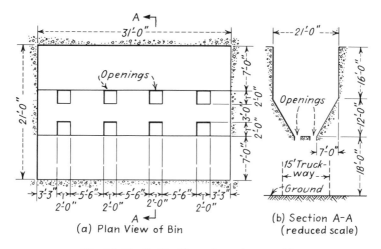

Fig. 15-17 Inside dimensions of an ore bin.

Problem 15-10. Figure 15-18 pictures a two-compartment tank elevated above ground. It is generally filled at A and overflows the baffle into the right-hand compartment and thence out at B. The chemical can be drawn off through C and D when desired. Outlets E

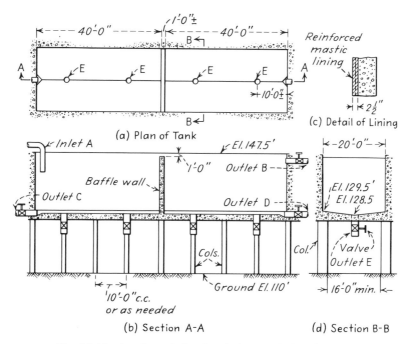

Fig. 15-18 A twin tank for chemicals at an industrial plant.

are used for disposing of sludge by emptying into tank trucks run under them.

Assume that the liquid weighs 70 pcf and that the compartments are filled up to the top. Design the tank and its supports for both compartments filled and for the left one full and the right one empty. The soil is safe for a bearing pressure of 4 tons/ft². The dimensions shown are for the concrete. Assume that the 2½-in. mastic lining covers all interior surfaces and weighs 25 psf.

Problem 15-11. Figure 15-19 shows a portion of an industrial plant. The upper part *DEGF* of this 25- by 64-ft structure contains electrical equipment. Assume a live load of 200 psf for the top, 150 psf for the control-room floor, and 200 psf for the pump room. The load on crane columns *A* may be assumed as 90 kips each. Columns *B* and *C* support a bin as well as the superstructure. Assume *B* = 200 kips each and *C* = 160 kips each. The soil can safely support a load of 8,000 psf. The access to the pump room *FGJH* is through a 10-ft square centrally located door in wall *FH*.

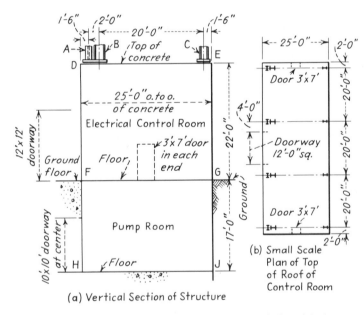

(a) Vertical Section of Structure

Fig. **15-19** Control and pump rooms at a large industrial plant.

Problem 15-12. Figure 15-20 shows the plans for a conveyor junction house at a mine. It is under a very large embankment as indicated in (*a*). The crane rails are to be supported upon corbels projecting from the side walls.

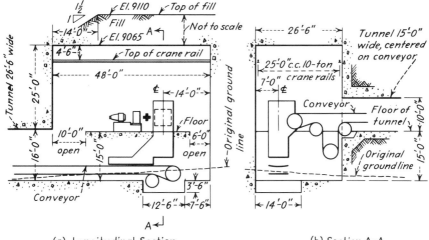

(a) Longitudinal Section   (b) Section A-A

Fig. 15-20   A conveyor junction house under a deep fill.

Assume that the fill weighs 100 pcf, and that $\phi = 35°$. Assume $f'_c = 4,000$ psi, $f_s = 24,000$ psi, and other stresses as specified by the Code. Design the roof, bottom, side walls, and tunnels for the earth loads and lateral pressures only. The intermediate floor can be assumed to be a 12-in. slab with four beams across the 26.5-ft width. Assume that the maximum wheel loads of the crane at one end are two reactions of 18,000 lb at 9 ft c.c. The soil below the structure is hard gravel that can safely resist a pressure of 12,000 psf.

This structure involves some very difficult problems. Both bending and shear are very large. The locations and details of construction joints need special attention. The junction of the 15-ft tunnel at the right of sketch (b) may crack if the tunnel settles on the fill under it. Probably both tunnels should be cut loose from the main structure.

APPENDIX

Explanation of Data. The tables and diagrams here are intended to be of special assistance to the designer of reinforced-concrete structures. They are not to take the place of the extensive information that is given in handbooks. The following instructions are given for each one:

Table 1. This contains data pertaining to the latest type of deformed bars, and it also gives information regarding the large square bars that were formerly prevalent and that may occasionally be encountered. The $\frac{1}{4}$-in. round bars are furnished in plain rods only unless by special order.

Table 1   **Table of Areas, Perimeters, Weights, and Other Data
for Reinforcing Bars**
All bars are deformed bars except $\frac{1}{4}$-in. rounds, which are plain.

Type of bar	Bar No. or size*	Nominal dimensions			Weight, plf	Deformations		
		Diameter, in.	Net area, in.^{2}	Perimeter, in.		Max avg spacing, in.	Min height, in.	Max gap, in.†
ASTM Designation: A-305-50T‡	3	0.375	0.11	1.178	0.376	0.262	0.015	0.143
	4	0.500	0.20	1.571	0.668	0.350	0.020	0.191
	5	0.625	0.31	1.963	1.043	0.437	0.028	0.239
	6	0.750	0.44	2.356	1.502	0.525	0.038	0.286
	7	0.875	0.60	2.749	2.044	0.612	0.044	0.334
	8	1.000	0.79	3.142	2.670	0.700	0.050	0.383
	9	1.128	1.00	3.544	3.400	0.790	0.056	0.431
	10	1.270	1.27	3.990	4.303	0.889	0.064	0.487
	11	1.410	1.56	4.430	5.313	0.987	0.071	0.540
ASTM A408	S14	1.693	2.25	5.32	7.65	1.185	0.085	0.648
	S18	2.257	4.00	7.09	13.60	1.58	0.102	0.864

   * Numbers 3 to 11, inclusive, are based on number of $\frac{1}{8}$ in. in nominal diameter of the bar section.

   † Chord of $12\frac{1}{2}$ per cent of nominal perimeter.

   ‡ Data given by American Society for Testing Materials. The ASTM Standards are subject to revision from time to time. ASTM headquarters in Philadelphia has copies and information on latest editions.

Table 2 Areas and Perimeters of Reinforcing Bars per Foot of Slab

Size of bar

Spacing, in.	2		3		4		5		6		7		8		9		10		11	
	¼ in. φ		⅜ in. φ		½ in. φ		⅝ in. φ		¾ in. φ		⅞ in. φ		1 in. φ		1⅛ in. + φ		1¼ in. + φ		1 7/16 in. − φ	
	A_s	Σo	A_s	Σo	A_s	Σo	A_s	Σo	A_s	Σo	A_s	Σo	A_s	Σo	A_s	Σo	A_s	Σo	A_s	Σo
2	0.30	4.68	0.66	7.08	1.20	9.42	1.86	11.76												
2½	0.24	3.74	0.53	5.66	0.96	7.54	1.49	9.41	2.11	11.33										
3	0.20	3.12	0.44	4.72	0.80	6.28	1.24	7.84	1.76	9.44	2.40	11.00	3.16	12.56	4.00	14.18				
3½	0.17	2.67	0.38	4.05	0.69	5.39	1.06	6.72	1.51	8.09	2.06	9.43	2.71	10.77	3.43	12.15	4.35	13.68		
4	0.15	2.34	0.33	3.54	0.60	4.71	0.93	5.88	1.32	7.08	1.80	8.25	2.37	9.42	3.00	10.63	3.81	11.97	4.68	13.29
4½	0.13	2.08	0.29	3.15	0.53	4.19	0.83	5.23	1.17	6.29	1.60	7.33	2.11	8.37	2.67	9.45	3.39	10.64	4.16	11.81
5	0.12	1.87	0.26	2.83	0.48	3.77	0.74	4.70	1.06	5.66	1.44	6.60	1.90	7.54	2.40	8.51	3.05	9.58	3.74	10.63
5½	0.11	1.70	0.24	2.57	0.44	3.43	0.68	4.28	0.96	5.15	1.31	6.00	1.72	6.85	2.18	7.73	2.77	8.70	3.40	9.66
6	0.10	1.56	0.22	2.36	0.40	3.14	0.62	3.92	0.88	4.72	1.20	5.50	1.58	6.28	2.00	7.09	2.54	7.98	3.12	8.86
6½	0.09	1.44	0.20	2.18	0.37	2.90	0.57	3.62	0.81	4.36	1.11	5.08	1.46	5.80	1.85	6.54	2.34	7.37	2.88	8.18
7	0.09	1.34	0.19	2.02	0.34	2.69	0.53	3.36	0.75	4.05	1.03	4.71	1.35	5.38	1.71	6.08	2.18	6.84	2.67	7.59
7½	0.08	1.25	0.18	1.89	0.32	2.51	0.50	3.14	0.70	3.78	0.96	4.40	1.26	5.02	1.60	5.67	2.03	6.38	2.50	7.09
8	0.08	1.17	0.16	1.77	0.30	2.36	0.46	2.94	0.66	3.54	0.90	4.12	1.18	4.71	1.50	5.32	1.90	5.98	2.34	6.64
9	0.07	1.04	0.15	1.57	0.27	2.09	0.41	2.61	0.59	3.15	0.80	3.67	1.05	4.19	1.33	4.72	1.69	5.32	2.08	5.91
10	0.06	0.94	0.13	1.42	0.24	1.88	0.37	2.35	0.53	2.83	0.72	3.30	0.95	3.77	1.20	4.25	1.52	4.79	1.87	5.32
11	0.05	0.85	0.12	1.29	0.22	1.71	0.34	2.14	0.48	2.57	0.65	3.00	0.86	3.43	1.09	3.87	1.39	4.35	1.70	4.83
12	0.05	0.78	0.11	1.18	0.20	1.57	0.31	1.96	0.44	2.36	0.60	2.75	0.79	3.14	1.00	3.54	1.27	3.99	1.56	4.43

Table 3 Areas and Perimeters of Reinforcing Bars

Size of bar

No. of bars	2 $\frac{1}{4}$ in. ϕ A_s	Σo	3 $\frac{3}{8}$ in. ϕ A_s	Σo	4 $\frac{1}{2}$ in. ϕ A_s	Σo	5 $\frac{5}{8}$ in. ϕ A_s	Σo	6 $\frac{3}{4}$ in. ϕ A_s	Σo	7 $\frac{7}{8}$ in. ϕ A_s	Σo	8 1 in. ϕ A_s	Σo	9 $1\frac{1}{8}$ in. $+ \phi$ A_s	Σo	10 $1\frac{1}{4}$ in. $+ \phi$ A_s	Σo	11 $1\frac{7}{16}$ in. $- \phi$ A_s	Σo
1	0.05	0.78	0.11	1.18	0.20	1.57	0.31	1.96	0.44	2.36	0.60	2.75	0.79	3.14	1.00	3.54	1.27	3.99	1.56	4.43
2	0.10	1.56	0.22	2.36	0.40	3.14	0.62	3.92	0.88	4.72	1.20	5.50	1.58	6.28	2.00	7.09	2.54	7.98	3.12	8.86
3	0.15	2.34	0.33	3.54	0.60	4.71	0.93	5.88	1.32	7.08	1.80	8.25	2.37	9.42	3.00	10.6	3.81	12.0	4.68	13.3
4	0.20	3.12	0.44	4.72	0.80	6.28	1.24	7.84	1.76	9.44	2.40	11.0	3.16	12.6	4.00	14.2	5.08	16.0	6.24	17.7
5	0.25	3.90	0.55	5.90	1.00	7.85	1.55	9.80	2.20	11.8	3.00	13.8	3.95	15.7	5.00	17.7	6.35	20.0	7.80	22.2
6	0.30	4.68	0.66	7.08	1.20	9.42	1.86	11.8	2.64	14.2	3.60	16.5	4.74	18.8	6.00	21.3	7.62	23.9	9.36	26.6
7	0.35	5.46	0.77	8.26	1.40	11.0	2.17	13.7	3.08	16.5	4.20	19.2	5.53	22.0	7.00	24.8	8.89	27.9	10.9	31.0
8	0.40	6.24	0.88	9.44	1.60	12.6	2.48	15.7	3.52	18.9	4.80	22.0	6.32	25.1	8.00	28.4	10.2	31.9	12.5	35.4
9	0.45	7.02	0.99	10.6	1.80	14.1	2.79	17.6	3.96	21.2	5.40	24.8	7.11	28.3	9.00	31.9	11.4	35.9	14.0	39.9
10	0.50	7.80	1.10	11.8	2.00	15.7	3.10	19.6	4.40	23.6	6.00	27.5	7.90	31.4	10.0	35.4	12.7	39.9	15.6	44.3
11	0.55	8.58	1.21	13.0	2.20	17.3	3.41	21.6	4.84	26.0	6.60	30.2	8.69	34.5	11.0	39.0	14.0	43.9	17.2	48.7
12	0.60	9.36	1.32	14.2	2.40	18.8	3.72	23.5	5.28	28.3	7.20	33.0	9.48	37.7	12.0	42.5	15.2	47.9	18.7	53.2
13	0.65	10.1	1.43	15.3	2.60	20.4	4.03	25.5	5.72	30.7	7.80	35.8	10.3	40.8	13.0	46.1	16.5	51.9	20.3	57.6
14	0.70	10.9	1.54	16.5	2.80	22.0	4.34	27.4	6.16	33.0	8.40	38.5	11.1	44.0	14.0	49.6	17.8	55.9	21.8	62.0
15	0.75	11.7	1.65	17.7	3.00	23.6	4.65	29.4	6.60	35.4	9.00	41.2	11.8	47.1	15.0	53.2	19.0	59.8	23.4	66.4
16	0.80	12.5	1.76	18.9	3.20	25.1	4.96	31.4	7.04	37.8	9.60	44.0	12.6	50.2	16.0	56.7	20.3	63.8	25.0	70.9
17	0.85	13.3	1.87	20.1	3.40	26.7	5.27	33.3	7.48	40.1	10.2	46.8	13.4	53.4	17.0	60.2	21.6	67.8	26.5	75.3
18	0.90	14.0	1.98	21.2	3.60	28.3	5.58	35.3	7.92	42.5	10.8	49.5	14.2	56.5	18.0	63.8	22.9	71.8	28.1	79.7
19	0.95	14.8	2.09	22.4	3.80	29.8	5.89	37.2	8.36	44.8	11.4	52.2	15.0	59.7	19.0	67.3	24.1	75.8	29.6	84.2
20	1.00	15.6	2.20	23.6	4.00	31.4	6.20	39.2	8.80	47.2	12.0	55.0	15.8	62.8	20.0	70.9	25.4	79.8	31.2	88.6

Table 3 Areas and Perimeters of Reinforcing Bars (*Continued*)

Size of bar

No. of bars	2 $\frac{1}{4}$ in. ϕ		3 $\frac{3}{8}$ in. ϕ		4 $\frac{1}{2}$ in. ϕ		5 $\frac{5}{8}$ in. ϕ		6 $\frac{3}{4}$ in. ϕ		7 $\frac{7}{8}$ in. ϕ		8 1 in. ϕ		9 $1\frac{1}{8}$ in. $+ \phi$		10 $1\frac{1}{4}$ in. $+ \phi$		11 $1\frac{7}{16}$ in. $- \phi$	
	A_s	Σo	A_s	Σo	A_s	Σo	A_s	Σo	A_s	Σo	A_s	Σo	A_s	Σo	A_s	Σo	A_s	Σo	A_s	Σo
21	1.05	16.4	2.31	24.8	4.20	33.0	6.51	41.2	9.24	49.6	12.6	57.8	16.6	65.9	21.0	74.4	26.7	83.8	32.8	93.0
22	1.10	17.2	2.42	26.0	4.40	34.5	6.82	43.1	9.68	51.9	13.2	60.5	17.4	69.1	22.0	78.0	27.9	87.8	34.3	97.5
23	1.15	17.9	2.53	27.1	4.60	36.1	7.13	45.1	10.1	54.3	13.8	63.2	18.2	72.2	23.0	81.5	29.2	91.8	35.9	101.9
24	1.20	18.7	2.64	28.3	4.80	37.7	7.44	47.0	10.6	56.6	14.4	66.0	19.0	75.4	24.0	85.1	30.5	95.8	37.4	106.3
25	1.25	19.5	2.75	29.5	5.00	39.2	7.75	49.0	11.0	59.0	15.0	68.8	19.8	78.5	25.0	88.6	31.8	99.8	39.0	110.8
26	1.30	20.3	2.86	30.7	5.20	40.8	8.06	51.0	11.4	61.4	15.6	71.5	20.5	81.6	26.0	92.1	33.0	103.7	40.6	115.2
27	1.35	21.1	2.97	31.9	5.40	42.4	8.37	52.9	11.9	63.7	16.2	74.2	21.3	84.8	27.0	95.7	34.3	107.7	42.1	119.6
28	1.40	21.8	3.08	33.0	5.60	44.0	8.68	54.9	12.3	66.1	16.8	77.0	22.1	87.9	28.0	99.2	35.6	111.7	43.7	124.0
29	1.45	22.6	3.19	34.2	5.80	45.5	8.99	56.8	12.8	68.4	17.4	79.8	22.9	91.1	29.0	102.8	36.8	115.7	45.2	128.5
30	1.50	23.4	3.30	35.4	6.00	47.1	9.30	58.8	13.2	70.8	18.0	82.5	23.7	94.2	30.0	106.3	38.1	119.7	46.8	132.9

Table 2. This contains information regarding cross-sectional areas and perimeters for use in designing and analyzing slabs. Except for the $\frac{1}{4}$-in. rounds, all bars are the modern A 305 type.

Table 3. This is a multiplication table to save time in computing areas and perimeters of groups of bars.

Table 4. This information is prepared to show the length of bar needed to develop various tensile unit stresses at lapped splices and anchorages of reinforcement by means of the allowable bond stresses specified in the Code for the strengths of concrete shown.

Table 4 **Theoretical Minimum Embedment Length to Develop Bond—A 305 Reinforcing Bars**

Bar No.	$f_s = 18,000$			$f_s = 20,000$			$f_s = 22,000$		
	f'_c								
	2,500	3,000	4,000	2,500	3,000	4,000	2,500	3,000	4,000
	Length for top bars, in.								
3	5	5	5	6	6	6	6	6	6
4	7	7	7	8	8	8	9	8	8
5	11	10	9	11	11	10	13	12	11
6	15	14	12	17	15	13	18	17	15
7	20	19	16	23	21	18	25	23	20
8	27	25	21	30	28	24	33	31	26
9	34	31	27	38	35	30	42	38	33
10	43	40	34	47	44	38	52	49	42
11	53	49	42	59	54	47	65	60	52
	Length for bottom bars, in.								
3	4	4	4	5	4	4	5	5	5
4	5	5	5	6	6	6	6	6	6
5	8	7	6	9	8	7	9	9	8
6	11	10	9	12	11	10	13	12	11
7	15	13	12	16	15	13	18	16	14
8	19	17	15	21	19	17	23	22	19
9	24	23	19	27	24	21	29	27	25
10	30	28	24	34	31	27	37	35	30
11	37	34	30	42	38	33	46	42	37

Figure 1.   This contains data showing the bending moments at the ends of beams of constant section with both ends fixed and for some of the most common loading conditions.   It is to be used when one must estimate the approximate bending in the interior spans of continuous beams.

$$M = m \times W \times l$$
m = Coefficient taken from diagram
W = Total load on beam
l = Length of beam
a = Length in terms of l

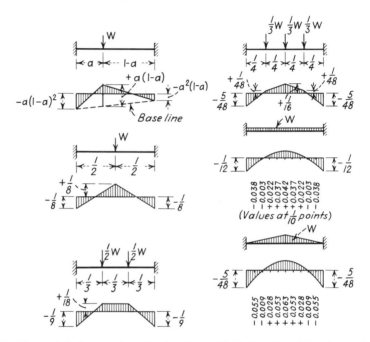

Fig. 1   Moments in beams of constant section with fixed ends.   (*Based upon data by Hardy Cross and the Portland Cement Association.*)

Figure 2. This information is for use in estimating the bending moments in the end spans of continuous beams.

$$M = m \times W \times l$$
$$m = \text{Coefficient taken from diagram}$$
$$W = \text{Total load on beam}$$
$$l = \text{Length of beam}$$

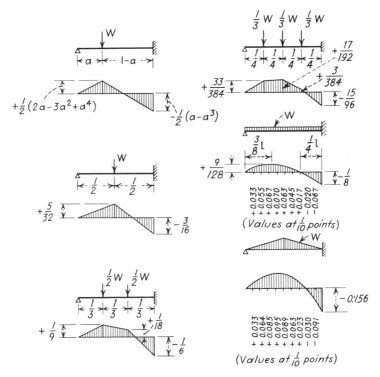

Fig. 2 Moments in beams of constant section with one end fixed and the other end simply supported.

Figure 3.   By interpolating from this diagram, one can determine the value of the coefficient in Coulomb's formula for the active pressure of earth against retaining walls when the soil is level or sloping behind the wall.   The values are generally sufficiently accurate for the purpose.

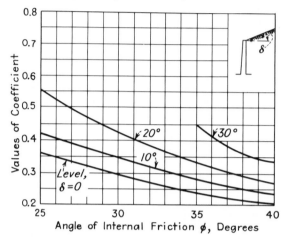

Fig. 3   Values of $\cos \phi/(1 + \sqrt{2 \sin^2 \phi - 2 \sin \phi \cos \phi \tan \delta})^2$ for Coulomb's equation for the magnitude of the active earth pressure inclined at angle ϕ when the surface is sloped at various angles δ.

Figures 4 *to* 9, *inclusive.* The analysis of rectangular beams can be greatly expedited by the use of these diagrams; they are also applicable for the analysis of T beams if the neutral axis lies within the flange or very close to it. In any case, the values of nA_s/b and $[(n-1)/b]A'_s$ per inch of width of the beam can be found, using b as the divisor for T beams as well as for rectangular ones when the bending moment is positive, or using b' for T beams when the bending moment is negative.

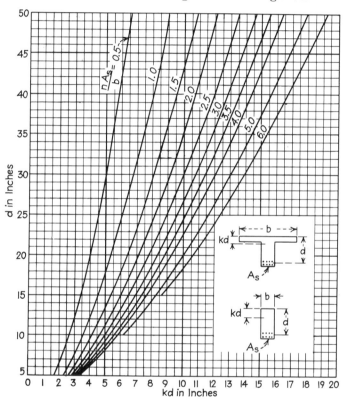

Fig. 4 Location of neutral axis of beam with tensile steel only.

The procedure in using the diagrams is as follows: When there is little or no steel in compression, use Figs. 4 and 5; otherwise use the curves that are prepared for the nearest value of $[(n-1)/b]A_s'$; with d as the ordinate, cross horizontally to the proper (or interpolated) value of nA_s/b, then read the corresponding magnitudes of kd and S_c/b; multiply S_c/b by b (or b') to find S_c, then compute $S_s = S_c(kd)/n(d-kd)$ from the quantities already found.   Results from two diagrams may be used for interpolation if greater accuracy is desired.

The advantage of these diagrams is the fact that they enable one to find the section moduli and thereby compute f_c and f_s.   Of course, important members should be checked analytically after the diagrams have been used for approximate analysis.

In using these diagrams, it will generally be sufficient to assume that

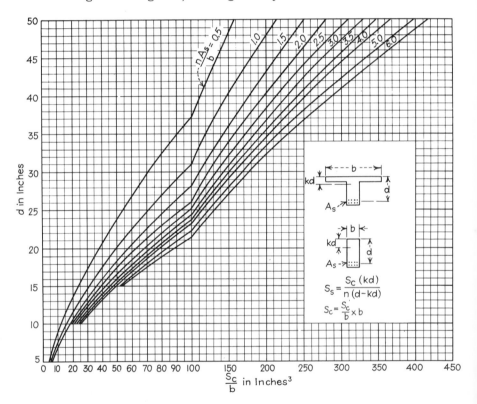

Fig. 5   Section modulus of a 1-in. width of beam with tensile steel only.

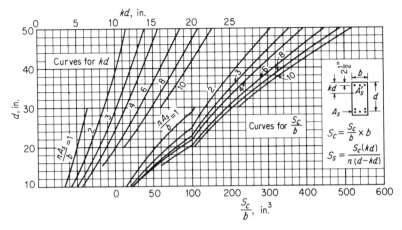

***Fig.* 6** Location of neutral axis and magnitude of section modulus of a 1-in. width of a beam, when $[(n-1)/b]A'_s = 1.0$.

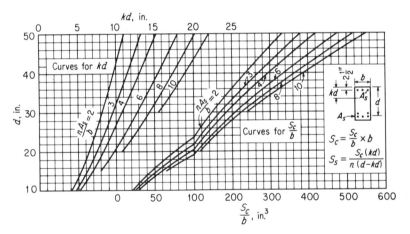

***Fig.* 7** Location of neutral axis and magnitude of section modulus of a 1-in. width of a beam, when $[(n-1)/b]A'_s = 2.0$.

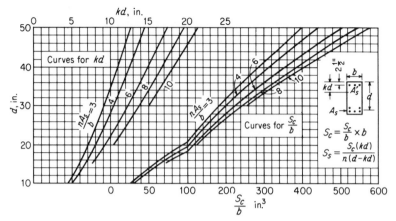

Fig. 8 Location of neutral axis and magnitude of section modulus of a 1-in. width of beam, when $[(n-1)/b]A_s' = 3.0$.

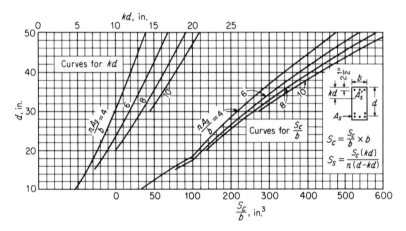

Fig. 9 Location of neutral axis and magnitude of section modulus of a 1-in. width of beam, when $[(n-1)/b]A_s' = 4.0$.

Tables 5 and 6. These tables are prepared for use in preliminary design to obtain the theoretically balanced design for a beam, the results then being modified by practical considerations if necessary.

Table 5 Coefficients p, k, j, and K for Rectangular Sections for Balanced Designs

$$k = \frac{1}{1 + (f_s/nf_c)} \qquad j = 1 - \tfrac{1}{3}k$$

$$p = \frac{A_s}{bd} = \frac{f_c}{2f_s} \times k \qquad K = \frac{f_c}{2}kj = \frac{M}{bd^2}$$

f'_c and n	f_c, psi	$f_s = 18,000$ psi				$f_s = 20,000$ psi			
		p	k	j	K	p	k	j	K
2,500 10	900	0.0083	0.333	0.889	133	0.0070	0.310	0.897	125
	975	0.0095	0.351	0.883	151	0.0080	0.328	0.891	142
	1,050	0.0107	0.368	0.877	169	0.0090	0.344	0.885	160
	1,125	0.0120	0.385	0.872	189	0.0101	0.360	0.880	178
3,000 9	1,000	0.0094	0.333	0.889	148	0.0078	0.311	0.896	139
	1,100	0.0108	0.355	0.882	172	0.0091	0.331	0.890	162
	1,200	0.0125	0.375	0.875	197	0.0105	0.350	0.883	185
	1,350	0.0151	0.403	0.866	236	0.0127	0.377	0.874	222
4,000 8	1,000	0.0086	0.308	0.897	138	0.0072	0.286	0.905	129
	1,100	0.0100	0.328	0.891	161	0.0084	0.306	0.898	151
	1,200	0.0116	0.348	0.884	185	0.0097	0.324	0.892	173
	1,300	0.0132	0.366	0.878	209	0.0111	0.342	0.886	197
	1,400	0.0149	0.384	0.872	234	0.0126	0.359	0.880	221
	1,600	0.0185	0.416	0.861	287	0.0156	0.390	0.870	272
	1,800	0.0222	0.444	0.852	340	0.0189	0.420	0.860	325
5,000 7	1,200	0.0106	0.318	0.894	170	0.0089	0.296	0.901	160
	1,400	0.0137	0.353	0.882	218	0.0115	0.329	0.890	205
	1,600	0.0170	0.383	0.872	267	0.0144	0.359	0.880	253
	1,800	0.0206	0.412	0.863	330	0.0174	0.387	0.871	303
	2,000	0.0244	0.438	0.854	374	0.0206	0.412	0.863	356
	2,100	0.0262	0.450	0.850	402	0.0222	0.424	0.859	382
	2,250	0.0292	0.467	0.844	444	0.0248	0.440	0.853	422

Table 6 Coefficients p, k, j, and K for Rectangular Sections for Balanced Designs

$$k = \frac{1}{1 + (f_s/nf_c)} \qquad j = 1 - \tfrac{1}{3}k$$

$$p = \frac{A_s}{bd} = \frac{f_c}{2f_s} \times k \qquad K = \frac{f_c}{2}kj = \frac{M}{bd^2}$$

f'_c and n	f_c, psi	$f_s = 22{,}000$ psi				$f_s = 24{,}000$ psi			
		p	k	j	K	p	k	j	K
2,500 10	900	0.0059	0.290	0.903	118	0.0051	0.272	0.909	111
	975	0.0068	0.307	0.898	134	0.0059	0.289	0.904	127
	1,050	0.0077	0.323	0.892	151	0.0066	0.304	0.899	143
	1,125	0.0086	0.338	0.887	169	0.0075	0.319	0.894	160
3,000 9	1,000	0.0066	0.290	0.903	131	0.0056	0.272	0.909	124
	1,100	0.0078	0.311	0.896	153	0.0067	0.292	0.903	145
	1,200	0.0090	0.329	0.890	176	0.0078	0.310	0.897	167
	1,350	0.0109	0.356	0.881	212	0.0094	0.336	0.888	201
4,000 8	1,000	0.0061	0.267	0.911	122	0.0052	0.250	0.917	115
	1,100	0.0072	0.286	0.905	142	0.0062	0.268	0.911	135
	1,200	0.0083	0.304	0.899	164	0.0072	0.286	0.905	155
	1,300	0.0095	0.321	0.893	186	0.0082	0.302	0.899	177
	1,400	0.0107	0.337	0.888	209	0.0093	0.318	0.894	199
	1,600	0.0134	0.368	0.877	258	0.0116	0.348	0.884	246
	1,800	0.0162	0.395	0.868	308	0.0141	0.375	0.875	295
5,000 7	1,200	0.0075	0.276	0.908	150	0.0065	0.259	0.914	142
	1,400	0.0098	0.308	0.897	193	0.0085	0.290	0.903	183
	1,600	0.0122	0.337	0.888	239	0.0106	0.318	0.894	227
	1,800	0.0149	0.364	0.879	288	0.0129	0.345	0.885	275
	2,000	0.0177	0.389	0.870	338	0.0153	0.368	0.877	323
	2,100	0.0191	0.400	0.867	364	0.0166	0.380	0.873	348
	2,250	0.0212	0.418	0.861	405	0.0186	0.396	0.868	386

Figures 10 *and* 11. These diagrams are for the analysis of members. If k for any T beam lies below and to the right of the straight line in Fig. 11(a), it means that the neutral axis is within the flange and that it can be located by means of Fig. 10.

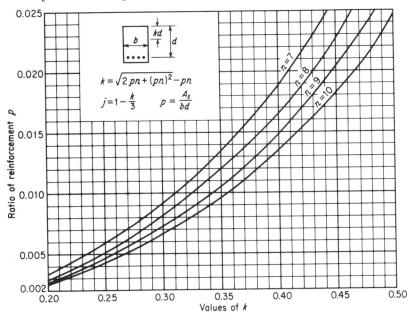

Fig. 10 Location of neutral axis of rectangular beams with tensile steel only.

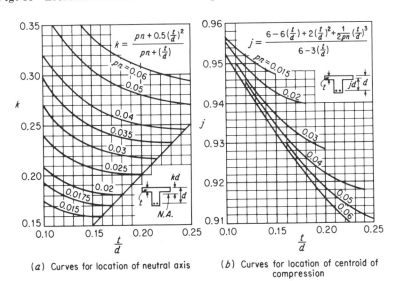

(*a*) Curves for location of neutral axis   (*b*) Curves for location of centroid of compression

Fig. 11 Magnitudes of k and j for T beams.

Figure 12 *and Table* 7.　Knowing the longitudinal shear to be with-stood by the stirrups per inch of length of the beam, varying combinations of sizes and spacings of stirrups can be secured from Fig. 12.　It is then important to check the chosen size of stirrup in Table 7 to make sure that the beam is deep enough to provide adequate bond to develop the stirrups in the upper (or compression) half of the beam.

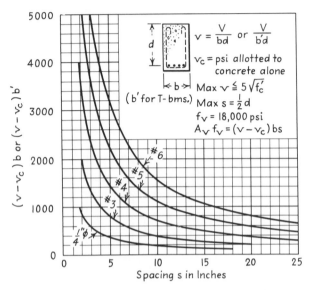

Fig. 12　Maximum spacing of vertical U-shaped stirrups.

Table 7　Recommended Minimum Effective Depth of Beam in Inches to Develop Single Vertical U-shaped Stirrups in One-half Depth Plus $1\frac{1}{2}$ In. of Cover, Using Standard Hooks

Size of stirrup, No.	f_c', psi				
	2,000	2,500	3,000	3,500	4,000
2	25	20	16	14	12
3	23	16	12	11	11
4	26	20	16	14	13
5	32	24	18	15	15
6	38	28	22	20	18

f_v = 18,000 psi.

Tables 8 *and* 8*A*.   Although the Code permits the use of narrower beams than those shown in Table 8, the data given there have been prepared to provide generous space for thorough encasement of the steel. Table 8*A* shows the minimum widths for proper encasement.

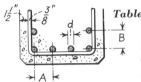

Table 8   **Preferred Widths of Beams**

Min $A = 3d$ or $d + 1\frac{1}{2} \times$ aggregate size.

Dimensions are in inches and are increased to nearest $\frac{1}{2}$ in.

Allow extra for splices.

	Bar No.	No. of longitudinal bars									Min A or B	Preferred B
		2	3	4	5	6	7	8	9	10		
$\frac{3}{4}$-in. aggregate	4	6	8	$9\frac{1}{2}$	11	$12\frac{1}{2}$	14	16	$17\frac{1}{2}$	19	$1\frac{5}{8}$	2
	5	$6\frac{1}{2}$	$8\frac{1}{2}$	10	12	14	16	$17\frac{1}{2}$	$19\frac{1}{2}$	$21\frac{1}{2}$	$1\frac{7}{8}$	$2\frac{1}{2}$
	6	7	9	$11\frac{1}{2}$	$13\frac{1}{2}$	16	18	$20\frac{1}{2}$	$22\frac{1}{2}$	25	$2\frac{1}{4}$	$2\frac{1}{2}$
	7	$7\frac{1}{2}$	10	$12\frac{1}{2}$	$15\frac{1}{2}$	18	$20\frac{1}{2}$	23	26	$28\frac{1}{2}$	$2\frac{5}{8}$	3
	8	8	11	14	17	20	23	26	29	32	3	3
	9	$8\frac{1}{2}$	12	15	$18\frac{1}{2}$	22	$25\frac{1}{2}$	$28\frac{1}{2}$	32	$35\frac{1}{2}$	$3\frac{3}{8}$	$3\frac{1}{2}$
	10	9	$12\frac{1}{2}$	$16\frac{1}{2}$	20	24	$27\frac{1}{2}$	$31\frac{1}{2}$	35	39	$3\frac{3}{4}$	4
	11	$9\frac{1}{2}$	14	$18\frac{1}{2}$	$22\frac{1}{2}$	27	31	$35\frac{1}{2}$	40	44	$4\frac{1}{4}$	$4\frac{1}{2}$
$1\frac{1}{2}$-in. aggregate	4	7	10	$12\frac{1}{2}$	$15\frac{1}{2}$	18	21	$23\frac{1}{2}$	$26\frac{1}{2}$	29	$2\frac{3}{4}$	3
	5	$7\frac{1}{2}$	$10\frac{1}{2}$	13	16	19	22	$24\frac{1}{2}$	$27\frac{1}{2}$	$30\frac{1}{2}$	$2\frac{7}{8}$	3
	6	$7\frac{1}{2}$	$10\frac{1}{2}$	$13\frac{1}{2}$	$16\frac{1}{2}$	$19\frac{1}{2}$	$22\frac{1}{2}$	$25\frac{1}{2}$	$28\frac{1}{2}$	$31\frac{1}{2}$	3	3
	7	8	11	14	$17\frac{1}{2}$	$20\frac{1}{2}$	$23\frac{1}{2}$	$26\frac{1}{2}$	30	33	$3\frac{1}{8}$	$3\frac{1}{2}$
	8	8	$11\frac{1}{2}$	$14\frac{1}{2}$	18	21	$24\frac{1}{2}$	$27\frac{1}{2}$	31	34	$3\frac{1}{4}$	$3\frac{1}{2}$
	9	$8\frac{1}{2}$	12	15	$18\frac{1}{2}$	22	$25\frac{1}{2}$	$28\frac{1}{2}$	32	$35\frac{1}{2}$	$3\frac{3}{8}$	4
	10	9	$12\frac{1}{2}$	$16\frac{1}{2}$	20	24	$27\frac{1}{2}$	$31\frac{1}{2}$	35	39	$3\frac{3}{4}$	4
	11	$9\frac{1}{2}$	14	$18\frac{1}{2}$	$22\frac{1}{2}$	27	31	$35\frac{1}{2}$	40	44	$4\frac{1}{4}$	$4\frac{1}{2}$

Table 8A   Minimum Widths of Beams*

Min $A = 2d$ or $d + 1\frac{1}{3} \times$ aggregate size.
Dimensions are in inches and are increased to nearest $\frac{1}{4}$ in.
Allow extra for splices.

	Bar No.	No. of longitudinal bars									Min A or B	Preferred B
		2	3	4	5	6	7	8	9	10		
$\frac{3}{4}$-in. aggregate	4	$5\frac{3}{4}$	$7\frac{1}{4}$	$8\frac{3}{4}$	$10\frac{1}{4}$	$11\frac{3}{4}$	$13\frac{1}{4}$	$14\frac{3}{4}$	$16\frac{1}{4}$	$17\frac{3}{4}$	$1\frac{1}{2}$	2
	5	6	$7\frac{3}{4}$	$9\frac{1}{4}$	11	$12\frac{1}{2}$	$14\frac{1}{4}$	$15\frac{3}{4}$	$17\frac{1}{2}$	19	$1\frac{5}{8}$	2
	6	$6\frac{1}{4}$	8	$9\frac{3}{4}$	$11\frac{1}{2}$	$13\frac{1}{4}$	15	$16\frac{3}{4}$	$18\frac{1}{2}$	$20\frac{1}{4}$	$1\frac{3}{4}$	$2\frac{1}{4}$
	7	$6\frac{1}{2}$	$8\frac{1}{2}$	$10\frac{1}{4}$	$12\frac{1}{4}$	14	16	$17\frac{3}{4}$	$19\frac{3}{4}$	$21\frac{1}{2}$	$1\frac{7}{8}$	$2\frac{1}{4}$
	8	$6\frac{3}{4}$	$8\frac{3}{4}$	$10\frac{3}{4}$	$12\frac{3}{4}$	$14\frac{3}{4}$	$16\frac{3}{4}$	$18\frac{3}{4}$	$20\frac{3}{4}$	$22\frac{3}{4}$	2	$2\frac{1}{2}$
	9	$7\frac{1}{4}$	$9\frac{1}{2}$	$11\frac{3}{4}$	14	$16\frac{1}{4}$	$18\frac{1}{2}$	$20\frac{3}{4}$	23	$25\frac{1}{4}$	$2\frac{1}{4}$	$2\frac{1}{2}$
	10	$7\frac{3}{4}$	$10\frac{1}{4}$	$12\frac{3}{4}$	$15\frac{1}{4}$	$17\frac{3}{4}$	$20\frac{1}{4}$	23	$25\frac{3}{4}$	$28\frac{1}{4}$	$2\frac{5}{8}$	3
	11	8	11	$13\frac{3}{4}$	$16\frac{1}{2}$	$19\frac{1}{2}$	$22\frac{1}{4}$	25	28	$30\frac{3}{4}$	$2\frac{7}{8}$	3
$1\frac{1}{2}$-in. aggregate	4	$6\frac{3}{4}$	$9\frac{1}{4}$	$11\frac{3}{4}$	$14\frac{1}{4}$	$16\frac{3}{4}$	$19\frac{1}{4}$	$21\frac{3}{4}$	$24\frac{1}{4}$	$26\frac{3}{4}$	$2\frac{1}{2}$	3
	5	7	$9\frac{3}{4}$	$12\frac{1}{4}$	15	$17\frac{1}{2}$	$20\frac{1}{4}$	$22\frac{3}{4}$	$25\frac{1}{4}$	28	$2\frac{5}{8}$	3
	6	$7\frac{1}{4}$	10	$12\frac{3}{4}$	$15\frac{1}{2}$	$18\frac{1}{4}$	21	$23\frac{3}{4}$	$26\frac{1}{2}$	$29\frac{1}{4}$	$2\frac{3}{4}$	3
	7	$7\frac{1}{2}$	$10\frac{1}{2}$	$13\frac{1}{4}$	$16\frac{1}{4}$	19	22	$24\frac{3}{4}$	$27\frac{3}{4}$	$30\frac{1}{2}$	$2\frac{7}{8}$	$3\frac{1}{4}$
	8	$7\frac{3}{4}$	$10\frac{3}{4}$	$13\frac{3}{4}$	$16\frac{3}{4}$	$19\frac{3}{4}$	$22\frac{3}{4}$	$25\frac{3}{4}$	$28\frac{3}{4}$	$31\frac{3}{4}$	3	$3\frac{1}{4}$
	9	8	$11\frac{1}{4}$	$14\frac{1}{4}$	$17\frac{1}{2}$	$20\frac{1}{2}$	$23\frac{3}{4}$	$26\frac{3}{4}$	30	33	$3\frac{1}{8}$	$3\frac{1}{4}$
	10	$8\frac{1}{4}$	$11\frac{1}{2}$	$14\frac{3}{4}$	18	$21\frac{1}{4}$	$24\frac{1}{2}$	$27\frac{3}{4}$	33	$34\frac{1}{4}$	$3\frac{1}{4}$	$3\frac{1}{2}$
	11	$8\frac{3}{4}$	$12\frac{1}{4}$	$15\frac{3}{4}$	$19\frac{1}{4}$	$22\frac{3}{4}$	$26\frac{1}{4}$	$29\frac{3}{4}$	$33\frac{1}{4}$	$36\frac{3}{4}$	$3\frac{1}{2}$	4

* Based in part on "Manual of Standard Practice for Reinforced Concrete Structures" (ACI 315-51).

Figures 13A *and* 13B.   These diagrams are to assist in roughly check-ing the safe loads for short columns or in obtaining approximate sizes for design purposes.   They are prepared for square tied columns and for round spirally reinforced ones.

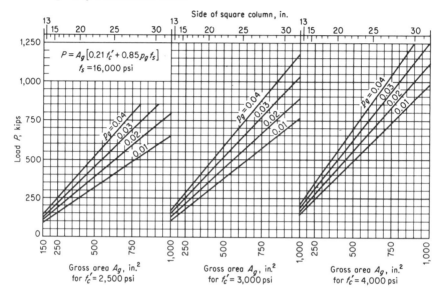

Fig. 13A   Safe loads P on short tied columns.

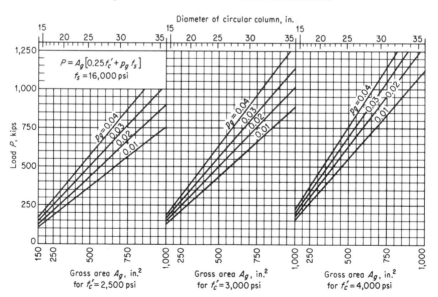

Fig. 13B   Safe loads P on short circular spirally reinforced columns.

Table 9 *and Figure* 14.   Table 9 gives some suggested size and pitch for spirals in round and square columns.   It is for a general guide.   It is based upon data formerly given in the American Concrete Institute's "Reinforced Concrete Handbook."   On the other hand, Fig. 14 gives the permissible size and pitch of spirals for any value of p_s (volume of spiral/volume of core).

Table 9   Size and Pitch of Spirals*

$p_s \geq 0.0112$; hot-rolled round bars; cover $= 1\frac{1}{2}$ in.

Pitch in inches

Column size, in.	Core size, in.	Square columns f'_c			Round columns f'_c		
		2,500	3,000	3,750	2,500	3,000	3,750
14	11				No. 3—$1\frac{3}{4}$	No. 3—$1\frac{3}{4}$	
15	12	No. 4—2			No. 3—2	No. 3—$1\frac{3}{4}$	No. 4—2
16	13	No. 4—2			No. 3—2	No. 3—$1\frac{3}{4}$	No. 4—2
17	14	No. 4—$2\frac{1}{4}$			No. 3—$2\frac{1}{4}$	No. 3—$1\frac{3}{4}$	No. 4—$2\frac{1}{4}$
18	15	No. 4—$2\frac{1}{4}$	No. 5—$2\frac{1}{2}$	No. 5—$2\frac{1}{4}$	No. 3—$2\frac{1}{2}$	No. 3—$1\frac{3}{4}$	No. 4—$2\frac{1}{2}$
19	16	No. 4—$2\frac{1}{4}$	No. 5—$2\frac{1}{2}$	No. 5—$2\frac{1}{4}$	No. 3—$2\frac{1}{4}$	No. 3—$1\frac{3}{4}$	No. 4—$2\frac{1}{2}$
20	17	No. 4—2	No. 5—$2\frac{3}{4}$	No. 5—$2\frac{1}{4}$	No. 3—$2\frac{1}{4}$	No. 3—2	No. 4—$2\frac{3}{4}$
21	18	No. 4—2	No. 5—$2\frac{3}{4}$	No. 5—2	No. 3—2	No. 3—2	No. 4—$2\frac{3}{4}$
22	19	No. 4—2	No. 5—$2\frac{3}{4}$	No. 5—2	No. 3—2	No. 3—2	No. 4—$2\frac{3}{4}$
23	20	No. 4—2	No. 5—$2\frac{1}{2}$	No. 5—2	No. 3—$1\frac{3}{4}$	No. 3—$1\frac{3}{4}$	No. 4—$2\frac{3}{4}$
24	21	No. 4—2	No. 5—$2\frac{1}{2}$	No. 5—2	No. 4—$3\frac{1}{4}$	No. 4—$3\frac{1}{4}$	No. 4—$2\frac{3}{4}$
25	22	No. 4—2	No. 5—$2\frac{1}{2}$	No. 5—2	No. 4—$3\frac{1}{4}$	No. 4—$3\frac{1}{4}$	No. 4—$2\frac{3}{4}$
26	23	No. 5—3	No. 5—$2\frac{1}{2}$	No. 5—2	No. 4—3	No. 4—3	No. 4—$2\frac{3}{4}$
27	24	No. 5—3	No. 5—$2\frac{1}{2}$	No. 5—2	No. 4—$2\frac{3}{4}$	No. 4—$2\frac{3}{4}$	No. 4—$2\frac{3}{4}$
28	25	No. 5—$2\frac{3}{4}$	No. 5—$2\frac{1}{4}$		No. 4—$2\frac{3}{4}$	No. 4—$2\frac{3}{4}$	No. 4—$2\frac{3}{4}$
29	26	No. 5—$2\frac{3}{4}$	No. 5—$2\frac{1}{4}$		No. 4—$2\frac{3}{4}$	No. 4—$2\frac{3}{4}$	No. 4—$2\frac{3}{4}$
30	27	No. 5—$2\frac{3}{4}$	No. 5—$2\frac{1}{4}$		No. 4—$2\frac{1}{2}$	No. 4—$2\frac{1}{2}$	No. 4—$2\frac{1}{2}$
31	28	No. 5—$2\frac{3}{4}$	No. 5—$2\frac{1}{4}$		No. 4—$2\frac{1}{2}$	No. 4—$2\frac{1}{2}$	No. 4—$2\frac{1}{2}$
32	29	No. 5—$2\frac{3}{4}$	No. 5—$2\frac{1}{4}$		No. 4—$2\frac{1}{4}$	No. 4—$2\frac{1}{4}$	No. 4—$2\frac{1}{4}$
33	30	No. 5—$2\frac{1}{2}$	No. 5—$2\frac{1}{4}$		No. 4—$2\frac{1}{4}$	No. 4—$2\frac{1}{4}$	No. 4—$2\frac{1}{4}$

* ACI "Reinforced Concrete Design Handbook."

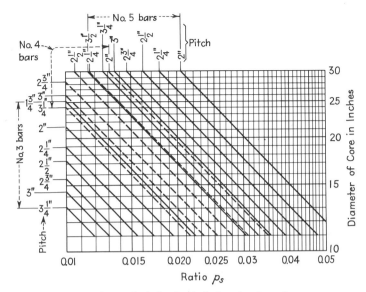

Fig. 14 Size and pitch of spirals for circular columns.

Table 10.   This gives the recommended dimensions for standard hooks and 90° bends of bars.   It is suitable for hard-grade bars as well as for structural and intermediate grades.   If necessary, the diameter of pin for structural and intermediate grades may be reduced, but not below 5*d*.

Table 10 Bending Details for Bars

Figure	Dim. \ d, No.	2	3	4	5	6	7	8	9	10	11	14S	18S
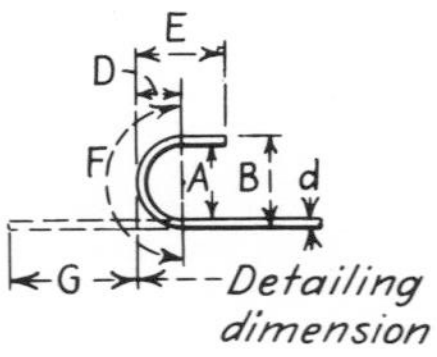	A	$1\frac{1}{4}$	$1\frac{7}{8}$	$2\frac{1}{2}$	$3\frac{1}{8}$	$4\frac{1}{2}$	$5\frac{1}{4}$	6	9	$10\frac{3}{16}$	$11\frac{1}{4}$	$1'5''$	$1'10\frac{1}{2}''$
	B	$1\frac{3}{4}$	$2\frac{5}{8}$	$3\frac{1}{2}$	$4\frac{3}{8}$	6	7	8	$11\frac{1}{4}$	$1'0\frac{11}{16}''$	$1'2\frac{1}{16}''$	$1'8\frac{3}{8}''$	$2'3''$
	C	$2\frac{1}{2}$	$2\frac{1}{2}$	$2\frac{1}{2}$	$2\frac{1}{2}$	3	$3\frac{1}{2}$	4	$4\frac{1}{2}$	5	$5\frac{3}{4}$	$6\frac{3}{4}$	9
	D	$\frac{7}{8}$	$1\frac{5}{16}$	$1\frac{3}{4}$	$2\frac{3}{16}$	3	$3\frac{1}{2}$	4	$5\frac{5}{8}$	$6\frac{3}{8}$	$7\frac{1}{16}$	$9\frac{3}{8}$	$1'0\frac{15}{16}''$
	E	$3\frac{3}{8}$	$3\frac{13}{16}$	$4\frac{1}{4}$	$4\frac{11}{16}$	6	7	8	$10\frac{1}{8}$	$11\frac{3}{8}$	$1'0\frac{13}{16}''$	$1'4\frac{1}{8}''$	$1'9\frac{15}{16}''$
	F	$2\frac{3}{8}$	$3\frac{1}{2}$	$4\frac{3}{4}$	$5\frac{7}{8}$	$8\frac{1}{4}$	$9\frac{5}{8}$	$1'1\frac{1}{4}''$	$1'3\frac{7}{8}''$	$1'5\frac{7}{8}''$	$1'7\frac{7}{8}''$	$2'4\frac{3}{8}''$	$3'2''$
Hook	G	4	$4\frac{3}{4}$	$5\frac{1}{2}$	$6\frac{1}{4}$	8	$9\frac{5}{8}$	$1'1\frac{1}{4}''$	$1'2\frac{3}{4}''$	$1'4\frac{1}{2}''$	$1'6\frac{1}{2}''$	$2'1\frac{3}{4}''$	$2'10''$
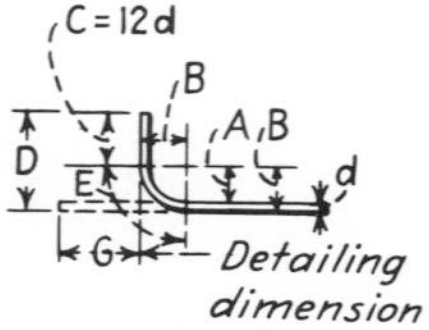	A	$\frac{5}{8}$	$\frac{15}{16}$	$1\frac{1}{4}$	$1\frac{9}{16}$	$2\frac{1}{4}$	$2\frac{5}{8}$	3	$4\frac{1}{2}$	$5\frac{1}{16}$	$5\frac{5}{8}$	$8\frac{1}{2}$	$11\frac{1}{4}$
	B	$\frac{7}{8}$	$1\frac{5}{16}$	$1\frac{3}{4}$	$2\frac{3}{16}$	3	$3\frac{1}{2}$	4	$5\frac{5}{8}$	$6\frac{5}{16}$	$7\frac{1}{16}$	$10\frac{3}{16}$	$1'1\frac{1}{2}''$
	C	3	$4\frac{1}{2}$	6	$7\frac{1}{2}$	9	$10\frac{1}{2}$	12	$1'1\frac{1}{2}''$	$1'3\frac{1}{4}''$	$1'5''$	$1'8\frac{1}{4}''$	$2'3''$
	D	$3\frac{7}{8}$	$5\frac{13}{16}$	$7\frac{3}{4}$	$9\frac{11}{16}$	$1'0''$	$1'2''$	$1'4''$	$1'7\frac{1}{8}''$	$1'9\frac{9}{16}''$	$2'0\frac{1}{16}''$	$2'6\frac{7}{16}''$	$3'4\frac{1}{2}''$
	E	$1\frac{3}{16}$	$1\frac{3}{4}$	$2\frac{3}{8}$	$2\frac{15}{16}$	$4\frac{1}{8}$	$4\frac{13}{16}$	$6\frac{5}{8}$	$6\frac{15}{16}$	$8\frac{15}{16}$	$9\frac{15}{16}$	$1'2\frac{3}{16}''$	$1'7''$
90° Bend	G	$3\frac{1}{4}$	5	$6\frac{5}{8}$	$8\frac{1}{4}$	$10\frac{1}{8}$	$11\frac{3}{4}$	$1'2\frac{5}{8}''$	$1'2\frac{3}{4}''$	$1'5\frac{7}{8}''$	$1'7\frac{7}{8}''$	$2'0\frac{1}{4}''$	$2'8\frac{1}{2}''$

Diameter of pin for bend: $5d$ for Nos. 2, 3, 4, and 5; $6d$ for Nos. 6, 7, and 8; $8d$ for Nos. 9, 10, and 11; $10d$ for Nos. 14S and 18S. G = approximate length to add to detailing dimension in order to make hook or bend. Rounded off to $\frac{1}{16}$-in. dimensions. Dimensions are in inches except as shown.

No. 2 bars are $\frac{1}{4}$-in. plain bars. (Deformed if by special order.)

Tables 11 *and* 12. These are tables giving recommendations for the maximum number of longitudinal bars to be used in columns. Table 12 shows the closest spacing for round columns. Table 11 allows more room and is suitable for most cases, even when the bars are spliced side by side instead of one inside the next bar.

Table 11 **Preferred Maximum Number of Longitudinal Bars in One Row in Columns**

Diam. or side of core, in.	Size of tie or spiral, No.	Round spirally reinforced ($p_g = 0.01$ min, 0.08 max) Size of bar, No.							Square tied ($p_g = 0.01$ min, 0.04 max) Size of bar, No.						
		5	6	7	8	9	10	11	5	6	7	8	9	10	11
10	3	9	8	7	7	6			12	8	8	8	6*		
11	3	11	9	9	8	7	6		12	12	8	8	6*		
12	3	12	11	10	9	8	7	6	12	12	12	8	8	6*	
13	3	13	12	11	10	8	7	6	16	12	12	12	8	8	6*
14	3	14	13	11	11	9	8	7	16	16	12	12	10*	8	6*
15	3	15	14	12	11	10	9	8	16	16	16	12	12	10*	8
16	4	16	14	13	12	11	9	8	20	16	16	16	12	10*	8
17	4	17	15	14	13	11	10	9	20	20	16	16	12	12	8
18	4	18	17	15	14	12	11	9	20	20	16	16	16	12	10*
19	4	19	18	16	15	13	12	10	24	20	20	16	16	12	12
20	4	21	19	17	16	14	12	10	24	24	20	20	16	12	12
21	4	22	20	18	17	15	13	11	28	24	20	20	16	16	12
22	4	23	21	19	18	16	14	12	28	24	24	20	20	16	12
23	5	24	22	20	18	16	14	12	28	28	24	24	20	16	16
24	5	25	23	21	19	17	15	13	32	28	24	24	20	16	16
25	5	26	24	22	20	18	16	13	32	28	28	24	20	20	16
26	5	27	25	23	21	18	16	14	32	32	28	24	24	20	16
27	5	28	26	24	22	19	17	15	36	32	28	28	24	20	16
28	5	29	27	25	23	20	17	15	36	32	28	28	24	20	20
29	5	31	28	26	24	21	18	16	36	36	32	28	24	20	20
30	5	32	29	27	24	22	19	17	40	36	32	32	28	24	24

Max size of aggregate = 1 in.
* Limited by max p_g. Arrange symmetrically but unequally.

Table 12 **Maximum Number of Longitudinal Bars in Round Columns***

Diameter of core, in.	Size of spiral, No.	Size of bar, No.						
		5	6	7	8	9	10	11
11	3	12	11	10	9	7	6	
12	3	13	12	11	10	8	7	6
13	3	15	13	12	11	9	8	6
14	3	16	15	14	12	11	9	7
15	3	18	16	15	14	12	10	8
16	3	19	18	16	15	13	11	9
17	3	21	19	18	16	14	12	10
18	4	22	20	19	17	15	13	11
19	4	23	22	20	18	16	14	11
20	4	25	23	21	20	17	15	12
21	4	26	24	22	21	18	16	13
22	4	28	26	24	22	19	17	14
23	4	29	27	25	23	20	18	15
24	4	31	28	26	25	21	19	16
25	5	32	30	28	26	22	20	17
26	5	33	31	29	27	23	21	17
27	5	35	32	30	28	25	22	18
28	5	36	34	31	29	26	23	19
29	5	38	35	33	31	27	24	20
30	5	39	37	34	32	28	25	21
31	5	41	38	35	33	29	26	22

* Based on "Manual of Standard Practice for Detailing Reinforced Concrete Structures" (ACI 315-51).

Tables 13 *and* 14. These show the allowable bond unit stresses for WSD and USD as specified in the Code (ACI 318-63). "Top" refers to horizontal bars having at least 12 in. of concrete below them. The unit stresses shown make an allowance for possible settlement of the wet con-

Table 13   **Maximum Allowable Bond Unit Stresses, psi, Type A 305, Working-stress Design**

		Size of bar																
	3		4		5		6		7		8		9		10		11	
f'_c	$\frac{3}{8}$ in. ϕ		$\frac{1}{2}$ in. ϕ		$\frac{5}{8}$ in. ϕ		$\frac{3}{4}$ in. ϕ		$\frac{7}{8}$ in. ϕ		1 in. ϕ		$1\frac{1}{8}$ in. $+\phi$		$1\frac{1}{4}$ in. $+\phi$		$1\frac{7}{16}$ in. $-\phi$	
	Top	Bottom	Top	Bottom	Top	Bottom	Top	Bottom	Top	Bottom	Top	Bottom	Top	Bottom	Top	Bottom	Top	Bottom
2,500	350	500	340	480	270	385	230	320	195	275	170	240	150	215	135	190	120	170
3,000	350	500	350	500	300	420	250	350	215	300	180	265	165	235	145	205	130	185
3,500	350	500	350	500	320	455	270	380	230	325	200	285	180	250	160	225	140	200
4,000	350	500	350	500	345	485	285	405	245	345	215	305	190	270	170	240	150	215
5,000	350	500	350	500	350	500	320	455	275	390	240	340	215	300	190	270	170	240

Note: Bond on plain bars = one-half that given above.

Table 14 Maximum Allowable Bond Unit Stresses, psi, Type A 305, Ultimate-strength Design

	3		4		5		6		7		8		9		10		11	
Size of bar	$\frac{3}{8}$ in. ϕ		$\frac{1}{2}$ in. ϕ		$\frac{5}{8}$ in. ϕ		$\frac{3}{4}$ in. ϕ		$\frac{7}{8}$ in. ϕ		1 in. ϕ		$1\frac{1}{8}$ in. $+ \phi$		$1\frac{1}{4}$in. $+ \phi$		$1\frac{7}{16}$ in. $- \phi$	
f'_c	Top	Bottom	Top	Bottom	Top	Bottom	Top	Bottom	Top	Bottom	Top	Bottom	Top	Bottom	Top	Bottom	Top	Bottom
2,500	560	800	560	800	535	760	445	635	385	545	335	475	295	420	265	375	235	335
3,000	560	800	560	800	560	800	490	695	420	595	365	520	325	460	290	410	260	370
3,500	560	800	560	800	560	800	530	750	455	645	395	560	350	500	310	440	280	400
4,000	560	800	560	800	560	800	560	800	485	685	425	600	375	535	335	475	300	425
5,000	560	800	560	800	560	800	560	800	540	770	475	670	420	595	375	530	335	475

Note: Bond on plain bars = one-half that given above.   See Table 16 in Appendix for safety provisions.

crete during compaction and vibration which might weaken the bond on the bottom portion of any bar. "Bottom" refers to horizontal bars in or near the bottom of a beam or in a vertical position where consolidation of the concrete will not affect the bond.

Table 15. This table gives the unit stresses allowed for bond on the special heavy bars 14S and 18S.

Table 15   **Bond Unit Stresses, psi, Type A 408**

f'_c	Working stress 14S and 18S bars		Ultimate strength* 14S and 18S bars	
	Top	Bottom	Top	Bottom
2,500	105	150	210	300
3,000	115	165	230	330
3,500	125	180	250	355
4,000	135	190	265	380
5,000	150	210	295	425

* See Table 16 in Appendix for safety provisions.

Table 16. The Code specifies these values for the safety provision ϕ for USD. They are to be used where applicable.

Table 16   **Safety Provisions for Ultimate-strength Design**

Coefficient ϕ	Structural type of action
0.90	Flexure
0.85	Diagonal tension
0.85	Bond and anchorage
0.75	Spirally reinforced compression members
0.70	Tied compression members

Table 17.   The first four columns apply to lightweight concrete.   The last column, with a unit weight of 145 pcf, is the one to use for ordinary stone concrete.   Notice that the figures are in ksi.

Table 17   **Modulus of Elasticity for Concrete, kips/in.²**

$$E_c = w^{1.5} \times 33 \sqrt{f'_c}$$

f'_c, psi	Unit weight w, lb/ft³				
	90	100	110	125	145
2,500	1,410	1,650	1,900	2,300	2,880
3,000	1,540	1,800	2,080	2,520	3,150
4,000	1,780	2,090	2,410	2,920	3,640
5,000	1,990	2,330	2,690	3,260	4,060

Figure 15.   This graph is for slabs with tensile reinforcement only. The reduction factor of 0.9 has already been included.

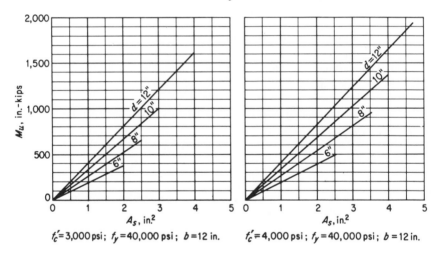

f'_c=3,000 psi; f_y=40,000 psi; b=12 in. f'_c=4,000 psi; f_y=40,000 psi; b=12 in.

Fig. 15 Ultimate resisting moments of slabs and small rectangular beams 12 in. wide.   (For $f_y = 50,000$ psi, multiply values of M_u by 1.25.)

Figure 16. This graph applies to rectangular beams with tensile reinforcement only. The steel is assumed to be the controlling factor.

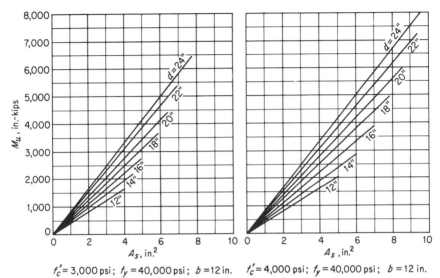

$f'_c = 3,000\,\text{psi};\ f_y = 40,000\,\text{psi};\ b = 12\,\text{in}.\qquad f'_c = 4,000\,\text{psi};\ f_y = 40,000\,\text{psi};\ b = 12\,\text{in}.$

Fig. 16 Ultimate resisting moments of rectangular beams 12 in. wide. (For $f_y = 50,000$ psi, multiply values of M_u by 1.25.)

Figure 17.   This figure is for use in making preliminary estimates and rough checks.   It applies to T beams with tensile reinforcement only.

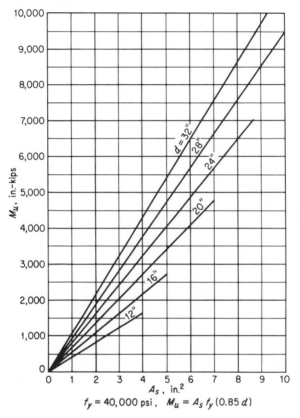

$f_y = 40,000$ psi, $M_u = A_s f_y (0.85\,d)$

Fig. 17 Approximate ultimate resisting moments of T beams.   (For $f_y = 50,000$ psi, multiply values of M_u by 1.25.)

INDEX

617